Student Solutions Manual to Accompany

Introduction to General, Organic, and Biochemistry

11th Edition

Morris Hein
Mount San Antonio College

Scott Pattison
Ball State University

Susan Arena
University of Illinois, Urbana–Champaign

WILEY

ISBN 978-1-118-50191-7

10 9 8 7 6 5 4 3 2 1

TO THE STUDENT

This Solutions Manual contains answers to all questions and solutions to all exercises at the end of each chapter in the text.

This book will be valuable to you if you use it properly. It is not intended to be a substitute for answering the questions and exercises by yourself. It is important that you go through the process of working out the answers before you see them in the solutions manual. Once you have seen the answer, much of the value of whether or not you have learned the material is lost. This does not mean that you cannot learn from the answers. If you absolutely cannot answer a question or an exercise, study the answer carefully to see where you are having difficulty. You should spend sufficient time answering each assigned exercise. Only after you have made a serious attempt to answer a question or exercise should you resort to the solutions manual.

Chemistry is one of the most challenging courses of study you will encounter. The core of the study of chemistry is problem solving. Skill at solving problems is achieved through effective and consistent practice. Watching and listening to others solve problems may be of some use but will not result in facility with chemistry problems. The methods used for solving problems in this manual are essentially the same as those used in the text. The basic steps in problem solving are fairly universal:

1. Read the problem carefully and determine the type of problem.
2. Develop a plan for solving the problem.
3. Write down the given information, including units, in an organized fashion.
4. Write a complete set-up and solution for the problem.
5. Use the solution manual to check the answer and solution.

Your solution might be correct and yet will vary from the manual. Usually, these differences are the result of completing operations in a different order, or using separate steps instead of a single line approach. If you make an error, take the time to analyze what went wrong. An error caused by an improper entry to the calculator can be frustrating, but it is not as serious as the inability to properly set up the problem. Do not become discouraged. Once you understand the steps in a problem, go back and rework it without looking at the answer.

The areas of organic and biochemistry present a different type of learning. Learning organic chemistry takes a great amount of practice. You need to put in a lot of time writing formulas, names, and reaction equations if you are to be successful. Initially there is a certain amount of memorization of material. For example, you must learn the names and formulas of the first ten aliphatic hydrocarbons: methane, ethane, propane, butane, pentane, hexane, heptane, octane, nonane, and decane. Say them to yourself several times daily; write the names together with their formulas over and over again, and soon you will know them very well. The names of many other classes of organic compounds are derived from the names of the aliphatic hydrocarbons. Remember, **practice** is the key word.

We have made every attempt to produce a manual with as few errors as possible. Please let us know of errors that you encounter so that they can be corrected. Allow this manual to help you have success as you begin the great adventure of studying chemistry.

MORRIS HEIN
SCOTT PATTISON
SUSAN ARENA

CONTENTS

CHAPTER 1

AN INTRODUCTION TO CHEMISTRY

SOLUTIONS TO REVIEW QUESTIONS

1. (a) A hypothesis is a tentative explanation of certain facts to provide a basis for further experimentation. A theory is an explanation of the general principles of certain phenomena with considerable evidence to support it.

   (b) A theory is an explanation of the general principles of certain phenomena with considerable evidence to support it. A scientific law is a simple statement of natural phenomena to which no exceptions are known under the given conditions.

2. (a) hypothesis
   (b) hypothesis
   (c) observation
   (d) theory
   (e) observation
   (f) scientific law

3. (a) A liquid has a definite volume but not a definite shape.
   (b) A gas has an indefinite volume and high compressibility.
   (c) A solid has a definite shape.
   (d) A liquid has an indefinite shape and slight compressibility.

4. A crystalline solid has a regular, repeating, three-dimensional, geometric pattern. An amorphous solid does not
   (a) A solid that has a regular, repeating pattern is a crystalline solid.
   (b) A plastic solid is amorphous.
   (c) A solid that has no regular repeating pattern is amorphous.
   (d) Glass is an amorphous solid.
   (e) Gold is a crystalline solid.

5. A phase is a homogeneous part of a system separated from other parts by a physical boundary.

6. There are six phases present.

7. Another name for a homogeneous mixture is solution.

8. Alcohol, mercury, and water are the only liquids in the table which are not mixtures. Mercury is an element; alcohol and water are compounds.

9. Air is the only gas mixture found in the table. The other gases are elements or compounds.

10. Three phases are present within the bottle; solid and liquid are observed visually, while gas is detected by the immediate odor.

11. The system is heterogeneous as three phases are present.

12. A system containing only one substance is not necessarily homogeneous. Two phases may be present. Example: ice in water.

13. A system containing two or more substances is not necessarily heterogeneous. In a solution only one phase is present. Examples: sugar dissolved in water, dilute sulfuric acid.

14. Homogeneous mixtures contain only one phase, while heterogeneous mixtures contain two or more phases.

15. (a) sugar, a compound and (c) gold, an element

16. Using the steps of the scientific method to help determine why your cell phone has stopped working.
    (a) Observation: My cell phone has stopped working.
    (b) Hypothesis: I think that the battery needs to be recharged.
    (c) Experiment: Plug in the phone to recharge the battery and allow sufficient time for the battery to fully recharge. Turn the phone back on. The phone now works again.
    (d) Theory: The battery in the phone has a limited charge time and needs to be recharged on a regular basis in order to keep it in working order.

SOLUTIONS TO EXERCISES

1. Two states are present; solid and gas.

2. Two states are present; solid and liquid.

3. The photo represents a heterogeneous mixture.

4. The maple leaf represents a heterogeneous mixture.

5. (a) homogeneous
   (b) heterogeneous
   (c) heterogeneous
   (d) heterogeneous

6. (a) homogeneous
   (b) homogeneous
   (c) heterogeneous
   (d) heterogeneous

7. Typical answers could be

Substance	Main or Active Ingredient
chocolate syrup	high fructose corn syrup
margarine	vegetable oil blend
non-dairy creamer	corn syrup solids
beef bouillon	salt
toothpaste	sodium fluoride
antibacterial soap	triclocarban
first aid spray	lidocaine HCl
sunblock lotion	ethylhexyl p-methoxycinnamate

8. The steps of the scientific method can be used to predict the outcome of the semester in the following way:

   (a) Collect the facts or data. They include the number of classes you have enrolled in, the amount of time you will give to each class, the number of hours required for your off campus job, and the amount of time for social activities.
   (b) Formulate a hypotheses. You predict that the amount of time you have allocated each week for class work will be sufficient to result in good grades at the end of the semester.
   (c) Plan and do additional experiments to test the hypothesis. In the first two weeks of the semester keep a record of your performance in each class, including grades on homework, quizzes, and exams.
   (d) Modify the hypothesis. If your grades in any class are not good, your hypothesis was incorrect. It will be necessary to increase the amount of time allocated to that class.

9. (a) water, a pure substance
   (b) chicken stock, a mixture
   (c) salt, a pure substance
   (d) mustard flour, a mixture

10. Hypothesis – The soup is a not a good source of nutrition because it contains 12 grams of fat per serving.

   Test – Search for the number of grams of fat per day recommended for an adult. Compare this number to the grams per fat on the label.

11. Observations –

   Calcite crystals have the ability to reflect light around an object rendering it "unseeable."

   "Invisibility cloaks" work only with laser light aimed directly at the crystal.

   Hypotheses –

   If larger crystals were used, larger objects could be hidden.

   These "invisibility cloaks" will improve in the future.

12. (a) two phases, solid and gas
    (b) two phases, liquid and gas
    (c) two phases, solid and liquid

CHAPTER 2

STANDARDS FOR MEASUREMENT

SOLUTIONS TO REVIEW QUESTIONS

1. The exponent will be positive for a large number and negative for a small number.

2. The exponent will decrease.

3. The last digit in a measurement is uncertain because if the quantity were to be measured multiple times, the last digit would vary.

4. It must be written in scientific notation as 6.420×10^5 g.

5. Zeroes are significant when they are between non-zero digits or at the end of a number that includes a decimal point.

6. Rule 1.   When the first digit after those you want to retain is 4 or less, that digit and all others to its right are dropped. The last digit retained is not changed.

    Rule 2.   When the first digit after those you want to retain is 5 or greater, that digit and all others to the right of it are dropped and the last digit retained is increased by one.

7. No, the number of significant digits in the calculated value may not be more than the number of significant figures in any of the measurements.

8. Yes, the number of significant digits depends on the precision of each of the individual measurements and the calculated value may have more or fewer significant figures than the original measurements as long as the precision is no greater than the measurement with the lowest precision. An example of a calculation with an increase in significant figures is in example 2.9.

9.
$$100 \text{ cm} = 1 \text{ m}$$
$$1,000,000,000 \text{ nm} = 1 \text{ m} \qquad (1 \text{ cm}) \left(\frac{1 \text{ m}}{100 \text{ cm}} \right) \left(\frac{1,000,000,000 \text{ nm}}{\text{m}} \right) = 10,000,000 \text{ nm}$$

10.
$$1000 \text{ mg} = 1 \text{ g}$$
$$1000 \text{ g} = 1 \text{ kg} \qquad (1 \text{ kg}) \left(\frac{1000 \text{ g}}{\text{kg}} \right) \left(\frac{1000 \text{ mg}}{\text{g}} \right) = 1,000,000 \text{ mg}$$

11. Weight is a measure of how much attraction the earth's gravity has for an object (or person). In this case, the farther the astronaut is from the earth the less gravitational force is pulling on him or her. Less gravitational attraction means the astronaut will weigh less. The mass of the astronaut is the amount of matter that makes up him or her. This does not change as the astronaut moves away from the earth.

12. They are equivalent units.

13. $(3.5 \text{ in.}) \left(\dfrac{2.54 \text{ cm}}{1 \text{ in.}} \right) = 8.9 \text{ cm}$

14. Heat is a form of energy, while temperature is a measure of the intensity of heat (how hot the system is).

15. The number of degrees between the freezing and boiling point of water are

    Fahrenheit 180°F
    Celsius   100°C
    Kelvin   100 K

16. The three materials would sort out according to their densities with the most dense (mercury) at the bottom and the least dense (glycerin) at the top. In the cylinder, the solid magnesium would sink in the glycerin and float on the liquid mercury.

17. Order of increasing density: ethyl alcohol, vegetable oil, salt, lead.

18. The density of ice must be less than 0.91 g/mL and greater than 0.789 g/mL.

19. Density is the ratio of the mass of a substance to the volume occupied by that mass. Density has the units of mass over volume. Specific gravity is the ratio (no units) of the density of a substance to the density of a reference substance (usually water at a specific temperature for solids and liquids). Specific gravity has no units.

20. The density of water is 1.0 g/mL at approximately 4°C. However, when water changes from a liquid to a solid at 0°C there is actually an increase in volume. The density of ice at 0°C is 0.917 g/mL. Therefore, ice floats in water because solid water is less dense than liquid water.

21. If you collect a container of oxygen gas, you should store it with the mouth up. Oxygen gas is denser than air.

22. density of gold = 19.3 g/mL   density of silver = 10.5 g/mL

    $(25 \text{ g gold}) \left(\dfrac{1 \text{ mL}}{19.3 \text{ g}} \right) = 1.3 \text{ mL}$

    $(25 \text{ g silver}) \left(\dfrac{1 \text{ mL}}{10.5 \text{ g}} \right) = 2.4 \text{ mL}$ 25 g of silver has the greater volume.

SOLUTIONS TO EXERCISES

1. (a) kilogram = 1000 grams
   (b) centimeter = 1/100 of a meter (0.01 m)
   (c) microliter = 1/1,000,000 of a liter (0.000001 L)
   (d) millimeter = 1/1000 of a meter (0.001 m)
   (e) deciliter = 1/10 of a liter (0.1 L)

2. (a) 1000 meters = 1 kilometer
   (b) 0.1 gram = 1 decigram
   (c) 0.000001 liter = 1 microliter
   (d) 0.01 meter = 1 centimeter
   (e) 0.001 liter = 1 milliliter

3. (a) gram = g
   (b) microgram = μg
   (c) centimeter = cm
   (d) micrometer = μm
   (e) milliliter = mL
   (f) deciliter = dL

4. (a) milligram = mg
   (b) kilogram = kg
   (c) meter = m
   (d) nanometer = nm
   (e) angstrom = Å
   (f) microliter = μL

5. (a) 2050   the first zero is significant; the last zero is not significant
   (b) 9.00×10^2 zeros are significant
   (c) 0.0530   the first two zeros are not significant; the last zero is significant
   (d) 0.075   zeros are not significant
   (e) 300.   zeros are significant
   (f) 285.00   zeros are significant

6. (a) 0.005   zeros are not significant
   (b) 1500   zeros are not significant
   (c) 250.   zero is significant
   (d) 10.000   zeros are significant
   (e) 6.070×10^4 zeros are significant
   (f) 0.2300   the first zero is not significant; the last two zeros are significant

7. Significant figures
   (a) 0.025   (2 sig. fig.)
   (b) 22.4   (3 sig. fig.)
   (c) 0.0404   (3 sig. fig.)
   (d) 5.50×10^3 (3 sig. fig.)

8. Significant figures
   (a) 40.0   (3 sig. fig.)
   (b) 0.081   (2 sig. fig.)
   (c) 129,042   (6 sig. fig.)
   (d) 4.090×10^{-3} (4 sig. fig.)

9. Round to three significant figures
   (a) 93.2
   (b) 0.0286
   (c) 4.64
   (d) 34.3

10. Round to three significant figures
    (a) 8.87
    (b) 21.3
    (c) 130. (1.30×10^2)
    (d) 2.00×10^6

11. Exponential notation
    (a) 2.9×10^6
    (b) 5.87×10^{-1}
    (c) 8.40×10^{-3}
    (d) 5.5×10^{-6}

12. Exponential notation
    (a) 4.56×10^{-2}
    (b) 4.0822×10^3
    (c) 4.030×10^1
    (d) 1.2×10^7

13. (a)
$$\begin{array}{r} 12.62 \\ 1.5 \\ 0.25 \\ \hline 14.37 \end{array} = 14.4$$

    (b) $(2.25 \times 10^3)(4.80 \times 10^4) = 10.8 \times 10^7 = 1.08 \times 10^8$
    (c) $\dfrac{(452)(6.2)}{14.3} = 195.97 = 2.0 \times 10^2$
    (d) $(0.0394)(12.8) = 0.504$
    (e) $\dfrac{0.4278}{59.6} = 0.00718 = 7.18 \times 10^{-3}$
    (f) $10.4 + (3.75)(1.5 \times 10^4) = 5.6 \times 10^4$

14. (a)
$$\begin{array}{r} 15.2 \\ -2.75 \\ 15.67 \\ \hline 28.1 \end{array}$$

    (b) $(4.68)(12.5) = 58.5$
    (c) $\dfrac{182.6}{4.6} = 40.$
    (d)
$$\begin{array}{r} 1986 \\ 23.84 \\ 0.012 \\ \hline 2009.852 \end{array} = 2010. = 2.010 \times 10^3$$

    (e) $\dfrac{29.3}{(284)(415)} = 2.49 \times 10^{-4}$
    (f) $(2.92 \times 10^{-3})(6.14 \times 10^5) = 1.79 \times 10^3$

15. Fractions to decimals (3 significant figures)
    (a) $\dfrac{5}{6} = 0.833$
    (b) $\dfrac{3}{7} = 0.429$
    (c) $\dfrac{12}{16} = 0.750$
    (d) $\dfrac{9}{18} = 0.500$

16. Decimals to fractions

    (a) $0.25 = \dfrac{1}{4}$

    (b) $0.625 = \dfrac{5}{8}$

    (c) $1.67 = 1\dfrac{2}{3}$ or $\dfrac{5}{3}$

    (d) $0.888 = \dfrac{8}{9}$

17. (a) $3.42x = 6.5$

    $x = \dfrac{6.5}{3.42} = 1.9$

    (b) $\dfrac{x}{12.3} = 7.05$

    $x = (7.05)(12.3) = 86.7$

    (c) $\dfrac{0.525}{x} = 0.25$

    $0.525 = 0.25x$

    $x = \dfrac{0.525}{0.25} = 2.1$

18. (a) $x = \dfrac{212 - 32}{1.8}$

    $x = 1.0 \times 10^2$

    (b) $8.9\dfrac{g}{mL} = \dfrac{40.90\,g}{x}$

    $\left(8.9\dfrac{g}{mL}\right)x = 40.90\,g$

    $x = \dfrac{40.90\,g}{8.9\dfrac{g}{mL}} = 4.6\,mL$

    (c) $72 = 1.8x + 32$

    $72 - 32 = 1.8x$

    $40. = 1.8x$

    $\dfrac{40.}{1.8} = x$

    $22 = x$

19. (a) $(28.0\,cm)\left(\dfrac{1\,m}{100\,cm}\right) = 0.280\,m$

    (b) $(1000.\,m)\left(\dfrac{1\,km}{1000\,m}\right) = 1.000\,km$

    (c) $(9.28\,cm)\left(\dfrac{10\,mm}{1\,cm}\right) = 92.8\,mm$

    (d) $(10.68\,g)\left(\dfrac{1000\,mg}{1\,g}\right) = 1.068 \times 10^4\,mg$

    (e) $\left(6.8 \times 10^4\,mg\right)\left(\dfrac{1\,g}{1000\,mg}\right)\left(\dfrac{1\,kg}{1000\,g}\right) = 6.8 \times 10^{-2}\,kg$

    (f) $(8.54\,g)\left(\dfrac{1\,kg}{1000\,g}\right) = 0.00854\,kg$

    (g) $(25.0\,mL)\left(\dfrac{1\,L}{1000\,mL}\right) = 2.50 \times 10^{-2}\,L$

    (h) $(22.4\,L)\left(\dfrac{10^6\,\mu L}{1\,L}\right) = 2.24 \times 10^7\,\mu L$

20.   (a) $(4.5 \text{ cm})\left(\dfrac{1 \text{ m}}{100 \text{ cm}}\right)\left(\dfrac{1 \text{ Å}}{10^{-10} \text{ m}}\right) = 4.5 \times 10^{8} \text{ Å}$

    (b) $(12 \text{ nm})\left(\dfrac{10^{-9} \text{ m}}{1 \text{ nm}}\right)\left(\dfrac{100 \text{ cm}}{1 \text{ m}}\right) = 1.2 \times 10^{-6} \text{ cm}$

    (c) $(8.0 \text{ km})\left(\dfrac{1000 \text{ m}}{1 \text{ km}}\right)\left(\dfrac{1000 \text{ mm}}{1 \text{ m}}\right) = 8.0 \times 10^{6} \text{ mm}$

    (d) $(164 \text{ mg})\left(\dfrac{1 \text{ g}}{1000 \text{ mg}}\right) = 0.164 \text{ g}$

    (e) $(0.65 \text{ kg})\left(\dfrac{1000 \text{ g}}{1 \text{ kg}}\right)\left(\dfrac{1000 \text{ mg}}{1 \text{ g}}\right) = 6.5 \times 10^{5} \text{ mg}$

    (f) $(5.5 \text{ kg})\left(\dfrac{1000 \text{ g}}{1 \text{ kg}}\right) = 5.5 \times 10^{3} \text{ g}$

    (g) $(0.468 \text{ L})\left(\dfrac{1000 \text{ mL}}{1 \text{ L}}\right) = 468 \text{ mL}$

    (h) $(9.0 \text{ μL})\left(\dfrac{1 \text{ L}}{10^{6} \text{ μL}}\right)\left(\dfrac{1000 \text{ mL}}{1 \text{ L}}\right) = 9.0 \times 10^{-3} \text{ mL}$

21.   (a) $(42.2 \text{ in.})\left(\dfrac{2.54 \text{ cm}}{1 \text{ in.}}\right) = 107 \text{ cm}$

    (b) $(0.64 \text{ mi})\left(\dfrac{5280 \text{ ft}}{1 \text{ mi}}\right)\left(\dfrac{12 \text{ in.}}{1 \text{ ft}}\right) = 4.1 \times 10^{4} \text{ in.}$

    (c) $(2.00 \text{ in.}^{2})\left(\dfrac{2.54 \text{ cm}}{1 \text{ in.}}\right)^{2} = 12.9 \text{ cm}^{2}$

    (d) $(42.8 \text{ kg})\left(\dfrac{2.205 \text{ lb}}{\text{kg}}\right) = 94.4 \text{ lb}$

    (e) $(3.5 \text{ qt})\left(\dfrac{946 \text{ mL}}{1 \text{ qt}}\right) = 3.3 \times 10^{3} \text{ mL}$

    (f) $(20.0 \text{ L})\left(\dfrac{1 \text{ qt}}{0.946 \text{ L}}\right)\left(\dfrac{1 \text{ gal}}{4 \text{ qt}}\right) = 5.29 \text{ gal}$

22.   (a)   The conversion is: m → cm → in. → ft

        $(35.6 \text{ m})\left(\dfrac{100 \text{ cm}}{1 \text{ m}}\right)\left(\dfrac{1 \text{ in.}}{2.54 \text{ cm}}\right)\left(\dfrac{1 \text{ ft}}{12 \text{ in.}}\right) = 117 \text{ ft}$

    (b) $(16.5 \text{ km})\left(\dfrac{1 \text{ mi}}{1.609 \text{ km}}\right) = 10.3 \text{ mi}$

    (c) $(4.5 \text{ in.}^{3})\left(\dfrac{2.54 \text{ cm}}{1 \text{ in.}}\right)^{3}\left(\dfrac{10 \text{ mm}}{1 \text{ cm}}\right)^{3} = 7.4 \times 10^{4} \text{ mm}^{3}$

    (d) $(95 \text{ lb})\left(\dfrac{453.6 \text{ g}}{1 \text{ lb}}\right) = 4.3 \times 10^{4} \text{ g}$

(e)　$(20.0 \text{ gal})\left(\dfrac{4 \text{ qt}}{1 \text{ gal}}\right)\left(\dfrac{0.946 \text{ L}}{1 \text{ qt}}\right) = 75.7 \text{ L}$

(f)　The conversion is: $\text{ft}^3 \rightarrow \text{in.}^3 \rightarrow \text{cm}^3 \rightarrow \text{m}^3$

$$(4.5 \times 10^4 \text{ ft}^3)\left(\dfrac{12 \text{ in.}}{1 \text{ ft}}\right)^3\left(\dfrac{2.54 \text{ cm}}{1 \text{ in.}}\right)^3\left(\dfrac{1 \text{ m}}{100 \text{ cm}}\right)^3 = 1.3 \times 10^3 \text{ m}^3$$

23.　The conversion is: $\dfrac{\text{mi}}{\text{min}} \rightarrow \dfrac{\text{km}}{\text{mi}} \rightarrow \dfrac{\text{min}}{\text{hr}}$

$$\left(\dfrac{15.2 \text{ mi}}{45 \text{ min}}\right)\left(\dfrac{1.609 \text{ km}}{1 \text{ mi}}\right)\left(\dfrac{60 \text{ min}}{1 \text{ hr}}\right) = 33 \dfrac{\text{km}}{\text{hr}}$$

24.　The conversion is: $\dfrac{\text{km}}{\text{s}} \rightarrow \dfrac{\text{mi}}{\text{km}} \rightarrow \dfrac{\text{s}}{\text{min}} \rightarrow \dfrac{\text{min}}{\text{hr}}$

$$\left(\dfrac{5.0 \text{ km}}{923 \text{ s}}\right)\left(\dfrac{1 \text{ mi}}{1.609 \text{ km}}\right)\left(\dfrac{60 \text{ s}}{1 \text{ min}}\right)\left(\dfrac{60 \text{ min}}{1 \text{ hr}}\right) = 12 \dfrac{\text{mi}}{\text{hr}}$$

25.　The conversion is: $\text{L} \rightarrow \text{mL} \rightarrow \text{mg} \rightarrow \text{g}$

$$(1 \text{ L})\left(\dfrac{1000 \text{ mL}}{1 \text{ L}}\right)\left(\dfrac{500. \text{ mg}}{100. \text{ mL}}\right)\left(\dfrac{1 \text{ g}}{1000 \text{ mg}}\right) = 5.00 \text{ g}$$

26.　The conversion is: $\text{tablet} \rightarrow \text{g} \rightarrow \text{mg} \rightarrow \text{grains}$

$$(1 \text{ tablet})\left(\dfrac{0.500 \text{ g}}{1 \text{ tablet}}\right)\left(\dfrac{1000 \text{ mg}}{1 \text{ g}}\right)\left(\dfrac{1 \text{ grain}}{60 \text{ mg}}\right) = 8.33 \text{ grains}$$

27.　The conversion is: $\text{hr} \rightarrow \text{min} \rightarrow \text{s} \rightarrow \text{m} \rightarrow \text{km} \rightarrow \text{mi}$

$$(5 \text{ hours})\left(\dfrac{60 \text{ min}}{1 \text{ hr}}\right)\left(\dfrac{60 \text{ s}}{1 \text{ min}}\right)\left(\dfrac{0.11 \text{ m}}{1 \text{ s}}\right)\left(\dfrac{1 \text{ km}}{1000 \text{ m}}\right)\left(\dfrac{1 \text{ mi}}{1.61 \text{ km}}\right) = 1.2 \text{ miles}$$

28.　The conversion is: $\text{cm} \rightarrow \text{in.} \rightarrow \text{ft} \rightarrow \text{yr} \rightarrow \text{day}$

$$(1 \text{ cm})\left(\dfrac{1 \text{ in.}}{2.54 \text{ cm}}\right)\left(\dfrac{1 \text{ ft}}{12 \text{ in.}}\right)\left(\dfrac{1 \text{ yr}}{3.38 \text{ ft}}\right)\left(\dfrac{365 \text{ day}}{1 \text{ yr}}\right) = 3.54 \text{ days}$$

29.　The conversion is: $\text{lb body} \rightarrow \text{lb fat} \rightarrow \text{kg fat}$

$$(225 \text{ lb body})\left(\dfrac{11.2 \text{ lb fat}}{100 \text{ lb body}}\right)\left(\dfrac{1 \text{ kg}}{2.205 \text{ lb}}\right) = 11.4 \text{ kg fat}$$

30.　The conversion is: $\text{carats} \rightarrow \text{mg} \rightarrow \text{g} \rightarrow \text{lb}$

$$(5.75 \text{ carats})\left(\dfrac{200 \text{ mg}}{1 \text{ carat}}\right)\left(\dfrac{1 \text{ g}}{1000 \text{ mg}}\right)\left(\dfrac{1 \text{ lb}}{453.6 \text{ g}}\right) = 2.54 \times 10^{-3} \text{ lb}$$

31.　The conversion is: $\dfrac{\text{yd}}{\text{s}} \rightarrow \dfrac{\text{mi}}{\text{yd}} \rightarrow \dfrac{\text{km}}{\text{mi}} \rightarrow \dfrac{\text{m}}{\text{km}} \rightarrow \dfrac{\text{s}}{\text{min}}$

$$\left(\dfrac{100. \text{ yd}}{52 \text{ s}}\right)\left(\dfrac{1 \text{ mi}}{1760 \text{ yd}}\right)\left(\dfrac{1.609 \text{ km}}{1 \text{ mi}}\right)\left(\dfrac{1000 \text{ m}}{1 \text{ km}}\right)\left(\dfrac{60 \text{ s}}{1 \text{ min}}\right) = 1.1 \times 10^2 \dfrac{\text{m}}{\text{min}}$$

32. The conversion is: $\dfrac{mi}{hr} \to \dfrac{km}{mi} \to \dfrac{m}{km} \to \dfrac{cm}{m} \to \dfrac{hr}{min} \to \dfrac{min}{s}$

$$\left(\dfrac{133\ mi}{1\ hr}\right)\left(\dfrac{1.609\ km}{1\ mi}\right)\left(\dfrac{1000\ m}{1\ km}\right)\left(\dfrac{100\ cm}{1\ m}\right)\left(\dfrac{1\ hr}{60\ min}\right)\left(\dfrac{1\ min}{60\ s}\right) = 5.94 \times 10^3\ \dfrac{cm}{s}$$

33. (a) $29{,}035\ ft - 21{,}002\ ft = 8033$ (total feet climbed)

$$(16\ hr)\left(\dfrac{60\ min}{hr}\right) = 960\ min \quad 960\ min + 42\ min = 1.0 \times 10^3\ min$$

$$\left(\dfrac{8003\ ft}{1.0 \times 10^3\ min}\right)\left(\dfrac{1\ mi}{5280\ ft}\right) = 1.5 \times 10^{-3}\ \dfrac{mi}{min}$$

(b) $$\left(\dfrac{8003\ ft}{1.0 \times 10^3\ min}\right)\left(\dfrac{1\ mi}{5280\ ft}\right)\left(\dfrac{1.609\ km}{mi}\right)\left(\dfrac{1000\ m}{km}\right)\left(\dfrac{1\ min}{60\ s}\right) = 4.1 \times 10^{-2}\ \dfrac{m}{s}$$

34. (a) The conversion is: $\dfrac{m}{hr} \to \dfrac{cm}{m} \to \dfrac{in.}{cm} \to \dfrac{ft}{in.} \to \dfrac{hr}{min}$

$$\left(\dfrac{4500\ m}{5\ hr}\right)\left(\dfrac{100\ cm}{m}\right)\left(\dfrac{1\ in.}{2.54\ cm}\right)\left(\dfrac{1\ ft}{12\ in.}\right)\left(\dfrac{1\ hr}{60\ min}\right) = 50\ \dfrac{ft}{min}$$

(b) The conversion is: $\dfrac{m}{hr} \to \dfrac{km}{m} \to \dfrac{hr}{min} \to \dfrac{min}{s}$

$$\left(\dfrac{4500\ m}{5\ hr}\right)\left(\dfrac{1\ km}{1000\ m}\right)\left(\dfrac{1\ hr}{60\ min}\right)\left(\dfrac{1\ min}{60\ s}\right) = 3 \times 10^{-4}\ \dfrac{km}{s}$$

35. The conversion is: gal $\to$ qt $\to$ L $\to$ mL $\to$ cup

$$(2010\ gal)\left(\dfrac{4\ qt}{1\ gal}\right)\left(\dfrac{1\ L}{1.06\ qt}\right)\left(\dfrac{1000\ mL}{1\ L}\right)\left(\dfrac{1\ cup}{473\ mL}\right) = 1.60 \times 10^4\ \text{cups coffee}$$

36. The conversion is: lb $\to$ g $\to$ number of tilapia

$$(475 \times 10^6\ lb)\left(\dfrac{454\ g}{1\ lb}\right)\left(\dfrac{1\ tilapia}{535\ g}\right) = 4.03 \times 10^8\ \text{tilapia}$$

37. The conversion is: gal $\to$ qt $\to$ mL $\to$ drops

$$(1.0\ gal)\left(\dfrac{4\ qt}{gal}\right)\left(\dfrac{946\ mL}{qt}\right)\left(\dfrac{20.\ drops}{mL}\right) = 7.6 \times 10^4\ \text{drops}$$

38. The conversion is: gal $\to$ qt $\to$ L

$$(42\ gal)\left(\dfrac{4\ qt}{gal}\right)\left(\dfrac{0.946\ L}{qt}\right) = 160\ L$$

39. The conversion is: $ft^3 \to in.^3 \to cm^3 \to mL$

$$(1.00\ ft^3)\left(\dfrac{12\ in.}{1\ ft}\right)^3\left(\dfrac{2.54\ cm}{1\ in.}\right)^3\left(\dfrac{1\ mL}{1\ cm^3}\right) = 2.83 \times 10^4\ mL$$

40. $V = A \times h$ $A = \text{area}$ $h = \text{height}$ $V = \text{volume}$

The conversion is: $\dfrac{cm^3}{nm} \rightarrow \dfrac{cm^3}{m} \rightarrow m^2$

$A = \dfrac{V}{h} = \left(\dfrac{200\,cm^3}{0.5\,nm}\right)\left(\dfrac{1\,nm}{10^{-9}\,m}\right)\left(\dfrac{1\,m}{100\,cm}\right)^3 = 4 \times 10^5\,m^2$

41. (a) $(27\,cm)\,(21\,cm)\,(4.4\,cm) = 2.5 \times 10^3\,cm^3$

(b) $2.5 \times 10^3\,cm^3$ is $2.5 \times 10^3\,mL\left(\dfrac{1\,L}{1000\,mL}\right) = 2.5\,L$

(c) $\left(2.5 \times 10^3\,cm^3\right)\left(\dfrac{1\,in.}{2.54\,cm}\right)^3 = 1.5 \times 10^2\,in.^3$

42. $(16\,in.)\,(8\,in.)\,(10\,in.)\left(\dfrac{2.54\,cm}{1\,in.}\right)^3\left(\dfrac{1\,L}{1000\,mL}\right)\left(\dfrac{1\,qt}{0.946\,L}\right)\left(\dfrac{1\,gal}{4\,qt}\right) = 6\,gal$

43. (a) $°F = 1.8°C + 32 = (1.8)\,(38.8) + 32 = 101.8°F$

(b) Yes, the child has a fever since $101.8°F > 98.6°F$

44. $°F = 1.8°C + 32$ $(1.8)\,(45) + 32 = 113°F$ Summer!

45. (a) $\dfrac{162 - 32}{1.8} = 72.2°C$ Remember to express the answer to the same precision as the original measurement.

(b) $°C + 273 = K$ $\dfrac{0.0 - 32}{1.8} + 273 = 255.2\,K$

(c) $1.8(-18) + 32 = -0.40°F$

(d) $212 - 273 = -61°C$

46. (a) $1.8(32) + 32 = 90.°F$

(b) $\dfrac{-8.6 - 32}{1.8} = -22.6°C$

(c) $273 + 273 = 546\,K$

(d) $°C = 100 - 273 = -173°C$

$(-173)(1.8) + 32 = -279°F = -300°F$ (1 significant figure in 100 K)

47. $°F = °C$

$°F = 1.8(°C) + 32$ substitute $°F$ for $°C$

$°F = 1.8(°F) + 32$

$-32 = 0.8(°F)$

$\dfrac{-32}{0.8} = °F$

$-40 = °F$

$-40°F = -40°C$

48.
$$°F = -°C$$
$$°F = 1.8(°C) + 32 \qquad \text{substitute } -°C \text{ for } °F$$
$$-°C = 1.8(°C) + 32$$
$$2.8(°C) = -32$$
$$°C = \frac{-32}{2.8}$$
$$°C = -11.4$$
$$-11.4°C = 11.4°F$$

49. $°F = 1.8°C + 32 \qquad (1.8)(460) + 32 = 860°F$

50. $°C = \dfrac{°F - 32}{1.8} \qquad \dfrac{(-244) - 32}{1.8} = -153°C$

51. $d = \dfrac{m}{v} = \dfrac{59.82\,g}{65.0\,mL} = 0.920\,\dfrac{g}{mL}$

52. $d = \dfrac{m}{v} = \dfrac{20.41\,g}{25.2\,mL} = 0.810\,\dfrac{g}{mL}$

53. $50.92\,g - 25.23\,g = 25.69\,g$ (mass of liquid)
$$d = \frac{m}{v} = \frac{25.69\,g}{25.0\,mL} = 1.03\,\frac{g}{mL}$$

54. $54.6\,mL - 50.0\,mL = 4.6\,mL$ (volume of zinc)
$$d = \frac{m}{v} = \frac{32.95\,g}{4.6\,mL} = 7.2\,\frac{g}{mL}$$

55. The conversion is g → mL
$$(15\,g)\left(\frac{1\,mL}{0.929\,g}\right) = 16\,mL$$

56. The conversion is g → mL
$$(75\,g)\left(\frac{1\,mL}{1.20\,g}\right) = 63\,mL$$

57. (a) report 10.01 grams
    (b) report 10.012 grams
    (c) report 10.0124 grams

58. (a) $(175\,\text{Skittles})\left(\dfrac{1.134\,g\,\text{Skittle}}{1\,\text{Skittle}}\right) = 198\,g\,\text{Skittles}$

    (b) $(175\,\text{Skittles})\left(\dfrac{5.3\,mL\,\text{Skittles}}{6\,\text{Skittles}}\right)\left(\dfrac{1\,L\,\text{Skittles}}{1000\,mL\,\text{Skittles}}\right) = 0.15\,L\,\text{Skittles}$

    (c) $(325.0\,g\,\text{Skittles})\left(\dfrac{1\,\text{Skittle}}{1.134\,g\,\text{Skittles}}\right) = 286.6\,\text{Skittles}$

(d) $(0.550 \text{ L Skittles}) \left(\dfrac{1000 \text{ mL Skittles}}{1 \text{ L Skittles}} \right) \left(\dfrac{6 \text{ Skittles}}{5.3 \text{ mL Skittles}} \right) = 620 \text{ Skittles}$

(e) The mass measurement is more precise. It is also more accurate. Using calculations similar to those above there should be 309 Skittles in 350 grams and the average value is 310 Skittles. There should be 367 Skittles in .325 L of Skittles and the average value is 384 Skittles.

59. A graduated cylinder would be the best choice for adding 100 mL of solvent to a reaction. While the volumetric flask is also labeled 100 mL, volumetric flasks are typically used for doing dilutions. The other three pieces of glassware could also be used, but they hold smaller volumes so it would take a longer time to measure out 100 mL. Also, because you would have to repeat the measurement many times using the other glassware, there is a greater chance for error.

60. The conversion is: $g \rightarrow mL$

$(21.5 \text{ g}) \left(\dfrac{1 \text{ mL}}{1.484 \text{ g}} \right) = 14.5 \text{ mL}$

61. The conversion is: $g \rightarrow mL$

$(25.27 \text{ g}) \left(\dfrac{1 \text{ mL}}{0.97 \text{ g}} \right) = 26 \text{ mL}$

62. The conversion is: $day \rightarrow cups \rightarrow mg \rightarrow g \rightarrow lb$

$(1 \text{ day}) \left(\dfrac{4.00 \times 10^8 \text{ cups}}{\text{day}} \right) \left(\dfrac{160 \text{ mg}}{\text{cup}} \right) \left(\dfrac{1 \text{ g}}{1000 \text{ mg}} \right) \left(\dfrac{1 \text{ lb}}{453.6 \text{ g}} \right) = 1 \times 10^5 \text{ lb}$

63. Mass of gear carried by paladin after drinking strength potion $= 115 \text{ lb} + 50.0 \text{ lb} = 165 \text{ lb}$
Amount of mass potion paladin can carry $= 165 \text{ lb} - 92 \text{ lb} = 73 \text{ lb}$

The conversion is: $lb \rightarrow g \rightarrow mL \rightarrow vials\ potion$

$(73 \text{ lb}) \left(\dfrac{454 \text{ g}}{1 \text{ lb}} \right) \left(\dfrac{1 \text{ ml}}{193 \text{ g}} \right) \left(\dfrac{1 \text{ vial potion}}{50.0 \text{ mL}} \right) = 3.43 \text{ vials potion}$

You would only be able to collect only 3 vials, because 4 would put you over your mass limit.

64. The conversion is: $sequins \rightarrow cm^3 \rightarrow g \rightarrow kg$

$(4560 \text{ sequins}) \left(\dfrac{0.0241 \text{ cm}^3}{1 \text{ sequin}} \right) \left(\dfrac{41.6 \text{ g sequins}}{1 \text{ cm}^3} \right) \left(\dfrac{1 \text{ kg}}{1000 \text{ g}} \right) = 4.57 \text{ kg sequins}$

The conversion is: $kg \rightarrow lb$

$(4.57 \text{ kg sequins}) \left(\dfrac{2.20 \text{ lb}}{1 \text{ kg}} \right) = 10.1 \text{ lb sequins}$

65. $V = \text{side}^3 = (0.50 \text{ m})^3 = 0.13 \text{ m}^3$ \qquad (volume of the cube)

$(0.13 \text{ m}^3) \left(\dfrac{100. \text{ cm}}{\text{m}} \right)^3 \left(\dfrac{1 \text{ L}}{1000 \text{ cm}^3} \right) = 130 \text{ L}$ \qquad (volume of cube)

Yes, the cube will hold the solution. $130\ L - 8.5\ L = 120\ L$ additional solution is necessary to fill the container.

66. The conversion is: $\dfrac{\mu g}{m^3} \to \dfrac{\mu g}{L} \to \dfrac{\mu g}{day}$

$$\left(\dfrac{180\ \mu g}{1\ m^3}\right)\left(\dfrac{1\ m^3}{1000\ L}\right)\left(2 \times 10^4\ \dfrac{L}{day}\right) = 4000\ \mu g\ \text{ingested/day} \qquad \text{(1 sig. figure)}$$

Yes, the nurse is at risk. This is well over the toxic limit.

67. (a) Convert 20.27 K to °C

$$K - 273.15 = {}^\circ C$$
$$20.27\ K - 273.15 = -252.88{}^\circ C$$

(b) Convert 20.27 K to °F

$${}^\circ F = (1.8 \times {}^\circ C) + 32$$
$${}^\circ F = (1.8 \times -252.88) + 32 = -455.18 + 32$$
$${}^\circ F = -423.18{}^\circ F$$

68. $^\circ F = 1.8^\circ C + 32$

Convert 36°C to °F $\qquad (1.8)(36) + 32 = 97{}^\circ F$

Convert 38°C to °F $\qquad (1.8)(38) + 32 = 100{}^\circ F$

Sauropods have a body temperature close to the body temperature of other warm-blooded mammals such as dogs, humans, and cows.

69. The conversion is: $L \to dL \to mg \to g$

$$(4.7\ L)\left(\dfrac{10\ dL}{1\ L}\right)\left(\dfrac{130\ mg}{1\ dL}\right)\left(\dfrac{1\ g}{1000\ mg}\right) = 6.1\ g$$

70. A sample of gold will sink to the bottom of the mercury and a sample of iron pyrite will float.

71. The conversion is: hands $\to$ in. $\to$ cm $\to$ m

$$(14.2\ \text{hands})\left(\dfrac{4\ \text{in.}}{1\ \text{hand}}\right)\left(\dfrac{2.54\ cm}{1\ \text{in.}}\right)\left(\dfrac{1\ m}{100\ cm}\right) = 1.44\ m$$

72. The conversion is: days $\to$ hr $\to$ gal $\to$ qt $\to$ L

$$(30\ \text{days})\left(\dfrac{24\ hr}{1\ day}\right)\left(\dfrac{22.5\ gal}{12\ hr}\right)\left(\dfrac{4\ qt}{1\ gal}\right)\left(\dfrac{0.946\ L}{1\ qt}\right) = 5.1 \times 10^3\ L$$

73. $d = \dfrac{m}{V}$ The cube with the largest volume has the lowest density. Use Table 2.5.

Cube A - lowest density   1.74 g/mL - magnesium
Cube B   2.70 g/mL - aluminum
Cube C - highest density   10.5 g/mL - silver

74. The conversion is: ft → in. → cm → m → nm → nanotubes

$$(40.0\ \text{ft})\left(\frac{12\ \text{in.}}{1\ \text{ft}}\right)\left(\frac{2.54\ \text{cm}}{1\ \text{in.}}\right)\left(\frac{1\ \text{m}}{100\ \text{cm}}\right)\left(\frac{10^9\ \text{nm}}{1\ \text{m}}\right)\left(\frac{1\ \text{nanotube}}{1.3\ \text{nm}}\right) = 9.4 \times 10^9\ \text{nanotubes}$$

75. The conversion is: acres → ft^2 → mi^2 → km^2

$$(125\ \text{acres})\left(\frac{43560\ \text{ft}^2}{1\ \text{acre}}\right)\left(\frac{1\ \text{mi}}{5280\ \text{ft}}\right)^2\left(\frac{1.609\ \text{km}}{1\ \text{mi}}\right)^2 = 0.506\ \text{km}^2$$

76. The volume of the aluminum cube is:

$$V = \frac{m}{d} = \frac{500.\ \text{g}}{2.70\ \dfrac{\text{g}}{\text{mL}}} = 185\ \text{mL} \quad \text{Density of Al is } 2.70\ \text{g/mL}$$

This is the same volume as the gold cube thus:

$$m = dV = (185\ \text{mL})(19.3\ \text{g/mL}) = 3.57 \times 10^3\ \text{g of gold} \quad \text{Density of Au is } 19.3\ \text{g/mL}$$

77. $d = \dfrac{m}{V} = \dfrac{24.12\ \text{g}}{25.0\ \text{mL}} = \dfrac{0.965\ \text{g}}{\text{mL}}$ (density of water at 90°C)

78. $150.50\ \text{g} - 88.25\ \text{g} = 62.25\ \text{g}$ (mass of liquid)

$$d = \frac{m}{V} \text{ thus } V = \frac{m}{d} = \frac{62.25\ \text{g}}{1.25\ \dfrac{\text{g}}{\text{mL}}} = 49.8\ \text{mL} \qquad \text{(volume of liquid)}$$

The container must hold at least 50 mL.

79. H$_2$O $\qquad \dfrac{50\ \text{g}}{1.0\ \dfrac{\text{g}}{\text{mL}}} = 50\ \text{mL}$

alcohol $\qquad \dfrac{50\ \text{g}}{0.789\ \dfrac{\text{g}}{\text{mL}}} = 60\ \text{mL}$

Ethyl alcohol has the greater volume due to its lower density.

80. The conversion is: g → lb → oz

$$(8.1\ \text{g})\left(\frac{1\ \text{lb}}{453.6\ \text{g}}\right)\left(\frac{16\ \text{oz}}{\text{lb}}\right) = 0.29\ \text{oz} \qquad \text{(mass of the coin)}$$

$$(0.29\ \text{oz})\left(\frac{3.5\ \%\ \text{Mn}}{100}\right) = 0.010\ \text{oz Mn}$$

81. Volume of sulfuric acid

$$\left(\frac{1\ \text{mL}}{1.84\ \text{g}}\right)(100.\ \text{g}) = 54.3\ \text{mL}$$

82. The conversion is: cup → oz → qt → L → mg

$$(2.00 \text{ cup coffee})\left(\frac{10 \text{ oz coffee}}{1 \text{ cup coffee}}\right)\left(\frac{1 \text{ qt coffee}}{32 \text{ oz coffee}}\right)\left(\frac{1 \text{ L coffee}}{1.057 \text{ qt coffee}}\right)\left(\frac{31.4 \text{ mg NMP}}{1 \text{ L coffee}}\right) = 18.6 \text{ mg NMP}$$

83. $(1.00 \text{ kg Pd})\left(\frac{1000 \text{ g}}{\text{kg}}\right)\left(\frac{1.00 \text{ mL}}{12.0 \text{ g}}\right) = 83.3 \text{ mL Pd at } 20°\text{C}$

$(1.00 \text{ kg Pd})\left(\frac{1000 \text{ g}}{\text{kg}}\right)\left(\frac{1.00 \text{ mL}}{11.0 \text{ g}}\right) = 90.9 \text{ mL Pd at } 1550°\text{C}$

90.9 mL − 83.3 mL = 7.6 mL change in volume

84.

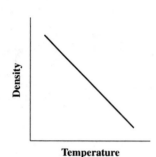

Since $d = \frac{m}{V}$, as the volume increases, the density decreases.

As solids are heated the density decreases due to an increase in the volume of the solid.

85. Original Apple computer

$$\left(\frac{\$9995}{5.0 \text{ Mbytes}}\right)\left(\frac{1 \text{ Mbyte}}{10^6 \text{ byte}}\right) = \frac{\$2.0 \times 10^{-3}}{\text{byte}}$$

iPod II

$$\left(\frac{\$699}{64 \text{ Gbytes}}\right)\left(\frac{1 \text{ Gbyte}}{10^9 \text{ byte}}\right) = \frac{\$1.1 \times 10^{-8}}{\text{byte}}$$

The iPod is definitely a better buy!

86. $V = (2.00 \text{ cm})(15.0 \text{ cm})(6.00 \text{ cm})\left(\frac{1 \text{ mL}}{1 \text{ cm}^3}\right) = 180.\text{mL}$ (volume of bar)

$d = \frac{m}{V} = 3300 \text{ g}/180.\text{ mL} = 18.3 \text{ g/mL}$

The density of pure gold is 19.3 g/mL (from Table 2.5), therefore, the gold bar is not pure gold, since its density is only 18.3 g/mL, or it is hollow inside.

87. $m = dV = (0.789 \text{ g/mL}) (35.0 \text{ mL}) = 27.6 \text{ g ethyl alcohol}$

27.6 g + 49.28 g = 76.9 g (mass of cylinder and alcohol)

88. The conversion is g → oz → gr → scruples

$$(695 \text{ g})\left(\frac{12 \text{ oz}}{373 \text{ g}}\right)\left(\frac{480 \text{ gr}}{\text{oz}}\right)\left(\frac{1 \text{ scruple}}{20 \text{ gr}}\right) = 537 \text{ scruples}$$

89. The conversion is: days → teaspoons → mL

$$(10 \text{ days})\left(\frac{2 \text{ teaspoons} \times 4}{\text{day}}\right)\left(\frac{5 \text{ mL}}{\text{teaspoon}}\right) = 400 \text{ mL}$$

Since you will need a total of 400 mL for the 10 days and the bottle contains 500 mL, you have purchased enough.

90. The density of lead is 11.34 g/mL. The density of aluminum is 2.70 g/mL. The density of silver is 10.5 g/mL. The density of the unknown piece of metal can be calculated from the mass (20.25 g) and the volume (57.5 mL − 50 mL = 7.5 mL) of the metal. Density of the unknown metal = 20.25 g/7.5 mL = 2.7 g/mL. The metal must be aluminum.

91. Volume of slug $\qquad\qquad$ 30.7 mL − 25.0 mL = 5.7 mL

Density of slug $\qquad\qquad d = \dfrac{m}{V} = \dfrac{15.454 \text{ g}}{5.7 \text{ mL}} = 2.7 \text{ g/mL}$

Mass of liquid, cylinder, and slug	125.934 g
Mass of slug(subtract)	−15.454 g
Mass of cylinder(subtract)	−89.450 g
Mass of the liquid	21.030 g

Density of liquid $\quad d = \dfrac{m}{V} = \dfrac{21.030 \text{ g}}{25.0 \text{ mL}} = 0.841 \text{ g/mL}$

92. The conversion is: km → m → cm → in. → ft → mi → hr → min → s → ns

$$(730.0 \text{ km})\left(\frac{1000 \text{ m}}{1 \text{ km}}\right)\left(\frac{100 \text{ cm}}{1 \text{ m}}\right)\left(\frac{1 \text{ in.}}{2.54 \text{ cm}}\right)\left(\frac{1 \text{ ft}}{12 \text{ in.}}\right)\left(\frac{1 \text{ mi}}{5280 \text{ ft}}\right)\left(\frac{1 \text{ hr}}{1.86 \times 10^8 \text{ mi}}\right)\left(\frac{60 \text{ min}}{1 \text{ hr}}\right)\left(\frac{60 \text{ s}}{1 \text{ min}}\right)$$

$$\left(\frac{10^9 \text{ ns}}{1 \text{ s}}\right) = 8.78 \times 10^6 \text{ ns}$$

8780000 ns is good to only 3 sig figs. You would need at least 6 significant digits to detect the difference between 8780000 ns and (8780000 + 60) ns. Sometimes experimenters do not see differences due to the precision of their measuring techniques.

CHAPTER 3

ELEMENTS AND COMPOUNDS

SOLUTIONS TO REVIEW QUESTIONS

1. Silicon   25.7%   Hydrogen   0.9%

    In 100 g $\dfrac{25.7\,\text{g Si}}{0.9\,\text{g H}}$ = 30 g Si/1 g H (1 sig. fig.)

    Si is 28 times heavier than H, thus since 30 > 28, there are more Si atoms than H atoms.

2. (a)   Mn   (c)   Na   (e)   Cl   (g)   Zn
    (b)   F   (d)   He   (f)   V   (h)   N

3. (a)   Iron   (c)   Carbon   (e)   Beryllium   (g)   Argon
    (b)   Magnesium   (d)   Phosphorus (f)   Cobalt   (h)   Mercury

4. The symbol of an element represents the element itself. It may stand for a single atom or a given quantity of the element.

5. Na   sodium   Ag   silver
    K   potassium   W   tungsten
    Fe   iron   Au   gold
    Sb   antimony   Hg   mercury
    Sn   tin   Pb   lead

6. H   hydrogen   S   sulfur
    B   boron   K   potassium
    C   carbon   V   vanadium
    N   nitrogen   Y   yttrium
    O   oxygen   I   iodine
    F   fluorine   W   tungsten
    P   phosphorus   U   uranium

7. 1 metal   0 metalloids   5 nonmetals

8. Hydrogen H_2 Chlorine Cl_2
    Nitrogen N_2 Bromine Br_2
    Oxygen O_2 Iodine I_2
    Fluorine F_2

9. (a)   CO – 1 atom of carbon and 1 atom of oxygen   Co – 1 atom of cobalt
    (b) H_2 – 1 molecule of hydrogen   2 H – 2 atoms of hydrogen
              (made of 2 atoms of hydrogen)
    (c) S_8 – 1 molecule of sulfur (made of 8 atoms of sulfur)   8 S – 8 atoms of sulfur
    (d)   CS – 1 atom of carbon and 1 atom of sulfur   Cs – 1 atom of cesium

10. In an element all atoms are alike, while a compound contains two or more elements (different atoms) which are chemically combined. Compounds can be decomposed into simpler substances while elements cannot.

11. 86 metals   7 metalloids   18 nonmetals (based on 111 elements)

12. 7 metals   1 metalloid   2 nonmetals

13. (a) iodine   (b) bromine

14. A compound is composed of two or more elements which are chemically combined in a definite proportion by mass. Its properties differ from those of its components. A mixture is the physical combining of two or more substances (not necessarily elements). The composition may vary, the substances retain their properties, and they may generally be separated by physical means.

15. Molecular compounds exist as molecules formed from two or more atoms of elements bonded together. Ionic compounds exist as cations and anions held together by electrical attractions.

16. Compounds are distinguished from one another by their characteristic physical and chemical properties.

17. Cations are positively charged, while anions are charged negatively.

SOLUTIONS TO EXERCISES

1. Diatomic molecules: (a) HCl, (b) O_2, (h) ClF

2. Diatomic molecules: (d) HI, (f) Cl_2, (g) CO

3. (a) Potassium, iodine
   (b) Sodium, carbon, oxygen
   (c) Aluminum, oxygen
   (d) Calcium, bromine
   (e) Hydrogen, carbon, oxygen

4. (a) Magnesium, bromine
   (b) Carbon, chlorine
   (c) Hydrogen, nitrogen, oxygen
   (d) Barium, sulfur, oxygen
   (e) Aluminum, phosphorus, oxygen

5. (a) ZnO
   (b) $KClO_3$
   (c) NaOH
   (d) C_2H_6O

6. (a) $AlBr_3$
   (b) CaF_2
   (c) $PbCrO_4$
   (d) C_6H_6

7. (a) $C_6H_{10}OS_2$
   (b) $C_{18}H_{27}NO_3$
   (c) $C_{10}H_{16}$

8. (a) $C_{55}H_{86}NO_{24}$
   (b) $C_{15}H_{14}O_6$
   (c) $C_{30}H_{48}O_3$

9. (a) 2 atoms iron, 3 atoms oxygen
   (b) 2 atoms calcium, 2 atoms nitrogen, 6 atoms oxygen
   (c) 1 atom cobalt, 4 atoms carbon, 6 atoms hydrogen, 4 atoms oxygen
   (d) 3 atoms carbon, 6 atoms hydrogen, 1 atom oxygen
   (e) 2 atoms potassium, 1 atom carbon, 3 atoms oxygen
   (f) 3 atoms copper, 2 atoms phosphorus, 8 atoms oxygen
   (g) 2 atoms carbon, 6 atoms hydrogen, 1 atom oxygen
   (h) 2 atoms sodium, 2 atoms chromium, 7 atoms oxygen

10. (a) 4 atoms hydrogen, 2 atoms carbon, 2 atoms oxygen
    (b) 3 atoms nitrogen, 12 atoms hydrogen, 1 atom phosphorus, 4 atoms oxygen
    (c) 1 atom magnesium, 2 atoms hydrogen, 2 atoms sulfur, 6 atoms oxygen
    (d) 1 atom zinc, 2 atoms chlorine
    (e) 1 atom nickel, 1 atom carbon, 3 atoms oxygen
    (f) 1 atom potassium, 1 atom manganese, 4 atoms oxygen
    (g) 4 atoms carbon, 10 atoms hydrogen
    (h) 1 atom lead, 1 atom chromium, 4 atoms oxygen

11. (a) 9 atoms (b) 14 atoms (c) 11 atoms (d) 45 atoms

12. (a) 9 atoms (b) 12 atoms (c) 12 atoms (d) 12 atoms

13. (a) 9 atoms H   (b) 8 atoms H   (c) 5 atoms H   (d) 10 atoms H

14. (a) 6 atoms O   (b) 4 atoms O   (c) 6 atoms O   (d) 21 atoms O

15. (a) mixture   (d) mixture
    (b) mixture   (e) pure substance
    (c) pure substance   (f) mixture

16. (a) mixture   (d) pure substance
    (b) mixture   (e) mixture
    (c) pure substance   (f) mixture

17. (c) compound   (e) element

18. (c) element   (d) compound

19. (a) mixture   (c) element
    (b) compound

20. (a) compound   (c) mixture
    (b) compound

21. Yes. The gaseous elements are all found on the extreme right of the periodic table. They are the entire last column and in the upper right corner of the table. Hydrogen is the exceptions and located at the upper left of the table.

22. No. The only common liquid elements (at room temperature) are mercury and bromine.

23. $\dfrac{18 \text{ metals}}{36 \text{ elements}} \times 100 = 50\% \text{ metals}$

24. $\dfrac{26 \text{ solids}}{36 \text{ elements}} \times 100 = 72\% \text{ solids}$

25. The formula for water is H_2O. There is one atom of oxygen for every two atoms of hydrogen. The molar mass of oxygen is 16.00 g and the molar mass of hydrogen is 1.008 g. For H_2O the mass of two hydrogen atoms is 2.016 g and the mass of one oxygen atom is 16.00 g. The ratio of hydrogen to oxygen is approximately 2:16 or 1:8. Therefore, there is 1 gram of hydrogen for every 8 grams of oxygen.

26. The formula for hydrogen peroxide is H_2O_2. There are two atoms of oxygen for every two atoms of hydrogen. The molar mass of oxygen is 16.00 g and the molar mass of hydrogen is 1.008 g. For hydrogen peroxide the total mass of hydrogen is 2.016 g and the total mass of oxygen is 32.00 g for a ratio of hydrogen to oxygen of approximately 2: 32 or 1:16. Therefore, there is 1 gram of hydrogen for every 16 grams of oxygen.

27. To a small sample of the mixture, add water and observe that the salt dissolves but the pepper does not. After the small trial, add water to the entire mixture to dissolve the salt. Separate the undissolved pepper from the salt solution by filtering the mixture. Coffee filters or strong paper towels would work well for this process.

28. The atoms that make up each ionic compound are on opposite ends of the periodic table from one another. An ionic compound is made up of a metal-nonmetal combination.

29. (a) 1 carbon atom and 1 oxygen atom, total number of atoms = 2
    (b) 1 boron atom and 3 fluorine atoms, total number of atoms = 4
    (c) 1 hydrogen atom, 1 nitrogen atom, 3 oxygen atoms, total number of atoms = 5
    (d) 1 potassium atom, 1 manganese atom, 4 oxygen atoms, total number of atoms = 6
    (e) 1 calcium atom, 2 nitrogen atoms, 6 oxygen atoms, total number of atoms = 9
    (f) 3 iron atoms, 2 phosphorus atoms, 8 oxygen atoms, total number of atoms = 13

30. (a) 181 atoms/module

    $$63\,C$$
    $$88\,N$$
    $$1\,Co$$
    $$14\,N$$
    $$14\,O$$
    $$\underline{1\,P}$$
    $$181\,atoms$$

    (b) $\dfrac{63\,C}{181\,atoms} \times 100 = 35\%\,C\,atoms$

    (c) $\dfrac{1\,Co}{181\,atoms} = \dfrac{1}{181}\,metals$

31. The conversion is: $cm^3 \rightarrow L \rightarrow mg \rightarrow g \rightarrow \$$

    $$\left(1 \times 10^{15}\,cm^3\right)\left(\frac{1\,L}{1000\,cm^3}\right)\left(\frac{4 \times 10^{-4}\,mg}{L}\right)\left(\frac{1\,g}{1000\,mg}\right)\left(\frac{\$19.40}{g}\right) = \$8 \times 10^6$$

32. $Ca(H_2PO_4)_2$

    $$(10\,formula\,units)\left(\frac{4\,atoms\,H}{formula\,unit}\right) = 40\,atoms\,H$$

33. $C_{145}H_{293}O_{168}$

    $$145\,C$$
    $$293\,H$$
    $$\underline{168\,O}$$
    $$606\,atoms/molecule$$

34. (a) magnesium, manganese, molybdenum, mendeleevium, mercury, meitnerium
    (b) carbon, phosphorus, sulfur, selenium, iodine, astatine, boron
    (c) sodium, potassium, iron, silver, tin, antimony

35. Add water to the mixture to dissolve the sugar. Filter the mixture to separate the sugar solution from the insoluble sand. Add another small amount of water to remove last traces of sugar. Filter. Allow the water to evaporate from the sugar solution to obtain crystals of sugar. Sand is the insoluble residue.

36. HNO_3 has 5 atoms/molecule

7 dozen = 84

(84 molecules)(5 atoms/molecule) = 420 atoms

or $(7 \text{ dz}) \left(\dfrac{12 \text{ molecules}}{\text{dz}} \right) \left(\dfrac{5 \text{ atoms}}{\text{molecule}} \right) = 420$ atoms

37.

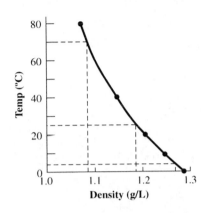

(a) As temperatures decreases, density increases.
(b) approximately 1.28 g/L at 5°C
     approximately 1.18 g/L at 25°C
     approximately 1.09 g/L at 70°C

38. Each represents eight units of sulfur. In 8 S the atoms are separate and distinct. In S_8 the atoms are joined as a unit (molecule).

39. (a) NaCl   (d) Fe_2S_3 (g) $C_6H_{12}O_6$
     (b) H_2SO_4 (e) K_3PO_4 (h) C_2H_5OH
     (c) K_2O (f) $Ca(CN)_2$ (i) $Cr(NO_3)_3$

40. Let X = grams sea water

$$\left(\frac{5.0 \times 10^{-8} \% \text{ I}_2}{100} \right)(X) = 1.0 \text{ g I}_2$$

$$X = \frac{(1.0 \text{ g I}_2)(100)}{(5.0 \times 10^{-8} \% \text{ I}_2)} = 2.0 \times 10^9 \text{ g sea water}$$

$$(2.0 \times 10^9 \text{g}) \left(\frac{1 \text{ kg}}{1000 \text{ g}} \right) = 2.0 \times 10^6 \text{ kg sea water}$$

41. (a) 12 carbons, 22 hydrogens, 11 oxygens
     (b) 7 carbons, 5 hydrogens, 3 oxygens, 1 nitrogen, 1 sulfur
     (c) 14 carbons, 18 hydrogens, 5 oxygens, 2 nitrogens
     (d) 4 carbons, 4 hydrogens, 3 oxygens, 1 nitrogen, 1 sulfur, 1 potassium
     (e) 12 carbons, 19 hydrogens, 8 oxygens, 3 chlorines

42. Cobalt should be written Co as CO is the formula for carbon monoxide.

43. $C_8N_4O_2H_{10}$

44. (a) NH_4Cl (d) FeF_2
    (b) H_2SO_4 (e) $Pb_3(PO_4)_2$
    (c) MgI_2 (f) Al_2O_3

45. Group 1A oxides: Li_2O, Na_2O, K_2O, Rb_2O, Cs_2O
    Group 2A oxides: BeO, MgO, CaO, SrO, BaO

46. (a)   Arachidic acid   20 carbons, 40 hydrogens, and 2 oxygens
            Arachidonic acid   20 carbons, 32 hydrogens, and 2 oxygens

    (b)   Stearic acid   18 carbons, 36 hydrogens, and 2 oxygens
            Linoleic acid   18 carbons, 32 hydrogens, and 2 oxygens

    (c)   Arachidic acid $\dfrac{40\,H}{20\,C}=\dfrac{2\,H}{C}$ Arachidonic acid $\dfrac{32\,H}{20\,C}=\dfrac{1.6\,H}{C}$

         Stearic acid $\dfrac{36\,H}{18\,C}=\dfrac{2\,H}{C}$ Linoleic acid $\dfrac{32\,H}{18\,C}=\dfrac{1.8\,H}{C}$

    (d)   Saturated molecules have more H's per C than unsaturated molecules. Saturated molecules must have more hydrogen atoms.

CHAPTER 4

PROPERTIES OF MATTER

SOLUTIONS TO REVIEW QUESTIONS

1. Gaseous state: 393 K = 120.0°C; boiling point of acetic acid is 118.0°C.

2. Liquid state: melting point of chlorine is −101.6°C and boiling point of chlorine is −34.6°C.

3. (a) $118.0°C + 273.15 = 391.2$ K
   (b) $(118.0°C)(1.8) + 32 = 244.4°F$

4. (a) $(16.7°C)(1.8) + 32 = 62.1°F$
   (b) $16.7°C + 273.15 = 289.9$ K

5. Small bubbles appear at each electrode, and a gas collects above each electrode. The system now contains water and two different gases.

6. A new substance is always formed during a chemical change, but never formed during physical changes.

7. Reading the problem carefully and writing down all of the important information including units is the critical first step in solving any chemistry problem.

8. No, the last step is always a check of the answer to make sure it makes sense.

9. Potential energy is the energy of position. By the position of an object, it has the potential of movement to a lower energy state. Kinetic energy is the energy matter possesses due to its motion.

10. A food Calorie is equal to 1000 cal or 1 kcal.

11. Iron requires more energy to heat because it has a higher specific heat.

12. A molecule of octane would produce more carbon dioxide because it contains more carbon atoms.

13. All hydrocarbons are composed only of the elements carbon and hydrogen.

14. Carbon

15. Fossil fuels produce carbon dioxide, a greenhouse gas. The supply of fossil fuels is also decreasing and mining or drilling for these fuels is harmful to the environment. Other sources of energy which may become important are wind and solar.

SOLUTIONS TO EXERCISES

1. (a) physical
   (b) chemical
   (c) physical
   (d) physical

   (e) chemical
   (f) physical
   (g) chemical
   (h) chemical

2. (a) physical
   (b) physical
   (c) chemical
   (d) physical

   (e) physical
   (f) physical
   (g) chemical
   (h) chemical

3. Although the appearance of the platinum wire changed during the heating, the original appearance was restored when the wire cooled. No change in the composition of the platinum could be detected.

4. A copper wire, like the platinum wire, changes to a glowing red color when heated (physical change). Upon cooling, the original appearance of the copper wire is not restored, but a new substance, black copper(II) oxide appears (chemical change).

5. Reactants: copper, oxygen
   Product: copper(II) oxide

6. Reactant: water
   Product: hydrogen, oxygen

7. (a) chemical
   (b) physical
   (c) chemical

   (d) chemical
   (e) chemical
   (f) physical

8. (a) physical
   (b) chemical
   (c) chemical

   (d) physical
   (e) chemical
   (f) chemical

9. (a) kinetic energy
   (b) kinetic energy
   (c) potential energy

   (d) potential energy
   (e) kinetic energy

10. (a) kinetic energy
    (b) potential energy
    (c) potential energy

    (d) kinetic energy
    (e) kinetic energy

11. The kinetic energy is converted to thermal energy (heat), chiefly in the brake system, and eventually dissipated into the atmosphere.

12. The transformation of kinetic energy to thermal energy (heat) is responsible for the fiery reentry of a space vehicle.

13. (a) +   (d) +
     (b) +   (e) −
     (c) −

14. (a) +   (d) −
     (b) −   (e) −
     (c) +

15. heat = (m)(specific heat)(Δt)
     $$= (125 \text{ g})(0.900 \text{ J/g}^\circ\text{C})(95.5^\circ\text{C} - 19.0^\circ\text{C})$$
     $$= 8.61 \times 10^3 \text{ J}$$

16. heat = (m)(specific heat)(Δt)
     $$= (65 \text{ g})(0.128 \text{ J/g}^\circ\text{C})(98.5^\circ\text{C} - 22.0^\circ\text{C})$$
     $$= 6.4 \times 10^2 \text{ J}$$

17. heat = (m)(specific heat)(Δt)
     $$= (25.0 \text{ g})(2.138 \text{ J/g}^\circ\text{C})(78.5^\circ\text{C} - 22.5^\circ\text{C})$$
     $$= 2.99 \times 10^3 \text{ J}$$

18. heat = (m)(specific heat)(Δt)
     $$= (35.0 \text{ g})(2.604 \text{ J/g}^\circ\text{C})(82.4^\circ\text{C} - 21.2^\circ\text{C})$$
     $$= 5.58 \times 10^3 \text{ J}$$

19. heat = (m)(specific heat)(Δt); change kJ to J

     $$\text{specific heat} = \frac{\text{heat}}{\text{m}(\Delta\text{t})} = \frac{2.50 \times 10^3 \text{ J}}{(135 \text{ g})(100.0^\circ\text{C} - 19.5^\circ\text{C})} = 0.230 \text{ J/g}^\circ\text{C}$$

20. heat = (m)(specific heat)(Δt); change kJ to J

     $$\text{specific heat} = \frac{\text{heat}}{\text{m}(\Delta\text{t})} = \frac{1.075 \times 10^4 \text{ J}}{(275 \text{ g})(327.5^\circ\text{C} - 21.2^\circ\text{C})} = 0.128 \text{ J/g}^\circ\text{C}$$

21. heat lost by copper = heat gained by water x = final temperature
     $$\text{(m)(specific heat)}(\Delta\text{t}) = \text{(m)(specific heat)}(\Delta\text{t})$$
     $$(155.0 \text{ g})(0.385 \text{ J/g}^\circ\text{C})(150.0^\circ\text{C} - x) = (250.0 \text{g})(4.184 \text{ J/g}^\circ\text{C})(x - 19.8\ ^\circ\text{C})$$
     $$8950 \text{ J} - 59.675x \text{ J}/^\circ\text{C} = 1046\, x \text{ J}/^\circ\text{C} - 20710 \text{ J}$$
     $$8950 \text{ J} + 20710 \text{ J} = 1046\, x \text{ J}/^\circ\text{C} + 59.675x \text{ J}/^\circ\text{C}$$
     $$29660 \text{ J} = 1106\, x \text{ J}/^\circ\text{C}$$
     $$26.8^\circ\text{C} = x$$

22. heat lost by copper = heat gained by water $\quad x$ = final temperature

$$(m)(\text{specific heat})(\Delta t) = (m)(\text{specific heat})(\Delta t)$$
$$(225.0\,\text{g})(0.900\,\text{J/g°C})(125.5°\text{C} - x) = (500.0\,\text{g})(4.184\,\text{J/g°C})(x - 22.5\,°\text{C})$$
$$25410\,\text{J} - 202.5x\,\text{J/°C} = 2090\,x\,\text{J/°C} - 47070\,\text{J}$$
$$25410\,\text{J} + 47070\,\text{J} = 2090\,x\,\text{J/°C} + 202.5x\,\text{J/°C}$$
$$72480\,\text{J} = 2290\,x\,\text{J/°C}$$
$$31.7°\text{C} = x$$

23. Physical properties of zeolites

   Crystalline solids, low density, porous, absorb water

   Chemical properties of zeolites

   Composed of silicon, aluminum, and oxygen, formed from reaction of volcanic rocks and ash with alkaline water

24. (a) heat lost by metal = heat gained by water

$$(m)(\text{specific heat})(\Delta t) = (m)(\text{specific heat})(\Delta t)$$
$$(110.0\,\text{g})(\text{specific heat})(92.0°\text{C} - 24.2°\text{C}) = (75.0\,\text{g})(4.184\,\text{J/g°C})(24.2°\text{C} - 21.0°\text{C})$$
$$(\text{specific heat})\,7458\,\text{g°C} = 1004.16\,\text{J}$$
$$\text{specific heat} = 0.13\,\text{J/g°C}$$

   (b) Iron has a specific heat of 0.47 J/g°C and lead has a specific heat of 0.13 J/g°C. The metal could possibly be lead, but further tests would be needed to determine this. The metal cannot be iron.

25. (a) heat = 262 Cal or 262 kcal or 262,000 cal

$$\text{Mass of man} = (165\,\text{lb})\left(\frac{453.6\,\text{g}}{1\,\text{lb}}\right) = 7.48 \times 10^4\,\text{g}$$
$$\text{specific heat} = \left(\frac{3.47\,\text{J}}{\text{g°C}}\right)\left(\frac{1\,\text{cal}}{4.184\,\text{J}}\right) = \frac{0.829\,\text{cal}}{\text{g°C}}$$
$$\text{heat} = (m)(\text{specific heat})(\Delta t)$$
$$\Delta t = \frac{\text{heat}}{(m)(\text{specific heat})} = \frac{(262000\,\text{cal})}{(7.48 \times 10^4\,\text{g})(0.829\,\text{cal/g°C})} = 4.22°\text{C}$$

   (b) heat = 25.0 Cal or 25.0 kcal or 25000 cal

$$\text{Mass of man} = (165\,\text{lb})\left(\frac{453.6\,\text{g}}{1\,\text{lb}}\right) = 7.48 \times 10^4\,\text{g}$$
$$\text{specific heat} = \left(\frac{3.47\,\text{J}}{\text{g°C}}\right)\left(\frac{1\,\text{cal}}{4.184\,\text{J}}\frac{0.829\,\text{cal}}{\text{g°C}}\right)$$
$$\text{heat} = (m)(\text{specific heat})(\Delta t)$$
$$\Delta t = \frac{\text{heat}}{(m)(\text{specific heat})} = \frac{(25000\,\text{cal})}{(7.48 \times 10^4\,\text{g})(0.829\,\text{cal/g°C})} = 0.403°\text{C}$$

26. $\text{heat} = (\text{m})(\text{specific heat})(\Delta t)$

$$\text{mass} = \frac{\text{heat}}{(\text{specific heat})(\Delta t)} = \frac{1.69 \times 10^4 \text{ J}}{(4.184 \text{ J/g°C})(100.0°\text{C} - 23.0°\text{C})} = 52.5 \text{ g H}_2\text{O}$$

27. $\text{heat} = (\text{m})(\text{specific heat})(\Delta t)$

$$\text{mass} = \frac{\text{heat}}{(\text{specific heat})(\Delta t)}; \text{ convert grams to pounds}$$

$$= \frac{3.25 \times 10^4 \text{ J}}{(0.131 \text{ J/g°C})(1064.4°\text{C} - 23.2°\text{C})} = 238 \text{ g gold}$$

$$(238 \text{ g})\left(\frac{1 \text{ lb}}{453.6 \text{ g}}\right) = 0.525 \text{ lb gold}$$

28. $\text{heat} = (\text{m})(\text{specific heat})(\Delta t)$ change kJ to J $x = $ final temperature

$4.00 \times 10^4 \text{ J} = (500.0 \text{ g})(4.184 \text{ J/g°C})(x - 10.0°\text{C})$

$4.00 \times 10^4 \text{ J} = 2092x \text{ J/°C} - 20920 \text{ J}$

$60{,}920 \text{ J} = 2092x \text{ J/°C}$

$60{,}920°\text{C} = 2092x$

$29.1°\text{C} = x$

29. heat lost by iron = heat gained by water $x = $ mass of iron in grams

$(\text{m})(\text{specific heat})(\Delta t) = (\text{m})(\text{specific heat})(\Delta t)$

$x(0.473 \text{ J/g°C})(125.0°\text{C} - 25.6°\text{C}) = (375 \text{ g})(4.184 \text{ J/g°C})(25.6°\text{C} - 19.8°\text{C})$

$47.0x \text{ J/g} = 9.1 \times 10^3 \text{ J}$

$$x = \frac{9.1 \times 10^3 \text{ J}}{47.0 \text{ J/g}} = 190 \text{ g Fe}$$

30. heat lost by copper = heat gained by water $x = $ mass of copper in grams

$(\text{m})(\text{specific heat})(\Delta t) = (\text{m})(\text{specific heat})(\Delta t)$

$x(0.385 \text{ J/g°C})(275.1°\text{C} - 29.7°\text{C}) = (272 \text{ g})(4.184 \text{ J/g°C})(29.7°\text{C} - 21.0°\text{C})$

$94.5x \text{ J/g} = 9.9 \times 10^3 \text{ J}$

$$x = \frac{9.9 \times 10^3 \text{ J}}{94.5 \text{ J/g}} = 1.0 \times 10^2 \text{ g Cu}$$

31. (a) $\text{heat} = (\text{m})(\text{specific heat})(\Delta t)$

$(100.0 \text{ g})(0.0921 \text{ cal/g°C})(100.0°\text{C} - 10.0°\text{C}) = 829 \text{ cal to heat Cu}$

(b)   let $x = $ temperature of Al after adding 829 cal

$829 \text{ cal} = (100.0 \text{ g})(0.215 \text{ cal/g°C})(x - 10.0°\text{C})$

$829 \text{ cal} = (21.5 \text{ cal/°C})x - 215 \text{ cal}$

$x = 48.6°\text{C}$ (final temperature for aluminum)

Therefore the copper gets hotter since it ended up at 100.0°C.

Note: You can figure this out without calculation if you consider the specific heats of the metals. Since the specific heat of copper is much less than aluminum the copper heats more easily.

32. heat lost by iron = heat gained by water $\qquad$ x = initial temperature of iron

$$(m)(\text{specific heat})(\Delta t) = (m)(\text{specific heat})(\Delta t)$$

$$(500.0\,\text{g})(0.473\,\text{J/g°C})(x - 90.0°\text{C}) = (400.0\,\text{g})(4.184\,\text{J/g°C})(90.0°\text{C} - 10.0°\text{C})$$

$$\left(237\,\frac{\text{J}}{°\text{C}}\right)x - 2.13 \times 10^4\,\text{J} = 1.34 \times 10^5\,\text{J}$$

$$\left(237\,\frac{\text{J}}{°\text{C}}\right)x = 1.55 \times 10^5\,\text{J°C}$$

$$x = \frac{1.55 \times 10^5\,\text{J°C}}{237\,\text{J}}$$

$$x = 654°\text{C}$$

33. heat lost by metal = heat gained by water $\qquad$ x = specific heat of metal

$$(m)(\text{specific heat})(\Delta t) = (m)(\text{specific heat})(\Delta t)$$

$$(20.0\,\text{g})(x)(203.°\text{C} - 29.0°\text{C}) = (100.0\,\text{g})(4.184\,\text{J/g°C})(29.0°\text{C} - 25.0°\text{C})$$

$$(3480\,\text{g°C})x = 1674\,\text{J}$$

$$x = 0.48\,\text{J}$$

34. heat lost = heat gained $\qquad$ x = final temperature

$$(m)(\text{specific heat})(\Delta t) = (m)(\text{specific heat})(\Delta t) \qquad (\text{specific heats are the same})$$

Let x = final temperature

$$(10.0\,\text{g})\left(\frac{\text{J}}{\text{g°C}}\right)(50.0°\text{C} - x) = (50.0\,\text{g})\left(\frac{\text{J}}{\text{g°C}}\right)(x - 10.0°\text{C})$$

$$500.\,\text{g°C} - 10.0\,\text{g}x = 50.0\,\text{g}x - 500\,\text{g°C}$$

$$1.00 \times 10^3\,\text{g°C} = 60.0\,\text{g}x$$

$$x = \frac{1.00 \times 10^3\,\text{g°C}}{60.0\,\text{g}} = 16.7°\text{C}$$

35. Specific heats for the metals are Fe: 0.473 J/g°C; Cu: 0.385 J/g°C; Al: 0.900 J/g°C. The metal with the lowest specific heat will warm most quickly, therefore, the copper pan heats fastest, and fries the egg fastest.

36. In order for the water to boil both the pan and water must reach 100.0°C. Specific heat for copper is 0.385 J/g°C.

$$(300.0\,\text{g})\left(0.385\,\frac{\text{J}}{\text{g°C}}\right)(100.°\text{C} - 25°\text{C}) + (800.0\,\text{g})\left(4.184\,\frac{\text{J}}{\text{g°C}}\right)(100.°\text{C} - 25°\text{C})$$

$$= 8.7 \times 10^3\,\text{J} + 2.5 \times 10^5\,\text{J}$$

$$= 2.6 \times 10^5\,\text{J needed to heat the pan and water to } 100°\text{C}$$

$$(2.6 \times 10^5\,\text{J})\left(\frac{1\,\text{s}}{628\,\text{J}}\right) = 414\,\text{s} = 6.9\,\text{min} = 6\,\text{min} + 54\,\text{s}$$

The water will boil at 6:06 and 54 s p.m.

37. Heat is transferred from the molecules of water on the surface to the air above them. As you blow you move the warmed air molecules away from the surface replacing them with cooler ones which are warmed by the coffee and cool it in a repeating cycle. Inserting a spoon into hot coffee cools the coffee by heat transfer as well. Heat is transferred from the coffee to the spoon lowering the temperature of the coffee and raising the temperature of the spoon.

38. The potatoes will cook at the same rate whether the water boils vigorously or slowly. Once the boiling point is reached the water temperature remains constant. The energy available is the same so the cooking time should be equal.

39. $(250\,\text{mL})(0.04) = 10\,\text{mL fat}$

    $(10\,\text{mL})(0.8\,\text{g/mL}) = 8\,\text{g fat in a glass of milk}$

40. (a) $\text{heat} = (m)(\text{specific heat})(\Delta t)$

    $m = 99.3\,\text{g}$

    $\text{specific heat} = \dfrac{1.33\,\text{J}}{\text{g}^\circ\text{C}}$

    $\Delta t = 15.3^\circ\text{F} \times \dfrac{100^\circ\text{C}}{180^\circ\text{F}} = 8.50^\circ\text{C}$

    $\text{heat} = (99.3\,\text{g})\left(\dfrac{1.33\,\text{J}}{\text{g}^\circ\text{C}}\right)(8.50^\circ\text{C}) = 1120\,\text{J or } 1.12\,\text{kJ}$

    (b) $\text{heat} = (m)(\text{specific heat})(\Delta t)$

    $m = 86.2\,\text{g}$

    $\Delta t = 25.9^\circ\text{F} \times \dfrac{100^\circ\text{C}}{180^\circ\text{F}} = 14.4^\circ\text{C}$

    $\text{heat} = 1.71\,\text{kJ} = 1710\,\text{J}$

    $\text{specific heat} = \dfrac{(\text{heat})}{(m)(\Delta t)} = \dfrac{(1710\,\text{J})}{(86.2\,\text{g})(14.4^\circ\text{C})} = \dfrac{1.38\,\text{J}}{\text{g}^\circ\text{C}}$

    The glove appears to be made of wool felt.

    (c) $\text{heat} = 1.65\,\text{kJ or } 1650\,\text{J}$

    $m = 50.0\,\text{g}$

    $\text{heat} = (m)(\text{specific heat})(\Delta t)$

    $\Delta t = \dfrac{\text{heat}}{(m)(\text{specific heat})}$

For wool felt specific heat $= \dfrac{1.38\,\text{J}}{\text{g}^{\circ}\text{C}}$

$$\Delta t = \frac{\text{heat}}{(\text{m})(\text{specific heat})} = \frac{(1650\,\text{cal})}{(50.0\,\text{g})(1.38\,\text{cal}/\text{g}^{\circ}\text{C})} = 23.9^{\circ}\text{C or } 43.0^{\circ}\text{F}$$

For cotton or paper specific heat $= \dfrac{1.33\,\text{J}}{\text{g}^{\circ}\text{C}}$

$$\Delta t = \frac{\text{heat}}{(\text{m})(\text{specific heat})} = \frac{(1650\,\text{cal})}{(50.0\,\text{g})(1.33\,\text{cal}/\text{g}^{\circ}\text{C})} = 24.8^{\circ}\text{C or } 44.7^{\circ}\text{F}$$

For rubber specific heat $= \dfrac{3.65\,\text{J}}{\text{g}^{\circ}\text{C}}$

$$\Delta t = \frac{\text{heat}}{(\text{m})(\text{specific heat})} = \frac{(1650\,\text{cal})}{(50.0\,\text{g})(3.65\,\text{cal}/\text{g}^{\circ}\text{C})} = 9.04^{\circ}\text{C or } 16.3^{\circ}\text{F}$$

For silicon specific heat $= \dfrac{1.46\,\text{J}}{\text{g}^{\circ}\text{C}}$

$$\Delta t = \frac{\text{heat}}{(\text{m})(\text{specific heat})} = \frac{(1650\,\text{cal})}{(50.0\,\text{g})(1.46\,\text{cal}/\text{g}^{\circ}\text{C})} = 22.6^{\circ}\text{C or } 40.7^{\circ}\text{F}$$

(d) Rubber

(e) The higher the specific heat the less the glove will increase in temperature as it absorbs heat. This means a high specific heat is good for an oven mitt.

41. Note that $100\,\text{MW} = \dfrac{100\,\text{MJ}}{\text{s}}$

$$(1\,\text{day})\left(\frac{7.50\,\text{hr}}{1\,\text{day}}\right)\left(\frac{60\,\text{min}}{1\,\text{hr}}\right)\left(\frac{60\,\text{s}}{1\,\text{min}}\right)\left(\frac{100\,\text{MJ}}{\text{s}}\right)\left(\frac{10^{6}\,\text{J}}{1\,\text{MJ}}\right)\left(\frac{1\,\text{GJ}}{10^{9}\,\text{J}}\right)\left(\frac{1\,\text{ton coal}}{26.6\,\text{GJ}}\right) = 102\,\text{tons of coal}$$

42. According to the law of conservation of energy the amount of potential energy the ball has initially should equal the amount of kinetic energy it has at the bottom of the hill. If the hill that the ball rolled down was frictionless then the ball should role up the other side until it has reached the same level as where it started. However, no hill is really frictionless. Therefore, some of the ball's potential energy is converted to kinetic energy of motion and some is lost by way of heat friction between the ball and the surface of the hill.

43. A chemical change has occurred. Hydrogen molecules and oxygen molecules have combined to form water molecules.

44. (a) As the water is heated, molecules that exist in the liquid state are changed into molecules in the gas gaseous state.
     (b) A physical change has occurred. No new substance was formed during the heating process. Only a change in state occurred.

45. Upon heating the substance, the appearance of the bright blue solid changes to black and a brownish orange gas is released. A chemical change has occurred.

CHAPTER 5

EARLY ATOMIC THEORY AND STRUCTURE

SOLUTIONS TO REVIEW QUESTIONS

1. • Elements are composed of indivisable particles called atoms.
   • Atoms of the same element have the same properties; atoms of different elements have different properties.
   • Compounds are composed of atoms joined together to form compounds.
   • Atoms combine in whole number ratios to form compounds.
   • Atoms may combine in different ratios to form more than one compound.

2. Dalton used Democritus' idea that all matter was composed of tiny indivisible particles or atomos when formulating his theory.

3. Dalton said that compounds could form only by combining whole atoms, not parts of atoms. Thus chemical formulas will always show whole numbers of atoms combining.

4. An atom is electrically neutral, containing equal numbers of protons and electrons. An ion has a charge resulting from an imbalance between the numbers of protons and electrons.

5. The force of attraction increases as the distance between the charged particles decreases.

6. Cations are ions with a positive charge and anions are ions with a negative charge.

7. The neutron is about 1840 times heavier than an electron.

8. | Particle | Charge | Mass |
   |----------|--------|------|
   | proton | +1 | 1 amu |
   | neutron | 0 | 1 amu |
   | electron | −1 | 0 |

9. | | Element | Atomic number |
   |---|---------|---------------|
   | (a) | copper | 29 |
   | (b) | nitrogen | 7 |
   | (c) | phosphorus | 15 |
   | (d) | radium | 88 |
   | (e) | zinc | 30 |

10. Isotopic notation $_Z^A E$

    Z represents the atomic number

    A represents the mass number

11. Isotopes contain the same number of protons and the same number of electrons. Isotopes have different numbers of neutrons and thus different atomic masses.

12. The mass number is equal to the sum of the number of protons and the number of neutrons in an atom. It is not possible to have a partial proton or neutron in an atom, thus the total number of nuclear particles will always be a whole number.

SOLUTIONS TO EXERCISES

1. N^{3-}, O^{2-}, and Te^{3-}

2. Mg^{2+}, Cr^{3+}, Ba^{2+}, Ca^{2+}, and Y^{3+}

3. Gold nuclei are very massive (compared to an alpha particle) and have a large positive charge. As the positive alpha particles approach the atom, some are deflected by this positive charge. Alpha particles approaching a gold nucleus directly are deflected backwards by the massive positive nucleus.

4. (a) The nucleus of the atom contains most of the mass since only a collision with a very dense, massive object would cause an alpha particle to be deflected back towards the source.

   (b) The deflection of the positive alpha particles from their initial flight indicates the nucleus of the atom is also positively charged.

   (c) Most alpha particles pass through the gold foil undeflected leading to the conclusion that the atom is mostly empty space.

5. In the atom, protons and neutrons are found within the nucleus. Electrons occupy the remaining space within the atom outside the nucleus.

6. The nucleus of an atom contains nearly all of its mass.

7. (a) Dalton contributed the concept that each element is composed of atoms which are unique, and can combine in ratios of small whole numbers.

   (b) Thomson discovered the electron, determined its properties, and found that the mass of a proton is 1840 times the mass of the electron. He developed the Thomson model of the atom.

   (c) Rutherford devised the model of a nuclear atom with the positive charge and mass concentrated in the nucleus. Most of the atom is empty space.

8. Electrons: Dalton – electrons are not part of his model
   Thomson – electrons are scattered throughout the positive mass of matter in the atom
   Rutherford – electrons are located out in space away from the central positive nucleus

   Positive matter: Dalton – no positive matter in his model
   Thomson – positive matter is distributed throughout the atom
   Rutherford – positive matter is concentrated in a small central nucleus

9. Atomic masses are not whole numbers because:
   (a) the neutron and proton do not have identical masses and neither is exactly 1 amu.
   (b) most elements exist in nature as a mixture of isotopes with different atomic masses due to different numbers of neutrons. The atomic mass given in the periodic table is the average mass of all these isotopes.

10. The isotope of $^{12}_{6}C$ with a mass of 12 is an exact number by definition. The mass of other isotopes, such as $^{63}_{29}Cu$, will not be an exact number for reasons given in Exercise 9.

11. The isotopes of hydrogen are protium, deuterium, and tritium.

12. All three isotopes of hydrogen have the same number of protons (1) and electrons (1). They differ in the number of neutrons (0, 1, and 2).

13. All five isotopes have nuclei that contain 32 protons. The numbers of neutrons are:

Isotope mass number	Neutrons
70	38
72	40
73	41
74	42
76	44

14. All five isotopes have nuclei that contain 30 protons. The numbers of neutrons are:

Isotope mass number	Neutrons
64	34
66	36
67	37
68	38
70	40

15. (a) $^{65}_{29}Cu$

    (b) $^{45}_{20}Ca$

    (c) $^{84}_{36}Kr$

16. (a) $^{109}_{47}Ag$

    (b) $^{18}_{8}O$

    (c) $^{57}_{26}Fe$

17. (a) 29 protons, 29 electrons, and 34 neutrons
    (b) 16 protons, 16 electrons, and 16 neutrons
    (c) 25 protons, 25 electrons, and 30 neutrons
    (d) 19 protons, 19 electrons, and 20 neutrons

18. (a) 26 protons, 26 electrons, and 28 neutrons
    (b) 11 protons, 11 electrons, and 12 neutrons
    (c) 35 protons, 35 electrons, and 44 neutrons
    (d) 15 protons, 15 electrons, and 16 neutrons

19. (a) 33
    (b) Arsenic, As

(c)   43

(d)   The charge is -3, this is an anion.

(e) $^{76}_{33}$As

20.   (a)   56

(b)   Barium, Ba

(c)   79

(d)   The charge is $+2$, this is a cation.

(e) $^{135}_{56}$Ba

21.   For each isotope:

(%)(amu) = that portion of the average atomic mass for that isotope.

Add together to obtain the average atomic mass.

(0.5145)(89.905 amu) + (0.1122)(90.906 amu) + (0.1715)(91.905 amu)

+ (0.1738)(93.906 amu) + (0.0280)(95.908 amu)

46.26 amu + 10.20 amu + 15.76 amu + 16.32 amu + 2.69 amu

= 91.23 amu = average atomic mass of Zr

22.   For each isotope:

(%)(amu) = that portion of the average atomic mass for that isotope.

The sum of the portions = the average atomic mass.

(0.080)(45.953) + (0.073)(46.952) + (0.738)(47.948) + (0.055)(48.948)

+ (1.000 − 0.946)x amu = 47.9 amu

= 3.7 amu + 3.4 amu + 35.4 amu + 2.7 amu + 0.054x amu = 47.9 amu

= 45.2 amu + 0.054x = 47.9 amu   0.054x = 47.9 amu − 45.2 amu

x amu = $\dfrac{2.7\ \text{amu}}{0.054}$ $x = 50.$ = mass of the fifth isotope of titanium

23.   (0.6917)(62.9296 amu) + (1.0000 − 0.6917)(64.9278 amu)

= 43.53 amu + 20.02 amu

= 63.55 amu = average atomic mass

The element is copper (see periodic table).

24.   (0.7577)(34.9689 amu) + (1.0000 − 0.7577)(36.9659 amu)

= 26.50 amu + 8.96 amu

= 35.46 amu = average atomic mass

The element is chlorine (see periodic table).

25. $V_{sphere} = \dfrac{4}{3}\pi r^3$ r_A = radius of atom, r_N = radius of nucleus

$$\frac{V_{atom}}{V_{nucleus}} = \frac{\frac{4}{3}\pi r_A^3}{\frac{4}{3}\pi r_N^3} = \frac{r_A^3}{r_N^3} = \frac{(1.0 \times 10^{-8})^3}{(1.0 \times 10^{-13})^3} = \frac{1.0 \times 10^{15}}{1.0}$$ (The ratio of atomic volume to nuclear volume is $1.0 \times 10^{15} : 1.0$.)

26. $\dfrac{3.0 \times 10^{-8}\ \text{cm}}{2.0 \times 10^{-13}\ \text{cm}} = \dfrac{1.5 \times 10^5}{1.0}$ (The ratio of the diameter of an Al atom to its nucleus diameter is $1.5 \times 10^5 : 1.0$.)

27. (a) In Rutherford's experiment the majority of alpha particles passed through the gold foil without deflection. This shows that the atom is mostly empty space and the nucleus is very small.
    (b) In Thomson's experiments with the cathode ray tube, rays were observed coming from both the anode and the cathode.
    (c) In Rutherford's experiment an alpha particle was occasionally dramatically deflected by the nucleus of a gold atom. The direction of deflection showed the nucleus to be positive. Positive charges repel each other.

28. Elements (a) and (c) are isotopes of phosphorus.

29. $(8.5 \text{ in.}) \left(\dfrac{2.54 \text{ cm}}{\text{in.}} \right) \left(\dfrac{1 \text{ atom Si}}{2.34 \times 10^{-8} \text{ cm}} \right) = 9.2 \times 10^8$ atoms Si

30. The properties of an element are related to the number of protons and electrons. If the number of neutrons differs, isotopes result. Isotopes of an element are still the same element even though the nuclear composition of the atoms are different.

31. ^{156}Dy has 90 neutrons; ^{160}Gd has 96 neutrons; ^{162}Er has 94 neutrons; ^{165}Ho has 98 neutrons.
    In order of increasing number of neutrons: Dy<Er<Gd<Ho
    On the periodic table, the order is based on increasing number of protons, so the order is Gd<Dy<Ho<Er.

32. percent of sample ^{60}Q $= x$

    percent of sample ^{63}Q $= 1 - x$

    $(x)(60. \text{ amu}) + (1 - x)(63 \text{ amu}) = 61.5 \text{ amu}$

    $60.x \text{ amu} + 63 \text{ amu} - 63x \text{ amu} = 61.5 \text{ amu}$

    $63 \text{ amu} - 61.5 \text{ amu} = 63x \text{ amu} - 60x \text{ amu}$

    $1.5 = 3x$

    $0.50 = x$

    ^{60}Q $= 50\%$

    ^{63}Q $= 50\%$

33. (a) Compare the mass of the unknown element to the mass of a carbon-12 atom.

    $(3.27 \times 10^{-19} \text{ mg unknown element}) \left(\dfrac{1 \text{ g}}{1000 \text{ mg}} \right) \left(\dfrac{12.0 \text{ amu}}{1.9927 \times 10^{-23} \text{ g}} \right) = 197 \text{ amu}$

    The atomic mass of the unknown element is 197 amu

    (b) The unknown element is Au, gold (see periodic table)

34. $(0.52 \text{ lb Ag}) \left(\dfrac{453.6 \text{ g}}{\text{lb}} \right) \left(\dfrac{1 \text{ atom Ag}}{1.79 \times 10^{-22} \text{ g Ag}} \right) = 1.3 \times 10^{24}$ atoms Ag

35. These are the elements that have the same number of protons, neutrons, and electrons.

	protons	neutrons	electrons
He	2	2	2
C	6	6	6
N	7	7	7
O	8	8	8
Ne	10	10	10
Mg	12	12	12
Si	14	14	14
S	16	16	16
Ca	20	20	20

36.

C^+	6 protons, 5 electrons
O^+	8 protons, 7 electrons
O^{2+}	8 protons, 6 electrons

37.

Mineral Supplement	Mineral use	Ion provided	Number protons	Number electrons
Calcium carbonate	Bones and Teeth	Ca^{2+}	20	18
Iron(II) sulfate	Hemoglobin	Fe^{2+}	26	24
Chromium(III) nitrate	Insulin	Cr^{3+}	24	21
Magnesium sulfate	Bones	Mg^{2+}	12	10
Zinc sulfate	Cellular metabolism	Zn^{2+}	30	28
Potassium iodide	Thyroid function	I^-	53	54

38.

Element	Symbol	Atomic #	Mass #	Protons	Neutrons	Electrons
Chlorine	^{36}Cl	17	36	17	19	17
Gold	^{197}Au	79	197	79	118	79
Barium	^{135}Ba	56	135	56	79	56
Argon	^{38}Ar	18	38	18	20	18
Nickel	^{58}Ni	28	58	28	30	28

39.

Element	Symbol	Atomic #	Mass #	Protons	Neutrons	Electrons
Xenon	^{134}Xe	54	134	54	80	54
Silver	^{107}Ag	47	107	47	60	47
Fluorine	^{19}F	9	19	9	10	9
Uranium	^{235}U	92	235	92	143	92
Potassium	^{41}K	19	41	19	22	19

40.

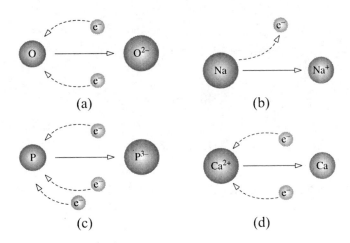

      (a)   (b)

      (c)   (d)

41.

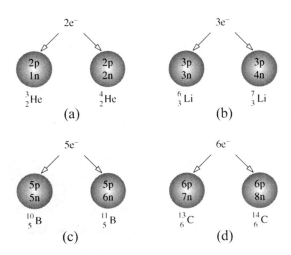

      (a)   (b)

      (c)   (d)

42. The mass of one electron is 9.110×10^{-28} grams.

    (a) Aluminum has 13 electrons. $\left(\dfrac{(13)(9.110 \times 10^{-28} \text{ g})}{4.480 \times 10^{-23} \text{ g}} \right)(100) = 0.02644\%$ electrons

(b) Phosphorus has 15 electrons. $\left(\dfrac{(15)(9.110 \times 10^{-28}\ g)}{5.143 \times 10^{-23}\ g}\right)(100) = 0.02657\%$ electrons

(c) Krypton has 36 electrons. $\left(\dfrac{(36)(9.110 \times 10^{-28}\ g)}{1.392 \times 10^{-22}\ g}\right)(100) = 0.02356\%$ electrons

(d) Platinum has 78 electrons. $\left(\dfrac{(78)(9.110 \times 10^{-28}\ g)}{3.240 \times 10^{-22}\ g}\right)(100) = 0.02193\%$ electrons

43. The mass of one proton is 1.673×10^{-24} grams.

(a) Selenium has 34 protons. $\left(\dfrac{(34)(1.673 \times 10^{-24}\ g)}{1.311 \times 10^{-22}\ g}\right)(100) = 43.39\%$ protons

(b) Xenon has 54 protons. $\left(\dfrac{(54)(1.673 \times 10^{-24}\ g)}{2.180 \times 10^{-22}\ g}\right)(100) = 41.44\%$ protons

(c) Chlorine has 17 protons. $\left(\dfrac{(17)(1.673 \times 10^{-24}\ g)}{5.887 \times 10^{-23}\ g}\right)(100) = 48.31\%$ protons

(d) Barium has 56 protons. $\left(\dfrac{(56)(1.673 \times 10^{-24}\ g)}{2.280 \times 10^{-22}\ g}\right)(100) = 41.09\%$ protons

44. The electron region is the area around the nucleus where electrons are most likely to be located.

45.

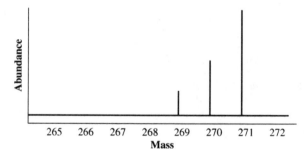

average atomic mass: $(0.1081)(269.14\ \text{amu}) = 29.09\ \text{amu}$

$\phantom{\text{average atomic mass: }}(0.3407)(270.51\ \text{amu}) = 92.16\ \text{amu}$

$\phantom{\text{average atomic mass: }}(0.5512)(271.23\ \text{amu}) = \underline{149.50}\ \text{amu}$

$\phantom{\text{average atomic mass: }(0.5512)(271.2)}\text{Total} \;= 270.75\ \text{amu}$

An atomic mass of 270.75 amu would come somewhere after Bohrium (mass = 264 amu). So, the atomic number of this new element would be greater than 107.

CHAPTER 6

NOMENCLATURE OF INORGANIC COMPOUNDS

SOLUTIONS TO REVIEW QUESTIONS

1. water (H_2O)–dihydrogen monoxide
   ammonia (NH_3)–nitrogen trihydride

2. An atom must gain one or more electrons to form an anion. An atom must lose one or more electrons to form a cation.

3. No, if elements combine in a one-to-one ratio the charges on their ions must be equal and opposite in sign. They could be $+1$, -1, or $+2$, -2 or $+3$, -3 etc.

4. The metal or cation is always named first.

5. (a) K_2S (d) H_3PO_4
   (b) $Co(BrO_3)_2$ (e) Fe_2O_3
   (c) NH_4NO_3 (f) $Mg(OH)_2$

6. The system for naming binary compounds composed of two nonmetals uses the name of the first element and the stem of the second element plus the suffix –ide. A prefix is attached to each element indicating the number of atoms of that element in the formula. Thus, N_2O_5 is dinitrogen pentoxide. Aluminum forms only one series of compounds in which the cation is always Al^{3+}. Thus the name for Al_2O_3, aluminum oxide, does not need to be distinguished from any other compound.

7. Barium forms only one series of compounds in which the cation is always Ba^{2+}. Thus the name for $BaCl_2$ (barium chloride) does not need to be distinguished from any other compound. Iron forms two series of compounds in which the iron ion is Fe^{2+} or Fe^{3+}. Thus the name iron chloride does not indicate which compound is in question. Therefore, $FeCl_2$ is called iron(II) chloride to indicate that the compound contains the Fe^{2+} ion. Neither uses prefixes because prefixes are used to name compounds composed of two nonmetals.

8. The mono prefix is generally not used for the first element.

9. The Stock System clearly indicates cation charges. The classical system works, but requires a chemist to memorize the common charges on all cations. It is much easier to use the Stock System and much more error free.

10. Nickel(II) compounds

   (a) $NiSO_4$ (d) $Ni(OH)_2$ (g) $NiCr_2O_7$ (j) $Ni(ClO)_2$
   (b) Ni_3P_2 (e) $Ni(IO_2)_2$ (h) $NiBr_2$
   (c) $NiCrO_4$ (f) $Ni(C_2H_3O_2)_2$ (i) $Ni(NO_3)_2$

11. (a)

HBrO	hypobromous acid
$HBrO_2$	bromous acid
$HBrO_3$	bromic acid
$HBrO_4$	perbromic acid

(b)

HIO	hypoiodous acid
HIO_2	iodous acid
HIO_3	iodic acid
HIO_4	periodic acid

12. Include hydrogen or dihydrogen in the name to tell how many hydrogens are left behind.

For example

Phosphate ion PO_4^{3-}

Hydrogen phosphate ion HPO_4^{2-}

Dihydrogen phosphate ion $H_2PO_4^{-}$

SOLUTIONS TO EXERCISES

1. Formulas of compounds:
   (a) Ba and S   BaS   (d) Mg and N Mg_3N_2
   (b) Cs and P Cs_3P (e) Ca and I CaI_2
   (c) Li and Br   LiBr   (f) H and Cl   HCl

2. Formulas of compounds:
   (a) Al and S Al_2S_3 (d) Sr and O   SrO
   (b) H and F   HF   (e) Cs and P Cs_3P
   (c) K and N K_3N (f) Al and Cl $AlCl_3$

3. (a) potassium K^+ (h) calcium Ca^{2+}
   (b) ammonium NH_4^+ (i) lead(II) Pb^{2+}
   (c) copper(I) Cu^+ (j) zinc Zn^{2+}
   (d) titanium(IV) Ti^{4+} (k) silver Ag^+
   (e) nickel(III) Ni^{3+} (l) hydrogen H^+
   (f) cesium Cs^+ (m) tin(II) Sn^{2+}
   (g) mercury(II) Hg^{2+} (n) iron(III) Fe^{3+}

4. (a) fluoride F^- (h) oxide O^{2-}
   (b) acetate $C_2H_3O_2^-$ (i) dichromate $Cr_2O_7^{2-}$
   (c) iodide I^- (j) hydrogen carbonate HCO_3^-
   (d) carbonate CO_3^{2-} (k) phosphate PO_4^{3-}
   (e) sulfide S^{2-} (l) sulfate SO_4^{2-}
   (f) nitrate NO_3^- (m) nitride N^{3-}
   (g) phosphide P^{3-} (n) chloride Cl^-

5. (a) sodium hydrogen carbonate   (d) acetic acid
   (b) mercury   (e) magnesium sulfate heptahydrate
   (c) calcium oxide   (f) sodium hydroxide

6. (a) sodium thiosulfate   (d) sodium chloride
   (b) dinitrogen monoxide   (e) magnesium hydroxide
   (c) aluminum oxide   (f) lead(II) sulfide

7.

Ion	Br^-	O^{2-}	NO_3^-	PO_4^{3-}	CO_3^{2-}
K^+	KBr	K_2O	KNO_3	K_3PO_4	K_2CO_3
Mg^{2+}	$MgBr_2$	MgO	$Mg(NO_3)_2$	$Mg_3(PO_4)_2$	$MgCO_3$
Al^{3+}	$AlBr_3$	Al_2O_3	$Al(NO_3)_3$	$AlPO_4$	$Al_2(CO_3)_3$
Zn^{2+}	$ZnBr_2$	ZnO	$Zn(NO_3)_2$	$Zn_3(PO_4)_2$	$ZnCO_3$
H^+	HBr	H_2O	HNO_3	H_3PO_4	H_2CO_3

8.

Ion	SO_4^{2-}	OH^-	AsO_4^{3-}	$C_2H_3O_2^-$	CrO_4^{2-}
NH_4^+	$(NH_4)_2SO_4$	NH_4OH	$(NH_4)_3AsO_4$	$NH_4C_2H_3O_2$	$(NH_4)_2CrO_4$
Ca^{2+}	$CaSO_4$	$Ca(OH)_2$	$Ca_3(AsO_4)_2$	$Ca(C_2H_3O_2)_2$	$CaCrO_4$
Fe^{3+}	$Fe_2(SO_4)_3$	$Fe(OH)_3$	$FeAsO_4$	$Fe(C_2H_3O_2)_3$	$Fe_2(CrO_4)_3$
Ag^+	Ag_2SO_4	$AgOH$	Ag_3AsO_4	$AgC_2H_3O_2$	Ag_2CrO_4
Cu^{2+}	$CuSO_4$	$Cu(OH)_2$	$Cu_3(AsO_4)_2$	$Cu(C_2H_3O_2)_2$	$CuCrO_4$

9. K^+ compounds: potassium bromide, potassium oxide, potassium nitrate, potassium phosphate, potassium carbonate.
Mg^{2+} compounds: magnesium bromide, magnesium oxide, magnesium nitrate, magnesium phosphate, magnesium carbonate.
Al^{3+} compounds: aluminum bromide, aluminum oxide, aluminum nitrate, aluminum phosphate, aluminum carbonate.
Zn^{2+} compounds: zinc bromide, zinc oxide, zinc nitrate, zinc phosphate, zinc carbonate.
H^+ compounds: hydrogen bromide (or hydrobromic acid), water, nitric acid, phosphoric acid, carbonic acid.

10. NH_4^+ compounds: ammonium sulfate, ammonium hydroxide, ammonium arsenate, ammonium acetate, ammonium chromate.
Ca^{2+} compounds: calcium sulfate, calcium hydroxide, calcium arsenate, calcium acetate, calcium chromate.
Fe^{3+} compounds: iron(III) sulfate, iron(III) hydroxide, iron(III) arsenate, iron(III) acetate, iron(III) chromate.
Ag^+ compounds: silver sulfate, silver hydroxide, silver arsenate, silver acetate, silver chromate.
Cu^{2+} compounds: copper(II) sulfate, copper(II) hydroxide, copper(II) arsenate, copper(II) acetate, copper(II) chromate.

11. Nonmetal binary compound formulas:
(a) diphosphorus pentoxide, P_2O_5
(b) carbon dioxide, CO_2
(c) tribromine octoxide, Br_3O_8
(d) sulfur hexachloride, SCl_6
(e) carbon tetrachloride, CCl_4
(f) dichlorine heptoxide, Cl_2O_7
(g) boron trifluoride, BF_3
(h) tetranitrogen hexasulfide, N_4S_6

12. Metal-nonmetal binary compound formulas:
(a) potassium nitride, K_3N
(b) barium oxide, BaO
(c) iron(II) oxide, FeO
(d) strontium phosphide, Sr_3P_2
(e) calcium nitride, Ca_3N_2
(f) cesium bromide, $CsBr$
(g) manganese(III) iodide, MnI_3
(h) sodium selenide, Na_2Se

13. Naming binary metal-nonmetal compounds:
(a) BaO, barium oxide
(b) K_2S, potassium sulfide
(c) $CaCl_2$, calcium chloride
(d) Cs_2S, cesium sulfide
(e) Al_2O_3, aluminum oxide
(f) $CaBr_2$, calcium bromide
(g) SrI_2, strontium iodide
(h) Mg_3N_2, magnesium nitride

14. Naming binary nonmetal compounds:
    (a) PBr_5, phosphorus pentabromide
    (b) I_4O_9, tetraiodine nonoxide
    (c) N_2S_5, dinitrogen pentasulfide
    (d) S_2F_{10}, disulfur decafluoride
    (e) $SiCl_4$, silicon tetrachloride
    (f) ClO_2, chlorine dioxide
    (g) P_4S_7, tetraphosphorus heptasulfide
    (h) IF_6, iodine hexafluoride

15. (a) $CuCl_2$ copper(II) chloride
    (b) $FeCl_2$ iron(II) chloride
    (c) $Fe(NO_3)_2$ iron(II) nitrate
    (d) $FeCl_3$ iron(III) chloride
    (e) SnF_2 tin(II) fluoride
    (f) VPO_4 vanadium(III) phosphate

16. Formulas:
    (a) tin(IV) bromide $SnBr_4$
    (b) copper(I) sulfate Cu_2SO_4
    (c) nickel(II) borate $Ni_3(BO_3)_2$
    (d) mercury(II) nitrite $Hg(NO_2)_2$
    (e) cobalt(III) carbonate $Co_2(CO_3)_3$
    (f) iron(II) acetate $Fe(C_2H_3O_2)_2$

17. Formulas of acids:
    (a) hydrochloric acid, HCl
    (b) chloric acid, $HClO_3$
    (c) nitric acid, HNO_3
    (d) carbonic acid, H_2CO_3
    (e) sulfurous acid, H_2SO_3
    (f) phosphoric acid, H_3PO_4

18. Formulas of acids:
    (a) acetic acid, $HC_2H_3O_2$
    (b) hydrofluoric acid, HF
    (c) hydrosulfuric acid, H_2S
    (d) boric acid, H_3BO_3
    (e) nitrous acid, HNO_2
    (f) hypochlorous acid, $HClO$

19. Naming acids:
    (a) HNO_2, nitrous acid
    (b) H_2SO_4, sulfuric acid
    (c) $H_2C_2O_4$, oxalic acid
    (d) HBr, hydrobromic acid
    (e) H_3PO_3, phosphorous acid
    (f) $HC_2H_3O_2$, acetic acid
    (g) HF, hydrofluoric acid
    (h) $HBrO_3$, bromic acid
    (i) HIO_4, periodic acid

20. Naming acids:
    (a) H_3PO_4, phosphoric acid
    (b) H_2CO_3 carbonic acid
    (c) HIO_3, iodic acid
    (d) HCl, hydrochloric acid
    (e) $HClO$, hypochlorous acid
    (f) HNO_3, nitric acid
    (g) HI, hydroiodic acid
    (h) $HClO_4$ perchloric acid
    (i) H_2SO_3, sulfurous acid

21. Formulas for:
    (a) silver sulfite Ag_2SO_3
    (b) cobalt(II) bromide $CoBr_2$
    (c) tin(II) hydroxide $Sn(OH)_2$
    (d) aluminum sulfate $Al_2(SO_4)_3$
    (e) lead(II) chloride $PbCl_2$
    (f) ammonium carbonate $(NH_4)_2CO_3$
    (g) chromium(III) oxide Cr_2O_3
    (h) cupric chloride $CuCl_2$
    (i) potassium permanganate $KMnO_4$

(j)   arsenic(V) sulfite $As_2(SO_3)_5$
(k)   sodium peroxide Na_2O_2
(l)   iron(II) sulfate $FeSO_4$
(m) potassium dichromate $K_2Cr_2O_7$
(n)   bismuth(III) chromate $Bi_2(CrO_4)_3$

22. Formulas for:

(a)   sodium chromate Na_2CrO_4
(b)   magnesium hydride MgH_2
(c)   nickel(II) acetate $Ni(C_2H_3O_2)_2$
(d)   calcium chlorate $Ca(ClO_3)_2$
(e)   magnesium bromate $Mg(BrO_3)_2$
(f)   potassium dihydrogen phosphate KH_2PO_4
(g)   manganese(II) hydroxide $Mn(OH)_2$
(h)   cobalt(II) hydrogen carbonate $Co(HCO_3)_2$
(i)   sodium hypochlorite $NaClO$
(j)   barium perchlorate $Ba(ClO_4)_2$
(k)   chromium(III) sulfite $Cr_2(SO_3)_3$
(l)   antimony(III) sulfate $Sb_2(SO_4)_3$
(m) sodium oxalate $Na_2C_2O_4$
(n)   potassium thiocyanate $KSCN$

23. Names for:

(a) $ZnSO_4$ zinc sulfate
(b) Hg_2S mercury(I) sulfide
(c) $CuCO_3$ copper(II) carbonate
(d) $Cd(NO_3)_2$ cadmium nitrate
(e) $Al(C_2H_3O_2)_3$ aluminum acetate
(f) CoF_2 cobalt(II) fluoride
(g) $Cr(ClO_3)_3$ chromium(III) chlorate
(h) Ag_3PO_4 silver phosphate
(i) MnS manganese(II) sulfide
(j) $BaCrO_4$ barium chromate

24. Names for:

(a) $Ca(HSO_4)_2$ calcium hydrogen sulfate
(b) $As_2(SO_3)_3$ arsenic(III) sulfite
(c) $Sn(NO_2)_2$ tin(II) nitrite
(d) CuI copper(I) iodide
(e) $KHCO_3$ potassium hydrogen carbonate
(f) $BiAsO_4$ bismuth(III) arsenate
(g) $(NH_4)_2CO_3$ ammonium carbonate
(h) $(NH_4)_2HPO_4$ ammonium monohydrogen phosphate
(i) $NaClO$ sodium hypochlorite
(j) $KMnO_4$ potassium permanganate

25. Formulas for:
    (a) slaked lime, $Ca(OH)_2$
    (b) fool's gold, FeS_2
    (c) washing soda, $Na_2CO_3 \cdot 10\,H_2O$
    (d) calcite, $CaCO_3$
    (e) cane sugar, $C_{12}H_{22}O_{11}$
    (f) borax, $Na_2B_4O_7 \cdot 10\,H_2O$
    (g) wood alcohol, CH_3OH
    (h) acetylene, C_2H_2

26. Formulas for:
    (a) grain alcohol, C_2H_5OH
    (b) cream of tartar, $KHC_4H_4O_6$
    (c) gypsum, $CaSO_4 \cdot 2\,H_2O$
    (d) brimstone, S
    (e) muriatic acid, HCl
    (f) plaster of paris, $CaSO_4 \cdot \frac{1}{2}H_2O$
    (g) lye, NaOH
    (h) laughing gas, N_2O

27. (a) $K \rightarrow K^+ + e^-$
    (b) $I + e^- \rightarrow I^-$
    (c) $Br + e^- \rightarrow Br^-$
    (d) $Fe \rightarrow Fe^{2+} + 2e^-$
    (e) $Ca \rightarrow Ca^{2+} + 2e^-$
    (f) $O + 2e^- \rightarrow O^{2-}$

28. (a) sulfate
    (b) phosphate
    (c) nitrate
    (d) chlorate
    (e) hydroxide
    (f) carbonate

29. (a) $CaSO_4$
    (b) $Ca_3(PO_4)_2$
    (c) $Ca(NO_3)_2$
    (d) $Ca(ClO_3)_2$
    (e) $Ca(OH)_2$
    (f) $CaCO_3$

30. (a) K_2SO_4
    (b) K_3PO_4
    (c) KNO_3
    (d) $KClO_3$
    (e) KOH
    (f) K_2CO_3

31. (a) CBr_4 carbon tetrabromide
    (b) BF_3 boron trifluoride
    (c) PCl_5 phosphorus pentachloride

32. Formula: KCl   Name: potassium chloride

33. *ide:* suffix is used to indicate a binary compound except for hydroxides, cyanides, and ammonium compounds.

    ous: used as a suffix to name an acid that has a lower oxygen content than the *-ic* acid (e.g., HNO_2, nitrous acid, and HNO_3, nitric acid). Also used as a suffix to name the lower ionic charge of a metal with more than one possible charge (e.g., Fe^{2+}, ferrous, and Fe^{3+}, ferric).

    ic: used as a suffix to name an acid that has a higher oxygen content than the *-ous* acid (e.g., HNO_3 nitric acid, and HNO_2 nitrous acid). Also used as a suffix to name the higher ionic charge of a metal with more than one possible charge (e.g., Fe^{2+}, ferrous, and Fe^{3+}, ferric).

    hypo: used as a prefix in naming an acid that has a lower oxygen content than the *-ous* acid when there are more than two oxyacids with the same elements (e.g., HClO, hypochlorous acid, and $HClO_2$, chlorous acid).

    per: used as a prefix in naming an acid that has a higher oxygen content than the *-ic* acid when there are more than two oxyacids with the same elements (e.g., $HClO_4$, perchloric acid, and $HClO_3$, chloric acid).

ite: the suffix used in naming a salt derived from an *-ous* acid.
For example, HNO_2 (nitrous acid); $NaNO_2$ (sodium nitrite).

ate: the suffix used in naming a salt derived from an *-ic* acid.
For example, H_2SO_4 (sulfuric acid); $CaSO_4$ (calcium sulfate).

Roman numerals: In the Stock System Roman numerals are used in naming compounds that contain metals that may exist as more than one type of cation. The charge of a metal is indicated by a Roman numeral written in parenthesis immediately after the name of the metal. For example, $FeCl_2$ [iron (II) chloride].

34. (a) Calcium carbonate or $CaCO_3$ is the principal ingredient in lime, chalk, and marble. It is harmless, though it does leave a residue on plumbing. There should be no concern regarding this spill.
    (b) Acetic acid or $HC_2H_3O_2$ is found in vinegar. This is not a particularly hazardous spill, though it will be widely apparent due to the characteristic odor of vinegar. (Perhaps it could be cleaned up with lots of lettuce?)
    (c) Dihydrogen oxide or H_2O is commonly known as water and it may be safely dumped into the sewer.

35. (a) $50\,e^-$, $50\,p$ (b) $48\,e^-$, $50\,p$ (c) $46\,e^-$, $50\,p$

36. The formula for a compound must be electrically neutral. Therefore $X = +3$ and $Y = -2$ since in X_2Y_3 this would give $2(+3) + 3(-2) = 0$.

37. $Li_3Fe(CN)_6$
    $AlFe(CN)_6$
    $Zn_3[Fe(CN)_6]_2$

38. (a) N^{3-} nitride   One has oxygen the other does not, charges on the ions differ.
        NO_2^- nitrite
    (b) NO_2^- nitrite   The number of oxygens differ, but the charge is the same.
        NO_3^- nitrate
    (c) HNO_2 nitrous acid   The number of oxygens in the compounds differ but they both have only
        HNO_3 nitric acid   one hydrogen and one nitrogen atom.

39. (a) Potassium chloride KCl
    (b) Calcium carbonate $CaCO_3$
    (c) Ferrous sulfate $FeSO_4$
    (d) Zinc oxide ZnO
    (e) Manganese sulfate $MnSO_4$ and $Mn_2(SO_4)_3$ would both be described by this name. Better names would be Manganese(II) sulfate for $MnSO_4$ and Manganese(III) sulfate for $Mn_2(SO_4)_3$
    (f) Copper sulfate $CuSO_4$ and Cu_2SO_4 would both be described by this name. Better names would be Copper(II) sulfate and cupric sulfate for $CuSO_4$ or Copper(I) sulfate or cuprous sulfate for Cu_2SO_4
    (g) Manganous oxide MnO
    (h) Potassium iodide KI
    (i) Cobalt carbonate $CoCO_3$ and $Co_2(CO_3)_3$ would both be described by this name. Better names would be Cobalt(II) carbonate for $CoCO_3$ and Cobalt(III) carbonate for $Co_2(CO_3)_3$
    (j) Sodium chloride $NaCl$

CHAPTER 7

QUANTITATIVE COMPOSITION OF COMPOUNDS

SOLUTIONS TO REVIEW QUESTIONS

1. A mole is an amount of substance containing the same number of atoms as there are atoms in exactly 12 g of carbon-12.

    It is Avogadro's number (6.022×10^{23}) of anything (atoms, molecules, ping-pong balls, etc.).

2. A mole of gold (197.0 g) has a higher mass than a mole of potassium (39.10 g).

3. Both samples (Au and K) contain the same number of atoms. (6.022×10^{23}).

4. A mole of gold atoms contains more electrons than a mole of potassium atoms, as each Au atom has 79 e^-, while each K atom has only 19 e^-.

5. 6.022×10^{23}

6. There are Avogadro's number of particles in one mole of substance.

7. 1 mole of ozone has the greater number of oxygen atoms.

8. The molar mass of an element is the mass of one mole (or 6.022×10^{23} atoms) of that element.

9. No. Avogadro's number is a constant. The mole is defined as Avogadro's number of C-12 atoms. Changing the atomic mass to 50 amu would change only the size of the atomic mass unit, not Avogadro's number.

10. (a) A mole of oxygen atoms (O) contains **6.022×10^{23}** atoms.
    (b) A mole of oxygen molecules (O_2) contains **6.022×10^{23}** molecules.
    (c) A mole of oxygen molecules (O_2) contains **1.204×10^{24}** atoms.
    (d) A mole of oxygen atoms (O) has a mass of **16.00** grams.
    (e) A mole of oxygen molecules (O_2) has a mass of **32.00** grams.

11. 6.022×10^{23} molecules in one molar mass of H_2SO_4.

    4.215×10^{24} atoms in one molar mass of H_2SO_4.

12. Either the chemical formula or experimental data giving the mass of the component elements in a sample.

13. There is 56.2% oxygen. (Remember the total percentage of all components must equal 100%.)

14. $\left(\dfrac{\text{total mass of the element}}{\text{molar mass of the element}}\right)(100) = \text{mass percent of the element}$

15. Choosing 100.0 g of a compound allows us to simply drop the % sign and use grams instead of percent.

16. C_4H_6 and C_8H_{12} both have the empirical formula C_2H_3.

17. The molecular formula represents the total number of atoms of each element in a molecule. The empirical formula represents the lowest number ratio of atoms of each element in a molecule.

18. The molar mass is the most useful additional information that can be used to determine the molecular formula of a compound from its empirical formula.

19. This formula tells us the number of units of the empirical formula found in the molecule.

SOLUTIONS TO EXERCISES

1. Molar masses

    (a) KBr   1   K   39.10 g
                      1   Br   79.90 g
                                    119.0

    (b) Na$_2$SO$_4$ 2   Na   45.98 g
                      1   S   32.07 g
                      4   O   64.00 g
                                    142.1 g

    (c) Pb(NO$_3$)$_2$ 1   Pb   207.2 g
                      2   N   28.02 g
                      6   O   96.00 g
                                    331.2 g

    (d) C$_2$H$_5$OH   2   C   24.02 g
                      6   H   6.048 g
                      1   O   16.00 g
                                    46.07 g

    (e) HC$_2$H$_3$O$_2$ 4   H   4.032 g
                      2   C   24.02 g
                      2   O   32.00 g
                                    60.05 g

    (f) Fe$_3$O$_4$ 3   Fe   167.6 g
                      4   O   64.00 g
                                    231.6 g

    (g) C$_{12}$H$_{22}$O$_{11}$ 12   C   144.1 g
                      22   H   22.18 g
                      11   O   176.0 g
                                    342.3 g

    (h) Al$_2$(SO$_4$)$_3$ 2   Al   53.96 g
                      3   S   96.21 g
                      12   O   192.0 g
                                    342.2 g

    (i) (NH$_4$)$_2$HPO$_4$ 9   H   9.072 g
                      2   N   28.02 g
                      1   P   30.97 g
                      4   O   64.00 g
                                    132.1 g

2. Molar masses

(a) NaOH

1	Na	22.99	g
1	O	16.00	g
1	H	1.008	g
		40.00	g

(b) Ag_2CO_3

2	Ag	215.8	g
1	C	12.01	g
3	O	48.00	g
		275.8	g

(c) Cr_2O_3

2	Cr	104.0	g
3	O	48.00	g
		152.0	g

(d) $(NH_4)_2CO_3$

2	N	28.02	g
8	H	8.064	g
1	C	12.01	g
3	O	48.00	g
		96.09	g

(e) $Mg(HCO_3)_2$

1	Mg	24.31	g
2	H	2.016	g
2	C	24.02	g
6	O	96.00	g
		146.3	g

(f) C_6H_5COOH

7	C	84.07	g
6	H	6.048	g
2	O	32.00	g
		122.1	g

(g) $C_6H_{12}O_6$

6	C	72.06	g
12	H	12.10	g
6	O	96.00	g
		180.2	g

(h) $K_4Fe(CN)_6$

4	K	156.4	g
1	Fe	55.85	g
6	C	72.06	g
6	N	84.06	g
		368.4	g

(i) $BaCl_2 \cdot 2\,H_2O$

1	Ba	137.3	g
2	Cl	70.90	g
4	H	4.032	g
2	O	32.00	g
		244.2	g

3. Moles of atoms.

(a) $(22.5 \text{ g Zn})\left(\dfrac{1 \text{ mol Zn}}{65.39 \text{ g Zn}}\right) = 0.344 \text{ mol Zn}$

(b) $(0.688 \text{ g Mg})\left(\dfrac{1 \text{ mol Mg}}{24.31 \text{ g Mg}}\right) = 2.83 \times 10^{-2} \text{ mol Mg}$

(c) $(4.5 \times 10^{22} \text{ atoms Cu})\left(\dfrac{1 \text{ mol Cu}}{6.022 \times 10^{23} \text{ atoms Cu}}\right) = 7.5 \times 10^{-2} \text{ mol Cu}$

(d) $(382 \text{ g Co})\left(\dfrac{1 \text{ mol Co}}{58.93 \text{ g Co}}\right) = 6.48 \text{ mol Co}$

(e) $(0.055 \text{ g Sn})\left(\dfrac{1 \text{ mol Sn}}{118.7 \text{ g Sn}}\right) = 4.6 \times 10^{-4} \text{ mol Sn}$

(f) $(8.5 \times 10^{24} \text{ molecules N}_2)\left(\dfrac{2 \text{ atoms N}}{1 \text{ molecule N}_2}\right)\left(\dfrac{1 \text{ mol N atoms}}{6.022 \times 10^{23} \text{ atoms N}}\right) = 28 \text{ mol N atoms}$

4. Number of moles.

(a) $(25.0 \text{ g NaOH})\left(\dfrac{1 \text{ mol NaOH}}{40.00 \text{ g NaOH}}\right) = 0.625 \text{ mol NaOH}$

(b) $(44.0 \text{ g Br}_2)\left(\dfrac{1 \text{ mol Br}_2}{159.8 \text{ g Br}_2}\right) = 0.275 \text{ mol Br}_2$

(c) $(0.684 \text{ g MgCl}_2)\left(\dfrac{1 \text{ mol MgCl}_2}{95.21 \text{ g MgCl}_2}\right) = 7.18 \times 10^{-3} \text{ mol MgCl}_2$

(d) $(14.8 \text{ g CH}_3\text{OH})\left(\dfrac{1 \text{ mol CH}_3\text{OH}}{32.04 \text{ g CH}_3\text{OH}}\right) = 0.462 \text{ mol CH}_3\text{OH}$

(e) $(2.88 \text{ g Na}_2\text{SO}_4)\left(\dfrac{1 \text{ mol Na}_2\text{SO}_4}{142.1 \text{ g Na}_2\text{SO}_4}\right) = 2.03 \times 10^{-2} \text{ mol Na}_2\text{SO}_4$

(f) $(4.20 \text{ lb ZnI}_2)\left(\dfrac{453.6 \text{ g}}{1 \text{ lb}}\right)\left(\dfrac{1 \text{ mol ZnI}_2}{319.2 \text{ g ZnI}_2}\right) = 5.97 \text{ mol ZnI}_2$

5. Number of grams.

(a) $(0.550 \text{ mol Au})\left(\dfrac{197.0 \text{ g Au}}{1 \text{ mol Au}}\right) = 108 \text{ g Au}$

(b) $(15.8 \text{ mol H}_2\text{O})\left(\dfrac{18.02 \text{ g H}_2\text{O}}{\text{mol H}_2\text{O}}\right) = 285 \text{ g H}_2\text{O}$

(c) $(12.5 \text{ mol Cl}_2)\left(\dfrac{70.90 \text{ g Cl}_2}{\text{mol Cl}_2}\right) = 886 \text{ g Cl}_2$

(d) $(3.15 \text{ mol NH}_4\text{NO}_3)\left(\dfrac{80.05 \text{ g NH}_4\text{NO}_3}{\text{mol NH}_4\text{NO}_3}\right) = 252 \text{ g NH}_4\text{NO}_3$

6. Number of grams.

(a) $\left(4.25 \times 10^{-4} \text{ mol H}_2\text{SO}_4\right)\left(\dfrac{98.09 \text{ g H}_2\text{SO}_4}{\text{mol H}_2\text{SO}_4}\right) = 0.0417 \text{ g H}_2\text{SO}_4$

(b) $\left(4.5 \times 10^{22} \text{ molecules CCl}_4\right)\left(\dfrac{1 \text{ mol}}{6.022 \times 10^{23} \text{ molecules}}\right)\left(\dfrac{153.8 \text{ g CCl}_4}{\text{mol CCl}_4}\right) = 11 \text{ g CCl}_4$

(c) $\left(0.00255 \text{ mol Ti}\right)\left(\dfrac{47.87 \text{ g Ti}}{\text{mol Ti}}\right) = 0.122 \text{ g Ti}$

(d) $\left(1.5 \times 10^{16} \text{ atoms S}\right)\left(\dfrac{32.07 \text{ g S}}{6.022 \times 10^{23} \text{atoms S}}\right) = 8.0 \times 10^{-7} \text{g S}$

7. Number of molecules

(a) $\left(2.5 \text{ mol S}_8\right)\left(\dfrac{6.022 \times 10^{23} \text{ molecules}}{\text{mol}}\right) = 1.5 \times 10^{24} \text{ molecules S}_8$

(b) $\left(7.35 \text{ mol NH}_3\right)\left(\dfrac{6.022 \times 10^{23} \text{ molecules}}{\text{mol}}\right) = 4.43 \times 10^{24} \text{ molecules NH}_3$

(c) $\left(17.5 \text{ g C}_2\text{H}_5\text{OH}\right)\left(\dfrac{6.022 \times 10^{23} \text{ molecules}}{46.07 \text{ g C}_2\text{H}_5\text{OH}}\right) = 2.29 \times 10^{23} \text{ molecules C}_2\text{H}_5\text{OH}$

(d) $\left(225 \text{ g Cl}_2\right)\left(\dfrac{6.022 \times 10^{23} \text{ molecules}}{70.90 \text{ g Cl}_2}\right) = 1.91 \times 10^{24} \text{ molecules Cl}_2$

8. Number of molecules

(a) $\left(9.6 \text{ mol C}_2\text{H}_4\right)\left(\dfrac{6.022 \times 10^{23} \text{ molecules}}{\text{mol}}\right) = 5.8 \times 10^{24} \text{ molecules C}_2\text{H}_4$

(b) $\left(2.76 \text{ mol N}_2\text{O}\right)\left(\dfrac{6.022 \times 10^{23} \text{ molecules}}{\text{mol}}\right) = 1.66 \times 10^{24} \text{ molecules N}_2\text{O}$

(c) $\left(23.2 \text{ g CH}_3\text{OH}\right)\left(\dfrac{6.022 \times 10^{23} \text{ molecules}}{32.04 \text{ g CH}_3\text{OH}}\right) = 4.36 \times 10^{23} \text{ molecules CH}_3\text{OH}$

(d) $\left(32.7 \text{ g CCl}_4\right)\left(\dfrac{6.022 \times 10^{23} \text{ molecules}}{153.8 \text{ g CCl}_4}\right) = 1.28 \times 10^{23} \text{ molecules CCl}_4$

9. Number of atoms

(a) $\left(25 \text{ molecules P}_2\text{O}_5\right)\left(\dfrac{7 \text{ atoms}}{1 \text{ molecule P}_2\text{O}_5}\right) = 1.8 \times 10^{2} \text{ atoms}$

(b) $\left(3.62 \text{ mol O}_2\right)\left(\dfrac{6.022 \times 10^{23} \text{ molecules}}{\text{mol O}_2}\right)\left(\dfrac{2 \text{ atoms}}{1 \text{ molecule}}\right) = 4.36 \times 10^{24} \text{ atoms}$

(c) $\left(12.2 \text{ mol CS}_2\right)\left(\dfrac{6.022 \times 10^{23} \text{ molecules}}{\text{mol CS}_2}\right)\left(\dfrac{3 \text{ atoms}}{1 \text{ molecule}}\right) = 2.20 \times 10^{25} \text{ atoms}$

(d) $(1.25 \text{ g Na})\left(\dfrac{6.022 \times 10^{23} \text{ atoms}}{22.99 \text{ g Na}}\right) = 3.27 \times 10^{22} \text{ atoms}$

(e) $(2.7 \text{ g CO}_2)\left(\dfrac{6.022 \times 10^{23} \text{ molecules}}{44.01 \text{ g CO}_2}\right)\left(\dfrac{3 \text{ atoms}}{1 \text{ molecule}}\right) = 1.1 \times 10^{23} \text{ atoms}$

(f) $(0.25 \text{ g CH}_4)\left(\dfrac{6.022 \times 10^{23} \text{ molecules}}{16.04 \text{ g CH}_4}\right)\left(\dfrac{5 \text{ atoms}}{1 \text{ molecule}}\right) = 4.7 \times 10^{22} \text{ atoms}$

10. Number of atoms.

(a) $(2 \text{ molecules CH}_3\text{COOH})\left(\dfrac{8 \text{ atoms}}{1 \text{ molecule}}\right) = 16 \text{ atoms}$

(b) $(0.75 \text{ mol C}_2\text{H}_6)\left(\dfrac{6.022 \times 10^{23} \text{ molecules CH}_3\text{COOH}}{\text{mol C}_2\text{H}_6}\right)\left(\dfrac{8 \text{ atoms}}{1 \text{ molecule}}\right) = 3.6 \times 10^{24} \text{ atoms}$

(c) $(25 \text{ mol H}_2\text{O})\left(\dfrac{6.022 \times 10^{23} \text{ molecules}}{\text{mol}}\right)\left(\dfrac{3 \text{ atoms}}{1 \text{ molecule H}_2\text{O}}\right) = 4.5 \times 10^{25} \text{ atoms}$

(d) $(92.5 \text{ g Au})\left(\dfrac{6.022 \times 10^{23} \text{ atoms}}{197.0 \text{ g Au}}\right) = 2.83 \times 10^{23} \text{ atoms}$

(e) $(75 \text{ g PCl}_3)\left(\dfrac{6.022 \times 10^{23} \text{ molecules}}{137.3 \text{ g PCl}_3}\right)\left(\dfrac{4 \text{ atoms}}{1 \text{ molecule}}\right) = 1.3 \times 10^{24} \text{ atoms}$

(f) $(15 \text{ g C}_6\text{H}_{12}\text{O}_6)\left(\dfrac{6.022 \times 10^{23} \text{ molecules}}{180.2 \text{ g C}_6\text{H}_{12}\text{O}_6}\right)\left(\dfrac{24 \text{ atoms}}{1 \text{ molecule}}\right) = 1.2 \times 10^{24} \text{ atoms}$

11. Number of grams.

(a) $(1 \text{ atom He})\left(\dfrac{4.003 \text{ g}}{6.022 \times 10^{23} \text{ atoms}}\right) = 6.647 \times 10^{-24} \text{ g He}$

(b) $(15 \text{ atoms C})\left(\dfrac{12.01 \text{ g}}{6.022 \times 10^{23} \text{ atoms}}\right) = 2.991 \times 10^{-22} \text{ g C}$

(c) $(4 \text{ molecules N}_2\text{O}_5)\left(\dfrac{108.0 \text{ g}}{6.022 \times 10^{23} \text{ molecules}}\right) = 7.175 \times 10^{-22} \text{ g N}_2\text{O}_5$

(d) $(11 \text{ molecules C}_6\text{H}_5\text{NH}_2)\left(\dfrac{93.13 \text{ g}}{6.022 \times 10^{23} \text{ molecules}}\right) = 1.701 \times 10^{-21} \text{ g C}_6\text{H}_5\text{NH}_2$

12. Number of grams.

(a) $(1 \text{ atom Xe})\left(\dfrac{131.3 \text{ g}}{6.022 \times 10^{23} \text{ atoms}}\right) = 2.180 \times 10^{-22} \text{ g Xe}$

(b) $(22 \text{ atoms Cl})\left(\dfrac{35.45 \text{ g}}{6.022 \times 10^{23} \text{ atoms}}\right) = 1.295 \times 10^{-21} \text{ g Cl}$

(c) $\left(9 \text{ molecules } CH_3COOH \left(\dfrac{60.05 \text{ g}}{6.022 \times 10^{23} \text{ molecules}}\right) = 8.975 \times 10^{-22} \text{ g } CH_3COOH\right.$

(d) $\left(15 \text{ molecules } C_4H_4O_2(NH_2)_2\right) \left(\dfrac{116.1 \text{ g}}{6.022 \times 10^{23} \text{ molecules}}\right) = 2.892 \times 10^{-21} \text{ g } C_4H_4O_2(NH_2)_2$

13. (a) $(25 \text{ kg } CO_2) \left(\dfrac{1000 \text{ g}}{\text{kg}}\right) \left(\dfrac{1 \text{ mol } CO_2}{44.01 \text{ g } CO_2}\right) = 5.7 \times 10^2 \text{ mol } CO_2$

(b) $(5 \text{ atoms Pb}) \left(\dfrac{1 \text{ mol Pb}}{6.022 \times 10^{23} \text{ atoms Pb}}\right) = 8 \times 10^{-24} \text{ mol Pb}$

(c) $(6 \text{ mol } O_2) \left(\dfrac{6.022 \times 10^{23} \text{ molecules } O_2}{\text{mol } O_2}\right) \left(\dfrac{2 \text{ atoms O}}{1 \text{ molecule } O_2}\right) = 7 \times 10^{24} \text{ atoms O}$

(d) $(25 \text{ molecules } P_4) \left(\dfrac{123.9 \text{ g } P_4}{6.022 \times 10^{23} \text{ molecules } P_4}\right) = 5.1 \times 10^{-21} \text{ g } P_4$

14. (a) $(275 \text{ atoms W}) \left(\dfrac{1 \text{ mol W}}{6.022 \times 10^{23} \text{ atoms W}}\right) = 4.57 \times 10^{-22} \text{ mol W}$

(b) $(95 \text{ mol } H_2O) \left(\dfrac{18.02 \text{ g } H_2O}{\text{mol } H_2O}\right) \left(\dfrac{1 \text{ kg}}{1000 \text{ g}}\right) = 1.7 \text{ kg } H_2O$

(c) $(12 \text{ molecules } SO_2) \left(\dfrac{64.07 \text{ g } SO_2}{6.022 \times 10^{23} \text{ molecules } SO_2}\right) = 1.277 \times 10^{-21} \text{ g } SO_2$

(d) $(25 \text{ mol } Cl_2) \left(\dfrac{6.022 \times 10^{23} \text{ molecules } Cl_2}{\text{mol } Cl_2}\right) \left(\dfrac{2 \text{ atoms Cl}}{1 \text{ molecule } Cl_2}\right) = 3.0 \times 10^{25} \text{ atoms Cl}$

15. One molecule of tetraphosphorus decoxide (P_4O_{10}) contains:

(a) $(1 \text{ molecule } P_4O_{10}) \left(\dfrac{1 \text{ mol } P_4O_{10}}{6.022 \times 10^{23} \text{ molecules } P_4O_{10}}\right) = 1.661 \times 10^{-24} \text{ mol } P_4O_{10}$

(b) $\left(1.661 \times 10^{-24} \text{ mol } P_4O_{10}\right) \left(\dfrac{283.9 \text{ g } P_4O_{10}}{\text{mol } P_4O_{10}}\right) = 4.716 \times 10^{-22} \text{ g } P_4O_{10}$

(c) $(1 \text{ molecule } P_4O_{10}) \left(\dfrac{4 \text{ P atoms}}{1 \text{ molecule } P_4O_{10}}\right) = 4 \text{ atoms P}$

(d) $(1 \text{ molecule } P_4O_{10}) \left(\dfrac{10 \text{ atoms O}}{1 \text{ molecule } P_4O_{10}}\right) = 10 \text{ atoms O}$

(e) $(1 \text{ molecule } P_4O_{10}) \left(\dfrac{14 \text{ atoms}}{1 \text{ molecule } P_4O_{10}}\right) = 14 \text{ total atoms}$

16. 125 grams of disulfur decofluoride (S_2F_{10}) contains:

    (a) $(125 \text{ g } S_2F_{10})\left(\dfrac{1 \text{ mol } S_2F_{10}}{254.1 \text{ g } S_2F_{10}}\right) = 0.492 \text{ mol } S_2F_{10}$

    (b) $(0.492 \text{ mol } S_2F_{10})\left(\dfrac{6.022 \times 10^{23} \text{ molecules}}{\text{mol}}\right) = 2.96 \times 10^{23} \text{ molecules } S_2F_{10}$

    (c) $(0.492 \text{ mol } S_2F_{10})\left(\dfrac{6.022 \times 10^{23} \text{ molecules}}{\text{mol}}\right)\left(\dfrac{12 \text{ atoms}}{1 \text{ molecule } S_2F_{10}}\right) = 3.56 \times 10^{24} \text{ total atoms}$

    (d) $(0.492 \text{ mol } S_2F_{10})\left(\dfrac{6.022 \times 10^{23} \text{ molecules}}{\text{mol}}\right)\left(\dfrac{2 \text{ S atoms}}{1 \text{ molecule } S_2F_{10}}\right) = 5.93 \times 10^{23} \text{ atoms S}$

    (e) $(0.492 \text{ mol } S_2F_{10})\left(\dfrac{6.022 \times 10^{23} \text{ molecules}}{\text{mol}}\right)\left(\dfrac{10 \text{ F atoms}}{1 \text{ molecule } S_2F_{10}}\right) = 2.96 \times 10^{24} \text{ atoms F}$

17. Atoms of hydrogen in:

    (a) $(25 \text{ molecules } C_6H_5CH_3)\left(\dfrac{8 \text{ H atoms}}{\text{molecule } C_6H_5CH_3}\right) = 2.0 \times 10^2 \text{ atoms H}$

    (b) $(3.5 \text{ mol } H_2CO_3)\left(\dfrac{6.022 \times 10^{23} \text{ molecules}}{\text{mol}}\right)\left(\dfrac{2 \text{ H atoms}}{\text{molecule } H_2CO_3}\right) = 4.2 \times 10^{24} \text{ atoms H}$

    (c) $(36 \text{ g } CH_3CH_2OH)\left(\dfrac{6.022 \times 10^{23} \text{ molecules}}{46.07 \text{ g}}\right)\left(\dfrac{6 \text{ H atoms}}{\text{molecule } CH_3CH_2OH}\right) = 2.8 \times 10^{24} \text{ atoms H}$

18. Atoms of hydrogen in:

    (a) $(23 \text{ molecules } CH_3CH_2COOH)\left(\dfrac{6 \text{ atoms H}}{1 \text{ molecule } CH_3CH_2COOH}\right) = 138 \text{ atoms H}$

    (b) $(7.4 \text{ mol } H_3PO_4)\left(\dfrac{6.022 \times 10^{23} \text{ molecules}}{\text{mol}}\right)\left(\dfrac{3 \text{ atoms H}}{1 \text{ molecule } H_3PO_4}\right) = 1.3 \times 10^{25} \text{ atoms H}$

    (c) $(57 \text{ g } C_6H_5ONH_2)\left(\dfrac{6.022 \times 10^{23} \text{ molecules}}{109.1 \text{ g}}\right)\left(\dfrac{7 \text{ atoms H}}{1 \text{ molecule } C_6H_5ONH_2}\right) = 2.2 \times 10^{24} \text{ atoms H}$

19. The number of grams of:

    (a) silver in 25.0 g AgBr

    $(25.0 \text{ g AgBr})\left(\dfrac{107.9 \text{ g Ag}}{187.8 \text{ g AgBr}}\right) = 14.4 \text{ g Ag}$

    (b) nitrogen in 6.34 mol $(NH_4)_3PO_4$

    $(6.34 \text{ mol } (NH_4)_3PO_4)\left(\dfrac{42.03 \text{ g N}}{\text{mol } (NH_4)_3PO_4}\right) = 266 \text{ g N}$

(c) oxygen in 8.45×10^{22} molecules SO_3

The conversion is: molecules $SO_3 \longrightarrow$ mol $SO_3 \longrightarrow$ g O

$$\left(8.45 \times 10^{22} \text{ molecules } SO_3\right)\left(\frac{1 \text{ mol}}{6.022 \times 10^{23} \text{ molecules}}\right)\left(\frac{48.00 \text{ g O}}{\text{mol } SO_3}\right) = 6.74 \text{ g O}$$

20. The number of grams of:
    (a) chlorine in 5.00 g $PbCl_2$

$$\left(5.00 \text{ g } PbCl_2\right)\left(\frac{70.90 \text{ g Cl}}{278.1 \text{ g } PbCl_2}\right) = 1.27 \text{ g Cl}$$

    (b) hydrogen in 4.50 mol H_2SO_4

$$\left(4.50 \text{ mol } H_2SO_4\right)\left(\frac{2.016 \text{ g H}}{1 \text{ mol } H_2SO_4}\right) = 9.07 \text{ g H}$$

    (c) hydrogen in 5.45×10^{22} molecules NH_3

The conversion is: molecules $NH_3 \longrightarrow$ moles $NH_3 \longrightarrow$ g H

$$\left(5.45 \times 10^{22} \text{ molecules } NH_3\right)\left(\frac{1 \text{ mol}}{6.022 \times 10^{23} \text{ molecules}}\right)\left(\frac{3.024 \text{ g H}}{\text{mol } NH_3}\right) = 0.274 \text{ g H}$$

21. Percent composition

    (a) NaBr   Na   22.99 g $\left(\dfrac{22.99 \text{ g}}{102.9 \text{ g}}\right)(100) = 22.34\%$ Na

                  Br $\underline{79.90}$ g

                         102.9 g $\left(\dfrac{79.90 \text{ g}}{102.9 \text{ g}}\right)(100) = 77.65\%$ Br

    (b) KHCO$_3$ K   39.10 g $\left(\dfrac{39.10 \text{ g}}{100.1 \text{ g}}\right)(100) = 39.06\%$ K

                      H   1.008 g

                    3 O   48.00 g $\left(\dfrac{1.008 \text{ g}}{100.1 \text{ g}}\right)(100) = 1.007\%$ H

                      C $\underline{12.01}$ g

                            100.1 g $\left(\dfrac{12.01 \text{ g}}{100.1 \text{ g}}\right)(100) = 12.00\%$ C

                                        $\left(\dfrac{48.00 \text{ g}}{100.1 \text{ g}}\right)(100) = 47.95\%$ O

(c) $FeCl_3$ Fe   55.85 g

                   3 Cl $\underline{106.4\ \ g}$

                          162.3   g

$$\left(\frac{55.85\ g}{162.3\ g}\right)(100) = 34.41\%\ Fe$$

$$\left(\frac{106.4\ g}{162.3\ g}\right)(100) = 65.56\%\ Cl$$

(d) $SiCl_4$ Si   28.09 g

                   4 Cl $\underline{141.8\ \ g}$

                          169.9   g

$$\left(\frac{28.09\ g}{169.9\ g}\right)(100) = 16.53\%\ Si$$

$$\left(\frac{141.8\ g}{169.9\ g}\right)(100) = 83.46\%\ Cl$$

(e) $Al_2(SO_4)_3$ 2 Al   53.96 g

                      3 S   96.21 g

                      12 O $\underline{192.0\ \ g}$

                          342.2   g

$$\left(\frac{53.96\ g}{342.2\ g}\right)(100) = 15.77\%\ Al$$

$$\left(\frac{96.21\ g}{342.2\ g}\right)(100) = 28.12\%\ S$$

$$\left(\frac{192.0\ g}{342.2\ g}\right)(100) = 56.11\%\ O$$

(f) $AgNO_3$ Ag   107.9   g

                      N   14.01 g

                      3 O $\underline{48.00\ g}$

                          169.9   g

$$\left(\frac{107.9\ g}{169.9\ g}\right)(100) = 63.51\%\ Ag$$

$$\left(\frac{14.01\ g}{169.9\ g}\right)(100) = 8.246\%\ N$$

$$\left(\frac{48.00\ g}{169.9\ g}\right)(100) = 28.25\%\ O$$

22.   Percent composition

(a) $ZnCl_2$ Zn   65.39 g

                   2 Cl $\underline{70.90\ g}$

                          136.3   g

$$\left(\frac{65.39\ g}{136.3\ g}\right)(100) = 47.98\%\ Zn$$

$$\left(\frac{70.90\ g}{136.3\ g}\right)(100) = 52.02\%\ Cl$$

(b) $NH_4C_2H_3O_2$ N   14.01   g

                         7 H   7.056 g

                         2 C   24.02   g

                         2 O $\underline{32.00\ \ g}$

                           77.09   g

$$\left(\frac{14.01\ g}{77.09\ g}\right)(100) = 18.17\%\ N$$

$$\left(\frac{7.056\ g}{77.09\ g}\right)(100) = 9.153\%\ H$$

$$\left(\frac{24.02\ g}{77.09\ g}\right)(100) = 31.16\%\ C$$

$$\left(\frac{32.00\ g}{77.09\ g}\right)(100) = 41.51\%\ O$$

(c) MgP_2O_7 Mg   24.31 g $\left(\dfrac{24.31\ g}{198.3\ g}\right)(100) = 12.26\%\ Mg$

                       2 P   61.94 g

                       7 O $\underline{112.0}$ g $\left(\dfrac{61.94\ g}{198.3\ g}\right)(100) = 31.24\%\ P$

                                 198.3 g

                                                   $\left(\dfrac{112.0\ g}{198.3\ g}\right)(100) = 56.48\%\ O$

(d) $(NH_4)_2SO_4$ 2 N   28.02 g $\left(\dfrac{28.02\ g}{132.2\ g}\right)(100) = 21.20\%\ N$

                       8 H   8.064 g

                         S   32.07 g $\left(\dfrac{8.064\ g}{132.2\ g}\right)(100) = 6.100\%\ H$

                       4 O $\underline{64.00}$ g

                               132.2 g $\left(\dfrac{32.07\ g}{132.2\ g}\right)(100) = 24.26\%\ S$

                                                   $\left(\dfrac{64.00\ g}{132.2\ g}\right)(100) = 48.41\%\ O$

(e) $Fe(NO_3)_3$ Fe   55.85 g $\left(\dfrac{55.85\ g}{241.9\ g}\right)(100) = 23.09\%\ Fe$

                       3 N   42.03 g

                       9 O $\underline{144.0}$ g $\left(\dfrac{42.03\ g}{241.9\ g}\right)(100) = 17.37\%\ N$

                               241.9 g

                                                  $\left(\dfrac{144.0\ g}{241.9\ g}\right)(100) = 59.53\%\ O$

(f) ICl_3 I   126.9 g $\left(\dfrac{126.9\ g}{233.3\ g}\right)(100) = 54.39\%\ I$

                     3 Cl $\underline{106.4}$ g

                           233.3 g $\left(\dfrac{106.4\ g}{233.3\ g}\right)(100) = 45.61\%\ Cl$

23.   Percent of iron

(a)   FeO   Fe   55.85 g $\left(\dfrac{55.85\ g}{71.85\ g}\right)(100) = 77.73\%\ Fe$

                     O $\underline{16.00}$ g

                           71.85 g

(b) Fe_2O_3 2 Fe   111.7 g $\left(\dfrac{111.7\ g}{159.7\ g}\right)(100) = 69.94\%\ Fe$

                       3 O $\underline{48.00}$ g

                           159.7 g

(c) Fe_3O_4 3 Fe   167.6 g $\left(\dfrac{167.6\ g}{231.6\ g}\right)(100) = 72.37\%\ Fe$

                       4 O $\underline{64.00}$ g

                           231.6 g

(d) $K_4Fe(CN)_6$ Fe   55.85 g $\left(\dfrac{55.85\ g}{368.4\ g}\right)(100) = 15.16\%\ Fe$

                       4 K   156.4 g

$$
\begin{array}{lr}
\text{6 C} & 72.06 \text{ g} \\
\text{6 N} & \underline{84.06 \text{ g}} \\
& 368.4 \ \ \text{g}
\end{array}
$$

24. Percent chlorine

(a) KCl

$$
\begin{array}{lr}
\text{K} & 39.10 \text{ g} \\
\text{Cl} & \underline{35.45 \text{ g}} \\
& 74.55 \text{ g}
\end{array}
$$
$\left(\dfrac{35.45 \text{ g}}{74.55 \text{ g}}\right)(100) = 47.55\% \text{ Cl}$

(b) $BaCl_2$

$$
\begin{array}{lr}
\text{Ba} & 137.3 \ \ \text{g} \\
\text{2 Cl} & \underline{70.90 \text{ g}} \\
& 208.2 \ \ \text{g}
\end{array}
$$
$\left(\dfrac{70.90 \text{ g}}{208.2 \text{ g}}\right)(100) = 34.05\% \text{ Cl}$

(c) $SiCl_4$

$$
\begin{array}{lr}
\text{Si} & 28.09 \ \ \text{g} \\
\text{4 Cl} & \underline{141.8 \ \ \text{g}} \\
& 169.9 \ \ \text{g}
\end{array}
$$
$\left(\dfrac{141.8 \text{ g}}{169.9 \text{ g}}\right)(100) = 83.46\% \text{ Cl}$

(d) LiCl

$$
\begin{array}{lr}
\text{Li} & 6.941 \text{ g} \\
\text{Cl} & \underline{35.45 \ \ \text{g}} \\
& 42.39 \ \ \text{g}
\end{array}
$$
$\left(\dfrac{35.45 \text{ g}}{42.39 \text{ g}}\right)(100) = 83.63\% \text{ Cl}$

Highest % Cl is LiCl; lowest % Cl is in $BaCl_2$

25. Percent composition

$$
\begin{array}{l}
73.16 \text{ g barium silicide} \\
\underline{-33.62 \text{ g Ba}} \\
39.54 \text{ g Si}
\end{array}
$$

$\left(\dfrac{39.54 \text{ g Si}}{73.16 \text{ g}}\right)(100) = 54.05\% \text{ Si}$

$\left(\dfrac{33.62 \text{ g Ba}}{73.16 \text{ g}}\right)(100) = 45.95\% \text{ Ba}$

26. Percent composition

$$
\begin{array}{ll}
7.52 & \text{g ajoene} \\
-3.09 & \text{g S} \\
-0.453 & \text{g H} \\
\underline{-0.513} & \text{g O} \\
3.46 & \text{g C}
\end{array}
$$

$\left(\dfrac{3.09 \text{ g S}}{7.52 \text{ g}}\right)(100) = 41.1 \ \% \text{ S}$

$\left(\dfrac{0.453 \text{ g H}}{7.52 \text{ g}}\right)(100) = 6.02 \ \% \text{ H}$

$\left(\dfrac{0.513 \text{ g O}}{7.52 \text{ g}}\right)(100) = 6.82 \ \% \text{ O}$

$\left(\dfrac{3.46 \text{ g C}}{7.52 \text{ g}}\right)(100) = 46.0 \ \% \text{ C}$

27. (a) H_2O has the higher percent Hydrogen
    (b) N_2O_3 has the lower percent Nitrogen
    (c) Both have the same percent Oxygen

28. (a) $KClO_3$ is lower. (Because a K atom has more mass than a Na atom.)
    (b) $KHSO_4$ is higher. (Because a H atom has less mass than a K atom.)
    (c) Na_2CrO_4 is lower. (Because only one Cr atom is present.)

29. Empirical formulas from percent composition.
    (a) Step 1. Express each element as grams/100 g material.

$$63.6\% \text{ N} = 63.6 \text{ g N}/100 \text{ g material}$$
$$36.4\% \text{ O} = 36.4 \text{ g O}/100 \text{ g material}$$

Step 2. Calculate the relative moles of each element.

$$(63.6 \text{ g N})\left(\frac{1 \text{ mol N}}{14.01 \text{ g N}}\right) = 4.54 \text{ mol N}$$

$$(36.4 \text{ g O})\left(\frac{1 \text{ mol O}}{16.00 \text{ g O}}\right) = 2.28 \text{ mol O}$$

Step 3. Change these moles to whole numbers by dividing each by the smaller number.

$$\frac{4.54 \text{ mol N}}{2.28} = 1.99 \text{ mol N}$$

$$\frac{2.28 \text{ mol O}}{2.28} = 1.00 \text{ mol O}$$

The simplest ratio of N:O is 2:1. The empirical formula, therefore, is N_2O.

(b) 46.7% N, 53.3% O

$$(46.7 \text{ g N})\left(\frac{1 \text{ mol N}}{14.01 \text{ g N}}\right) = 3.33 \text{ mol N} \qquad \frac{3.33 \text{ mol N}}{3.33} = 1.00 \text{ mol N}$$

$$(53.3 \text{ g N})\left(\frac{1 \text{ mol O}}{16.00 \text{ g O}}\right) = 3.33 \text{ mol O} \qquad \frac{3.33 \text{ mol O}}{3.33} = 1.00 \text{ mol O}$$

The empirical formula is NO.

(c) 25.9% N, 74.1% O

$$(25.9 \text{ g N})\left(\frac{1 \text{ mol N}}{14.01 \text{ g N}}\right) = 1.85 \text{ mol N} \qquad \frac{1.85 \text{ mol N}}{1.85} = 1.00 \text{ mol N}$$

$$(74.1 \text{ g O})\left(\frac{1 \text{ mol O}}{16.00 \text{ g O}}\right) = 4.63 \text{ mol O} \qquad \frac{4.63 \text{ mol O}}{1.85} = 2.50 \text{ mol O}$$

Since these values are not whole numbers, multiply each by 2 to change them to whole numbers.
$(1.00 \text{ mol N})(2) = 2.00 \text{ mol N}; \quad (2.5 \text{ mol O})(2) = 5.00 \text{ mol O}$
The empirical formula is N_2O_5.

(d)   43.4% Na, 11.3% C, 45.3% O

$$(43.4 \text{ g Na})\left(\frac{1 \text{ mol Na}}{22.99 \text{ g Na}}\right) = 1.89 \text{ mol Na} \qquad \frac{1.89 \text{ mol Na}}{0.941} = 2.01 \text{ mol Na}$$

$$(11.3 \text{ g C})\left(\frac{1 \text{ mol C}}{12.01 \text{ g C}}\right) = 0.941 \text{ mol C} \qquad \frac{0.941 \text{ mol C}}{0.941} = 1.00 \text{ mol C}$$

$$(45.3 \text{ g O})\left(\frac{1 \text{ mol O}}{16.00 \text{ g O}}\right) = 2.83 \text{ mol O} \qquad \frac{2.83 \text{ mol O}}{0.941} = 3.00 \text{ mol O}$$

The empirical formula is Na_2CO_3.

(e)   18.8% Na, 29.0% Cl, 52.3% O

$$(18.8 \text{ g Na})\left(\frac{1 \text{ mol Na}}{22.99 \text{ g Na}}\right) = 0.818 \text{ mol Na} \qquad \frac{0.818 \text{ mol Na}}{0.818} = 1.00 \text{ mol Na}$$

$$(29.0 \text{ g Cl})\left(\frac{1 \text{ mol Cl}}{35.45 \text{ g Cl}}\right) = 0.818 \text{ mol Cl} \qquad \frac{0.818 \text{ mol Cl}}{0.818} = 1.00 \text{ mol Cl}$$

$$(52.3 \text{ g O})\left(\frac{1 \text{ mol O}}{16.00 \text{ g O}}\right) = 3.27 \text{ mol O} \qquad \frac{3.27 \text{ mol O}}{0.818} = 4.00 \text{ mol O}$$

The empirical formula is $NaClO_4$.

(f)   72.02% Mn, 27.98% O

$$(72.02 \text{ g Mn})\left(\frac{1 \text{ mol Mn}}{54.94 \text{ g Mn}}\right) = 1.311 \text{ mol Mn} \qquad \frac{1.311 \text{ mol Mn}}{1.311} = 1.000 \text{ mol Mn}$$

$$(27.98 \text{ g O})\left(\frac{1 \text{ mol O}}{16.00 \text{ g O}}\right) = 1.749 \text{ mol O} \qquad \frac{1.749 \text{ mol O}}{1.311} = 1.334 \text{ mol O}$$

Multiply both values by 3 to give whole numbers.
$(1.000 \text{ mol Mn})(3) = 3.000 \text{ mol Mn}; \ (1.334 \text{ mol O})(3) = 4.002 \text{ mol O}$
The empirical formula is Mn_3O_4.

30. Empirical formulas from percent composition.

(a)   64.1% Cu, 35.9% Cl

$$(64.1 \text{ g Cu})\left(\frac{1 \text{ mol Cu}}{63.55 \text{ g Cu}}\right) = 1.01 \text{ mol Cu} \qquad \frac{1.01 \text{ mol Cu}}{1.01} = 1.00 \text{ mol Cu}$$

$$(35.9 \text{ g Cl})\left(\frac{1 \text{ mol Cl}}{35.45 \text{ g Cl}}\right) = 1.01 \text{ mol Cl} \qquad \frac{1.01 \text{ mol Cl}}{1.01} = 1.00 \text{ mol Cl}$$

The empirical formula is CuCl.

(b)   47.2% Cu, 52.8% Cl

$$(47.2 \text{ g Cu})\left(\frac{1 \text{ mol Cu}}{63.55 \text{ g Cu}}\right) = 0.743 \text{ mol Cu} \qquad \frac{0.743 \text{ mol Cu}}{0.743} = 1.00 \text{ mol Cu}$$

$$(52.8 \text{ g Cl})\left(\frac{1 \text{ mol Cl}}{35.45 \text{ g Cl}}\right) = 1.49 \text{ mol Cl} \qquad \frac{1.49 \text{ mol Cl}}{0.743} = 2.01 \text{ mol Cl}$$

The empirical formula is $CuCl_2$.

(c)   51.9% Cr, 48.1% S

$$(51.9 \text{ g Cr})\left(\frac{1 \text{ mol Cr}}{52.00 \text{ g Cr}}\right) = 0.998 \text{ mol Cr} \qquad \frac{0.998 \text{ mol Cr}}{0.998} = 1.00 \text{ mol Cr}$$

$$(48.1 \text{ g S})\left(\frac{1 \text{ mol S}}{32.07 \text{ g S}}\right) = 1.50 \text{ mol S} \qquad \frac{1.50 \text{ mol S}}{0.998} = 1.50 \text{ mol S}$$

Multiply both values by 2 to give whole numbers.
$(1.00 \text{ mol Cr})(2) = 2.00 \text{ mol Cr}; \ (1.50 \text{ mol S})(2) = 3.00 \text{ mol S}$
The empirical formula is Cr_2S_3.

(d)   55.3% K, 14.6% P, 30.1% O

$$(55.3 \text{ g K})\left(\frac{1 \text{ mol K}}{39.10 \text{ g K}}\right) = 1.41 \text{ mol K} \qquad \frac{1.41 \text{ mol K}}{0.471} = 2.99 \text{ mol K}$$

$$(14.6 \text{ g P})\left(\frac{1 \text{ mol P}}{30.97 \text{ g P}}\right) = 0.471 \text{ mol P} \qquad \frac{1.471 \text{ mol P}}{0.471} = 1.00 \text{ mol P}$$

$$(30.1 \text{ g O})\left(\frac{1 \text{ mol O}}{16.00 \text{ g O}}\right) = 1.88 \text{ mol O} \qquad \frac{1.88 \text{ mol O}}{0.471} = 3.99 \text{ mol O}$$

The empirical formula is K_3PO_4.

(e)   38.9% Ba, 29.4% Cr, 31.7% O

$$(38.9 \text{ g Ba})\left(\frac{1 \text{ mol Ba}}{137.3 \text{ g Ba}}\right) = 0.283 \text{ mol Ba} \qquad \frac{0.283 \text{ mol Ba}}{0.283} = 1.00 \text{ mol Ba}$$

$$(29.4 \text{ g Cr})\left(\frac{1 \text{ mol Cr}}{52.00 \text{ g Cr}}\right) = 0.565 \text{ mol Cr} \qquad \frac{0.565 \text{ mol Cr}}{0.283} = 2.00 \text{ mol Cr}$$

$$(31.7 \text{ g O})\left(\frac{1 \text{ mol O}}{16.00 \text{ g O}}\right) = 1.98 \text{ mol O} \qquad \frac{1.98 \text{ mol O}}{0.283} = 7.00 \text{ mol O}$$

The empirical formula is $BaCr_2O_7$.

(f) 3.99% P, 82.3% Br, 13.7% Cl

$$(3.99 \text{ g P})\left(\frac{1 \text{ mol P}}{30.97 \text{ g P}}\right) = 0.129 \text{ mol P} \qquad \frac{0.129 \text{ mol P}}{0.129} = 1.00 \text{ mol P}$$

$$(82.3 \text{ g Br})\left(\frac{1 \text{ mol Br}}{79.90 \text{ g Br}}\right) = 1.03 \text{ mol Br} \qquad \frac{1.03 \text{ mol Br}}{0.129} = 7.98 \text{ mol Br}$$

$$(13.7 \text{ g Cl})\left(\frac{1 \text{ mol Cl}}{35.45 \text{ g Cl}}\right) = 0.386 \text{ mol Cl} \qquad \frac{0.386 \text{ mol Cl}}{0.129} = 2.99 \text{ mol Cl}$$

The empirical formula is PBr_8Cl_3.

31. Empirical formula:

(a) $$(26.08 \text{ g Zn})\left(\frac{1 \text{ mol Zn}}{65.39 \text{ g}}\right) = 0.3988 \text{ mol Zn} \qquad \frac{0.3988 \text{ mol Zn}}{0.3988} = 1.00 \text{ mol Zn}$$

$$(4.79 \text{ g C})\left(\frac{1 \text{ mol C}}{12.01 \text{ g}}\right) = 0.399 \text{ mol C} \qquad \frac{0.399 \text{ mol C}}{0.3988} = 1.00 \text{ mol C}$$

$$(19.14 \text{ g O})\left(\frac{1 \text{ mol O}}{16.00 \text{ g}}\right) = 1.196 \text{ mol O} \qquad \frac{1.196 \text{ mol O}}{0.3988} = 2.999 \text{ mol O}$$

The empirical formula is $ZnCO_3$

(b) 150.0 g compound
$$\begin{array}{r} -57.66 \text{ g C} \\ -7.26 \text{ g H} \\ \hline 85.1 \text{ g Cl} \end{array}$$

$$(57.66 \text{ g C})\left(\frac{1 \text{ mol C}}{12.01 \text{ g C}}\right) = 4.801 \text{ mol C} \qquad \frac{4.801 \text{ mol C}}{2.40} = 2.000 \text{ mol C}$$

$$(7.26 \text{ g H})\left(\frac{1 \text{ mol H}}{1.008 \text{ g H}}\right) = 7.20 \text{ mol H} \qquad \frac{7.20 \text{ mol H}}{2.40} = 3.00 \text{ mol H}$$

$$(85.1 \text{ g Cl})\left(\frac{1 \text{ mol Cl}}{35.45 \text{ g}}\right) = 2.40 \text{ mol Cl} \qquad \frac{2.40 \text{ mol Cl}}{2.40} = 1.00 \text{ mol Cl}$$

The empirical formula is C_2H_3Cl

(c)   75.0 g Oxide − 42.0 g V = 33.0 g O

$$(42.0 \text{ g V})\left(\frac{1 \text{ mol V}}{50.94 \text{ g V}}\right) = 0.824 \text{ mol V} \qquad \frac{0.824 \text{ mol V}}{0.824} = 1.00 \text{ mol V}$$

$$(33.0 \text{ g O})\left(\frac{1 \text{ mol O}}{16.00 \text{ g O}}\right) = 2.06 \text{ mol O} \qquad \frac{2.06 \text{ mol O}}{0.824} = 2.50 \text{ mol O}$$

Multiplying both by 2 gives the empirical formula V_2O_5

(d) $$(67.35 \text{ g Ni})\left(\frac{1 \text{ mol Ni}}{58.69 \text{ g Ni}}\right) = 1.148 \text{ mol Ni} \qquad \frac{1.148 \text{ mol Ni}}{0.7649} = 1.501 \text{ mol Ni}$$

$$(48.96 \text{ g O})\left(\frac{1 \text{ mol O}}{16.00 \text{ g O}}\right) = 3.060 \text{ mol O} \qquad \frac{3.060 \text{ mol O}}{0.7649} = 4.001 \text{ mol O}$$

$$(23.69 \text{ g P})\left(\frac{1 \text{ mol P}}{30.97 \text{ g P}}\right) = 0.7649 \text{ mol P} \qquad \frac{0.7649 \text{ mol P}}{0.7649} = 1.000 \text{ mol P}$$

Multiplying all by 2 gives the empirical formula $Ni_3O_8P_2$

32.   Empirical formula

(a) $$(55.08 \text{ g C})\left(\frac{1 \text{ mol C}}{12.01 \text{ g C}}\right) = 4.586 \text{ mol C} \qquad \frac{4.586 \text{ mol C}}{0.7643 \text{ mol}} = 6.000 \text{ mol C}$$

$$(3.85 \text{ g H})\left(\frac{1 \text{ mol H}}{1.008 \text{ g H}}\right) = 3.82 \text{ mol H} \qquad \frac{3.82 \text{ mol H}}{0.7643} = 5.00 \text{ mol H}$$

$$(61.07 \text{ g Br})\left(\frac{1 \text{ mol Br}}{79.90 \text{ g Br}}\right) = 0.7643 \text{ mol Br} \qquad \frac{0.7643 \text{ mol Br}}{0.7643} = 1.000 \text{ mol Br}$$

The empirical formula is C_6H_5Br

(b)   65.2 g compound − 36.8 g Ag − 12.1 g Cl = 16.3 g O

$$(36.8 \text{ g Ag})\left(\frac{1 \text{ mol Ag}}{107.9 \text{ g Ag}}\right) = 0.341 \text{ mol Ag} \qquad \frac{0.341 \text{ mol Ag}}{0.341} = 1.00 \text{ mol Ag}$$

$$(12.1 \text{ g Cl})\left(\frac{1 \text{ mol Cl}}{35.45 \text{ g Cl}}\right) = 0.341 \text{ mol Cl} \qquad \frac{0.341 \text{ mol Cl}}{0.341} = 1.00 \text{ mol Cl}$$

$$(16.3 \text{ g O})\left(\frac{1 \text{ mol O}}{16.00 \text{ g O}}\right) = 1.02 \text{ mol O} \qquad \frac{1.02 \text{ mol O}}{0.341} = 2.99 \text{ mol O}$$

The empirical formula is $AgClO_3$

(c) 25.25 g sulfide $-$ 12.99 g V = 12.26 g S

$$(12.99 \text{ g V})\left(\frac{1 \text{ mol V}}{50.94 \text{ g V}}\right) = 0.2550 \text{ mol V} \qquad \frac{0.2550 \text{ mol V}}{0.2550} = 1.000 \text{ mol V}$$

$$(12.26 \text{ g S})\left(\frac{1 \text{ mol S}}{32.07 \text{ g S}}\right) = 0.3823 \text{ mol S} \qquad \frac{0.3823 \text{ mol S}}{0.2550} = 1.499 \text{ mol S}$$

Multiplying both by 2 gives the empirical formula V_2S_3

(d) $$(38.0 \text{ g Zn})\left(\frac{1 \text{ mol Zn}}{65.39 \text{ g}}\right) = 0.581 \text{ mol Zn} \qquad \frac{0.581 \text{ mol Zn}}{0.387} = 1.50 \text{ mol Zn}$$

$$(12.0 \text{ g P})\left(\frac{1 \text{ mol P}}{30.97 \text{ g}}\right) = 0.387 \text{ mol P} \qquad \frac{0.387 \text{ mol P}}{0.387} = 1.00 \text{ mol P}$$

Multiplying both by 2 gives the empirical formula Zn_3P_2

33. 15.267 g sulfide $-$ 12.272 g Au = 2.995 g S

$$(12.272 \text{ g Au})\left(\frac{1 \text{ mol Au}}{197.0 \text{ g Au}}\right) = 0.06229 \text{ mol Au} \qquad \frac{0.06229 \text{ mol Au}}{0.06229} = 1.000 \text{ mol Au}$$

$$(2.995 \text{ g S})\left(\frac{1 \text{ mol S}}{32.07 \text{ g S}}\right) = 0.09339 \text{ mol S} \qquad \frac{0.09339 \text{ mol S}}{0.06229} = 1.499 \text{ mol S}$$

Multiplying both by 2 gives the empirical formula Au_2S_3

34. 10.724 g oxide $-$ 7.143 g Ti = 3.581 g O

$$(7.143 \text{ g Ti})\left(\frac{1 \text{ mol Ti}}{47.88 \text{ g Ti}}\right) = 0.1492 \text{ mol Ti} \qquad \frac{0.1492 \text{ mol Ti}}{0.1492} = 1.000 \text{ mol Ti}$$

$$(3.581 \text{ g O})\left(\frac{1 \text{ mol O}}{16.00 \text{ g O}}\right) = 0.2238 \text{ mol O} \qquad \frac{0.2238 \text{ mol O}}{0.1492} = 1.500 \text{ mol O}$$

Multiplying both by 2 gives the empirical formula Ti_2O_3

35. 5.000 g compound $-$ (0.6375 g C + 0.1070 g H) = 4.256 g S

$$(0.6375 \text{ g C})\left(\frac{1 \text{ mol C}}{12.01 \text{ g C}}\right) = 0.05308 \text{ mol C} \qquad \left(\frac{0.05308 \text{ mol C}}{0.05308 \text{ mol}}\right) = 1.000 \text{ mol C}$$

$$(0.1070 \text{ g H})\left(\frac{1 \text{ mol H}}{1.008 \text{ g H}}\right) = 0.1062 \text{ mol H} \qquad \left(\frac{0.1062 \text{ mol H}}{0.05308 \text{ mol}}\right) = 2.001 \text{ mol H}$$

$$(4.256 \text{ g S})\left(\frac{1 \text{ mol S}}{32.07 \text{ g S}}\right) = 0.1327 \text{ mol S} \qquad \left(\frac{0.1327 \text{ mol S}}{0.05308 \text{ mol}}\right) = 2.500 \text{ mol S}$$

Multiplying by 2 gives the empirical formula $C_2H_4S_5$

36. Empirical formula

5.276 g compound − 3.898 g Hg = 1.378 g Cl

$$(3.898 \text{ g Hg})\left(\frac{1 \text{ mol Hg}}{200.6 \text{ g Hg}}\right) = 0.01943 \text{ mol Hg} \qquad \frac{0.01943 \text{ mol Hg}}{0.01943} = 1.000 \text{ mol Hg}$$

$$(1.378 \text{ g Cl})\left(\frac{1 \text{ mol Cl}}{35.45 \text{ g Cl}}\right) = 0.03887 \text{ mol Cl} \qquad \frac{0.03887 \text{ mol Cl}}{0.01943} = 2.001 \text{ mol Cl}$$

The empirical formula is $HgCl_2$.

37. Empirical and molecular formula of traumatic acid.
    63.13% C, 8.830% H, 28.03% O; molar mass = 228 g

$$(63.13 \text{ g C})\left(\frac{1 \text{ mol C}}{12.01 \text{ g C}}\right) = 5.256 \text{ mol C} \qquad \left(\frac{5.256 \text{ mol C}}{1.752}\right) = 3.000 \text{ mol C}$$

$$(8.830 \text{ g H})\left(\frac{1 \text{ mol H}}{1.008 \text{ g H}}\right) = 8.760 \text{ mol H} \qquad \left(\frac{8.760 \text{ mol H}}{1.752}\right) = 5.000 \text{ mol H}$$

$$(28.03 \text{ g O})\left(\frac{1 \text{ mol O}}{16.00 \text{ g O}}\right) = 1.752 \text{ mol O} \qquad \left(\frac{1.752 \text{ mol O}}{1.752 \text{ mol}}\right) = 1.000 \text{ mol O}$$

The empirical formula for traumatic acid is C_3H_5O. The empirical formula mass is 57 g.

$$\frac{\text{molar mass}}{\text{empirical formula mass}} = \frac{228}{57} = 4$$

The molecular formula is four times that of the empirical formula.
Molecular formula is $(C_3H_5O)_4 = C_{12}H_{20}O_4$.

38. Empirical and molecular formulas of dixanthogen.

29.73% C, 4.16% H, 13.20% O, 52.91% S; molar mass = 242.4 g

$$(29.73 \text{ g C})\left(\frac{1 \text{ mol C}}{12.01 \text{ g C}}\right) = 2.475 \text{ mol C} \qquad \frac{2.475 \text{ mol C}}{0.8250} = 3.000 \text{ mol C}$$

$$(4.16 \text{ g H})\left(\frac{1 \text{ mol H}}{1.008 \text{ g H}}\right) = 4.13 \text{ mol H} \qquad \frac{4.13 \text{ mol H}}{0.8250} = 5.01 \text{ mol H}$$

$$(13.20 \text{ g O})\left(\frac{1 \text{ mol O}}{16.00 \text{ g O}}\right) = 0.8250 \text{ mol O} \qquad \frac{0.8250 \text{ mol O}}{0.8250} = 1.000 \text{ mol O}$$

$$(52.91 \text{ g S})\left(\frac{1 \text{ mol S}}{32.07 \text{ g S}}\right) = 1.650 \text{ mol S} \qquad \frac{1.650 \text{ mol S}}{0.8250} = 2.000 \text{ mol S}$$

The empirical formula is $C_3H_5OS_2$. The empirical formula mass is 121.2 g.

$$\frac{\text{molar mass}}{\text{empirical formula mass}} = \frac{242.4 \text{ g}}{121.2 \text{ g}} = 2$$

The molecular formula is twice that of the empirical formula
Molecular formula is $(C_3H_5OS_2)_2 = C_6H_{10}O_2S_4$

39. Molecular formula of oxalic acid (ethanedioic acid)
26.7% C, 2.24% H, 71.1% O; molar mass = 90.04

$$26.7 \text{ g C}\left(\frac{1 \text{ mol C}}{12.01 \text{ g C}}\right) = 2.22 \text{ mol C} \qquad \frac{2.22 \text{ mol C}}{2.2} = 1.0 \text{ mol C}$$

$$2.2 \text{ g H}\left(\frac{1 \text{ mol H}}{1.008 \text{ g H}}\right) = 2.2 \text{ mol H} \qquad \frac{2.2 \text{ mol H}}{2.2} = 1.0 \text{ mol H}$$

$$71.1 \text{ g O}\left(\frac{1 \text{ mol O}}{16.00 \text{ g O}}\right) = 4.44 \text{ mol O} \qquad \frac{4.44 \text{ mol O}}{2.2} = 2.0 \text{ mol O}$$

The empirical formula is CHO_2, making the empirical formula mass 45.02 g.

$$\frac{\text{molar mass}}{\text{mass of empirical formula}} = \frac{90.04 \text{ g}}{45.02 \text{ g}} = 2$$

The molecular formula is twice that of the empirical formula.
Molecular formula = $(CHO_2)_2 = C_2H_2O_4$

40. Molecular formula of butyric acid

54.5% C, 9.2% H, 36.3% O; molar mass = 88.11

$$(54.5 \text{ g C})\left(\frac{1 \text{ mol C}}{12.01 \text{ g C}}\right) = 4.54 \text{ mol C} \qquad \frac{4.54 \text{ mol C}}{2.27} = 2.00 \text{ mol C}$$

$$(9.2 \text{ g H})\left(\frac{1 \text{ mol H}}{1.008 \text{ g H}}\right) = 9.1 \text{ mol H} \qquad \frac{9.1 \text{ mol H}}{2.27} = 4.0 \text{ mol H}$$

$$(36.3 \text{ g O})\left(\frac{1 \text{ mol O}}{16.00 \text{ g O}}\right) = 2.27 \text{ mol O} \qquad \frac{2.27 \text{ mol O}}{2.27} = 1.0 \text{ mol O}$$

The empirical formula is C_2H_4O, making the empirical formula mass 44.05 g.

$$\frac{\text{molar mass}}{\text{mass of empirical formula}} = \frac{88.11 \text{ g}}{44.05 \text{ g}} = 2$$

The molecular formula is twice that of the empirical formula.
Molecular formula = $(C_2H_4O)_2 = C_4H_8O_2$

41. $\%$ nitrogen $= \dfrac{12.04\ g}{39.54\ g}(100) = 30.45\%$

$\%$ oxygen $= \dfrac{39.54\ g - 12.04\ g}{39.54\ g}(100) = 69.55\%$

empirical formula: moles of nitrogen $= \dfrac{12.04\ g\ N}{14.01\ g/mol} = 0.8594\ mol\ N$

moles of oxygen $= \dfrac{27.50\ g\ O}{16.00\ g/mol\ O} = 1.719\ mol\ O$

relative number of nitrogen atoms $= \dfrac{0.8594\ mol}{0.8594\ mol} = 1.000$

relative number of oxygen atoms $= \dfrac{1.719\ mol}{0.8594\ mol} = 2.000$

empirical formula $= NO_2$

molecular formula: (molar mass of NO_2) $x = 92.02\ g$, $46.01\ x = 92.02$, $x = 2$
The molecular formula is twice the empirical formula.
molecular formula $= N_2O_4$

42. Total mass of $C + H + O = 30.21\ g + 5.08\ g + 40.24\ g = 75.53\ g$

$\%$ carbon $= \dfrac{30.21\ g}{75.53\ g}(100) = 40.0\%$

$\%$ hydrogen $= \dfrac{5.08\ g}{75.53\ g}(100) = 6.73\%$

$\%$ oxygen $= \dfrac{40.24\ g}{75.53\ g}(100) = 53.3\%$

empirical formula: moles of carbon $= \dfrac{30.21\ g\ C}{12.01\ g/mol} = 2.515\ mol\ C$

moles of hydrogen $= \dfrac{5.080\ g\ H}{1.008\ g/mol} = 5.03\ mol\ H$

moles of oxygen $= \dfrac{40.24\ g\ O}{16.00\ g/mol} = 2.515\ mol\ O$

relative number of carbon atoms $= \dfrac{2.515\ mol}{2.515\ mol} = 1.000$

relative number of hydrogen atoms $= \dfrac{5.03\ mol}{2.515\ mol} = 2.00$

relative number of oxygen atoms $= \dfrac{2.515\ mol}{2.515\ mol} = 1.000$

empirical formula $= CH_2O$

molecular formula: (molar mass of CH_2O) $x = 180.18\ g/mol$,

$(30.03\ g/mol)x = 180.18\ g/mol$,

$x = \dfrac{180.18\ g/mol}{30.03\ g/mol} = 6$

The molecular formula is six times the empirical formula.
molecular formula $= C_6H_{12}O_6$

43. What is compound XYZ_3

   X: $(0.4004)(100.09 \text{ g}) = 40.08 \text{ g (calcium)}$

   Y: $(0.1200)(100.09 \text{ g}) = 12.01 \text{ g (carbon)}$

   Z: $(0.4796)(100.09 \text{ g}) = 48.00 \text{ g}; \dfrac{48.00 \text{ g}}{3} = 16.00 \text{ g (oxygen)}$

   Elements determined from atomic masses in the periodic table.
   $XYZ_3 = CaCO_3$

44. What is compound $X_2(YZ_3)_3$

   X: $(0.1912)(282.23 \text{ g}) = \dfrac{53.96 \text{ g}}{2} = 26.98 \text{ g (aluminum)}$

   Y: $(0.2986)(282.23 \text{ g}) = \dfrac{84.27 \text{ g}}{3} = 28.09 \text{ g (silicon)}$

   Z: $(0.5102)(282.23 \text{ g}) = \dfrac{143.99 \text{ g}}{9} = 16.00 \text{ g (oxygen)}$

   Elements determined from atomic masses in the periodic table.
   $X_2(YZ_3)_3 = Al_2(SiO_3)_3$

45. $(0.350 \text{ mol } P_4)\left(\dfrac{6.022 \times 10^{23} \text{ molecules}}{\text{mol}}\right)\left(\dfrac{4 \text{ atoms P}}{\text{molecule } P_4}\right) = 8.43 \times 10^{23} \text{ atoms P}$

46. $(10.0 \text{ g K})\left(\dfrac{1 \text{ mol K}}{39.10 \text{ g K}}\right)\left(\dfrac{1 \text{ mol Na}}{1 \text{ mol K}}\right)\left(\dfrac{22.99 \text{ g Na}}{\text{mol Na}}\right) = 5.88 \text{ g Na}$

47. $\left(\dfrac{3.27 \times 10^{-22} \text{ g}}{1 \text{ molecule}}\right)\left(\dfrac{6.022 \times 10^{23} \text{ molecules}}{1 \text{ mol}}\right) = \dfrac{197 \text{ g}}{\text{mol}}$

48. $(5 \text{ lb } C_{12}H_{22}O_{11})\left(\dfrac{453.6 \text{ g}}{1 \text{ lb}}\right)\left(\dfrac{6.022 \times 10^{23} \text{ molecules}}{342.3 \text{ g}}\right) = 4 \times 10^{24} \text{ molecules } C_{12}H_{22}O_{11}$

49. $(6.022 \times 10^{23} \text{ sheets})\left(\dfrac{4.60 \text{ cm}}{500 \text{ sheets}}\right)\left(\dfrac{1 \text{ m}}{100 \text{ cm}}\right) = 5.54 \times 10^{19} \text{ m}$

50. $\left(\dfrac{6.022 \times 10^{23} \text{ dollars}}{7.0 \times 10^9 \text{ people}}\right) = 8.6 \times 10^{13} \text{ dollars/person}$

51. The conversion is: $\text{mi}^3 \longrightarrow \text{ft}^3 \longrightarrow \text{in.}^3 \longrightarrow \text{cm}^3 \longrightarrow \text{drops}$

   (a) $(1 \text{ mi}^3)\left(\dfrac{5280 \text{ ft}}{\text{mile}}\right)^3\left(\dfrac{12.0 \text{ in.}}{\text{ft}}\right)^3\left(\dfrac{2.54 \text{ cm}}{\text{inch}}\right)^3\left(\dfrac{20 \text{ drops}}{1.0 \text{ cm}^3}\right) = 8 \times 10^{16} \text{ drops}$

   (b) $(6.022 \times 10^{23} \text{ drops})\left(\dfrac{1 \text{ mi}^3}{8 \times 10^{16} \text{ drops}}\right) = 8 \times 10^6 \text{ mi}^3$

52. 1 mol Ag = 107.9 g Ag

(a) $(107.9 \text{ g Ag})\left(\dfrac{1 \text{ cm}^3}{10.5 \text{ g}}\right) = 10.3 \text{ cm}^3 \text{(volume of cube)}$

(b) $10.3 \text{ cm}^3 = \text{volume of cube} = (\text{side})^3$

$\text{side} = \sqrt[3]{10.3 \text{ cm}^3} = 2.18 \text{ cm}$

53. (a) Determine the molar mass of each compound.
CO_2, 44.01 g; O_2, 32.00 g; H_2O, 18.02 g; CH_3OH, 32.04 g. The 1.00 gram sample with the lowest molar mass will contain the most molecules. Thus, H_2O will contain the most molecules.

(b) $(1.00 \text{ g } H_2O)\left(\dfrac{1 \text{ mol}}{18.02 \text{ g}}\right)\left(\dfrac{(3)(6.022 \times 10^{23} \text{atoms})}{\text{mol}}\right) = 1.00 \times 10^{23} \text{atoms}$

$(1.00 \text{ g } CH_3OH)\left(\dfrac{1 \text{ mol}}{32.04 \text{ g}}\right)\left(\dfrac{(6)(6.022 \times 10^{23} \text{atoms})}{\text{mol}}\right) = 1.13 \times 10^{23} \text{atoms}$

$(1.00 \text{ g } CO_2)\left(\dfrac{1 \text{ mol}}{44.01 \text{ g}}\right)\left(\dfrac{(3)(6.022 \times 10^{23} \text{atoms})}{\text{mol}}\right) = 4.10 \times 10^{22} \text{atoms}$

$(1.00 \text{ g } O_2)\left(\dfrac{1 \text{ mol}}{32.00 \text{ g}}\right)\left(\dfrac{(2)(6.022 \times 10^{23} \text{atoms})}{\text{mol}}\right) = 3.76 \times 10^{22} \text{atoms}$

The 1.00 g sample of CH_3OH contains the most atoms

54. 1 mol Fe_2S_3 = 207.9 g Fe_2S_3 = 6.022 × 10²³ formula units

$(6.022 \times 10^{23} \text{atoms})\left(\dfrac{1 \text{ formula unit}}{5 \text{ atoms}}\right)\left(\dfrac{207.9 \text{ g } Fe_2S_3}{6.022 \times 10^{23} \text{formula units}}\right) = 41.58 \text{ g } Fe_2S_3$

55. The conversion is g P $\longrightarrow$ mol P $\longrightarrow$ mol Ca $\longrightarrow$ g Ca

$(1.00 \text{ g P})\left(\dfrac{1 \text{ mol P}}{30.97 \text{ g P}}\right)\left(\dfrac{3 \text{ mol Ca}}{2 \text{ mol P}}\right)\left(\dfrac{40.08 \text{ g Ca}}{1 \text{ mol Ca}}\right) = 1.94 \text{ g Ca}$

1.94 g Ca combines with 1.00 g P.

56. Grams of Fe per ton of ore that contains 5% $FeSO_4$.

The conversion is: ton $\longrightarrow$ lb $\longrightarrow$ g $\longrightarrow$ g $FeSO_4$ $\longrightarrow$ g Fe

$(1.0 \text{ ton})\left(\dfrac{2000 \text{ lb}}{\text{ton}}\right)\left(\dfrac{453.6 \text{ g}}{\text{lb}}\right)(0.05 \text{ } FeSO_4)\left(\dfrac{55.85 \text{ g Fe}}{151.9 \text{ g } FeSO_4}\right) = 2 \times 10^4 \text{g Fe}$

1.0 ton of iron ore contains 2×10^4 g Fe.

57. From the formula, 2 Li (13.88 g) combine with 1 S (32.07 g).

$\left(\dfrac{13.88 \text{ g Li}}{32.07 \text{ g S}}\right)(20.0 \text{ g S}) = 8.66 \text{ g Li}$

58. (a) $HgCO_3$

	Hg	200.6 g	
	C	12.01 g	$\left(\dfrac{200.6 \text{ g Hg}}{260.6 \text{ g}}\right)(100) = 76.98\% \text{ Hg}$
	3 O	48.00 g	
		260.6 g	

(b) $Ca(ClO_3)_2$

	6 O	96.00 g	
	2 Cl	70.90 g	$\left(\dfrac{96.00 \text{ g O}}{207.0 \text{ g}}\right)(100) = 46.38\% \text{ O}$
	Ca	40.08 g	
		207.0 g	

(c) $C_{10}H_{14}N_2$

	2 N	28.02 g	
	10 C	120.1 g	$\left(\dfrac{28.02 \text{ g N}}{162.6 \text{ g}}\right)(100) = 17.27\% \text{ N}$
	14 H	14.11 g	
		162.2 g	

(d) $C_{55}H_{72}MgN_4O_5$

	Mg	24.31 g	
	55 C	660.55 g	$\left(\dfrac{24.31 \text{ g Mg}}{893.5 \text{ g}}\right)(100) = 2.721\% \text{ Mg}$
	72 H	72.58 g	
	4 N	56.04 g	
	5 O	80.00 g	
		893.5 g	

59. According to the formula, 1 mol (65.39 g) Zn combines with 1 mol (32.07 g) S.

$$(19.5 \text{ g Zn})\left(\dfrac{32.07 \text{ g S}}{65.39 \text{ g Zn}}\right) = 9.56 \text{ g S}$$

19.5 g Zn require 9.56 g S for complete reaction. Therefore, there is not sufficient S present (9.40 g) to react with the Zn.

60. Percent composition of $C_{15}H_{20}O_6$

15 C	180.2 g	$\left(\dfrac{180.2 \text{ g C}}{296.4 \text{ g}}\right)(100) = 60.80 \% \text{ C}$
20 H	20.16 g	$\left(\dfrac{20.16 \text{ g H}}{296.4 \text{ g}}\right)(100) = 6.802 \% \text{ H}$
6 O	96.00 g	$\left(\dfrac{96.00 \text{ g O}}{296.4 \text{ g}}\right)(100) = 32.39 \% \text{ O}$
	296.4 g	

61. Percent composition of $C_{17}H_{21}NO \cdot HCl$

17 C	204.2 g
22 H	22.18 g
N	14.01 g
O	16.00 g
Cl	35.45 g
	291.8 g

$$\left(\frac{204.2 \text{ g C}}{291.8 \text{ g}}\right)(100) = 69.98\% \text{ C}$$

$$\left(\frac{22.18 \text{ g H}}{291.8 \text{ g}}\right)(100) = 7.60\% \text{ H}$$

$$\left(\frac{14.01 \text{ g N}}{291.8 \text{ g}}\right)(100) = 4.80\% \text{ N}$$

$$\left(\frac{16.00 \text{ g O}}{291.8 \text{ g}}\right)(100) = 5.48\% \text{ O}$$

$$\left(\frac{35.45 \text{ g Cl}}{291.8 \text{ g}}\right)(100) = 12.15\% \text{ Cl}$$

62. Percent composition of sucrose

12 C	144.1 g
22 H	22.18 g
11 O	176.0 g
	342.3 g

$$\left(\frac{144.1 \text{ g C}}{342.3 \text{ g}}\right)(100) = 42.10\% \text{ C}$$

$$\left(\frac{22.18 \text{ g H}}{342.3 \text{ g}}\right)(100) = 6.480\% \text{ H}$$

$$\left(\frac{176.0 \text{ g O}}{342.3 \text{ g}}\right)(100) = 51.42\% \text{ O}$$

63. Molecular formula of aspirin
60.0% C, 4.48% H, 35.5% O; molar mass of aspirin = 180.2

$$(60.0 \text{ g C})\left(\frac{1 \text{ mol C}}{12.01 \text{ g C}}\right) = 5.00 \text{ mol C} \qquad \frac{5.00 \text{ mol C}}{2.22} = 2.25 \text{ mol C}$$

$$(4.48 \text{ g H})\left(\frac{1 \text{ mol H}}{1.008 \text{ g H}}\right) = 4.44 \text{ mol H} \qquad \frac{4.44 \text{ mol H}}{2.22} = 2.00 \text{ mol H}$$

$$(35.5 \text{ g O})\left(\frac{1 \text{ mol O}}{16.00 \text{ g O}}\right) = 2.22 \text{ mol O} \qquad \frac{2.22 \text{ mol O}}{2.22} = 1.00 \text{ mol O}$$

Multiplying each by 4 give the empirical formula $C_9H_8O_4$. The empirical formula mass is 180.2 g. Since the empirical formula mass equals the molar mass, the molecular formula is the same as the empirical formula, $C_9H_8O_4$.

64. First calculate the percent oxygen in $Al_2(SO_4)_3$.

2 Al	53.96 g
3 S	96.21 g
12 O	192.0 g
	342.2 g

$$\left(\frac{192.0 \text{ g}}{342.2 \text{ g}}\right)(100) = 56.11\% \text{ O}$$

Second calculate grams of oxygen in 8.50 g of $Al_2(SO_4)_3$.

Now take 56.11% of 8.50 g
$$(8.50 \text{ g Al}_2(SO_4)_3)(0.5611) = 4.77 \text{ g O}$$

65. Empirical formula of gallium arsenide; 48.2% Ga, 51.8% As

$$(48.2 \text{ g Ga})\left(\frac{1 \text{ mol Ga}}{69.72 \text{ g Ga}}\right) = 0.691 \text{ mol Ga} \qquad \frac{0.691 \text{ mol Ga}}{0.691} = 1.00 \text{ mol Ga}$$

$$(51.8 \text{ g As})\left(\frac{1 \text{ mol As}}{74.92 \text{ g As}}\right) = 0.691 \text{ mol As} \qquad \frac{0.691 \text{ mol As}}{0.691} = 1.00 \text{ mol As}$$

The empirical formula is GaAs.

66. Empirical formula of calcium tartrate; 25.5% C, 2.1% H, 21.3% Ca, 51.0% O.

$$(25.5 \text{ g C})\left(\frac{1 \text{ mol C}}{12.01 \text{ g C}}\right) = 2.12 \text{ mol C} \qquad \frac{2.212 \text{ mol C}}{0.531} = 3.99 \text{ mol C}$$

$$(2.1 \text{ g H})\left(\frac{1 \text{ mol H}}{1.008 \text{ g H}}\right) = 2.1 \text{ mol H} \qquad \frac{2.1 \text{ mol H}}{0.531} = 4.0 \text{ mol H}$$

$$(21.2 \text{ g Ca})\left(\frac{1 \text{ mol C}}{40.08 \text{ g Ca}}\right) = 0.531 \text{ mol Ca} \qquad \frac{0.529 \text{ mol Ca}}{0.531} = 1.00 \text{ mol Ca}$$

$$(51.0 \text{ g O})\left(\frac{1 \text{ mol O}}{16.00 \text{ g O}}\right) = 3.19 \text{ mol O} \qquad \frac{3.19 \text{ mol O}}{0.531} = 6.01 \text{ mol O}$$

The empirical formula is $C_4H_4CaO_6$

67. (a) 7.79% C, 92.21% Cl

$$(7.79 \text{ g C})\left(\frac{1 \text{ mol C}}{12.01 \text{ g C}}\right) = 0.649 \text{ mol C} \qquad \frac{0.649 \text{ mol C}}{0.649} = 1.00 \text{ mol C}$$

$$(92.21 \text{ g Cl})\left(\frac{1 \text{ mol Cl}}{35.45 \text{ g Cl}}\right) = 2.601 \text{ mol Cl} \qquad \frac{2.601 \text{ mol Cl}}{0.649} = 4.01 \text{ mol Cl}$$

The empirical formula is CCl_4. The empirical formula mass is 153.8 which equals the molar mass, therefore the molecular formula is CCl_4.

(b) 10.13% C, 89.87% Cl

$$(10.13 \text{ g C})\left(\frac{1 \text{ mol C}}{12.01 \text{ g C}}\right) = 0.8435 \text{ mol C} \qquad \frac{0.8435 \text{ mol C}}{0.8435} = 1.000 \text{ mol C}$$

$$(89.87 \text{ g Cl})\left(\frac{1 \text{ mol Cl}}{35.45 \text{ g Cl}}\right) = 2.535 \text{ mol Cl} \qquad \frac{2.535 \text{ mol Cl}}{0.8435} = 3.005 \text{ mol Cl}$$

The empirical formula is CCl_3. The empirical formula mass is 118.4 g.

$$\frac{\text{molar mass}}{\text{empirical formula mass}} = \frac{236.7 \text{ g}}{118.4 \text{ g}} = 1.999$$

The molecular formula is twice that of the empirical formula.
Molecular formula = C_2Cl_6.

(c)   25.26% C, 74.74% Cl

$$(25.26 \text{ g C})\left(\frac{1 \text{ mol C}}{12.01 \text{ g C}}\right) = 2.103 \text{ mol C} \qquad \frac{2.103 \text{ mol C}}{2.103} = 1.000 \text{ mol C}$$

$$(74.74 \text{ g Cl})\left(\frac{1 \text{ mol Cl}}{35.45 \text{ g Cl}}\right) = 2.108 \text{ mol Cl} \qquad \frac{2.108 \text{ mol Cl}}{2.103} = 1.002 \text{ mol Cl}$$

The empirical formula is CCl. The empirical formula mass is 47.46 g.

$$\frac{\text{molar mass}}{\text{empirical formula mass}} = \frac{284.8 \text{ g}}{47.46 \text{ g}} = 6.000$$

The molecular formula is six times that of the empirical formula.
Molecular formula = C_6Cl_6.

(d)   11.25% C, 88.75% Cl

$$(11.25 \text{ g C})\left(\frac{1 \text{ mol C}}{12.01 \text{ g C}}\right) = 0.9367 \text{ mol C} \qquad \frac{0.9367 \text{ mol C}}{0.9367} = 1.000 \text{ mol C}$$

$$(88.75 \text{ g Cl})\left(\frac{1 \text{ mol Cl}}{35.45 \text{ g Cl}}\right) = 2.504 \text{ mol Cl} \qquad \frac{2.504 \text{ mol Cl}}{0.9367} = 2.673 \text{ mol Cl}$$

Multiplying each by 3 gives the empirical formula C_3Cl_8. The empirical formula mass is 319.6. Since the molar mass is also 319.6 the molecular formula is C_3Cl_8.

68.   The conversion is: s $\longrightarrow$ min $\longrightarrow$ hr $\longrightarrow$ day $\longrightarrow$ yr

$$(6.022 \times 10^{23} \text{s})\left(\frac{1 \text{ min}}{60 \text{ s}}\right)\left(\frac{1 \text{ hr}}{60 \text{ min}}\right)\left(\frac{1 \text{ day}}{24 \text{ hr}}\right)\left(\frac{1 \text{ year}}{365 \text{ days}}\right) = 1.910 \times 10^{16} \text{ years}$$

69.   The conversion is: g $\longrightarrow$ mol $\longrightarrow$ atom

$$(2.5 \text{ g Cu})\left(\frac{1 \text{ mol Cu}}{63.55 \text{ g Cu}}\right)\left(\frac{6.022 \times 10^{23} \text{atoms}}{\text{mol}}\right) = 2.4 \times 10^{22} \text{ atoms Cu}$$

70.   The conversion is: molecules $\longrightarrow$ mol $\longrightarrow$ g   1 trillion = 10^{12}

$$\left(1000. \times 10^{12} \text{ molecules C}_3\text{H}_8\text{O}_3\right)\left(\frac{1 \text{ mol}}{6.022 \times 10^{23} \text{ molecules}}\right)\left(\frac{92.09 \text{ g C}_3\text{H}_8\text{O}_3}{\text{mol C}_3\text{H}_8\text{O}_3}\right)$$
$$= 1.529 \times 10^{-7} \text{g C}_3\text{H}_8\text{O}_3$$

71. $(7.0 \times 10^9 \text{ people})\left(\dfrac{1 \text{ mol people}}{6.022 \times 10^{23} \text{ people}}\right) = 1.2 \times 10^{-14} \text{ mol of people}$

72. Empirical formula 23.3% Co, 25.3% Mo, 51.4% Cl

$$(23.3 \text{ g Co})\left(\frac{1 \text{ mol Co}}{58.93 \text{ g Co}}\right) = 0.935 \text{ mol Co} \qquad \frac{0.395 \text{ mol Co}}{0.264} = 1.50 \text{ mol Co}$$

$$(25.3 \text{ g Mo})\left(\frac{1 \text{ mol Mo}}{95.94 \text{ g Mo}}\right) = 0.264 \text{ mol Mo} \qquad \frac{0.264 \text{ mol Mo}}{0.264} = 1.00 \text{ mol Mo}$$

$$(51.4 \text{ g Cl})\left(\frac{1 \text{ mol Cl}}{35.45 \text{ g Cl}}\right) = 1.45 \text{ mol Cl} \qquad \frac{1.45 \text{ mol Cl}}{0.264} = 5.49 \text{ mol Cl}$$

Multiplying by 2 gives the empirical formula $Co_3Mo_2Cl_{11}$.

73. The conversion is: g Al $\longrightarrow$ mol Al $\longrightarrow$ mol Mg $\longrightarrow$ g Mg

$$(18 \text{ g Al})\left(\frac{1 \text{ mol Al}}{26.98 \text{ g Al}}\right)\left(\frac{2 \text{ mol Mg}}{1 \text{ mol Al}}\right)\left(\frac{24.31 \text{ g Mg}}{\text{mol Mg}}\right) = 32 \text{ g Mg}$$

74. $(10.0 \text{ g compound}) (0.177) = 1.77 \text{ g N}$

$$(1.77 \text{ g N})\left(\frac{1 \text{ mol N}}{14.01 \text{ g N}}\right) = 0.126 \text{ mol N}$$

$$(3.8 \times 10^{23} \text{ atoms H})\left(\frac{1 \text{ mol}}{6.022 \times 10^{23} \text{ atoms}}\right) = 0.63 \text{ mol H}$$

To determine the mol C, first find grams of H and subtract the grams of H and N from the grams of the sample.

$$(0.63 \text{ mol H})\left(\frac{1.008 \text{ g H}}{\text{mol H}}\right) = 0.64 \text{ g H}$$

$$
\begin{array}{r}
10.0 \text{ g sample} \\
-1.77 \text{ g N} \\
-0.64 \text{ g H} \\
\hline
7.6 \text{ g C}
\end{array}
$$

$$(7.6 \text{ g C})\left(\frac{1 \text{ mol C}}{12.01 \text{ g C}}\right) = 0.63 \text{ mol C}$$

Now determine the empirical formula from the moles of C, H, and N.

N $\quad \dfrac{0.126 \text{ mol N}}{0.126} = 1.00 \text{ mol N}$

H $\quad \dfrac{0.63 \text{ mol H}}{0.126} = 5.0 \text{ mol H}$

C $\quad \dfrac{0.63 \text{ mol C}}{0.126} = 5.0 \text{ mol C}$

The empirical formula is C_5H_5N

75. Let x = molar mass of A_2O

$$0.400x = 16.00 \text{ g O (since } A_2O \text{ has only one mol of O atoms)}$$

$$x = 40.0 \text{ g O/mol } A_2O$$

$$40.0 = 16.00 + 2y \quad y = \text{molar mass of A}$$

$$40.0 - 16.00 = 2y$$

$$12.0\frac{g}{mol} = y$$

Look in the periodic table for the element that has 12.0 g/mol.

The element is carbon. The mystery element is carbon.

76. (a) CH_2O (divide the molecular formula by 6)
    (b) C_4H_9 (divide the molecular formula by 2)
    (c) CH_2O (divide the molecular formula by 3)
    (d) $C_{25}H_{52}$ (divide the molecular formula by 1)
    (e) $C_6H_2Cl_2O$ (divide the molecular formula by 2)

77. $(1 \text{ L water})\left(\dfrac{9.0 \text{ μg Cu}^{2+} \text{ ions}}{1 \text{ L water}}\right)\left(\dfrac{1 \text{ g}}{10^6 \text{ μg}}\right)\left(\dfrac{1 \text{ mol Cu}^{2+} \text{ ions}}{63.55 \text{ g Cu}^{2+} \text{ ions}}\right)\left(\dfrac{6.022 \times 10^{23} \text{ Cu}^{2+} \text{ ions}}{1 \text{ mol Cu}^{2+} \text{ ions}}\right)$

$$= 8.5 \times 10^{16} \text{ Cu}^{2+} \text{ ions}$$

78. $(3.0 \text{ g H}_2)\left(\dfrac{1 \text{ mol H}_2}{2.016 \text{ g H}_2}\right)\left(\dfrac{6.022 \times 10^{23} \text{ molecules H}_2}{1 \text{ mol H}_2}\right)\left(\dfrac{1 \text{ molecule O}_2}{1.0 \times 10^6 \text{ molecule H}_2}\right)$

$$= 9.0 \times 10^{17} \text{ molecules O}_2$$

79. $(1 \text{ day})\left(\dfrac{10.75 \text{ hr}}{1 \text{ day}}\right)\left(\dfrac{60 \text{ min}}{1 \text{ hr}}\right)\left(\dfrac{60 \text{ s}}{1 \text{ min}}\right)\left(\dfrac{250.0 \text{ kg H}_2O}{1 \text{ s}}\right)\left(\dfrac{1000 \text{ g H}_2O}{1 \text{ kg H}_2O}\right)\left(\dfrac{1 \text{ mol H}_2O}{18.02 \text{ g H}_2O}\right)$

$$= 5.369 \times 10^8 \text{ mol H}_2O$$

80. Empirical formula of 38.65% C, 9.74% H, 51.61% S

$$(38.65 \text{ g C})\left(\frac{1 \text{ mol C}}{12.01 \text{ g C}}\right) = 3.218 \text{ mol C} \quad \left(\frac{3.218 \text{ mol C}}{1.609}\right) = 2.000 \text{ mol C}$$

$$(9.74 \text{ g H})\left(\frac{1 \text{ mol H}}{1.008 \text{ g H}}\right) = 9.66 \text{ mol H} \quad \left(\frac{9.66 \text{ mol H}}{1.609}\right) = 6.00 \text{ mol H}$$

$$(51.61 \text{ g S})\left(\frac{1 \text{ mol S}}{32.07 \text{ g S}}\right) = 1.609 \text{ mol S} \quad \left(\frac{1.609 \text{ mol S}}{1.609}\right) = 1.000 \text{ mol S}$$

The empirical formula is C_2H_6S

81. First determine the elements in compound $A(BC)_3$:

    A: $(0.3459)(78.01\ g) = 26.98\ g$ (aluminum)

    B: $(0.6153)(78.01\ g) = \dfrac{48.00\ g}{3} = 16.00\ g$ (oxygen)

    C: $(0.0388)(78.01\ g) = \dfrac{3.03\ g}{3} = 1.01\ g$ (hydrogen)

    Element determined from atomic masses in the periodic table.
    $A(BC)_3 = Al(OH)_3$
    Then compound $A_2B_3 = Al_2O_3$ with a molar mass of

    $2(26.98\ g) + 3(16.00\ g) = 102.0\ g$

    $\% \ Al = \dfrac{2(26.98\ g)}{102.0\ g}(100) = 52.90\%$

    $\% \ O = \dfrac{3(16.00)}{102.0}(100) = 47.06\%$

82. (a) Percent composition of the original unknown compound.

    Convert g CO_2 to g C and g H_2O to g H

    $(4.776\ g\ CO_2)\left(\dfrac{12.01\ g\ C}{44.01\ g\ CO_2}\right) = 1.303\ g\ C$

    $(2.934\ g\ H_2O)\left(\dfrac{2.016\ g\ H}{18.02\ g\ H_2O}\right) = 0.3282\ g\ H$

    $2.500\ g$ compound $-1.303\ g\ C - 0.3282\ g\ H = 0.8688\ g\ O$

    $\left(\dfrac{1.303\ g\ C}{2.500\ g}\right)(100) = 52.12\%\ C$

    $\left(\dfrac{0.3282\ g\ H}{2.500\ g}\right)(100) = 13.13\%\ H$

    $\left(\dfrac{0.8688\ g\ O}{2.500\ g}\right)(100) = 34.75\%\ O$

    (b) Empirical formula of unknown compound; 52.12% C, 13.13% H, 34.76% O.

    $(52.12\ g\ C)\left(\dfrac{1\ mol\ C}{12.01\ g}\right) = 4.340$ $\dfrac{4.340\ mol\ C}{2.173} = 1.997\ mol\ C$

    $(13.13\ g\ H)\left(\dfrac{1\ mol\ H}{1.008\ g\ H}\right) = 13.03$ $\dfrac{13.03\ mol\ H}{2.173} = 5.996\ mol\ H$

    $(34.76\ g\ O)\left(\dfrac{1\ mol\ C}{16.00\ g}\right) = 2.173$ $\dfrac{2.173\ mol\ O}{2.173} = 1.000\ mol\ O$

    The empirical formula is C_2H_6O

83. $1.00 \text{ pg C}_{10}\text{H}_{16} \left(\dfrac{1 \text{ g}}{10^{12} \text{ pg}} \right) \left(\dfrac{1 \text{ mol C}_{10}\text{H}_{16}}{136.2 \text{ g C}_{10}\text{H}_{16}} \right) \left(\dfrac{6.022 \times 10^{23} \text{ molecules}}{1 \text{ mol}} \right)$

$\left(\dfrac{1 \text{ photon}}{1 \text{ molecule C}_{10}\text{H}_{16}} \right) \left(\dfrac{1 \text{ s}}{2.64 \times 10^{18} \text{ photons}} \right) = 1.67 \times 10^{-9} \text{ s}$

1 gram $= 10^{12}$ picogram (pg)

CHAPTER 8

CHEMICAL EQUATIONS

SOLUTIONS TO REVIEW QUESTIONS

1. The coefficients in a balanced chemical equation represent the number of moles (or molecules or formula units) of each of the chemical species in the reaction.

2. The physical state of a substance may be a solid, a liquid, or a gas. The symbols indicate whether a substance is a solid, a liquid, a gas, or is in an aqueous solution. A solid is indicated by (*s*), a liquid by (*l*), a gas by (*g*) and an aqueous solution by (*aq*).

3. The purpose of balancing chemical equations is to conform to the Law of Conservation of Mass. Ratios of reactants and products can then be easily determined.

4. (a) Yes. It is necessary to conserve atoms to follow the Law of Conservation of Mass.
    (b) No. Molecules can be taken apart and rearranged to form different molecules in reactions.
    (c) Moles of molecules are not conserved (b). Moles of atoms are conserved (a).

5. These charts help to keep track of which elements are balanced in a chemical equation. The charts are one way of keeping track of the number of atoms of each element on the reactant side of a chemical equation and on the product side of an equation. The top row in a chart gives the number and types of atoms on the reactant side and the bottom row gives the number and types of atoms on the product side of a chemical equation. Using a chart may make it easier to see where coefficients are needed in a reaction and what number that coefficient should be.

6. Chemical equations can only be balanced by changing the number of each substance reacted or produced. If the subscripts are changed then the identity of the reactants and products is also changed. Equations must be balanced using the actual reactants and products.

7. A combustion reaction is an exothermic process (usually burning) done in the presence of oxygen.

8. The activity series given in Table 8.2 shows the relative activity of certain metals and halogens. As you move up the table starting with gold (Au) and ending with potassium (K) the activity increases. The same is true as you move up from iodine (I_2) to fluorine (F_2). The table is useful for predicting the products of some reactions because an element in the series will replace any element given below it. For example, hydrogen can replace copper, silver, mercury, or gold in a chemical reaction.

9. The major types of chemical reactions are combination reactions, decomposition reactions, single displacement reactions, and double displacement reactions.

10. Indicators that a chemical reaction has occurred are
    *   Formation of a precipitate
    *   Formation of a gas
    *   A temperature change

11. A chemical change that absorbs heat energy is said to be an *endothermic* reaction. The products are at a higher energy level than the reactants. A chemical change that liberates heat energy is said to be an *exothermic* reaction. The products are at a lower energy level than the reactants.

12. Although an exothermic reaction will liberate more heat than it absorbs, some heat is still necessary to get the reaction started. The energy required to start a reaction is the activation energy.

13. Methane, carbon dioxide and water

14. Carbon dioxide levels fall in the spring as plants grow and incorporate carbon dioxide into their cells. Carbon dioxide levels rise in the fall as plants begin to decay back into the soil releasing carbon dioxide back into the air.

SOLUTIONS TO EXERCISES

1. (a) exothermic   (d) exothermic
   (b) endothermic   (e) endothermic
   (c) exothermic

2. (a) endothermic   (d) exothermic
   (b) exothermic   (e) exothermic
   (c) endothermic

3. (a) $2 H_2 + O_2 \longrightarrow 2 H_2O$ combination

   (b) $3 N_2H_4(l) \longrightarrow 4 NH_3(g) + N_2(g)$ decomposition

   (c) $H_2SO_4 + 2 NaOH \longrightarrow 2 H_2O + Na_2SO_4$ double displacement

   (d) $Al_2(CO_3)_3 \overset{\Delta}{\longrightarrow} Al_2O_3 + 3 CO_2$ decomposition

   (e) $2 NH_4I + Cl_2 \longrightarrow 2 NH_4Cl + I_2$ single displacement

4. (a) $H_2 + Br_2 \longrightarrow 2 HBr$ combination

   (b) $BaO_2(s) + H_2SO_4(aq) \longrightarrow BaSO_4(s) + H_2O_2(aq)$ double displacement

   (c) $Ba(ClO_3)_2 \overset{\Delta}{\longrightarrow} BaCl_2 + 3 O_2$ decomposition

   (d) $CrCl_3 + 3 AgNO_3 \longrightarrow Cr(NO_3)_3 + 3 AgCl$ double displacement

   (e) $2 H_2O_2 \longrightarrow 2 H_2O + O_2$ decomposition

5. There are many ways to form oxides. For example: (1) some metals plus oxygen, (2) some nonmetals plus oxygen, (3) some metals plus water (steam), (4) combustion of hydrocarbons

6. A metal and a nonmetal can react to form a salt; the reaction of an acid and a base can form a salt.

7. (a) $2 MnO_2 + CO \longrightarrow Mn_2O_3 + CO_2$

   (b) $Cu_2O(s) + C(s) \longrightarrow 2 Cu(s) + CO(g)$

   (c) $4 C_3H_5(NO_3)_3 \longrightarrow 12 CO_2 + 10 H_2O + 6 N_2 + O_2$

   (d) $4 FeS + 7 O_2 \longrightarrow 2 Fe_2O_3 + 4 SO_2$

   (e) $2 Cu(NO_3)_2 \longrightarrow 2 CuO + 4 NO_2 + O_2$

   (f) $3 NO_2 + H_2O \longrightarrow 2 HNO_3 + NO$

   (g) $2 Fe(s) + 3 S(l) \longrightarrow Fe_2S_3(s)$

   (h) $4 HCN + 5 O_2 \longrightarrow 2 N_2 + 4 CO_2 + 2 H_2O$

   (i) $2 B_5H_9 + 12 O_2 \longrightarrow 5 B_2O_3 + 9 H_2O$

8. (a) $2\,SO_2 + O_2 \longrightarrow 2\,SO_3$

   (b) $Li_2O(s) + H_2O(l) \longrightarrow 2\,LiOH(aq)$

   (c) $2\,Na + 2\,H_2O \longrightarrow 2\,NaOH + H_2$

   (d) $2\,AgNO_3 + Ni \longrightarrow Ni(NO_3)_2 + 2\,Ag$

   (e) $Bi_2S_3 + 6\,HCl \longrightarrow 2\,BiCl_3 + 3\,H_2S$

   (f) $2\,PbO_2 \xrightarrow{\Delta} 2\,PbO + O_2$

   (g) $Hg_2(C_2H_3O_2)_2(aq) + 2\,KCl(aq) \longrightarrow Hg_2Cl_2(s) + 2\,KC_2H_3O_2(aq)$

   (h) $2\,KI + Br_2 \longrightarrow 2\,KBr + I_2$

   (i) $2\,K_3PO_4 + 3\,BaCl_2 \longrightarrow 6\,KCl + Ba_3(PO_4)_2$

9. (a) $Mg(s) + 2\,HBr(aq) \longrightarrow H_2(g) + MgBr_2(aq)$

   (b) $Ca(ClO_3)_2(s) \xrightarrow{\Delta} CaCl_2(s) + 3\,O_2(g)$

   (c) $4\,Li(s) + O_2(g) \longrightarrow 2\,Li_2O(s)$

   (d) $3\,Ba(BrO_3)_2(aq) + 2\,Na_3PO_4(aq) \longrightarrow Ba_3(PO_4)_2(s) + 6\,NaBrO_3(aq)$

   (e) $2\,HC_2H_3O_2(aq) + Na_2CO_3(aq) \longrightarrow 2\,NaC_2H_3O_2(aq) + CO_2(g) + H_2O(l)$

   (f) $3\,AgNO_3(aq) + AlI_3(aq) \longrightarrow 3\,AgI(s) + Al(NO_3)_3(aq)$

10. (a) $MgCO_3(s) \xrightarrow{\Delta} MgO(s) + CO_2(g)$

    (b) $Ca(OH)_2(s) + 2\,HClO_3(aq) \longrightarrow Ca(ClO_3)_2(aq) + 2\,H_2O(l)$

    (c) $Fe_2(SO_4)_3(aq) + 6\,NaOH(aq) \longrightarrow 2\,Fe(OH)_3(s) + 3\,Na_2SO_4(aq)$

    (d) $Zn(s) + 2\,HC_2H_3O_2(aq) \longrightarrow H_2(g) + Zn(C_2H_3O_2)_2(aq)$

    (e) $SO_3(g) + H_2O(l) \longrightarrow H_2SO_4(aq)$

    (f) $Na_2CO_3(aq) + CoCl_2(aq) \longrightarrow CoCO_3(s) + 2\,NaCl(aq)$

11. (a) $H_2SO_4(aq) + 2\,NaOH(aq) \longrightarrow 2\,H_2O(l) + Na_2SO_4(aq) + \text{heat}$

    (b) $Pb(NO_3)_2(aq) + 2\,KBr(aq) \longrightarrow PbBr_2(s) + 2\,KNO_3(aq)$

    (c) $NH_4Cl(aq) + AgNO_3(aq) \longrightarrow AgCl(s) + NH_4NO_3(aq)$

    (d) $CaCO_3(s) + 2\,HC_2H_3O_2(aq) \longrightarrow Ca(C_2H_3O_2)_2(aq) + H_2O(l) + CO_2(g)$

12. (a) $CuSO_4(aq) + 2\,KOH(aq) \longrightarrow Cu(OH)_2(s) + K_2SO_4(aq)$

    (b) $H_3PO_4(aq) + 3\,NaOH(aq) \longrightarrow Na_3PO_4(aq) + 3\,H_2O(l) + \text{heat}$

    (c) $3\,NaHCO_3(s) + H_3PO_4(aq) \longrightarrow Na_3PO_4(aq) + 3\,H_2O(l) + 3\,CO_2(g)$

    (d) $2\,AlCl_3(aq) + 3\,Pb(NO_3)_2 \longrightarrow 3\,PbCl_2(s) + 2\,Al(NO_3)_3(aq)$

13. (a) $2\,Ca(s) + 2\,H_2O(l) \longrightarrow Ca(OH)_2(aq) + H_2(g)$

    (b) $Br_2(l) + 2\,KI(aq) \longrightarrow I_2(s) + 2\,KBr(aq)$

    (c) $Cu(s) + HCl(aq) \longrightarrow$ no reaction

    (d) $2\,Al(s) + 3\,H_2SO_4(aq) \longrightarrow 3\,H_2(g) + Al_2(SO_4)_3(aq)$

14. (a) $Cu(s) + NiCl_2(aq) \longrightarrow$ no reaction

    (b) $2\,Rb(s) + 2\,H_2O(l) \longrightarrow H_2(g) + 2\,RbOH(aq)$

    (c) $I_2(s) + CaCl_2(aq) \longrightarrow$ no reaction

    (d) $3\,Mg(s) + 2\,Al(NO_3)_3(aq) \longrightarrow 2\,Al(s) + 3\,Mg(NO_3)_2(aq)$

15. (a) $Sr(s) + 2\,H_2O(l) \longrightarrow H_2(g) + Sr(OH)_2(s)$

    (b) $BaCl_2(aq) + 2\,AgNO_3(aq) \longrightarrow Ba(NO_3)_2(aq) + 2\,AgCl(s)$

    (c) $Mg(s) + ZnBr_2(aq) \longrightarrow Zn(s) + MgBr_2(aq)$

    (d) $2\,K(s) + Cl_2(g) \longrightarrow 2\,KCl(s)$

16. (a) $Li_2O(s) + H_2O \longrightarrow 2\,LiOH(aq)$

    (b) $Na_2SO_4(aq) + Pb(NO_3)_2(aq) \longrightarrow 2\,NaNO_3(aq) + PbSO_4(s)$

    (c) $Zn(s) + CuSO_4(aq) \longrightarrow Cu(s) + ZnSO_4(aq)$

    (d) $4\,Al(s) + 3\,O_2(g) \longrightarrow 2\,Al_2O_3(s)$

17. (a) $2\,Ba + O_2 \longrightarrow 2\,BaO$

    (b) $2\,NaHCO_3 \xrightarrow{\Delta} Na_2CO_3 + H_2O + CO_2$

    (c) $Ni + CuSO_4 \longrightarrow NiSO_4 + Cu$

    (d) $MgO + 2\,HCl \longrightarrow MgCl_2 + H_2O$

    (e) $H_3PO_4 + 3\,KOH \longrightarrow K_3PO_4 + 3\,H_2O$

18. (a) $C + O_2 \longrightarrow CO_2$

    (b) $2\,Al(ClO_3)_3 \xrightarrow{\Delta} 9\,O_2 + 2\,AlCl_3$

    (c) $CuBr_2 + Cl_2 \longrightarrow CuCl_2 + Br_2$

    (d) $2\,SbCl_3 + 3\,(NH_4)_2S \longrightarrow Sb_2S_3 + 6\,NH_4Cl$

    (e) $2\,NaNO_3 \xrightarrow{\Delta} 2\,NaNO_2 + O_2$

19. (a) One mole of $MgBr_2$ reacts with two moles of $AgNO_3$ to yield one mole of $Mg(NO_3)_2$ and two moles of AgBr.

    (b) One mole of N_2 reacts with three moles of H_2 to produce two moles of NH_3.

    (c) Two moles of C_3H_7OH react with nine moles of O_2 to form six moles of CO_2 and eight moles of H_2O.

20. (a) Two moles of Na react with one mole of Cl_2 to produce two moles of NaCl and release 822 kJ of energy. The reaction is exothermic.
    (b) One mole of PCl_5 absorbs 92.9 kJ of energy to produce one mole of PCl_3 and one mole of Cl_2. The reaction is endothermic.
    (c) 1 mol of solid sulfur will react with 2 moles of gaseous carbon monoxide to produce 1 mol of gaseous sulfur dioxide, 2 moles of solid carbon, and 76 kilojoules of energy in this exothermic reaction.

21. (a) $2\,HgO(s) + 182\,kJ \longrightarrow 2\,Hg(l) + O_2(g)$
    (b) $2\,H_2(g) + O_2(g) \longrightarrow 2\,H_2O(l) + 571.6\,kJ$

22. (a) $Ca(s) + 2\,H_2O(l) \longrightarrow Ca(OH)_2(aq) + H_2(g) + 635.1\,kJ$
    (b) $2\,BrF_3 + 601.6\,kJ \longrightarrow Br_2 + 3\,F_2$

23.
    (a) decomposition reaction, $2\,AgClO_3(s) \xrightarrow{\Delta} 2\,AgCl(s) + 3\,O_2(g)$
    (b) single-displacement, $Fe(s) + H_2SO_4(aq) \longrightarrow H_2(g) + FeSO_4(aq)$
    (c) combination reaction, $Zn(s) + Cl_2(g) \longrightarrow ZnCl_2(s)$
    (d) double-displacement, $HBr(aq) + KOH(aq) \longrightarrow KBr(aq) + H_2O(l)$

24. (a) single-displacement, $Ni(s) + Pb(NO_3)_2(aq) \longrightarrow Pb(s) + Ni(NO_3)_2(aq)$
    (b) combination, $MgO(s) + H_2O(l) \longrightarrow Mg(OH)_2(s)$
    (c) decomposition, $2\,HgO(s) \longrightarrow 2\,Hg(l) + O_2(g)$
    (d) double-displacement, $PbCl_2(aq) + (NH_4)_2CO_3(aq) \longrightarrow PbCO_3(s) + 2\,NH_4Cl(aq)$

25. Combinations that form a precipitate:

    $Ca(NO_3)_2(aq) + (NH_4)_2SO_4(aq) \longrightarrow CaSO_4(s) + 2\,NH_4NO_3(aq)$

    $Ca(NO_3)_2(aq) + (NH_4)_2CO_3(aq) \longrightarrow CaCO_3(s) + 2\,NH_4NO_3(aq)$

    $AgNO_3(aq) + NH_4Cl(aq) \longrightarrow AgCl(s) + NH_4NO_3(aq)$

    $2\,AgNO_3(aq) + (NH_4)_2SO_4(aq) \longrightarrow Ag_2SO_4(s) + 2\,NH_4NO_3(aq)$

    $2\,AgNO_3(aq) + (NH_4)_2CO_3(aq) \longrightarrow Ag_2CO_3(s) + 2\,NH_4NO_3(aq)$

26. $P_4O_{10} + 12\,HClO_4 \longrightarrow 6\,Cl_2O_7 + 4\,H_3PO_4$

    $\quad 10\,O + 12(4\,O) \qquad\quad 6(7\,O) + 4(4\,O)$

    $\quad 10\,O + 48\,O \qquad\qquad 42\,O + 16\,O$

    $\qquad 58\,O \qquad\qquad\quad 58\,O$

27. In $5\,Ni_3(PO_4)_2$ there are:
    (a) 15 atoms of N   (c) 40 atoms of O
    (b) 10 atoms of P   (d) 65 total atoms

28. $CaCO_3(s) + 2 HC_2H_3O_2(aq) \longrightarrow Ca(C_2H_3O_2)_2(aq) + CO_2(g) + H_2O(l)$

29.

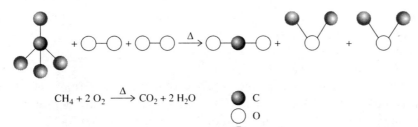

$$CH_4 + 2 O_2 \xrightarrow{\Delta} CO_2 + 2 H_2O$$

● C

○ O

● H

30. (a) $C(s) + O_2(g) \longrightarrow CO_2(g)$

(b) $CO_2(g) + C(s) \longrightarrow 2 CO(g)$

(c) $Fe_2O_3(s) + 3 CO(g) \xrightarrow{\Delta} 2 Fe(l) + 3 CO_2(g)$

$Fe_3O_4(s) + 4 CO(g) \xrightarrow{\Delta} 3 Fe(l) + 4 CO_2(g)$

31. $2 B + 3 G_2 \rightarrow 2 BG_3$

32. The metals that should be chosen are (b) zinc, (c) aluminum, and (e) calcium. These metals are more active than nickel, therefore will react in a solution of nickel(II) chloride; (a) copper and (d) lead are less active and will not react with nickel(II) chloride solution (see Table 8.2).

Equations

$Zn + NiCl_2 \longrightarrow Ni + ZnCl_2$

$2 Al + 3 NiCl_2 \longrightarrow 3 Ni + 2 AlCl_3$

$Ca + NiCl_2 \longrightarrow Ni + CaCl_2$

33. $Ti + Ni(NO_3)_2 \longrightarrow$ Reaction occurs

$Ti + Pb(NO_3)_2 \longrightarrow$ Reaction occurs

$Ti + Mg(NO_3)_2 \longrightarrow$ no reaction

Ti is above Ni and Pb in the activity series since both react. Ti is below Mg in the series since it will not replace Mg. From the printed activity series in the chapter Ni lies above Pb so the order is:

Mg

Ti

Ni

Pb

34. (a) $4 Cs + O_2 \longrightarrow 2 Cs_2O$

(b) $2 Al + 3 S \longrightarrow Al_2S_3$

(c) $SO_3 + H_2O \longrightarrow H_2SO_4$

(d) $Na_2O + H_2O \longrightarrow 2 NaOH$

35. (a) $2 H_2(g) + O_2(g) \longrightarrow 2 H_2O(g)$

(b) $CH_4(g) + H_2O(g) \longrightarrow 3 H_2(g) + CO(g)$
$CO(g) + H_2O(g) \longrightarrow H_2(g) + CO_2(g)$

(c) $CH_4(g) + 2 O_2(g) \longrightarrow 2 H_2O(g) + CO_2(g)$

(d) One molecule of carbon dioxide is produced from one molecule of methane regardless of whether it is converted into hydrogen gas or burned in oxygen. If the goal is to reduce carbon dioxide emission, there is no advantage in using hydrogen synthesized from methane as a fuel.

36. (a) $Pb(NO_3)_2(aq) + K_2CrO_4(aq) \longrightarrow 2 KNO_3(aq) + PbCrO_4(s)$ Lead(II) chromate

(b) $CdCl_2(aq) + Li_2S(aq) \longrightarrow CdS(s) + 2 LiCl(aq)$ Cadmium sulfide

(c) $Pb(C_2H_3O_2)_2(aq) + H_2CO_3(aq) \longrightarrow PbCO_3(s) + 2 HC_2H_3O_2(aq)$ Lead(II) carbonate

(d) $Hg(NO_3)_2(aq) + Na_2S(aq) \longrightarrow HgS(s) + 2 NaNO_3(aq)$ Mercury(II) sulfide

(e) $4 Ti(s) + 3 O_2(g) \longrightarrow 2 Ti_2O_3(s)$ Titanium(III) oxide

(f) $4 Fe(s) + 3 O_2(g) \longrightarrow 2 Fe_2O_3(s)$ Iron(III) oxide

37. (a) $2 ZnO \xrightarrow{\Delta} 2 Zn + O_2$

(b) $SnO_2 \xrightarrow{\Delta} Sn + O_2$ or $2 SnO_2 \xrightarrow{\Delta} 2 SnO + O_2$

(c) $Na_2CO_3 \xrightarrow{\Delta} Na_2O + CO_2$

(d) $Mg(ClO_3)_2 \xrightarrow{\Delta} MgCl_2 + 3 O_2$

38. (a) $Mg(s) + 2 HCl(aq) \longrightarrow H_2(g) + MgCl_2(aq)$

(b) $2 NaBr(aq) + Cl_2(g) \longrightarrow 2 NaCl(aq) + Br_2(l)$

(c) $3 Zn(s) + 2 Fe(NO_3)_3(aq) \longrightarrow 2 Fe(s) + 3 Zn(NO_3)_2(aq)$

(d) $2 Al(s) + 3 Cu(NO_3)_2(aq) \longrightarrow 3 Cu(s) + 2 Al(NO_3)_3(aq)$

39. (a) $2 (NH_4)_3PO_4(aq) + 3 Ba(NO_3)_2(aq) \longrightarrow Ba_3(PO_4)_2(s) + 6 NH_4NO_3(aq)$

(b) $Na_2S(aq) + Pb(C_2H_3O_2)_2(aq) \longrightarrow PbS(s) + 2 NaC_2H_3O_2(aq)$

(c) $CuSO_4(aq) + Ca(ClO_3)_2(aq) \longrightarrow CaSO_4(s) + Cu(ClO_3)_2(aq)$

(d) $Ba(OH)_2(aq) + H_2C_2O_4(aq) \longrightarrow BaC_2O_2(aq) + 2 H_2O(l)$

(e) $H_3PO_4(aq) + 3 KOH(aq) \longrightarrow K_3PO_4(aq) + 3 H_2O(l)$

(f) $H_2SO_4(aq) + Na_2CO_3(aq) \longrightarrow Na_2SO_4(aq) + H_2O(l) + CO_2(g)$

40. (a) $K_2SO_4(aq) + Ba(C_2H_3O)_2(aq) \longrightarrow 2 KC_2H_3O_2(aq) + BaSO_4(s)$

(b) $H_2SO_4(aq) + 2 LiOH(aq) \longrightarrow Li_2SO_4(aq) + 2 H_2O(l)$

(c) $(NH_4)_3PO_4(aq) + NaBr(aq) \longrightarrow$ no reaction

(d) $CaI_2(aq) + 2 AgNO_3(aq) \longrightarrow Ca(NO_3)_2(aq) + 2 AgI(s)$

(e) $2 HNO_3(aq) + Sr(OH)_2(aq) \longrightarrow Sr(NO_3)_2(aq) + 2 H_2O(l)$

(f) $CsNO_3(aq) + Ca(OH)_2(aq) \longrightarrow$ No reaction

41. (a) $CH_4 + 2\,O_2 \longrightarrow CO_2 + 2\,H_2O$
    (b) $2\,H_2 + O_2 \longrightarrow 2\,H_2O$

42. (a) $CH_4 + 2\,O_2 \longrightarrow CO_2 + 2\,H_2O$
    (b) $2\,C_3H_6 + 9\,O_2 \longrightarrow 6\,CO_2 + 6\,H_2O$
    (c) $C_6H_5CH_3 + 9\,O_2 \longrightarrow 7\,CO_2 + 4\,H_2O$

43. 1. combustion of fossil fuels
    2. destruction of the rain forests by burning
    3. increased population

44. Carbon dioxide, methane, and water are all considered to be greenhouse gases. They each act to trap the heat near the surface of the earth in the same manner in which a greenhouse is warmed.

45. The effects of global warming can be reduced by:
    1. developing new energy sources (not dependant on fossil fuels)
    2. conservation of energy resources
    3. recycling
    4. decreased destruction of the rain forests and other forests

46. About half the carbon dioxide released into the atmosphere remains in the air. The rest is absorbed by plants and used in photosynthesis or is dissolved in the oceans.

47. In the Northern Hemisphere, the concentration of CO_2 peaks once in May, dropping as plants use the CO_2 to produce growth, until October when the second peak occurs as a result of the fallen leaves decaying.

48. (a) Ag^+, Co^{2+}, Ba^{2+}, Zn^{2+}, Sn^{2+},

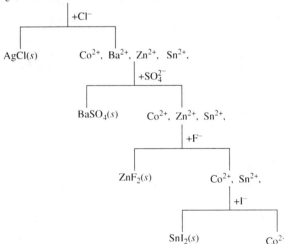

    (b) All the anionic salts are sodium salts because you need to use soluble reagents and all sodium salts are soluble.

CHAPTER 9

CALCULATIONS FROM CHEMICAL EQUATIONS

SOLUTIONS TO REVIEW QUESTIONS

1. A mole ratio is the ratio between the mole amounts of two atoms and/or molecules involved in a chemical reaction.

2. In order to convert grams to moles the molar mass of the compound under consideration needs to be determined.

3. The balanced equation is

$$Ca_3P_2 + 6\,H_2O \rightarrow 3\,Ca(OH)_2 + 2\,PH_3$$

(a) Correct: $(1\text{ mol Ca}_3\text{P}_2)\left(\dfrac{2\text{ mol PH}_3}{1\text{ mol Ca}_3\text{P}_2}\right) = 2\text{ mol PH}_3$

(b) Incorrect: 1 g Ca_3P_2 would produce 0.4 g PH_3

$$(1\text{ g Ca}_3\text{P}_2)\left(\dfrac{1\text{ mol}}{182.2\text{ g}}\right)\left(\dfrac{2\text{ mol PH}_3}{1\text{ mol Ca}_3\text{P}_2}\right)\left(\dfrac{33.99\text{ g}}{\text{mol}}\right) = 0.4\text{ g PH}_3$$

(c) Correct: see equation
(d) Correct: see equation
(e) Incorrect: 2 mol Ca_3P_2 requires 12 mol H_2O to produce 4.0 mol PH_3.

$$(2\text{ mol Ca}_3\text{P}_2)\left(\dfrac{6\text{ mol H}_2\text{O}}{1\text{ mol Ca}_3\text{P}_2}\right) = 12\text{ mol H}_2\text{O}$$

(f) Correct: 2 mol Ca_3P_2 will react with 12 mol H_2O (3 mol H_2O are present in excess) and 6 mol $Ca(OH)_2$ will be formed.

$$(2\text{ mol Ca}_3\text{P}_2)\left(\dfrac{3\text{ mol Ca(OH)}_2}{1\text{ mol Ca}_3\text{P}_2}\right) = 6\text{ mol Ca(OH)}_2$$

(g) Incorrect: $(200.\text{ g Ca}_3\text{P}_2)\left(\dfrac{1\text{ mol}}{182.2\text{ g}}\right)\left(\dfrac{6\text{ mol H}_2\text{O}}{1\text{ mol Ca}_3\text{ P}_2}\right)\left(\dfrac{18.02\text{ g}}{\text{mol}}\right) = 119\text{ g H}_2\text{O}$

The amount of water present (100. g) is less than needed to react with 200. g $Ca_3P_2 \cdot H_2O$ is the limiting reactant.

(h) Incorrect: water is the limiting reactant.

$$(100.\text{ g H}_2\text{O})\left(\dfrac{1\text{ mol}}{18.02\text{ g}}\right)\left(\dfrac{2\text{ mol PH}_3}{6\text{ mol H}_2\text{O}}\right)\left(\dfrac{33.99\text{ g}}{\text{mol}}\right) = 62.9\text{ g PH}_3(\text{theoretical})$$

4. The balanced equation is

$$2\,CH_4 + 3\,O_2 + 2\,NH_3 \rightarrow 2\,HCN + 6\,H_2O$$

(a) Correct

(b) Incorrect: $(16\,mol\,O_2)\left(\dfrac{2\,mol\,HCN}{3\,mol\,O_2}\right) = 10.7\,mol\,HCN$ (not 12 mol HCN)

(c) Correct

(d) Incorrect: $(12\,mol\,HCN)\left(\dfrac{6\,mol\,H_2O}{2\,mol\,HCN}\right) = 36\,mol\,H_2O$ (not 4 mol H_2O)

(e) Correct

(f) Incorrect: O_2 is the limiting reactant

$$(3\,mol\,O_2)\left(\dfrac{2\,mol\,HCN}{3\,mol\,O_2}\right) = 2\,mol\,HCN \text{ (not 3 mol HCN)}$$

5. The molar mass of the product is also needed to convert moles of product into grams.

6. $g\,H_2 \rightarrow mol\,H_2 \rightarrow mol\,NH_3 \rightarrow g\,NH_3$

7. The theoretical yield of a chemical reaction is the maximum amount of product that can be produced based on a balanced equation. The actual yield of a reaction is the actual amount of product obtained.

8. You can calculate the percent yield of a chemical reaction by dividing the actual yield by the theoretical yield and multiplying by one hundred.

SOLUTIONS TO EXERCISES

1. (a) $(25.0 \text{ g KNO}_3)\left(\dfrac{1 \text{ mol}}{101.1 \text{ g}}\right) = 0.247 \text{ mol KNO}_3$

   (b) $(56 \text{ mmol NaOH})\left(\dfrac{1 \text{ mol}}{1000 \text{ mmol}}\right) = 0.056 \text{ mol NaOH}$

   (c) $(5.4 \times 10^2 \text{ g (NH}_4)_2\text{C}_2\text{O}_4)\left(\dfrac{1 \text{ mol}}{124.1 \text{ g}}\right) = 4.4 \text{ mol (NH}_4)_2\text{C}_2\text{O}_4$

   (d) The conversion is: mL sol $\rightarrow$ g sol $\rightarrow$ g H_2SO_4 $\rightarrow$ mol H_2SO_4

   $(16.8 \text{ mL solution})\left(\dfrac{1.727 \text{ g}}{\text{mL}}\right)\left(\dfrac{0.800 \text{ g H}_2\text{SO}_4}{\text{g solution}}\right)\left(\dfrac{1 \text{ mol}}{98.09 \text{ g}}\right) = 0.237 \text{ mol H}_2\text{SO}_4$

2. (a) $(2.10 \text{ kg NaHCO}_3)\left(\dfrac{1000 \text{ g}}{\text{kg}}\right)\left(\dfrac{1 \text{ mol}}{84.01 \text{ g}}\right) = 25.0 \text{ mol NaHCO}_3$

   (b) $(525 \text{ mg ZnCl}_2)\left(\dfrac{1 \text{ g}}{1000 \text{ mg}}\right)\left(\dfrac{1 \text{ mol}}{136.3 \text{ g}}\right) = 3.85 \times 10^{-3} \text{ mol ZnCl}_2$

   (c) $(9.8 \times 10^{24} \text{ molecules CO}_2)\left(\dfrac{1 \text{ mol}}{6.022 \times 10^{23} \text{ molecules}}\right) = 16 \text{ mol CO}_2$

   (d) $(250 \text{ mL C}_2\text{H}_5\text{OH})\left(\dfrac{0.789 \text{ g}}{\text{mL}}\right)\left(\dfrac{1 \text{ mol}}{46.07 \text{ g}}\right) = 4.3 \text{ mol C}_2\text{H}_5\text{OH}$

3. (a) $(2.55 \text{ mol Fe(OH)}_3)\left(\dfrac{106.9 \text{ g}}{\text{mol}}\right) = 273 \text{ g Fe(OH)}_3$

   (b) $(125 \text{ kg CaCO}_3)\left(\dfrac{1000 \text{ g}}{\text{kg}}\right) = 1.25 \times 10^5 \text{ g CaCO}_3$

   (c) $(10.5 \text{ mol NH}_3)\left(\dfrac{17.03 \text{ g}}{\text{mol}}\right) = 179 \text{ g NH}_3$

   (d) $(72 \text{ mmol HCl})\left(\dfrac{1 \text{ mol}}{1000 \text{ mmol}}\right)\left(\dfrac{36.46 \text{ g}}{\text{mol}}\right) = 2.6 \text{ g HCl}$

   (e) $(500.0 \text{ mL Br}_2)\left(\dfrac{3.119 \text{ g}}{\text{mL}}\right) = 1.560 \times 10^3 \text{ g Br}_2$

4. (a) $(0.00844 \text{ mol NiSO}_4)\left(\dfrac{154.8 \text{ g}}{\text{mol}}\right) = 1.31 \text{ g NiSO}_4$

   (b) $(0.0600 \text{ mol HC}_2\text{H}_3\text{O}_2)\left(\dfrac{60.05 \text{ g}}{\text{mol}}\right) = 3.60 \text{ g HC}_2\text{H}_3\text{O}_2$

   (c) $(0.725 \text{ mol Bi}_2\text{S}_3)\left(\dfrac{514.2 \text{ g}}{\text{mol}}\right) = 373 \text{ g Bi}_2\text{S}_3$

(d) $(4.50 \times 10^{21} \text{ molecules } C_6H_{12}O_6)\left(\dfrac{1 \text{ mol}}{6.022 \times 10^{23} \text{ molecules}}\right)\left(\dfrac{180.2 \text{ g}}{\text{mol}}\right) = 1.35 \text{ g } C_6H_{12}O_6$

(e) $(75 \text{ mL solution})\left(\dfrac{1.175 \text{ g}}{\text{mL}}\right)\left(\dfrac{0.200 \text{ g } K_2CrO_4}{\text{g solution}}\right) = 18 \text{ g } K_2CrO_4$

5. Larger number of molecules: $10.0 \text{ g } H_2O$ or $10.0 \text{ g } H_2O_2$

   Water has a lower molar mass than hydrogen peroxide. 10.0 grams of water has a lower molar mass, contains more moles, and therefore more molecules than 10.0 g of H_2O_2.

6. Larger number of molecules: $25.0 \text{ g } HCl$ or $85 \text{ g } C_6H_{12}O_6$

   $(25.0 \text{ g } HCl)\left(\dfrac{1 \text{ mol}}{36.46 \text{ g}}\right)\left(\dfrac{6.022 \times 10^{23} \text{ molecules}}{\text{mol}}\right) = 4.13 \times 10^{23} \text{ molecules } HCl$

   $(85.0 \text{ g } C_6H_{12}O_6)\left(\dfrac{1 \text{ mol}}{180.2 \text{ g}}\right)\left(\dfrac{6.022 \times 10^{23} \text{ molecules}}{\text{mol}}\right) = 2.84 \times 10^{23} \text{ molecules } C_6H_{12}O_6$

   HCl contains more molecules

7. Mole Ratios

   $12 \, CO_2 + 11 \, H_2O \rightarrow C_{12}H_{22}O_{11} + 12 \, O_2$

   (a) $\dfrac{12 \text{ mol } CO_2}{11 \text{ mol } H_2O}$ (d) $\dfrac{1 \text{ mol } C_{12}H_{22}O_{11}}{12 \text{ mol } CO_2}$

   (b) $\dfrac{11 \text{ mol } H_2O}{1 \text{ mol } C_{12}H_{22}O_{11}}$ (e) $\dfrac{11 \text{ mol } H_2O}{12 \text{ mol } O_2}$

   (c) $\dfrac{12 \text{ mol } O_2}{12 \text{ mol } CO_2}$ (f) $\dfrac{12 \text{ mol } O_2}{1 \text{ mol } C_{12}H_{22}O_{11}}$

8. Mole ratios

   $C_4H_9OH + 6 \, O_2 \rightarrow 4 \, CO_2 + 5 \, H_2O$

   (a) $\dfrac{6 \text{ mol } O_2}{1 \text{ mol } C_4H_9OH}$ (d) $\dfrac{1 \text{ mol } C_4H_9OH}{4 \text{ mol } CO_2}$

   (b) $\dfrac{5 \text{ mol } H_2O}{6 \text{ mol } O_2}$ (e) $\dfrac{5 \text{ mol } H_2O}{1 \text{ mol } C_4H_9OH}$

   (c) $\dfrac{4 \text{ mol } CO_2}{5 \text{ mol } H_2O}$ (f) $\dfrac{4 \text{ mol } CO_2}{6 \text{ mol } O_2}$

9. The balanced equation is $CO_2 + 4 \, H_2 \rightarrow CH_4 + 2 \, H_2O$

   (a) $(25 \text{ mol } CO_2)\left(\dfrac{2 \text{ mol } H_2O}{1 \text{ mol } CO_2}\right) = 50. \text{ mol } H_2O$

   (b) $(12 \text{ mol } H_2O)\left(\dfrac{1 \text{ mol } CH_4}{2 \text{ mol } H_2O}\right) = 6.0 \text{ mol } CH_4$

10. The balanced equation is $H_2SO_4 + 2\,NaOH \rightarrow Na_2SO_4 + 2\,H_2O$

    (a) $(17\,\text{mol}\,H_2SO_4)\left(\dfrac{2\,\text{mol}\,NaOH}{1\,\text{mol}\,H_2SO_4}\right) = 34\,\text{mol}\,NaOH$

    (b) $(21\,\text{mol}\,NaOH)\left(\dfrac{1\,\text{mol}\,Na_2SO_4}{2\,\text{mol}\,NaOH}\right) = 11\,\text{mol}\,Na_2SO_4$

11. The balanced equation is

    $MnO_2(s) + 4\,HCl(aq) \rightarrow Cl_2(g) + MnCl_2(aq) + 2\,H_2O(l)$

    (a) $(1.05\,\text{mol}\,MnO_2)\left(\dfrac{4\,\text{mol}\,HCl}{1\,\text{mol}\,MnO_2}\right) = 4.20\,\text{mol}\,HCl$

    (b) $(1.25\,\text{mol}\,H_2O)\left(\dfrac{1\,\text{mol}\,MnCl_2}{2\,\text{mol}\,H_2O}\right) = 0.625\,\text{mol}\,MnCl_2$

    (c) $(3.28\,\text{mol}\,MnO_2)\left(\dfrac{1\,\text{mol}\,Cl_2}{1\,\text{mol}\,MnO_2}\right)\left(\dfrac{70.90\,g}{1\,\text{mol}}\right) = 233\,g\,Cl_2$

    (d) $(15.0\,\text{kg}\,MnCl_2)\left(\dfrac{1000\,g}{1\,kg}\right)\left(\dfrac{1\,\text{mol}}{125.8\,g}\right)\left(\dfrac{4\,\text{mol}\,HCl}{1\,\text{mol}\,MnCl_2}\right) = 477\,\text{mol}\,HCl$

12. $Al_4C_3 + 12\,H_2O \rightarrow 4\,Al(OH)_3 + 3\,CH_4$

    (a) $(100.\,g\,Al_4C_3)\left(\dfrac{1\,\text{mol}}{144.0\,g}\right)\left(\dfrac{12\,\text{mol}\,H_2O}{1\,\text{mol}\,Al_4C_3}\right) = 8.33\,\text{mol}\,H_2O$

    (b) $(0.600\,\text{mol}\,CH_4)\left(\dfrac{4\,\text{mol}\,Al(OH)_3}{3\,\text{mol}\,CH_4}\right) = 0.800\,\text{mol}\,Al(OH)_3$

    (c) $(275\,g\,Al_4C_3)\left(\dfrac{1\,\text{mol}}{144.0\,g}\right)\left(\dfrac{3\,\text{mol}\,CH_4}{1\,\text{mol}\,Al_4C_3}\right) = 5.73\,\text{mol}\,CH_4$

    (d) $(4.22\,\text{mol}\,Al(OH)_3)\left(\dfrac{12\,\text{mol}\,H_2O}{4\,\text{mol}\,Al(OH)_3}\right)\left(\dfrac{18.02\,g}{1\,\text{mol}}\right) = 228\,g\,H_2O$

13. Grams of $CaCl_2$

    $CaCO_3 + 2\,HCl \rightarrow CaCl_2 + H_2O + CO_2$

    The conversion is: $g\,CaCO_3 \rightarrow \text{mol}\,CaCO_3 \rightarrow \text{mol}\,CaCl_2 \rightarrow g\,CaCl_2$

    $(50.0\,g\,CaCO_3)\left(\dfrac{1\,\text{mol}}{100.1\,g}\right)\left(\dfrac{1\,\text{mol}\,CaCl_2}{1\,\text{mol}\,CaCO_3}\right)\left(\dfrac{111.0\,g}{\text{mol}}\right) = 55.4\,g\,CaCl_2$

14. Grams of $AlBr_3$

    $2\,Al + 6\,HBr \rightarrow 2\,AlBr_3 + 3\,H_2$

    The conversion is: $g\,Al \rightarrow \text{mol}\,Al \rightarrow \text{mol}\,AlBr_3 \rightarrow g\,AlBr_3$

    $(25.2\,g\,Al)\left(\dfrac{1\,\text{mol}}{26.98\,g}\right)\left(\dfrac{2\,\text{mol}\,AlBr_3}{2\,\text{mol}\,Al}\right)\left(\dfrac{266.7\,g}{\text{mol}}\right) = 249\,g\,AlBr_3$

15. The balanced equation is $Fe_2O_3 + 3\,C \rightarrow 2\,Fe + 3\,CO$

    The conversion is: $kg\,Fe_2O_3 \rightarrow kmol\,Fe_2O_3 \rightarrow kmol\,Fe \rightarrow kg\,Fe$

    $$(125\,kg\,Fe_2O_3)\left(\frac{1\,kmol}{159.7\,kg}\right)\left(\frac{2\,kmol\,Fe}{1\,kmol\,Fe_2O_3}\right)\left(\frac{55.85\,kg}{kmol}\right) = 87.4\,kg\,Fe$$

16. The balanced equation is $3\,Fe + 4\,H_2O \rightarrow Fe_3O_4 + 4\,H_2$

    Calculate the grams of both H_2O and Fe to produce $375\,g\,Fe_3O_4$

    $$(375\,g\,Fe_3O_4)\left(\frac{1\,mol}{231.6\,g}\right)\left(\frac{4\,mol\,H_2O}{1\,mol\,Fe_3O_4}\right)\left(\frac{18.02\,g}{mol}\right) = 117\,g\,H_2O$$

    $$(375\,g\,Fe_3O_4)\left(\frac{1\,mol}{231.6\,g}\right)\left(\frac{3\,mol\,Fe}{1\,mol\,Fe_3O_4}\right)\left(\frac{55.85\,g}{mol}\right) = 271\,g\,Fe$$

17. The balanced equation is: $2\,C_{12}H_4Cl_6 + 23\,O_2 + 2\,H_2O \rightarrow 24\,CO_2 + 12\,HCl$

    (a) $$(10.0\,mol\,O_2)\left(\frac{2\,mol\,H_2O}{23\,mol\,O_2}\right) = 0.870\,mol\,H_2O$$

    (b) $$(15.2\,mol\,H_2O)\left(\frac{12\,mol\,HCl}{2\,mol\,H_2O}\right)\left(\frac{36.46\,g}{mol}\right) = 3.33 \times 10^3\,g\,HCl$$

    (c) $$(76.5\,g\,HCl)\left(\frac{1\,mol}{36.46\,g}\right)\left(\frac{24\,mol\,CO_2}{12\,mol\,HCl}\right) = 4.20\,mol\,CO_2$$

    (d) $$(100.25\,g\,CO_2)\left(\frac{1\,mol}{44.01\,g}\right)\left(\frac{2\,mol\,C_{12}H_4Cl_6}{24\,mol\,CO_2}\right)\left(\frac{360.9\,g}{mol}\right) = 68.51\,g\,C_{12}H_4Cl_6$$

    (e) $$(2.5\,kg\,C_{12}H_4Cl_6)\left(\frac{1000\,g}{1\,kg}\right)\left(\frac{1\,mol}{360.9\,g}\right)\left(\frac{12\,mol\,HCl}{2\,mol\,C_{12}H_4Cl_6}\right)\left(\frac{36.46\,g}{mol}\right) = 1.5 \times 10^3\,g\,HCl$$

18. $4\,HgS + 4\,CaO \rightarrow 4\,Hg + 3\,CaS + CaSO_4$

    (a) $$(2.5\,mol\,CaO)\left(\frac{3\,mol\,CaS}{4\,mol\,CaO}\right) = 1.9\,mol\,CaS$$

    (b) $$(9.75\,mol\,CaSO_4)\left(\frac{4\,mol\,Hg}{1\,mol\,CaSO_4}\right)\left(\frac{200.6\,g}{mol}\right) = 7.82 \times 10^3\,g\,Hg$$

    (c) $$(97.25\,g\,HgS)\left(\frac{1\,mol}{232.7\,g}\right)\left(\frac{4\,mol\,CaO}{4\,mol\,HgS}\right) = 0.4179\,mol\,CaO$$

    (d) $$(87.6\,g\,HgS)\left(\frac{1\,mol}{232.7\,g}\right)\left(\frac{4\,mol\,Hg}{4\,mol\,HgS}\right)\left(\frac{200.6\,g}{mol}\right) = 75.5\,g\,Hg$$

    (e) $$(9.25\,kg\,Hg)\left(\frac{1000\,g}{1\,kg}\right)\left(\frac{1\,mol}{200.6\,g}\right)\left(\frac{3\,mol\,CaS}{4\,mol\,Hg}\right)\left(\frac{72.15\,g}{1\,mol}\right) = 2.50 \times 10^3\,g\,CaS$$

19. (a) ○ Hydrogen ● Oxygen
Hydrogen is the limiting reactant.

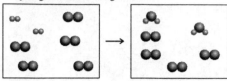

(b) ○ Hydrogen ● Bromine
Bromine is the limiting reactant.

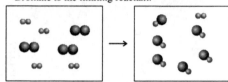

20. (a) ○ Lithium ● Iodine
No limiting reactant.

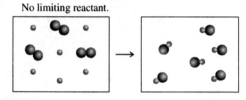

(b) ● Silver ○ Chlorine
Silver is the limiting reactant.

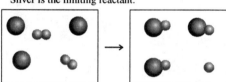

Note that because chlorine gas exists as a diatomic molecule, only a single atom of chlorine would remain if all of the silver were to react. In actual reactions we count in terms of moles and a ½ mole of chlorine gas would remain if 3 moles of silver reacted with 2 moles of chlorine gas.

21. (a) ○ Potassium ● Chlorine
Potassium is the limiting reactant.

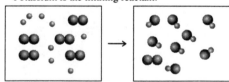

(b) ○ Aluminum ● Oxygen
Oxygen is the limiting reactant.

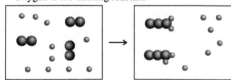

22. (a) ◦ Nitrogen ● Oxygen

Oxygen is the limiting reactant.

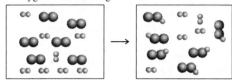

(b) ◦ Iron ◦ Hydrogen ● Oxygen

Water is the limiting reactant.

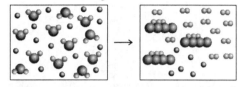

23. (a)
$$KOH \quad + \quad HNO_3 \quad \rightarrow \quad KNO_3 \quad + \quad H_2O$$
16.0 g   12.0 g

Choose one of the products and calculate its mass that would be produced from each given reactant. Using KNO_3 as the product:

$$(16.0\text{ g KOH})\left(\frac{1\text{ mol}}{56.10\text{ g}}\right)\left(\frac{1\text{ mol KNO}_3}{1\text{ mol KOH}}\right)\left(\frac{101.1\text{ g}}{\text{mol}}\right) = 28.8\text{ g KNO}_3$$

$$(12.0\text{ g HNO}_3)\left(\frac{1\text{ mol}}{63.02\text{ g}}\right)\left(\frac{1\text{ mol KNO}_3}{1\text{ mol KOH}}\right)\left(\frac{101.1\text{ g}}{\text{mol}}\right) = 19.3\text{ g KNO}_3$$

Since HNO_3 produces less KNO_3, it is the limiting reactant and KOH is in excess.

(b)
$$2\,NaOH \quad + \quad H_2SO_4 \quad \rightarrow \quad Na_2SO_4 \quad + \quad 2\,H_2O$$
10.0 g   10.0 g

Choose one of the products and calculate its mass that would be produced from each given reactant. Using H_2O as the product:

$$(10.0\text{ g NaOH})\left(\frac{1\text{ mol}}{40.00\text{ g}}\right)\left(\frac{2\text{ mol H}_2\text{O}}{2\text{ mol NaOH}}\right)\left(\frac{18.02\text{ g}}{\text{mol}}\right) = 4.51\text{ g H}_2\text{O}$$

$$(10.0\text{ g H}_2\text{SO}_4)\left(\frac{1\text{ mol}}{98.09\text{ g}}\right)\left(\frac{2\text{ mol H}_2\text{O}}{1\text{ mol H}_2\text{SO}_4}\right)\left(\frac{18.02\text{ g}}{\text{mol}}\right) = 3.67\text{ g H}_2\text{O}$$

Since H_2SO_4 produces less H_2O, it is the limiting reactant and NaOH is in excess.

24. (a)
$$2\,Bi(NO_3)_3 \quad + \quad 3\,H_2S \quad \rightarrow \quad Bi_2S_3 \quad + \quad 6\,HNO_3$$
50.0 g   6.00 g

Choose one of the products and calculate its mass that would be produced from each given reactant. Using Bi_2S_3 as the product:

$$(50.0 \text{ g Bi(NO}_3)_3)\left(\frac{1 \text{ mol}}{395.0 \text{ g}}\right)\left(\frac{1 \text{ mol Bi}_2\text{S}_3}{2 \text{ mol Bi(NO}_3)_3}\right)\left(\frac{514.2 \text{ g}}{\text{mol}}\right) = 32.5 \text{ g Bi}_2\text{S}_3$$

$$(6.00 \text{ g H}_2\text{S})\left(\frac{1 \text{ mol}}{34.09 \text{ g}}\right)\left(\frac{1 \text{ mol Bi}_2\text{S}_3}{3 \text{ mol H}_2\text{S}}\right)\left(\frac{514.2 \text{ g}}{\text{mol}}\right) = 30.2 \text{ g Bi}_2\text{S}_3$$

Since H_2S produces less Bi_2S_3, it is the limiting reactant and $Bi(NO_3)_3$ is in excess.

(b) $3 \text{ Fe} \quad + \quad 4 \text{ H}_2\text{O} \quad \rightarrow \quad \text{Fe}_3\text{O}_4 \quad + \quad 4 \text{ H}_2$
    40.0 g   16.0 g

Choose one of the products and calculate its mass that would be produced from each given reactant. Using H_2 as the product:

$$(40.0 \text{ g Fe})\left(\frac{1 \text{ mol}}{55.85 \text{ g}}\right)\left(\frac{4 \text{ mol H}_2}{3 \text{ mol Fe}}\right)\left(\frac{2.016 \text{ g}}{\text{mol}}\right) = 1.93 \text{ g H}_2$$

$$(16.0 \text{ g H}_2\text{O})\left(\frac{1 \text{ mol}}{18.02 \text{ g}}\right)\left(\frac{4 \text{ mol H}_2}{4 \text{ mol H}_2\text{O}}\right)\left(\frac{2.016 \text{ g}}{\text{mol}}\right) = 1.79 \text{ g H}_2$$

Since H_2O produces less H_2, it is the limiting reactant and Fe is in excess.

25. Limiting reactant calculations

$$2 \text{ Al(OH)}_3 + 3 \text{ H}_2\text{SO}_4 \rightarrow \text{Al}_2(\text{SO}_4)_3 + 6 \text{ H}_2\text{O}$$

(a) Reaction between 35.0 g $Al(OH)_3$ and 35.0 g H_2SO_4.
Convert each amount to moles of $Al_2(SO_4)_3$.

$$(35.0 \text{ g Al(OH)}_3)\left(\frac{1 \text{ mol}}{78.00 \text{ g}}\right)\left(\frac{1 \text{ mol Al}_2(\text{SO}_4)_3}{2 \text{ mol Al(OH)}_3}\right) = 0.224 \text{ mol Al}_2(\text{SO}_4)_3$$

$$(35.0 \text{ g H}_2\text{SO}_4)\left(\frac{1 \text{ mol}}{98.09 \text{ g}}\right)\left(\frac{1 \text{ mol Al}_2(\text{SO}_4)_3}{3 \text{ mol H}_2\text{SO}_4}\right) = 0.119 \text{ mol Al}_2(\text{SO}_4)_3$$

H_2SO_4 is the limiting reactant. The yield is 0.119 mol $Al_2(SO_4)_3$

(b) Reaction between 45.0 g H_2SO_4 and 25.0 g $Al(OH)_3$.
Calculate the grams of $Al_2(SO_4)_3$ from each reactant

$$(45.0 \text{ g H}_2\text{SO}_4)\left(\frac{1 \text{ mol}}{98.09 \text{ g}}\right)\left(\frac{1 \text{ mol Al}_2(\text{SO}_4)_3}{3 \text{ mol H}_2\text{SO}_4}\right)\left(\frac{342.2 \text{ g}}{\text{mol}}\right) = 52.3 \text{ g Al}_2(\text{SO}_4)_3$$

$$(25.0 \text{ g Al(OH)}_3)\left(\frac{1 \text{ mol}}{78.09 \text{ g}}\right)\left(\frac{1 \text{ mol Al}_2(\text{SO}_4)_3}{2 \text{ mol Al(OH)}_3}\right)\left(\frac{342.2 \text{ g}}{\text{mol}}\right) = 54.8 \text{ g Al}_2(\text{SO}_4)_3$$

H_2SO_4 is the limiting reactant. The yield is 52.3 g $Al_2 (SO_4)_3$
$Al(OH)_3$ is the excess reactant.

(c) Reaction between 2.5 mol $Al(OH)_3$ and 5.5 mol H_2SO_4.
Convert each amount to moles of product.

$$(2.5 \text{ mol Al(OH)}_3)\left(\frac{1 \text{ mol Al}_2(\text{SO}_4)_3}{2 \text{ mol Al(OH)}_3}\right) = 1.3 \text{ mol Al}_2(\text{SO}_4)_3$$

$$(5.5 \text{ mol H}_2\text{SO}_4)\left(\frac{1 \text{ mol Al}_2(\text{SO}_4)_3}{3 \text{ mol H}_2\text{SO}_4}\right) = 1.8 \text{ mol Al}_2(\text{SO}_4)_3$$

Al(OH)$_3$ is the limiting reactant. 1.3 mol Al$_2$(SO$_4$)$_3$ produced

$$(2.5 \text{ mol Al(OH)}_3)\left(\frac{6 \text{ mol H}_2\text{O}}{2 \text{ mol Al(OH)}_3}\right) = 7.5 \text{ mol H}_2\text{O produced}$$

$$(2.5 \text{ mol Al(OH)}_3)\left(\frac{3 \text{ mol H}_2\text{SO}_4}{2 \text{ mol Al(OH)}_3}\right) = 3.8 \text{ mol H}_2\text{SO}_4 \text{ produced}$$

5.5 mol H$_2$SO$_4$ − 3.8 mol H$_2$SO$_4$ = 1.7 mol H$_2$SO$_4$ unreacted

When the reaction is complete, 1.3 mol Al$_2$(SO$_4$)$_3$, 7.5 mol H$_2$O, and 1.7 mol H$_2$SO$_4$ will be in the container.

26. The balanced equation is P$_4$ + 6 Cl$_2$ → 4 PCl$_3$
   (a) Reaction between 20.5 g P$_4$ and 20.5 g Cl$_2$
       Convert each amount to moles of PCl$_3$

$$(20.5 \text{ g P}_4)\left(\frac{1 \text{ mol}}{123.9 \text{ g}}\right)\left(\frac{4 \text{ mol PCl}_3}{1 \text{ mol P}_4}\right) = 0.662 \text{ mol PCl}_3$$

$$(20.5 \text{ g Cl}_2)\left(\frac{1 \text{ mol}}{70.90 \text{ g}}\right)\left(\frac{4 \text{ mol PCl}_3}{6 \text{ mol Cl}_2}\right) = 0.193 \text{ mol PCl}_3$$

Cl$_2$ is the limiting reactant. The yield is 0.193 mol PCl$_3$.
   (b) Reaction between 55 g Cl$_2$ and 25 g P$_4$.
       Calculate the grams of PCl$_3$ from each reactant.

$$(55 \text{ g Cl}_2)\left(\frac{1 \text{ mol}}{70.90 \text{ g}}\right)\left(\frac{4 \text{ mol PCl}_3}{6 \text{ mol Cl}_2}\right)\left(\frac{137.3 \text{ g}}{\text{mol}}\right) = 71 \text{ g PCl}_3$$

$$(25 \text{ g P}_4)\left(\frac{1 \text{ mol}}{123.9 \text{ g}}\right)\left(\frac{4 \text{ mol PCl}_3}{1 \text{ mol P}_4}\right)\left(\frac{137.3 \text{ g}}{\text{mol}}\right) = 1.1 \times 10^2 \text{ g PCl}_3$$

Cl$_2$ is the limiting reactant. The yield is 71 g PCl$_3$.
P$_4$ is the excess reactant.
   (c) Reaction between 15 mol P$_4$ and 35 mol Cl$_2$.
       Convert each amount to moles of product.

$$(15 \text{ mol P}_4)\left(\frac{4 \text{ mol PCl}_3}{1 \text{ mol P}_4}\right) = 60. \text{ mol PCl}_3$$

$$(35 \text{ mol Cl}_2)\left(\frac{4 \text{ mol PCl}_3}{6 \text{ mol Cl}_2}\right) = 23 \text{ mol PCl}_3 \text{ produced}$$

Cl$_2$ is the limiting reactant.

$$(35 \text{ mol Cl}_2)\left(\frac{1 \text{ mol P}_4}{6 \text{ mol Cl}_2}\right) = 5.8 \text{ mol P}_4 \text{ reacted}$$

15 mol P$_4$ − 5.8 mol P$_4$ = 9 mol P$_4$ left over

When the reaction is complete, 23 mol PCl$_3$ and 9 mol P$_4$ will be in the container.

27. $X_8 + 12\,O_2 \rightarrow 8\,XO_3$

    The conversion is: $g\,O_2 \rightarrow mol\,O_2 \rightarrow mol\,X_8$

    $$(120.0\,g\,O_2)\left(\frac{1\,mol}{32.00\,g}\right)\left(\frac{1\,mol\,X_8}{12\,mol\,O_2}\right) = 0.3125\,mol\,X_8 \qquad 80.0\,g\,X_8 = 0.3125\,mol\,X_8$$

    $$\frac{80.0\,g}{0.3125\,mol} = 256\,g/\,mol\,X_8$$

    $$\text{molar mass } X = \frac{256\,\dfrac{g}{mol}}{8} = 32.0\,\frac{g}{mol}$$

    Using the periodic table we find that the element with 32.0 g/mol is sulfur.

28. $X + 2\,HCl \rightarrow XCl_2 + H_2$

    The conversion is: $g\,H_2 \rightarrow mol\,H_2 \rightarrow mol\,X$

    $$(2.42\,g\,H_2)\left(\frac{1\,mol}{2.016\,g}\right)\left(\frac{1\,mol\,X}{1\,mol\,H_2}\right) = 1.20\,mol\,X \qquad 78.5\,g\,X = 1.20\,mol\,X$$

    $$\frac{78.5\,g}{1.20\,mol} = 65.4\,g/mol$$

    Using the periodic table we find that the element with atomic mass 65.4 is zinc.

29. Limiting reactant calculation and percent yield

    $SiO_2 + 2\,C \rightarrow Si + 2\,CO$

    $kg\,SiO_2 \rightarrow g\,SiO_2 \rightarrow mol\,SiO_2 \rightarrow mol\,Si \rightarrow g\,Si \rightarrow kg\,Si$

    $$(35.0\,kg\,SiO_2)\left(\frac{1000\,g}{1\,kg}\right)\left(\frac{1\,mol}{60.09\,g}\right)\left(\frac{1\,mol\,Si}{1\,mol\,SiO_2}\right)\left(\frac{28.09\,g}{1\,mol}\right)\left(\frac{1\,kg}{1000\,g}\right) = 16.4\,kg\,Si$$

    $kg\,C \rightarrow g\,C \rightarrow mol\,C \rightarrow mol\,Si \rightarrow g\,Si \rightarrow kg\,Si$

    $$(25.3\,kg\,C)\left(\frac{1000\,g}{1\,kg}\right)\left(\frac{1\,mol}{12.01\,g}\right)\left(\frac{1\,mol\,Si}{2\,mol\,C}\right)\left(\frac{28.09\,g}{1\,mol}\right)\left(\frac{1\,kg}{1000\,g}\right) = 29.6\,kg\,Si$$

    The limiting reactant is SiO_2. 16.4 kg Si is the theoretical yield.

    $$\text{Percent yield} = \left(\frac{\text{actual yield}}{\text{theoretical yield}}\right)(100) = \left(\frac{14.4\,kg}{16.4\,kg}\right)(100) = 87.8\%\ \text{yield of Si}$$

30. Limiting reactant calculation and percent yield

    $2\,Cr_2O_3 + 3\,Si \rightarrow 4\,Cr + 3\,SiO_2$

    $$(350.0\,g\,Cr_2O_3)\left(\frac{1\,mol}{152.0\,g}\right)\left(\frac{4\,mol\,Cr}{2\,mol\,Cr_2O_3}\right)\left(\frac{52.00\,g}{1\,mol}\right) = 239.5\,g\,Cr$$

    $$(235.0\,g\,Si)\left(\frac{1\,mol}{28.09\,g}\right)\left(\frac{4\,mol\,Cr}{3\,mol\,Si}\right)\left(\frac{52.00\,g}{1\,mol}\right) = 580.0\,g\,Cr$$

    The limiting reactant is Cr_2O_3. 239.5 g Cr is the theoretical yield.

    $$\text{Percent yield} = \left(\frac{\text{actual yield}}{\text{theoretical yield}}\right)(100) = \left(\frac{213.2\,g}{239.5\,g}\right)(100) = 89.02\,\%\ \text{yield}$$

31. $3\,Cu + 8\,HNO_3 \rightarrow 3\,Cu(NO_3)_2 + 4\,H_2O + 2\,NO$

    (a) Limiting reactant problem

$$(27.5\,g\,Cu)\left(\frac{1\,mol}{63.55\,g}\right)\left(\frac{3\,mol\,Cu(NO_3)_2}{3\,mol\,Cu}\right)\left(\frac{187.6\,g}{1\,mol}\right) = 81.2\,g\,Cu(NO_3)_2$$

$$(125\,g\,HNO_3)\left(\frac{1\,mol}{63.02\,g}\right)\left(\frac{3\,mol\,Cu(NO_3)_2}{8\,mol\,HNO_3}\right)\left(\frac{187.6\,g}{1\,mol}\right) = 140\,g\,Cu(NO_3)_2$$

    (b) The excess reactant is HNO_3

    (c) The percent yield is 87.3%. The actual yield is

$$(0.873)(81.2\,g) = 70.9\,g\,Cu(NO_3)_2\ actual\ yield$$

32. $Fe_2O_3(s) + 6\,HCl(aq) \rightarrow 2\,FeCl_3(aq) + 3\,H_2O(l)$

    (a) Limiting reactant problem.

$$(35\,g\,Fe_2O_3)\left(\frac{1\,mol}{159.7\,g}\right)\left(\frac{2\,mol\,FeCl_3}{1\,mol\,Fe_2O_3}\right) = 0.44\ mol\,FeCl_3$$

$$(35\,g\,HCl)\left(\frac{1\,mol}{36.46\,g}\right)\left(\frac{2\,mol\,FeCl_3}{6\,mol\,HCl}\right) = 0.32\ mol\,FeCl_3$$

HCl is the limiting reactant.

$$(35\,g\,HCl)\left(\frac{1\,mol}{36.46\,g}\right)\left(\frac{3\,mol\,H_2O}{6\,mol\,HCl}\right) = 0.48\ mol\,H_2O$$

0.32 mol $FeCl_3$ and 0.48 mol H_2O are the theoretical yields.

    (b) Fe_2O_3 is the excess reactant.

$$(35\,g\,HCl)\left(\frac{1\,mol}{36.46\,g}\right)\left(\frac{1\,mol\,Fe_2O_3}{6\,mol\,HCl}\right)\left(\frac{159.7\,g}{mol}\right) = 26\,g\,Fe_2O_3\ reacted$$

35 g Fe_2O_3 − 26 g Fe_2O_3 = 9 g Fe_2O_3 excess

    (c) The percent yield is 92.5%. The actual yield is $(0.925)(0.32\,mol\,FeCl_3)\left(\frac{162.2\,g}{mol}\right) = 48\,g\,FeCl_3$

33. No. There are not enough screwdrivers, wrenches, or pliers. 2400 screwdrivers, 3600 wrenches, and 1200 pliers are needed for 600 tool sets.

34. The balanced equation is $C_6H_{12}O_6 \rightarrow 2\,C_2H_5OH + 2\,CO_2$

$$(575\,lb\,C_6H_{12}O_6)\left(\frac{453.6\,g}{lb}\right)\left(\frac{1\,mol}{180.2\,g}\right)\left(\frac{2\,mol\,C_2H_5OH}{1\,mol\,C_6H_{12}O_6}\right)\left(\frac{46.07\,g}{mol}\right)\left(\frac{1\,mL}{0.789\,g}\right)\left(\frac{1\,L}{1000\,mL}\right) = 169\,L\,C_2H_5OH$$

35. Consider the reaction A → 2B and assume that you have 1 gram of A. This does not guarantee that you will produce 1 gram of B because A and B have different molar masses. One gram of A does not contain the same number of molecules as 1 gram of B. However, 1 mole of A does have the same number of molecules as one mole of B. (Remember, 1 mole = 6.022×10^{23} molecules always.) If you determine the number of moles in one gram of A and multiply by 2 to get the number of moles of B . . . then from that you can determine the grams of B using its molar mass. Equations are written in terms of moles not grams.

36. $4 KO_2 + 2 H_2O + 4 CO_2 \rightarrow 4 KHCO_3 + 3 O_2$

(a) $\left(\dfrac{0.85 \text{ g } CO_2}{\text{min}}\right)\left(\dfrac{1 \text{ mol}}{44.01 \text{ g}}\right)\left(\dfrac{4 \text{ mol } KO_2}{4 \text{ mol } CO_2}\right) = \dfrac{0.019 \text{ mol } KO_2}{\text{min}}$

$\left(\dfrac{0.019 \text{ mol } KO_2}{\text{min}}\right)(10.0 \text{ min}) = 0.19 \text{ mol } KO_2$

(b) The conversion is: $\dfrac{\text{g } CO_2}{\text{min}} \rightarrow \dfrac{\text{mol } CO_2}{\text{g } CO_2} \rightarrow \dfrac{\text{mol } O_2}{\text{mol } CO_2} \rightarrow \dfrac{\text{g } O_2}{\text{mol } O_2} \rightarrow \dfrac{\text{min}}{\text{hr}} \rightarrow \dfrac{\text{g } O_2}{\text{hr}}$

$\left(\dfrac{0.85 \text{ g } CO_2}{\text{min}}\right)\left(\dfrac{1 \text{ mol}}{44.01 \text{ g}}\right)\left(\dfrac{3 \text{ mol } O_2}{4 \text{ mol } CO_2}\right)\left(\dfrac{32.00 \text{ g}}{1.0 \text{ mol}}\right)\left(\dfrac{60 \text{ min}}{1.0 \text{ hr}}\right) = \dfrac{28 \text{ g } O_2}{\text{hr}}$

37. $C_{12}H_{22}O_{11} \xrightarrow{\text{H}_2\text{SO}_4} 12 C + 11 H_2O$

(a) $(2.0 \text{ lb } C_{12}H_{22}O_{11})\left(\dfrac{453.6 \text{ g}}{\text{lb}}\right)\left(\dfrac{1 \text{ mol}}{342.3 \text{ g}}\right)\left(\dfrac{12 \text{ mol } C}{1 \text{ mol } C_{12}H_{22}O_{11}}\right)\left(\dfrac{12.01 \text{ g}}{\text{mol}}\right) = 3.8 \times 10^2 \text{ g C}$

$(2.0 \text{ lb } C_{12}H_{22}O_{11})\left(\dfrac{453.6 \text{ g}}{\text{lb}}\right)\left(\dfrac{1 \text{ mol}}{342.3 \text{ g}}\right)\left(\dfrac{11 \text{ mol } H_2O}{1 \text{ mol } C_{12}H_{22}O_{11}}\right)\left(\dfrac{18.02 \text{ g}}{\text{mol}}\right) = 5.3 \times 10^2 \text{ g } H_2O$

From 2.0 lb $C_{12}H_{22}O_{11}$, 3.8×10^2 g C and 5.3×10^2 g H_2O are yielded.

(b) $(25.2 \text{ g } C_{12}H_{22}O_{11})\left(\dfrac{1 \text{ mol } C_{12}H_{22}O_{11}}{342.3 \text{ g}}\right)\left(\dfrac{11 \text{ mol } H_2O}{1 \text{ mol } C_{12}H_{22}O_{11}}\right)\left(\dfrac{18.02 \text{ g}}{\text{mol}}\right)\left(\dfrac{1 \text{ mL}}{0.994 \text{ g}}\right) = 14.7 \text{ mL } H_2O$

38. $4 C_3H_5(NO_3)_3(l) \rightarrow 12 CO_2(g) + 10 H_2O(g) + 6 N_2(g) + O_2(g)$

(a) $(45.0 \text{ g } C_3H_5(NO_3)_3)\left(\dfrac{1 \text{ mol}}{227.1 \text{ g}}\right)\left(\dfrac{12 \text{ mol } CO_2}{4 \text{ mol } C_3H_5(NO_3)_3}\right)\left(\dfrac{44.01 \text{ g}}{1 \text{ mol}}\right) = 26.2 \text{ g } CO_2$

$(45.0 \text{ g } C_3H_5(NO_3)_3)\left(\dfrac{1 \text{ mol}}{227.1 \text{ g}}\right)\left(\dfrac{10 \text{ mol } H_2O}{4 \text{ mol } C_3H_5(NO_3)_3}\right)\left(\dfrac{18.02 \text{ g}}{1 \text{ mol}}\right) = 8.93 \text{ g } H_2O$

$(45.0 \text{ g } C_3H_5(NO_3)_3)\left(\dfrac{1 \text{ mol}}{227.1 \text{ g}}\right)\left(\dfrac{6 \text{ mol } N_2}{4 \text{ mol } C_3H_5(NO_3)_3}\right)\left(\dfrac{28.02 \text{ g}}{1 \text{ mol}}\right) = 8.33 \text{ g } N_2$

$(45.0 \text{ g } C_3H_5(NO_3)_3)\left(\dfrac{1 \text{ mol}}{227.1 \text{ g}}\right)\left(\dfrac{1 \text{ mol } O_2}{4 \text{ mol } C_3H_5(NO_3)_3}\right)\left(\dfrac{32.00 \text{ g}}{1 \text{ mol}}\right) = 1.59 \text{ g } O_2$

From 45.0 g $C_5H_5(NO_3)_3$, 26.2 g CO_2, 8.93 g H_2O, 8.33 g N_2, and 1.59 g O_2 are yielded.

(b) $(45.0 \text{ g } C_3H_5(NO_3)_3)\left(\dfrac{1 \text{ mol}}{227.1 \text{ g}}\right)\left(\dfrac{29 \text{ mol gas}}{4 \text{ mol } C_3H_5(NO_3)_3}\right) = 1.44 \text{ mol gas}$

39. $2 CH_3OH + 3 O_2 \rightarrow 2 CO_2 + 4 H_2O$

The conversion is:

mL $CH_3OH \rightarrow$ g $CH_3OH \rightarrow$ mol $CH_3OH \rightarrow$ mol $O_2 \rightarrow$ g O_2

$(60.0 \text{ mL } CH_3OH)\left(\dfrac{0.787 \text{ g}}{\text{mL}}\right)\left(\dfrac{1 \text{ mol}}{32.04 \text{ g}}\right)\left(\dfrac{3 \text{ mol } O_2}{2 \text{ mol } CH_3OH}\right)\left(\dfrac{32.00 \text{ g}}{\text{mol}}\right) = 70.7 \text{ g } O_2$

40. The balanced equation is $7\,H_2O_2 + N_2H_4 \rightarrow 2\,HNO_3 + 8\,H_2O$

   The conversion is:

   (a) $(75\,kg\,N_2H_4)\left(\dfrac{1000\,g}{1\,kg}\right)\left(\dfrac{1\,mol}{32.05\,g}\right)\left(\dfrac{2\,mol\,HNO_3}{1\,mol\,N_2H_4}\right)\left(\dfrac{63.02\,g}{mol}\right) = 2.9 \times 10^5\,g\,HNO_3$

   (b) $(250\,L\,H_2O_2)\left(\dfrac{1000\,mL}{1\,L}\right)\left(\dfrac{1.41\,g}{1\,mL}\right)\left(\dfrac{1\,mol}{34.02\,g}\right)\left(\dfrac{8\,mol\,H_2O}{7\,mol\,H_2O_2}\right)\left(\dfrac{18.02\,g}{mol}\right) = 2.1 \times 10^5\,g\,H_2O$

   (c) $(725\,g\,H_2O_2)\left(\dfrac{1\,mol}{34.02\,g}\right)\left(\dfrac{1\,mol\,N_2H_4}{7\,mol\,H_2O_2}\right)\left(\dfrac{32.05\,g}{mol}\right) = 97.6\,g\,N_2H_4$

   (d) Reaction between $750\,g$ of N_2H_4 and $125\,g$ of H_2O_2.

   Convert each amount to grams of H_2O.

   $(750\,g\,N_2H_4)\left(\dfrac{1\,mol}{32.05\,g}\right)\left(\dfrac{8\,mol\,H_2O}{1\,mol\,N_2H_4}\right)\left(\dfrac{18.02\,g}{mol}\right) = 3.4 \times 10^3\,g\,H_2O$

   $(125\,g\,H_2O_2)\left(\dfrac{1\,mol}{34.02\,g}\right)\left(\dfrac{8\,mol\,H_2O}{7\,mol\,H_2O_2}\right)\left(\dfrac{18.02\,g}{mol}\right) = 75.7\,g\,H_2O$

   $75.7\,g\,H_2O$ can be produced.

   (e) Since H_2O_2 is the limiting reactant, N_2H_4 is in excess.

   $(125g\,H_2O_2)\left(\dfrac{1\,mol}{34.02\,g}\right)\left(\dfrac{1\,mol\,N_2H_4}{7\,mol\,H_2O_2}\right)\left(\dfrac{32.05\,g}{mol}\right) = 16.8\,g\,N_2H_4$ reacted

   $750\,g\,N_2H_4$ given $- 16.8\,g\,N_2H_4$ used $= 730\,g\,N_2H_4$ remaining

41. The balanced equation is

   $16\,HCl + 2\,KMnO_4 \rightarrow 5\,Cl_2 + 2\,KCl + 2\,MnCl_2 + 8\,H_2O$

   (a) Reaction between $25\,g\,KMnO_4$ and $85\,g\,HCl$. Convert each to moles of $MnCl_2$.

   $(25\,g\,KMnO_4)\left(\dfrac{1\,mol\,KMnO_4}{158.04\,g\,KMnO_4}\right)\left(\dfrac{2\,mol\,MnCl_2}{2\,mol\,KMnO_4}\right) = 0.16\,mol\,MnCl_2$

   $(85\,g\,HCl)\left(\dfrac{1\,mol}{36.46\,g}\right)\left(\dfrac{2\,mol\,MnCl_2}{16\,mol\,HCl}\right) = 0.29\,mol\,MnCl_2$

   $KMnO_4$ is the limiting reactant; $0.16\,mol\,MnCl_2$ produced.

   (b) $(75\,g\,KCl)\left(\dfrac{1\,mol}{74.55\,g}\right)\left(\dfrac{8\,mol\,H_2O}{2\,mol\,KCl}\right)\left(\dfrac{18.02\,g}{mol}\right) = 73\,g\,H_2O$

   (c) $(150\,g\,HCl)\left(\dfrac{1\,mol}{36.46\,g}\right)\left(\dfrac{5\,mol\,Cl_2}{16\,mol\,HCl}\right)\left(\dfrac{70.90\,g}{mol}\right) = 91\,g\,Cl_2$

   Theoretical yield is $91\,g\,Cl_2$; Percent yield: $\left(\dfrac{75\,g}{91\,g}\right)(100) = 82\%$ yield

(d)   Reaction between 25 g HCl and 25 g $KMnO_4$. Convert each amount to grams of Cl_2.

$$(25\,\text{g HCl})\left(\frac{1\,\text{mol}}{36.46\,\text{g}}\right)\left(\frac{5\,\text{mol Cl}_2}{16\,\text{mol HCl}}\right)\left(\frac{70.90\,\text{g}}{\text{mol}}\right) = 15\,\text{g Cl}_2$$

$$(25\,\text{g KMnO}_4)\left(\frac{1\,\text{mol}}{158.04\,\text{g}}\right)\left(\frac{5\,\text{mol Cl}_2}{2\,\text{mol KMnO}_4}\right)\left(\frac{70.90\,\text{g}}{\text{mol}}\right) = 28\,\text{g Cl}_2$$

HCl is the limiting; $KMnO_4$ is in excess; 15 g Cl_2 will be produced.

(e)   Calculate the mass of unreacted $KMnO_4$:

$$(25\,\text{g HCl})\left(\frac{1\,\text{mol}}{36.46\,\text{g}}\right)\left(\frac{2\,\text{mol KMnO}_4}{16\,\text{mol HCl}}\right)\left(\frac{158.0\,\text{g}}{\text{mol}}\right) = 14\,\text{g KMnO}_4 \text{ will react.}$$

Unreacted $KMnO_4 = 25\,\text{g} - 14\,\text{g} = 11\,\text{g KMnO}_4$ remain unreacted.

42.   The balanced equation is

$$4\,\text{Ag} + 2\,\text{H}_2\text{S} + \text{O}_2 \rightarrow 2\,\text{Ag}_2\text{S} + 2\,\text{H}_2\text{O}$$

(a) $$(1.1\,\text{g Ag})\left(\frac{1\,\text{mol}}{107.9\,\text{g}}\right)\left(\frac{2\,\text{mol Ag}_2\text{S}}{4\,\text{mol Ag}}\right)\left(\frac{247.9\,\text{g}}{\text{mol}}\right) = 1.3\,\text{g Ag}_2\text{S}$$

$$(0.14\,\text{g H}_2\text{S})\left(\frac{1\,\text{mol}}{34.09\,\text{g}}\right)\left(\frac{2\,\text{mol Ag}_2\text{S}}{2\,\text{mol H}_2\text{S}}\right)\left(\frac{247.9\,\text{g}}{\text{mol}}\right) = 1.0\,\text{g Ag}_2\text{S}$$

$$(0.080\,\text{g O}_2)\left(\frac{1\,\text{mol}}{32.00\,\text{g}}\right)\left(\frac{2\,\text{mol Ag}_2\text{S}}{1\,\text{mol O}_2}\right)\left(\frac{247.9\,\text{g}}{\text{mol}}\right) = 1.2\,\text{g Ag}_2\text{S}$$

H_2S is limiting; 1.0 g Ag_2S forms.

(b) $$(1.1\,\text{g Ag})\left(\frac{1\,\text{mol}}{107.9\,\text{g}}\right)\left(\frac{2\,\text{mol H}_2\text{S}}{4\,\text{mol Ag}}\right)\left(\frac{34.09\,\text{g}}{\text{mol}}\right) = 0.17\,\text{g H}_2\text{S reacts}$$

0.17 g H_2S − 0.14 g H_2S = 0.03 grams more H_2S needed to completely react Ag.

43.   Limiting reactant calculation and percent yield

$$C_3H_5(C_{15}H_{31}CO_2)_3 + 3\,\text{KOH} \rightarrow C_3H_5(OH)_3 + 3\,C_{15}H_{31}CO_2K$$

$$\left(500.0\,\text{g C}_3\text{H}_5(\text{C}_{15}\text{H}_{31}\text{CO}_2)_3\right)\left(\frac{1\,\text{mol}}{807.3\,\text{g}}\right)\left(\frac{3\,\text{mol C}_{15}\text{H}_{31}\text{CO}_2\text{K}}{1\,\text{mol C}_3\text{H}_5(\text{C}_{15}\text{H}_{31}\text{CO}_2)_3}\right)\left(\frac{294.5\,\text{g}}{1\,\text{mol}}\right) = 547.2\,\text{g C}_{15}\text{H}_{31}\text{CO}_2\text{K}$$

$$(500.0\,\text{g KOH})\left(\frac{1\,\text{mol}}{56.11\,\text{g}}\right)\left(\frac{3\,\text{mol C}_{15}\text{H}_{31}\text{CO}_2\text{K}}{3\,\text{mol KOH}}\right)\left(\frac{294.5\,\text{g}}{1\,\text{mol}}\right) = 2624\,\text{g C}_{15}\text{H}_{31}\text{CO}_2\text{K}$$

The limiting reactant is the triglyceride ($C_3H_5(C_{15}H_{31}CO_2)_3$). 547.2 g soap ($C_{15}H_{31}CO_2K$) is the theoretical yield.

$$\text{Percent yield} = \left(\frac{\text{actual yield}}{\text{theoretical yield}}\right)(100) = \left(\frac{361.7\,\text{g}}{547.2\,\text{g}}\right)(100) = 66.10\%\text{ yield of soap}$$

44. Equation: $CaO + H_2O \rightarrow Ca(OH)_2$

$$(35.55\,g\,CaO)\left(\frac{1\,mol}{56.08\,g}\right)\left(\frac{1\,mol\,Ca(OH)_2}{1\,mol\,CaO}\right)\left(\frac{74.10\,g}{1\,mol}\right) = 46.97\,g\,Ca(OH)_2$$

45. (a) $Pb(NO_3)_2(aq) + Na_2CrO_4(aq) \rightarrow 2\,NaNO_3(aq) + PbCrO_4(s)$

(b) $(26.41\,g\,Pb(NO_3)_2)\left(\frac{1\,mol}{331.2\,g}\right)\left(\frac{1\,mol\,PbCrO_4}{1\,mol\,Pb(NO_3)_2}\right)\left(\frac{323.2\,g}{1\,mol}\right) = 25.77\,g\,PbCrO_4$

$(18.33\,g\,Na_2CrO_4)\left(\frac{1\,mol}{162.0\,g}\right)\left(\frac{1\,mol\,PbCrO_4}{1\,mol\,Na_2CrO_4}\right)\left(\frac{323.2\,g}{1\,mol}\right) = 36.57\,g\,PbCrO_4$

Theoretical yield is 25.77 g PbCrO$_4$

$$Percent\ yield = \left(\frac{actual\ yield}{theoretical\ yield}\right)(100) = \left(\frac{21.23\,g}{25.77\,g}\right)(100) - 82.38\,\%\ yield\ of\ PbCrO_4$$

46. $C_2H_5OH + 3\,O_2 \rightarrow 2\,CO_2 + 3\,H_2O$

(a) $(2.5\,mol\,C_2H_5OH)\left(\frac{2\,mol\,CO_2}{1\,mol\,C_2H_5OH}\right) = 5.0\,mol\,CO_2$

$(7.5\,mol\,O_2)\left(\frac{2\,mol\,CO_2}{3\,mol\,O_2}\right) = 5.0\,mol\,CO_2$

Neither reactant is limiting.

$(2.5\,mol\,C_2H_5OH)\left(\frac{3\,mol\,H_2O}{1\,mol\,C_2H_5OH}\right) = 7.5\,mol\,H_2O$

When the reaction is complete, there will be 5.0 mol CO$_2$ and 7.5 mol H$_2$O.

(b) $(225\,g\,C_2H_5OH)\left(\frac{1\,mol}{46.07\,g}\right)\left(\frac{2\,mol\,CO_2}{1\,mol\,C_2H_5OH}\right)\left(\frac{44.01\,g}{mol}\right) = 430.\,g\,CO_2$

$(225\,g\,C_2H_5OH)\left(\frac{1\,mol}{46.07\,g}\right)\left(\frac{3\,mol\,H_2O}{1\,mol\,C_2H_5OH}\right)\left(\frac{18.02\,g}{mol}\right) = 264\,g\,H_2O$

47. The balanced equation is $Zn + 2\,HCl \rightarrow ZnCl_2 + H_2$

180.0 g Zn − 35 g Zn = 145 g Zn reacted with HCl

(a) $(145\,g\,Zn)\left(\frac{1\,mol}{65.39\,g}\right)\left(\frac{1\,mol\,H_2}{1\,mol\,Zn}\right)\left(\frac{2.016\,g}{mol}\right) = 4.47\,g\,H_2\ produced$

(b) $(145\,g\,Zn)\left(\frac{1\,mol}{65.39\,g}\right)\left(\frac{2\,mol\,HCl}{1\,mol\,Zn}\right)\left(\frac{36.46\,g}{mol}\right) = 162\,g\,HCl\ reacted$

(c) $(180.0\,g\,Zn)\left(\frac{1\,mol}{65.39\,g}\right)\left(\frac{2\,mol\,HCl}{1\,mol\,Zn}\right)\left(\frac{36.46\,g}{mol}\right) = 201\,g\,HCl\ reacts$

201 g − 162 g = 39 g more HCl needed to react with the 180.0 g Zn

48. $Fe(s) + CuSO_4(aq) \rightarrow Cu(s) + FeSO_4(aq)$
        2.0 mol   3.0 mol

(a)   2.0 mol Fe react with 2.0 mol $CuSO_4$ to yield 2.0 mol Cu and 2.0 mol $FeSO_4$. 1.0 mol $CuSO_4$ is unreacted. At the completion of the reaction, there will be 2.0 mol Cu, 2.0 mol $FeSO_4$, and 1.0 mol $CuSO_4$.

(b)   Determine which reactant is limiting and then calculate the g $FeSO_4$ produced from that reactant.

$$(20.0\,g\,Fe)\left(\frac{1\,mol}{55.85\,g}\right)\left(\frac{1\,mol\,Cu}{1\,mol\,Fe}\right)\left(\frac{63.55\,g}{mol}\right) = 22.8\,g\,Cu$$

$$(40.0\,g\,CuSO_4)\left(\frac{1\,mol}{159.6\,g}\right)\left(\frac{1\,mol\,Cu}{1\,mol\,CuSO_4}\right)\left(\frac{63.55\,g}{mol}\right) = 15.9\,g\,Cu$$

Since $CuSO_4$ produces less Cu, it is the limiting reactant. Determine the mass of $FeSO_4$. produced from 40.0 g $CuSO_4$.

$$(40.0\,g\,CuSO_4)\left(\frac{1\,mol}{159.6\,g}\right)\left(\frac{1\,mol\,FeSO_4}{1\,mol\,CuSO_4}\right)\left(\frac{151.9\,g}{mol}\right) = 38.1\,g\,FeSO_4\,produced$$

Calculate the mass of unreacted Fe.

$$(40.0\,g\,CuSO_4)\left(\frac{1\,mol}{159.6\,g}\right)\left(\frac{1\,mol\,Fe}{1\,mol\,FeSO_4}\right)\left(\frac{55.85\,g}{mol}\right) = 14.0\,g\,Fe\,will\,react$$

Unreacted Fe = 20.0 g $-$ 14.0 g = 6.0 g. Therefore, at the completion of the reaction, 15.9 g Cu, 38.1 g $FeSO_4$, 6.0 g Fe, and no $CuSO_4$ remain.

49.   Limiting reactant calculation

$CO(g) + 2\,H_2(g) \rightarrow CH_3OH(l)$

Reaction between 40.0 g CO and 10.0 g H_2: determine the limiting reactant by calculating the amount of CH_3OH that would be formed from each reactant.

$$(40.0\,g\,CO)\left(\frac{1\,mol}{28.01\,g}\right)\left(\frac{1\,mol\,CH_3OH}{1\,mol\,CO}\right)\left(\frac{32.04\,g}{mol}\right) = 45.8\,g\,CH_3OH$$

$$(10.0\,g\,H_2)\left(\frac{1\,mol}{2.016\,g}\right)\left(\frac{1\,mol\,CH_3OH}{2\,mol\,H_2}\right)\left(\frac{32.04\,g}{mol}\right) = 79.5\,g\,CH_3OH$$

CO is limiting; H_2 is in excess; 45.8 g CH_3OH will be produced.

Calculate the mass of unreacted H_2:

$$(40.0\,g\,CO)\left(\frac{1\,mol}{28.01\,g}\right)\left(\frac{2\,mol\,H_2}{1\,mol\,CO}\right)\left(\frac{2.016\,g}{mol}\right) = 5.76\,g\,H_2\,react$$

10.0 g H_2 $-$ 5.76 g H_2 = 4.2 g H_2 remain unreacted

50. The balanced equation is $C_6H_{12}O_6 \rightarrow 2\,C_2H_5OH + 2\,CO_2$
    (a) First calculate the theoretical yield.

$$(750\,g\,C_6H_{12}O_6)\left(\frac{1\,mol}{180.2\,g}\right)\left(\frac{2\,mol\,C_2H_5OH}{1\,mol\,C_6H_{12}O_6}\right)\left(\frac{46.07\,g}{mol}\right) = 3.8 \times 10^2\,g\,C_2H_5OH\ \text{(theoretical yield)}$$

Then take 84.6% of the theoretical yield to obtain the actual yield.

$$\text{actual yield} = \frac{(\text{theoretical yield})(84.6)}{100} = \frac{(3.8 \times 10^2\,g\,C_2H_5OH)(84.6)}{100}$$

$$= 3.2 \times 10^2\,g\,C_2H_5OH$$

    (b) 475 g C_2H_5OH represents 84.6% of the theoretical yield. Calculate the theoretical yield.

$$\text{theoretical yield} = \frac{475\,g}{0.846} = 561\,g\,C_2H_5OH$$

Now calculate the g $C_6H_{12}O_6$ needed to produce 561 g C_2H_5OH.

$$(561\,g\,C_2H_5OH)\left(\frac{1\,mol}{46.07\,g}\right)\left(\frac{1\,mol\,C_6H_{12}O_6}{2\,mol\,C_2H_5OH}\right)\left(\frac{180.2\,g}{mol}\right) = 1.10 \times 10^3\,g\,C_6H_{12}O_6$$

51. The balanced equations are:

$CaCl_2(aq) + 2\,AgNO_3(aq) \rightarrow Ca(NO_3)_2(aq) + 2\,AgCl(s)$

$MgCl_2(aq) + 2\,AgNO_3(aq) \rightarrow Mg(NO_3)_2(aq) + 2\,AgCl(s)$

1 mol of each salt will produce the same amount (2 mol) of AgCl. $MgCl_2$ has a higher percentage of Cl than $CaCl_2$ because Mg has a lower atomic mass than Ca. Therefore, on an equal mass basis, $MgCl_2$ will produce more AgCl than will $CaCl_2$.

Calculations show that 1.00 g $MgCl_2$ produces 3.01 g AgCl, and 1.00 g $CaCl_2$ produces 2.56 g AgCl.

52. The balanced equation is $Li_2O + H_2O \rightarrow 2\,LiOH$

The conversion is: g $H_2O \rightarrow$ mol $H_2O \rightarrow$ mol $Li_2O \rightarrow$ g $Li_2O \rightarrow$ kg Li_2O

$$\left(\frac{2500\,g\,H_2O}{\text{astronaut day}}\right)\left(\frac{1\,mol}{18.02\,g}\right)\left(\frac{1\,mol\,Li_2O}{1\,mol\,H_2O}\right)\left(\frac{29.88\,g}{mol}\right)\left(\frac{1\,kg}{1000\,g}\right) = \frac{4.1\,kg\,Li_2O}{\text{astronaut day}}$$

$$\left(\frac{4.1\,kg\,Li_2O}{\text{astronaut day}}\right)(30\,\text{days})(3\,\text{astronauts}) = 3.7 \times 10^2\,kg\,Li_2O$$

53. The balanced equation is

$H_2SO_4 + 2\,NaCl \rightarrow Na_2SO_4 + 2\,HCl$

First calculate the g HCl to be produced

$$(20.0\,L\,HCl\,\text{solution})\left(\frac{1000\,mL}{1\,L}\right)\left(\frac{1.20\,g}{1.00\,mL}\right)(0.420) = 1.01 \times 10^4\,g\,HCl$$

Then calculate the g H_2SO_4 required to produce the HCl

$$(1.01 \times 10^4 \text{ g HCl})\left(\frac{1 \text{ mol}}{36.46 \text{ g}}\right)\left(\frac{1 \text{ mol } H_2SO_4}{2 \text{ mol HCl}}\right) = 139 \text{ g } H_2SO_4$$

Finally, calculate the kg H_2SO_4 (96%)

$$(1.39 \text{ g } H_2SO_4)\left(\frac{1.00 \text{ g } H_2SO_4 \text{ solution}}{0.96 \text{ g } H_2SO_4}\right)\left(\frac{1 \text{ kg}}{1000 \text{ g}}\right) = 0.14 \text{ kg concentrated } H_2SO_4$$

54. $SiO_2 + 3 \text{ C} \rightarrow \text{SiC} + 2 \text{ CO}$

The conversion is kg $SiO_2 \rightarrow$ g $SiO_2 \rightarrow$ mol $SiO_2 \rightarrow$ mol SiC $\rightarrow$ g SiC $\rightarrow$ kg SiC

$$(250.0 \text{ kg } SiO_2)\left(\frac{1000 \text{ g}}{1 \text{ kg}}\right)\left(\frac{1 \text{ mol}}{60.09 \text{ g}}\right)\left(\frac{1 \text{ mol SiC}}{1 \text{ mol } SiO_2}\right)\left(\frac{40.10 \text{ g}}{1 \text{ mol}}\right)\left(\frac{1 \text{ kg}}{1000 \text{ g}}\right) = 166.8 \text{ kg SiC}$$

55. Percent yield of H_2SO_4

$$(100.0 \text{ g S})\left(\frac{1 \text{ mol}}{32.07 \text{ g}}\right) = 3.118 \text{ mol S to start with}$$

3.118 mol S $\rightarrow$ 3.118 mol SO_2 $-$ 10% $= 2.806$ mol SO_2

$$\begin{pmatrix} 3.118 \\ -0.3118 \\ \hline 2.8062 \end{pmatrix}$$

2.806 mol SO_2 $\rightarrow$ 2.806 mol SO_3 $-$ 10% $= 2.525$ mol SO_3

$$\begin{pmatrix} 2.806 \\ -0.2806 \\ \hline 2.5254 \end{pmatrix}$$

2.525 mol SO_3 $\rightarrow$ 2.525 mol H_2SO_4 $-$ 10% $= 2.273$ mol H_2SO_4

$$\begin{pmatrix} 2.525 \\ -0.2525 \\ \hline 2.2725 \end{pmatrix}$$

$$(2.273 \text{ mol } H_2SO_4)\left(\frac{98.09 \text{ g}}{\text{mol}}\right) = 223.0 \text{ g } H_2SO_4 \text{ formed}$$

$$(3.118 \text{ mol S})\left(\frac{1 \text{ mol } H_2SO_4}{1 \text{ mol S}}\right) = 3.118 \text{ mol } H_2SO_4 \text{(theoretical yield)}$$

$$\left(\frac{2.273 \text{ mol } H_2SO_4}{3.118 \text{ mol } H_2SO_4}\right)(100) = 72.90\% \text{ yield}$$

Alternate Solution:

Calculation of yield. There are three chemical steps to the formation of H_2SO_4. Each step has a 10% loss of yield.

Step 1:   100% yield $-$ 10% $=$ 90.00% yield
Step 2:   90.00% yield $-$ 10% $=$ 81.00% yield
Step 3:   81.00% yield $-$ 10% $=$ 72.90% yield

Now calculate the grams of product. One mole of sulfur will yield a maximum of 1 mol H_2SO_4. Therefore 3.118 mol S will give a maximum of 3.118 mol H_2SO_4.

$$(3.118 \text{ mol S})\left(\frac{1 \text{ mol } H_2SO_4}{1 \text{ mol S}}\right)\left(\frac{98.09 \text{ g}}{\text{mol}}\right)(0.7290) = 223.0 \text{ g } H_2SO_4 \text{ yield}$$

56. According to the equations, the moles of CO_2 come from both reactions and the moles H_2O come from only the first reaction.

So the mol $NaHCO_3 = 2 \times$ mol $H_2O = 2 \times 0.0357$ mol $= 0.0714$ mol $NaHCO_3$

$$(0.0714 \text{ mol } NaHCO_3)\left(\frac{84.01 \text{ g}}{\text{mol}}\right) = 6.00 \text{ g } NaHCO_3 \text{ in the sample}$$

$(10.00 \text{ g } NaHCO_3 + Na_2CO_3) - 6.00 \text{ g } NaHCO_3 = 4.00 \text{ g } Na_2CO_3$ in the sample

$$\left(\frac{6.00 \text{ g } NaHCO_3}{10.00 \text{ g}}\right)(100) - 60.0\% \text{ } NaHCO_3$$

$$\left(\frac{4.00 \text{ g } Na_2CO_3}{10.00 \text{ g}}\right)(100) = 40.0\% \text{ } Na_2CO_3$$

57. (a) ○ Hydrogen ◎ Carbon ● Oxygen

O_2 is the limiting reagent. Because each CH_4 requires 2 O_2 molecules, there will still be some oxygen gas left at the end of the reaction.

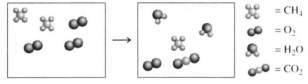

$= CH_4$
$= O_2$
$= H_2O$
$= CO_2$

58. The balanced equation is $2 \text{ } KClO_3 \rightarrow 2 \text{ } KCl + 3 \text{ } O_2$

12.82 g mixture $-$ 9.45 g residue $=$ 3.37 g O_2 lost by heating

Because the O_2 lost came only from $KClO_3$, we can use it to calculate the amount of $KClO_3$ in the mixture

The conversion is: g $O_2 \rightarrow$ mol $O_2 \rightarrow$ mol $KClO_3 \rightarrow$ g $KClO_3$

$$(3.37 \text{ g } O_2)\left(\frac{1 \text{ mol}}{32.00 \text{ g}}\right)\left(\frac{2 \text{ mol } KClO_3}{3 \text{ mol } O_2}\right)\left(\frac{122.6 \text{ g}}{\text{mol}}\right) = 8.61 \text{ g } KClO_3 \text{ in the mixture}$$

$$\left(\frac{8.61 \text{ g } KClO_3}{12.82 \text{ g sample}}\right)(100) = 67.2\% \text{ } KClO_3$$

59. The balanced equation is

$Al(OH)_3(s) + 3 \text{ } HCl(aq) \rightarrow AlCl_3(aq) + 3 \text{ } H_2O(l)$

The conversion is: L HCl $\rightarrow$ g HCl $\rightarrow$ mol HCl $\rightarrow$ mol $Al(OH)_3 \rightarrow$ g $Al(OH)_3$

$$\left(\frac{2.5 \text{ L}}{\text{day}}\right)\left(\frac{3.0 \text{ g } HCl}{\text{L}}\right)\left(\frac{1 \text{ mol}}{36.46 \text{ g}}\right)\left(\frac{1 \text{ mol } Al(OH)_3}{3 \text{ mol } HCl}\right)\left(\frac{78.00 \text{ g}}{\text{mol}}\right) = 5.3 \text{ g } Al(OH)_3/\text{day}$$

Now calculate the number of 400. mg tablets that can be made from 5.3 g Al(OH)$_3$

$$\left(\frac{5.3\,\text{g Al(OH)}_3}{\text{day}}\right)\left(\frac{1000\,\text{mg}}{\text{g}}\right)\left(\frac{1\,\text{tablet}}{400.\,\text{mg Al(OH)}_3}\right) = 13\,\text{tablets/day}$$

60. $4\,\text{P} + 5\,\text{O}_2 \rightarrow \text{P}_4\text{O}_{10}$

$\text{P}_4\text{O}_{10} + 6\,\text{H}_2\text{O} \rightarrow 4\,\text{H}_3\text{PO}_4$

In the first reaction:

$$(20.0\,\text{g P})\left(\frac{1\,\text{mol}}{30.97\,\text{g}}\right) = 0.646\,\text{mol P}$$

$$(30.0\,\text{g O}_2)\left(\frac{1\,\text{mol}}{32.00\,\text{g}}\right) = 0.938\,\text{mol O}_2$$

This is a ratio of $\dfrac{0.646\,\text{mol P}}{0.938\,\text{mol O}_2} = \dfrac{3.44\,\text{mol P}}{5.00\,\text{mol O}_2}$

Therefore, P is the limiting reactant and the P$_4$O$_{10}$ produced is:

$$(0.646\,\text{mol P})\left(\frac{1\,\text{mol P}_4\text{O}_{10}}{4\,\text{mol P}}\right) = 0.162\,\text{mol P}_4\text{O}_{10}$$

In the second reaction:

$$(15.0\,\text{g H}_2\text{O})\left(\frac{1\,\text{mol}}{18.02\,\text{g}}\right) = 0.832\,\text{mol H}_2\text{O}$$

and we have 0.162 mol P$_4$O$_{10}$. The ratio of $\dfrac{\text{H}_2\text{O}}{\text{P}_4\text{O}_{10}}$ is $\dfrac{0.832\,\text{mol}}{0.162\,\text{mol}} = \dfrac{5.14\,\text{mol}}{1.00\,\text{mol}}$

Therefore, H$_2$O is the limiting reactant and the H$_3$PO$_4$ produced is:

$$(0.832\,\text{mol H}_2\text{O})\left(\frac{4\,\text{mol H}_3\text{PO}_4}{6\,\text{mol H}_2\text{O}}\right)\left(\frac{97.99\,\text{g}}{\text{mol}}\right) = 54.4\,\text{g H}_3\text{PO}_4$$

61. Limiting reactant-like calculation

Calculate the number of bags of supplement that can be made from each starting material.

25 lb oats + 17.5 lb molasses + 30.0 lb alfalfa + 28.5 lb apples → 1 bag supplement

$$(4.20\,\text{tons oats})\left(\frac{2000\,\text{lb}}{1\,\text{ton}}\right)\left(\frac{1\,\text{bag}}{25\,\text{lb oats}}\right) = 336\,\text{bags supplement}$$

$$(2490\,\text{gal molasses})\left(\frac{3.785\,\text{L}}{1\,\text{gal}}\right)\left(\frac{1000\,\text{mL}}{1\,\text{L}}\right)\left(\frac{1.46\,\text{g}}{1\,\text{mL}}\right)\left(\frac{1\,\text{lb}}{453.6\,\text{g}}\right)\left(\frac{1\,\text{bag}}{17.5\,\text{lb molasses}}\right) = 1730\,\text{bags supplement}$$

$$(250.0 \text{ crates apple})\left(\frac{12.5 \text{ kg apple}}{1 \text{ crate}}\right)\left(\frac{1 \text{ lb}}{0.4536 \text{ kg}}\right)\left(\frac{1 \text{ bag}}{28.5 \text{ lb apple}}\right) = 242 \text{ bags supplement}$$

$$(350.0 \text{ bales})\left(\frac{62.8 \text{ kg alfalfa}}{1 \text{ bale}}\right)\left(\frac{1 \text{ lb}}{0.4536 \text{ kg}}\right)\left(\frac{1 \text{ bag}}{30.0 \text{ lb alfalfa}}\right) = 1620 \text{ bags supplement}$$

The limiting reactant is apples. 242 bags of feed supplement is the theoretical yield.

$$\text{Income} = (242 \text{ bag})\left(\frac{\$2.25}{1 \text{ bag}}\right) = \$545$$

CHAPTER 10

MODERN ATOMIC THEORY AND THE PERIODIC TABLE
SOLUTIONS TO REVIEW QUESTIONS

1. **Wavelength** is defined as the distance between consecutive peaks in a wave. It is generally symbolized by the Greek letter lambda, λ.
    Frequency is a measure of the number of waves that pass a specific point every second. It is generally symbolized by the Greek letter nu, ν.

2. Visible light ranges in wavelength from about 4×10^{-7} m to 7×10^{-7} m. Red light has a longer wavelength than blue light.

3. Photon

4. An electron orbital is a region in space around the nucleus of an atom where an electron is most probably found.

5. The electrons in the atom are located in the orbitals with the lowest energies.

6. The main difference is that the Bohr orbit has an electron traveling a specific path around the nucleus while an orbital is a region in space where the electron is most probably found.

7. Bohr's model was inadequate since it could not account for atoms more complex than hydrogen. It was modified by Schrödinger into the modern concept of the atom in which electrons exhibit wave and particle properties. The motion of electrons is determined only by probability functions as a region in space, or a cloud surrounding the nucleus.

8. Both 1s and 2s orbitals are spherical in shape and located symmetrically around the nucleus. The sizes of the spheres are different—the radius of the 2s orbital is larger than the 1s. The electrons in 2s orbitals are located further from the nucleus.

9. The letters used to designate the energy sublevels are s, p, d, and f.

10. 1s, 2s, 2p, 3s, 3p, 4s, 3d, 4p.

11. s–2 electrons per shell
    p–6 electrons per shell after the first energy level
    d–10 electrons per shell after the second energy level.

12. *s* orbital.

p orbitals

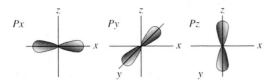

13. A second electron may enter an orbital already occupied by an electron if its spin is opposite that of the electron already in the orbital and all other orbitals of the same sublevel contain an electron.

14. The valence shell is the outermost energy level of an atom.

15. Valence electrons are the electrons located in the outermost energy level of an atom. Valence electrons are involved in bonding. They are important because ion formation involves the gain or loss of valence electrons. Covalent bonding involves sharing valence electrons.

16. 4 is the fourth principal energy level

    f indicates the energy sublevel

    3 indicates the number of electrons in the f sublevel

17. Ir, Zr, and Ag are not representative elements; they are transition elements.

18. Elements in the p-block all have one to six electrons in the p sublevel.

19.

Atomic #	Symbol
6	C
7	N
8	O
15	P
33	As

Elements with atomic numbers 7, 15, and 33 are all in the same group on the periodic table. They have an outermost electron structure of s^2p^3.

20. The first three elements that have six electrons in their outermost energy level are O, S, and Se.

21. The greatest number of elements in any period is 32. The 6th period has this number of elements.

22. The elements in Group A always have their last electrons in the outermost energy level, while the last electrons in Group B lie in an inner level.

23. Pairs of elements which are out of sequence with respect to atomic masses are: Ar and K; Co and Ni; Te and I; Th and Pa; U and Np; Pu and Am; Lr and Rf; Sg and Bh.

24. Dimitri Mendeleev, of Russia and Lothar Meyer, of Germany both independently published results that led to the current periodic table.

25. Dimitri Mendeleev is credited with being the father of the modern periodic table.

SOLUTIONS TO EXERCISES

1.

	Element	Total Electrons	Valence Electrons
(a)	Li	3	1
(b)	Mg	12	2
(c)	Ca	20	2
(d)	F	9	7

2.

	Element	Total Electrons	Valence Electrons
(a)	Na	11	1
(b)	As	33	5
(c)	P	15	5
(d)	Al	13	3

3. Electron configurations

    (a)   Sc $1s^2 2s^2 2p^6 3s^2 3p^6 4s^2 3d^1$

    (b)   Rb $1s^2 2s^2 2p^6 3s^2 3p^6 4s^2 3d^{10} 4p^6 5s^1$

    (c)   Br $1s^2 2s^2 2p^6 3s^2 3p^6 4s^2 3d^{10} 4p^5$

    (d)   S $1s^2 2s^2 2p^6 3s^2 3p^4$

4.   (a)   Mn $1s^2 2s^2 2p^6 3s^2 3p^6 4s^2 3d^5$

    (b)   Kr $1s^2 2s^2 2p^6 3s^2 3p^6 4s^2 3d^{10} 4p^6$

    (c)   Ga $1s^2 2s^2 2p^6 3s^2 3p^6 4s^2 3d^{10} 4p^1$

    (d)   B $1s^2 2s^2 2p^1$

5. The spectral lines of hydrogen are produced by energy emitted when the electron from a hydrogen atom, which has absorbed energy, falls from a higher energy level to a lower energy level (closer to the nucleus).

6. Bohr said that a number of orbits were available for electrons, each corresponding to an energy level. When an electron falls from a higher energy orbit to a lower energy orbit, energy is given off as a specific wavelength of light. Only those energies in the visible range are seen in the hydrogen spectrum. Each line corresponds to a change from one orbit to another.

7. 16 orbitals in the 4th principal energy level; 1 in s, 3 in p, 5 in d, and 7 in f. The s and p orbitals are in the 4th period, the d orbitals are in the 5th period, and the f orbitals are in the 6th period.

8. 18 electrons in third energy level; 2 in s, 6 in p, 10 in d

9. (a) $^{14}_{7}\text{N}$ [↑↓] [↑↓] [↑][↑][↑]

    (b) $^{35}_{17}\text{Cl}$ [↑↓] [↑↓] [↑↓][↑↓][↑↓] [↑↓] [↑↓][↑↓][↑]

    (c) $^{65}_{30}\text{Zn}$ [↑↓] [↑↓] [↑↓][↑↓][↑↓] [↑↓] [↑↓][↑↓][↑↓]
         [↑↓] [↑↓][↑↓][↑↓][↑↓][↑↓]

    (d) $^{91}_{40}\text{Zr}$ [↑↓] [↑↓] [↑↓][↑↓][↑↓] [↑↓] [↑↓][↑↓][↑↓]
         [↑↓] [↑↓][↑↓][↑↓][↑↓][↑↓] [↑↓][↑↓][↑↓][↑↓]
         [↑][↑][ ][ ][ ]

    (e) $^{127}_{53}\text{I}$ [↑↓] [↑↓] [↑↓][↑↓][↑↓] [↑↓] [↑↓][↑↓][↑↓]
         [↑↓] [↑↓][↑↓][↑↓][↑↓][↑↓] [↑↓][↑↓][↑↓][↑↓]
         [↑↓][↑↓][↑↓][↑↓][↑↓] [↑↓][↑↓][↑]

10. (a) $^{28}_{14}\text{Si}$ [↑↓] [↑↓] [↑↓][↑↓][↑↓] [↑↓] [↑][↑][ ]

    (b) $^{32}_{16}\text{S}$ [↑↓] [↑↓] [↑↓][↑↓][↑↓] [↑↓] [↑↓][↑][↑]

    (c) $^{40}_{18}\text{Ar}$ [↓↑] [↓↑] [↓↑][↓↑][↓↑] [↓↑] [↓↑][↓↑][↓↑]

    (d) $^{51}_{23}\text{V}$ [↑↓] [↑↓] [↑↓][↑↓][↑↓] [↑↓] [↑↓][↑↓][↑↓]
         [↑↓] [↑][↑][↑][ ][ ]

    (e) $^{31}_{15}\text{P}$ [↑↓] [↑↓] [↑↓][↑↓][↑↓] [↑↓] [↑][↑][↑]

11. (a) O $1s^2 2s^2 2p^4$

    (b) Ca $1s^2 2s^2 2p^6 3s^2 3p^6 4s^2$

    (c) Ar $1s^2 2s^2 2p^6 3s^2 3p^6$

    (d) Br $1s^2 2s^2 2p^6 3s^2 3p^6 4s^2 3d^{10} 4p^5$

    (e) Fe $1s^2 2s^2 2p^6 3s^2 3p^6 4s^2 3d^6$

12. (a) Li $1s^2 2s^1$

    (b) P $1s^2 2s^2 2p^6 3s^2 3p^3$

    (c) Zn $1s^2 2s^2 2p^6 3s^2 3p^6 4s^2 3d^{10}$

    (d) Na $1s^2 2s^2 2p^6 3s^1$

    (e) K $1s^2 2s^2 2p^6 3s^2 3p^6 4s^1$

13. (a) Incorrect – the 2 p sublevel should be completely filled before the 3s sublevel is populated.

[↑↓] [↑↓] [↑↓|↑↓|↑↓] [ ]

(b) Correct

(c) Correct

(d) Incorrect – electrons in the 3d sublevel should not be paired until all 3d orbitals are populated.

[↑↓] [↑↓] [↑↓|↑↓|↑↓] [↑↓] [↑↓|↑↓|↑↓] [↑↓] [↑|↑|↑|↑|↑]

14. (a) Correct

(b) Incorrect – the 3d sublevel should be populated before the 4p sublevel.

[↑↓] [↑↓] [↑↓|↑↓|↑↓] [↑↓] [↑↓|↑↓|↑↓] [↑↓] [↑↓|↑↓|↑↓|↑|↑]

(c) Incorrect – the second electron in the 4s orbital must be represented by a down arrow

[↑↓] [↑↓] [↑↓|↑↓|↑↓] [↑↓] [↑↓|↑↓|↑↓] [↑↓]

(d) Correct

15. (a) Neon   (c) Gallium

(b) Phosphorus   (d) Manganese

16. (a) Nitrogen   (c) Calcium

(b) Nickel   (d) Sulfur

17. (a) fluorine, F   (c) sulfur, S

(b) sodium, Na   (d) nickel, Ni

18. (a) boron, B   (c) lead, Pb

(b) silicon, Si   (d) tellurium, Te

19. (a) Titanium (Ti)

[↑↓] [↑↓] [↑↓|↑↓|↑↓] [↑↓] [↑↓|↑↓|↑↓] [↑↓]
[↑|↑| | |]

(b) Argon (Ar)

[↑↓] [↑↓] [↑↓|↑↓|↑↓] [↑↓] [↑↓|↑↓|↑↓]

(c) Arsenic (As)

[↑↓] [↑↓] [↑↓|↑↓|↑↓] [↑↓] [↑↓|↑↓|↑↓] [↑↓]
[↑↓|↑↓|↑↓|↑↓|↑↓] [↑|↑|↑]

(d) Bromine (Br)

[↑↓] [↑↓] [↑↓|↑↓|↑↓] [↑↓] [↑↓|↑↓|↑↓] [↑↓]
[↑↓|↑↓|↑↓|↑↓|↑↓] [↑↓|↑↓|↑]

(e) Manganese (Mn)

[↑↓] [↑↓] [↑↓|↑↓|↑↓] [↑↓] [↑↓|↑↓|↑↓] [↑↓] [↑↓]
[↑|↑|↑|↑|↑]

20. (a) Phosphorus (P)   [↑↓] [↑↓] [↑↓][↑↓][↑↓] [↑↓] [↑][↑][↑]

   (b) Zinc (Zn)   [↑↓] [↑↓] [↑↓][↑↓][↑↓] [↑↓] [↑↓][↑↓][↑↓] [↑↓]
                   [↑↓][↑↓][↑↓][↑↓][↑↓]

   (c) Calcium (Ca)   [↑↓] [↑↓] [↑↓][↑↓][↑↓] [↑↓] [↑↓][↑↓][↑↓] [↑↓]

   (d) Selenium (Se)   [↑↓] [↑↓] [↑↓][↑↓][↑↓] [↑↓] [↑↓][↑↓][↑↓] [↑↓]
                    [↑↓][↑↓][↑↓][↑↓][↑↓] [↑↓][↑][↑]

   (e) Potassium (K)   [↑↓] [↑↓] [↑↓][↑↓][↑↓] [↑↓] [↑↓][↑↓][↑↓] [↑]

21. (a) F   [↑↓] [↑↓] [↑↓][↑↓][↑]

   (b) S   [↑↓] [↑↓] [↑↓][↑↓][↑↓] [↑↓] [↑↓][↑][↑]

   (c) Co   [↑↓] [↑↓] [↑↓][↑↓][↑↓] [↑↓] [↑↓][↑↓][↑↓] [↑↓]
         [↑↓][↑↓][↑][↑][↑]

   (d) Kr   [↑↓] [↑↓] [↑↓][↑↓][↑↓] [↑↓] [↑↓][↑↓][↑↓] [↑↓]
         [↑↓][↑↓][↑↓][↑↓][↑↓] [↑↓][↑↓][↑↓]

   (e) Ru   [↑↓] [↑↓] [↑↓][↑↓][↑↓] [↑↓] [↑↓][↑↓][↑↓] [↑↓]
         [↑↓][↑↓][↑↓][↑↓][↑↓] [↑↓][↑↓][↑↓] [↑↓] [↑↓][↑][↑][↑][↑]

22. (a) Cl   [↑↓] [↑↓] [↑↓][↑↓][↑↓] [↑↓] [↑↓][↑↓][↑]

   (b) Mg   [↑↓] [↑↓] [↑↓][↑↓][↑↓] [↑↓]

   (c) Ni   [↑↓] [↑↓] [↑↓][↑↓][↑↓] [↑↓] [↑↓][↑↓][↑↓] [↑↓]
         [↑↓][↑↓][↑↓][↑][↑]

   (d) Cu   [↑↓] [↑↓] [↑↓][↑↓][↑↓] [↑↓] [↑↓][↑↓][↑↓] [↑↓]
         [↑↓][↑↓][↑↓][↑↓][↑]

   (e) Ba   [↑↓] [↑↓] [↑↓][↑↓][↑↓] [↑↓] [↑↓][↑↓][↑↓] [↑↓]
         [↑↓][↑↓][↑↓][↑↓][↑↓] [↑↓][↑↓][↑↓] [↑↓]
         [↑↓][↑↓][↑↓][↑↓][↑↓] [↑↓][↑↓][↑↓] [↑↓]

23. (a) $^{32}_{16}S$ (b) $^{60}_{28}Ni$

24. (a) (13p 14n) $2e^-8e^-3e^-$ $^{27}_{13}Al$

   (b) (22p 26n) $2e^-8e^-8e^-4e^-$ $^{48}_{22}Ti$

25. The eleventh electron of sodium is located in the third energy level because the first and second levels are filled. Also the properties of sodium are similar to the other elements in Group 1A.

26. The last electron in potassium is located in the fourth energy level because the 4s orbital is at a lower energy level than the 3d orbital. Also the properties of potassium are similar to the other elements in Group 1A.

27. Noble gases all have filled s and p orbitals in the outermost energy level.

28. Noble gases each have filled s and p orbitals in the outermost energy level.

29. Moving from left to right in any period of elements, the atomic number increases by one from one element to the next and the atomic radius generally decreases. Each period (except period 1) begins with an alkali metal and ends with a noble gas. There is a trend in properties of the elements changing from metallic to nonmetaliic from the beginning to the end of the period.

30. The elements in a group have the same number of outer energy level electrons. They are located vertically on the periodic table.

31.   (a)   4   (b)   6   (c)   1   (d)   7   (e)   3

32.   (a)   5   (b)   5   (c)   6   (d)   2   (e)   3

33. The outermost energy level contains one electron in an s orbital.

34. All of these elements have a s^2d^{10} electron configuration in their outermost energy levels.

35.   (a) and (g)

    (b) and (d)

36.   (a) and (f)

    (e) and (h)

37. 12, 38 since they are in the same group or family of elements.

38. 7, 33 since they are in the same group or family of elements.

39.   (a)   K, metal   (c)   S, nonmetal
    (b)   Pu, metal   (d)   Sb, metalloid

40.   (a)   I, nonmetal   (c)   Mo, metal
    (b)   W, metal   (d)   Ge, metalloid

41. Period 6, lanthanide series, contains the first element with an electron in an f orbital.

42. Period 4 Group 3B contains the first element with an electron in a d orbital.

43. Group 7A contain 7 valence electrons.
    Group 7B contain 2 electrons in the outermost level and 5 electrons in an inner d orbital.
    Group A elements are representative while Group B elements are transition elements.

44. Group 3A contain 3 valence electrons.
    Group 3B contain 2 electrons in the outermost level and one electron in an inner d orbital.
    Group A elements are representative while Group B elements are transition elements.

45. (a) arsenic   (c) lithium
    (b) cobalt   (d) chlorine

46. (a) lead   (c) gallium
    (b) samarium   (d) iridium

47. The valence energy level of an atom can be determined by looking at what period the element is in. Period 1 corresponds to valence energy level 1, period 2 to valence energy level 2 and so on. The number of valence electrons for element's 1–18 can be determined by looking at the group number. For example, boron is under Group 3A, therefore it has three valence shell electrons.

48. (a) Mg $\quad 1s^2 2s^2 2p^6 \boxed{3s^2}$

    (b) P $\quad 1s^2 2s^2 2p^6 \boxed{3s^2}\boxed{3p^3}$

    (c) K $\quad 1s^2 2s^2 2p^6 3s^2 3p^6 \boxed{4s^1}$

    (d) F $\quad 1s^2 \boxed{2s^2}\boxed{2p^5}$

    (e) Se $\quad 1s^2 2s^2 2p^6 3s^2 3p^6 \boxed{4s^2} 3d^{10}\boxed{4p^4}$

    (f) N $\quad 1s^2 \boxed{2s^2} \boxed{2p^3}$

49. (a) Na^+, (d) F^-, and (e) Ne have 8 valence electrons.

50. (a)   7A, Halogens   (d)   8A, Noble Gases
    (b)   2A, Alkaline Earth Metals   (e)   8A, Noble Gases
    (c)   1A, Alkali Metals   (f)   1A, Alkali Metals

51. (a) No, the electronic configuration predicted by the periodic table is $1s^2\, 2s^2\, 2p^6\, 3s^2\, 3p^6\, 4s^2\, 3d^4$.

    (b) $\quad (5.00\,\text{cm}^3)\left(\dfrac{7.19\,\text{g}}{1\,\text{cm}^3}\right)\left(\dfrac{1\,\text{mol}}{52.00\,\text{g}}\right)\left(\dfrac{6.022 \times 10^{23}\,\text{atoms}}{1\,\text{mol}}\right) = 4.16 \times 10^{23}\,\text{atoms Cr}$

    (c) $\quad V = \dfrac{4}{3}\pi r^3 = \dfrac{4}{3}\pi\left(1.40 \times 10^{-8}\right)^3 = 1.15 \times 10^{-23}\,\text{cm}^3$

    (d) $\quad (5.00\,\text{cm}^3)\left(\dfrac{1\,\text{atom}}{1.15 \times 10^{-23}\,\text{cm}^3}\right) = 4.35 \times 10^{23}\,\text{atoms Cr}$

52. Each of the different elements has a characteristic emission spectra which will be observed as different colors in the fireworks.

53. Sb $\quad 1s^2 2s^2 2p^6 3s^2 3p^6 4s^2 3d^{10} 4p^6 5s^2 4d^{10} 5p^3$ or [Kr] $5s^2 4d^{10} 5p^3$

54. Bi $\quad 1s^2 2s^2 2p^6 3s^2 3p^6 4s^2 3d^{10} 4p^6 5s^2 4d^{10} 5p^6 6s^2 4f^{14} 5d^{10} 6p^3$ or [Xe] $6s^2 4f^{14} 5d^{10} 6p^3$

55. (a) The four most abundant elements in the
earth's crust, seawater, and air are:
O: $1s^2 2s^2 2p^4$ Si: $1s^2 2s^2 2p^6 3s^2 3p^2$ Al: $1s^2 2s^2 2p^6 3s^2 3p^1$
Fe: $1s^2 2s^2 2p^6 3s^2 3p^6 4s^2 3d^6$

    (b) The five most abundant elements in the human body are:
O: $1s^2 2s^2 2p^4$ C: $1s^2 2s^2 2p^2$ H: $1s^1$ N: $1s^2 2s^2 2p^3$
Ca: $1s^2 2s^2 2p^6 3s^2 3p^6 4s^2$

56. Maximum number of electrons
    (a) Any orbital can hold a maximum of two electrons.
    (b) A d sublevel can hold a maximum of ten electrons.
    (c) The third principal energy level can hold two electrons in 3s, six electrons in 3p, and ten electrons in 3d for a total of eighteen electrons.
    (d) Any orbital can hold a maximum of two electrons.
    (e) An f sublevel can hold a maximum of fourteen electrons.

57. Name of elements

    (a) Magnesium   (b) Phosphorus   (c) Argon

58. Nitrogen has more valence electrons on more energy levels than hydrogen. More varied electron transitions are possible.

59. (a) Ne   (b) Ge   (c) F   (d) N

60. The outermost electron structure for the elements in 7A is $s^2 p^5$.

61. Transition elements are found in Groups 1B–8B, lanthanides and actinides.

62. In transition elements the last electron added is in a d or f orbital. The last electron added in a representative element is in an s or p orbital.

63. Elements 7, 15, 33, 51, and 83 all have 5 electrons in their valence shell.

64. Family names

    (a) Alkali Metals   (b) Alkaline Earth Metals   (c) Halogens

65. Sublevels

    (a) sublevel p   (b) sublevel d   (c) sublevel f

66. (a) Na   representative element   metal
    (b) N   representative element   nonmetal
    (c) Mo   transition element   metal
    (d) Ra   representative element   metal
    (e) As   representative element   metalloid
    (f) Ne   noble gas   nonmetal

67. If element 36 is a noble gas, 35 would be in periodic Group 7A and 37 would be in periodic Group 1A.

68. Answers will vary but should at least include a statement about: (1) Numbering of the elements and their relationship to atomic structure; (2) division of the elements into periods and groups; (3) division of the elements into metals, nonmetals, and metalloids; (4) identification and location of the representative and transition elements.

69. (a)   The two elements are isotopes.
    (b)   The two elements are adjacent to each other in the same period.

70. Most gases are located in the upper right part of the periodic table (H is an exception). They are nonmetals. Liquids show no pattern. Neither do solids, except the vast majority of solids are metals.

71. excited sulfur atom:
    electron configuration: $1s^2 2s^2 2p^6 3s^1 3p^5$
    orbital diagram:

    | ↑↓ | | ↑↓ | | ↑↓ | ↑↓ | ↑↓ | | ↑ | | ↑↓ | ↑↓ | ↑ |

72. Electrons are located in seven principal energy levels. The outermost energy level has one electron residing in a 7s orbital.

73. Metals are located on the left side of the periodic table. The elements in Group 1A have only one valence electron and those in Group 2A have only two valence electrons.
    All metals easily lose their valence electrons to obtain a Noble Gas configuration. Nonmetals are located on the right side of the periodic table where they are only a few electrons short of a noble gas configuration. Nonmetals gain valence electrons to obtain a noble gas configuration.

74. On the periodic table, the period number corresponds to the principal energy level in which the s and p sublevels are filling. The group number of the Main Representative elements corresponds to the number of electrons filling in the principal energy level. Groups 1A and 2A are known as the s-block elements and Groups 3A through 8A are known as the p-block elements.

CHAPTER 11

CHEMICAL BONDS: THE FORMATION
OF COMPOUNDS FROM ATOMS

SOLUTIONS TO REVIEW QUESTIONS

1. smallest Cl, Mg, Na, K, Rb largest.

2. More energy is required for neon because it has a very stable outer electron structure consisting of an octet of electrons in filled orbitals (noble gas electron structure). Sodium, an alkali metal, has a relatively unstable outer electron structure with a single electron in an unfilled orbital. The sodium electron is also farther away from the nucleus and is shielded by more inner electrons than are neon outer electrons.

3. When a third electron is removed from beryllium, it must come from a very stable electron structure corresponding to that of the noble gas, helium. In addition, the third electron must be removed from a +2 beryllium ion, which increases the difficulty of removing it.

4. The first ionization energy decreases from top to bottom because in the successive alkali metals, the outermost electron is farther away from the nucleus and is more shielded from the positive nucleus by additional electron energy levels.

5. The first ionization energy decreases from top to bottom because the outermost electrons in the successive noble gases are farther away from the nucleus and are more shielded by additional inner electron energy levels.

6. Helium has a much higher first ionization energy than does hydrogen because helium has a very stable outer electron structure, consisting of a filled principal energy level.

7. After the first electron is removed from an atom of lithium, much more energy would be required to remove a second electron since that one would come from a noble gas electron configuration, a filled principal energy level.

8. (a) Li > Be   (d) Cl > F
   (b) Rb > K   (e) Se > 0
   (c) Al > P   (f) As > Kr

9. The first element in each group has the smallest radius.

10. Atomic size increases down a column since each successive element has an additional energy level which contains electrons located farther from the nucleus.

11. By losing one electron, a potassium atom acquires a noble gas structure and becomes a K^+ ion. To become a K^{2+} ion requires the loss of a second electron and breaking into the noble gas structure of the K^+ ion. This requires too much energy to generally occur.

12. An aluminum ion has a +3 charge because it has lost 3 electrons in acquiring a noble gas electron structure.

13. A Lewis structure is a representation of the bonding in a compound. Valence electrons are responsible for bonding, therefore they are the only electrons that need to be shown in a Lewis structure.

14. Group   1A   2A   3A   4A   5A   6A   7A

         E·   E: E: ·E: ·E: ·E: ·E:

15. Lewis structure:

     Cs·   Ba: Tl: ·Pb: ·Po: ·At: :Rn:

     Each of these is a representative element and has the same number of electrons in its outer energy level as its periodic group.

16. Valence electrons are the electrons found in the outermost s and p energy levels of an atom.

17. Metals are less electronegative than nonmetals. Therefore, metals lose electrons more easily than nonmetals. So, metals will transfer electrons to nonmetals leaving the metals with a positive charge and the nonmetals with a negative charge.

18. The noble gases are the most stable of all the elements because they have a complete octet (8 electrons) in their valence level. When the elements in Groups 1A, 2A, 6A, and 7A form ions, they do so to establish a stable electron structure.
     (a) The elements in Group 1A lose an electron (obtain a positive charge) in order to achieve a noble gas electron configuration.
     (b) The elements in Group 2A lose electrons (obtain a positive charge) in order to achieve a noble gas electron configuration.
     (c) The elements in Group 6A gain electrons (obtain a negative charge) in order to achieve a noble gas electron configuration.
     (d) The elements in Group 7A gain an electron (obtain a negative charge) in order to achieve a noble gas electron configuration.

19. All compounds must be neutrally charged. This means the overall charge on an ionic compound must be zero.

20. The alkaline earth elements.

21. $Mg_3(PO_4)_2$, $Be_3(PO_4)_2$, $Sr_3(PO_4)_2$, and $Ba_3(PO_4)_2$. Note that the basic formula is the same for all of the elements in the same family when they form an ionic compound with a phosphate ion.

22. When magnesium loses two electrons it will achieve the same electron configuration as the noble gas neon. Elements tend to gain or lose electrons to achieve a noble gas electron configuration.

23. The term molecule is used to describe covalent compounds that have a distinct set of atoms in their structure. Ionic compounds are composed of collections of ions which come together in ratios that balance their charges, but there are no independent molecules formed.

24. A single covalent bond is composed of two electrons. A maximum of three covalent bonds may be formed between any two atoms.

25. A covalent bond is formed by the overlap of the orbitals on individual atoms. There is a sharing of electrons in the region where the orbitals overlap.

26. In a Lewis structure the dash represents a two electron covalent bond.

27. Not all molecules that contain polar bonds are polar due to dipoles that cancel each other by acting in equal and opposite direction.

28. The more electronegative atom in the bond between two atoms will more strongly attract electrons so it will have a partial negative charge ($\delta-$). The less electronegative atom will have a partial positive charge ($\delta+$) because the bonding pair of electrons has been pulled away by the more electronegative atom.

29. (a) Elements with the highest electronegativities are found in the upper right hand corner of the periodic table.
    (b) Elements with the lowest electronegativities are found in the lower left of the periodic table.

30. A Lewis structure is a visual representation of the arrangement of atoms and electrons in a molecule or an ion. It shows how the atoms in a molecule are bonded together.

31. The dots in a Lewis structure represent electrons. The lines represent bonding pairs of electrons.

32. There are times when two or more Lewis structures can be drawn for a single skeleton structure. These different structures all exist and are called resonance structures.

33. Resonance structures

34. The Lewis structure for an ion should be drawn inside of a set of square brackets and the charge of the entire ion specified at the upper right hand outer corner of the brackets. An example is [Na^+].

35. Na_2SO_4 and $CaCO_3$ are two of many possible examples of compounds with both ionic and covalent bonds.

36. The electron pair arrangement is the arrangement of both bonding and nonbonding electrons in a Lewis structure. The molecular shape of a molecule is the three dimensional arrangement of its atoms in space.

SOLUTIONS TO EXERCISES

1.

Drawing 1 K^+ ion   Both particles have the same number of protons, so
Drawing 2   K atom K^+ with one fewer electron is smaller.

2.

Drawing 1   Cl atom   Both particles have the same number of protons, so
Drawing 2 Cl^- ion Cl^- with one more electron is larger.

3. (a) A calcium atom is larger because it has electrons in the 4th shell, while a calcium ion does not. In addition, Ca^{2+} ion has 20 protons and 18 electrons, creating a charge imbalance and drawing the electrons in towards the positive nucleus.

   (b) A chloride ion is larger because it has one more electron than a chlorine atom, in its outer shell. Also, the ion has 17 protons and 18 electrons, creating a charge imbalance, resulting in a lessening of the attraction of the electrons towards the nucleus.

   (c) A magnesium ion is larger than an aluminum ion. Both ions will have 10 electrons in their electron shells, but the aluminium ion will have a greater charge imbalance since it has 13 protons and the magnesium ion has 12 protons. The charge imbalance draws the electrons in more closely to the nucleus.

   (d) A sodium atom is larger than a silicon atom. Sodium and silicon are both in period 3. Going across a period, the radii of atoms decrease.

   (e) A bromide ion is larger than a potassium ion. The bromide ion has 35 protons and 36 electrons, creating a charge imbalance that results in a lessening of the attraction of the electrons towards the nucleus. The potassium ion has 19 protons and 18 electrons, creating a charge imbalance that results in the electrons being drawn more closely to the nucleus.

4. (a) Fe^{2+} has one electron more than Fe^{3+}, so it will have a larger radius.

   (b) A potassium atom is larger than a potassium ion. They both have 19 protons. The potassium atom has 19 electrons. The potassium ion has 18 electrons, creating a charge imbalance that results in the electrons being drawn more closely to the nucleus.

   (c) A chloride ion is larger than a sodium ion. The chloride ion has 17 protons and 18 electrons, creating a charge imbalance that results in a lessening of the attraction of the electrons towards the nucleus. The sodium ion has 11 protons and 10 electrons, creating a charge imbalance that results in the electrons being drawn more closely to the nucleus.

   (d) A strontium atom is larger than an iodine atom. Strontium and iodine are both in period 5. Going across a period, the radii of atoms decrease.

   (e) A rubidium ion is larger than a strontium ion. Both ions will have 36 electrons in their electron shells, but the strontium ion will have a greater charge imbalance since it has 38 protons and the rubidium ion has 37 protons. The charge imbalance draws the electrons in more closely to the nucleus.

5.

	+	−		+	−
(a)	H	O	(d)	Pb	S
(b)	Rb	Cl	(e)	P	F
(c)	H	N	(f)	H	C

6.

		+	−			+	−
(a)	H		Cl	(d)	I		Br
(b)	S		O	(e)	Cs		I
(c)	C		Cl	(f)	O		F

7. (a) covalent   (b) ionic   (c) ionic   (d) covalent

8. (a) ionic   (b) covalent   (c) covalent   (d) covalent

9. (a) $Mg \rightarrow Mg^{2+} + 2\,e^-$ (b) $Br + 1\,e^- \rightarrow Br^-$

10. (a) $K \rightarrow K^+ + 1\,e^-$ (b) $S + 2e^- \rightarrow S^{2-}$

11. (a) $\longrightarrow$ Li_2O

    (b) $\longrightarrow$ K_3N

12. (a) $\longrightarrow$ K_2S

    (b) $\longrightarrow$ Ca_3N_2

13. (a) Se (6)   (b) P (5)   (c) Br (7)   (d) Mg (2)   (e) He (2)   (f) As (5)

14. (a) Pb (4)   (b) Li (1)   (c) O (6)   (d) Cs (1)   (e) Ga (3)   (f) Ar (8)

15. Noble gas structures:
    (a) potassium atom, lose 1 e^- (c) bromine atom, gain 1 e^-
    (b) aluminum ion, none   (d) selenium atom, gain 2 e^-

16. (a) sulfur atom, gain 2 e^- (c) nitrogen atom, gain 3e^-
    (b) calcium atom, lose 2 e^- (d) iodide ion, none

17. (a) ionic, NaCl   (c) ionic, $MgBr_2$ (e) molecular, CO_2
    (b) molecular, CH_4 (d) molecular, Br_2

18. (a) Two oxygen atoms will form a nonpolar covalent compound. The formula is O_2.
    (b) Hydrogen and bromine will form a polar covalent compound. The formula is HBr.

(c)   Oxygen and two hydrogen atoms will form a polar covalent compound. The formula is H_2O.

(d)   Two iodine atoms will form a nonpolar covalent compound, the formula is I_2.

19. (a) NaH, Na_2O (c) AlH_3, Al_2O_3
      (b) CaH_2, CaO (d) SnH_4, SnO_2

20. (a) SbH_3, Sb_2O_3 (c) HCl, Cl_2O_7
      (b) H_2Se, SeO_3 (d) CH_4, CO_2

21. Li_2SO_4 lithium sulfate K_2SO_4 potassium sulfate
     Rb_2SO_4 rubidium sulfate Cs_2SO_4 cesium sulfate
     Fr_2SO_4 francium sulfate

22. $BeBr_2$ beryllium bromide $BaBr_2$ barium bromide
     $MgBr_2$ magnesium bromide $RaBr_2$ radium bromide
     $SrBr_2$ strontium bromide

23. Lewis structures:

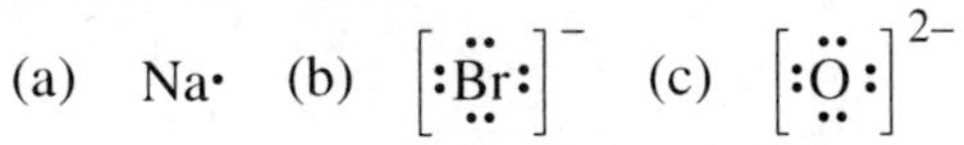

    (a) Na·   (b) $\left[:\overset{\cdot\cdot}{Br}:\right]^{-}$ (c) $\left[:\overset{\cdot\cdot}{\underset{\cdot\cdot}{O}}:\right]^{2-}$

24. (a) Ga: (b) $[Ga]^{3+}$ (c) $[Ca]^{2+}$

25. (a) covalent   (c) ionic
      (b) ionic   (d) covalent

26. (a) covalent   (c) covalent
      (b) ionic   (d) covalent

27. (a) covalent   (b) covalent   (c) covalent

28. (a) covalent   (b) covalent   (c) covalent

29. 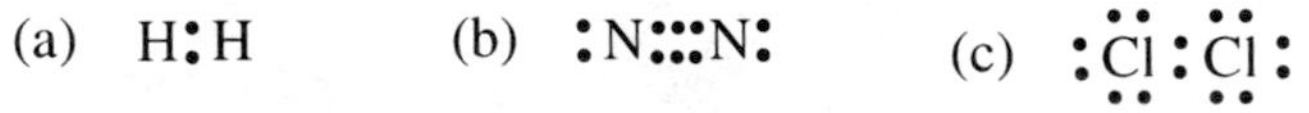 (a) H:H   (b) :N:::N: (c) :Cl:Cl:

30. 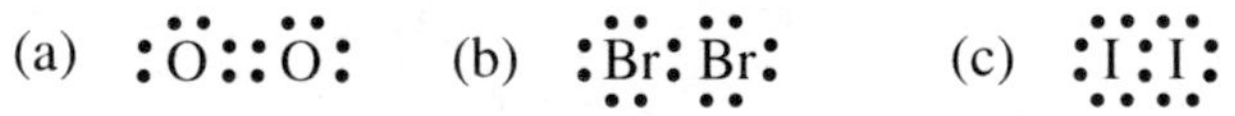 (a) :O::O: (b) :Br:Br: (c) :I:I:

31. 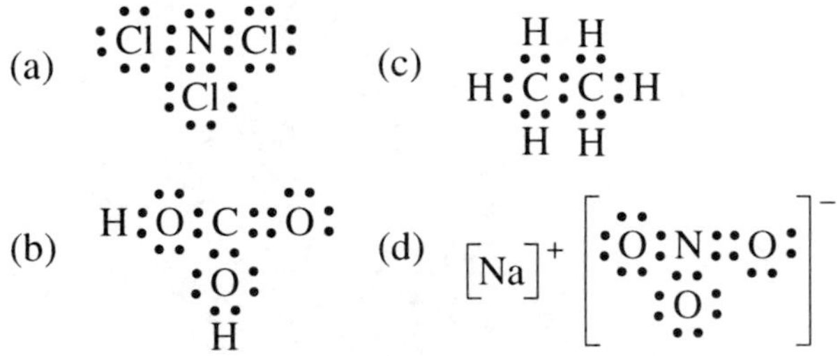 (a) :Cl:N:Cl: :Cl: (c) H:C:C:H with H above and below each C

     (b) H:O:C::O: with :O: and H below C   (d) $[Na]^{+}$ $\left[:O:N::O: :O:\right]^{-}$

32. 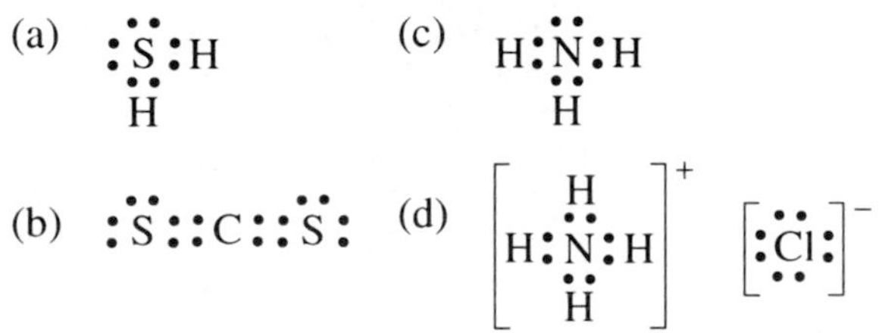 (a) :S:H with H below   (c) H:N:H with H below

     (b) :S::C::S: (d) $\left[H:N:H\right]^{+}$ with H above and below $[:Cl:]^{-}$

33. (a) $[Ba]^{2+}$

 (d) $\left[:C:::N: \right]^-$

 (b) $[Al]^{3+}$

 (e) 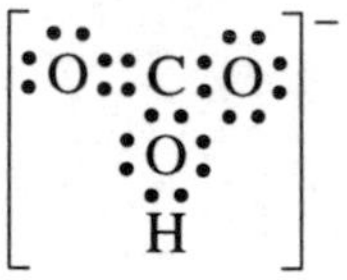 $\left[\begin{array}{c} :O::C:O: \\ :O: \\ H \end{array} \right]^-$

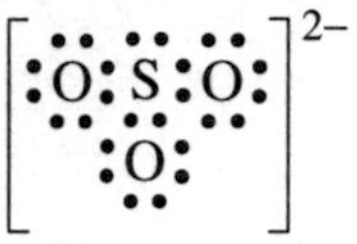

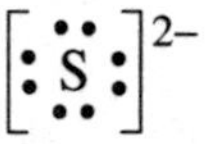

 (c) $\left[\begin{array}{c} :O:S:O: \\ :O: \end{array} \right]^{2-}$

34. (a) 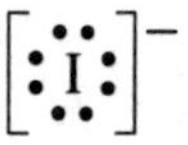 $\left[:I: \right]^-$

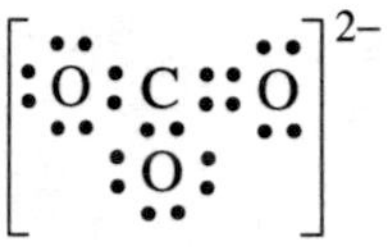

 (d) 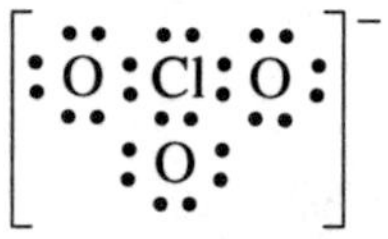 $\left[\begin{array}{c} :O:Cl:O: \\ :O: \end{array} \right]^-$

 (b) $\left[:S: \right]^{2-}$

 (e)  $\left[\begin{array}{c} :O:N::O: \\ :O: \end{array} \right]^-$

 (c) $\left[\begin{array}{c} :O:C::O: \\ :O: \end{array} \right]^{2-}$

35. (a) CH_3Cl, polar   (c) OF_2, polar
    (b) Cl_2, nonpolar   (d) PBr_3, polar

36. (a) H_2, nonpolar   (c) CH_3OH, polar
    (b) NI_3, polar   (d) CS_2, nonpolar

37. (a) 4 electron pairs, tetrahedral
    (b) 4 electron pairs, tetrahedral
    (c) 3 electron pairs, trigonal planar

38. (a) 3 electron pairs, trigonal planar
    (b) 4 electron pairs, tetrahedral
    (c) 4 electron pairs, tetrahedral

39. (a) tetrahedral   (b) trigonal pyramidal   (c) tetrahedral

40. (a) tetrahedral   (b) trigonal pyramidal   (c) tetrahedral

41. (a) tetrahedral   (b) trigonal pyramidal   (c) bent

42. (a) tetrahedral   (b) bent   (c) bent

43. Oxygen

44. Potassium

45.

1 2 3	
Atom 1	Argon
Atom 2	Sodium
Atom 3	Cesium

46.

1 2 3 4 5

Drawing 1	Sr^{2+}
Drawing 2	Rb^+
Drawing 3	Kr
Drawing 4	Br^- ion
Drawing 5	Se^{2-} ion

All of these particles are isoelectronic, meaning they have the same number of electrons. The more protons in their nucleus, the smaller the particle.

47. hydrazine N_2H_4 14 e⁻

$$\overset{\displaystyle H \quad H}{\underset{\displaystyle H \quad H}{:\!N\!-\!N\!:}}$$

hydrozoic acid HN_3 16 e⁻

$$H-\ddot{N}=N=\ddot{N}:$$

48. (a) NO_2^- 18 e⁻

$$\left[:\!\ddot{O}\!-\!N\!=\!\ddot{O}:\right]^-$$

bent

(b) SO_4^{2-} 32 e⁻

$$\left[\begin{array}{c}:\ddot{O}:\\ | \\ :\!\ddot{O}\!-\!S\!-\!\ddot{O}\!: \\ | \\ :\ddot{O}:\end{array}\right]^{2-}$$

tetrahedral

(c) $SOCl_2$ 26 e⁻

$$\overset{\displaystyle :\!\ddot{Cl}\!-\!\ddot{S}\!-\!\ddot{Cl}\!:}{\underset{\displaystyle :\ddot{O}:}{|}}$$

trigonal pyramidal

(d) Cl_2O 20 e⁻

$$\overset{\displaystyle \ddot{O}}{\underset{\displaystyle .\ddot{Cl}\qquad\ddot{Cl}:}{\diagup \diagdown}}$$

bent

49. (a) C_2H_6 14 e⁻

$$\overset{\displaystyle H \quad H}{\underset{\displaystyle H \quad H}{H\!-\!C\!-\!C\!-\!H}}$$

(b) C_2H_4 12 e⁻

$$\underset{\displaystyle H \quad H}{H\!-\!C\!=\!C\!-\!H}$$

(c) C_2H_2 10 e⁻ $H-C\equiv C-H$

50. (a) Be   (b) He   (c) K   (d) F   (e) Fr   (f) Ne

51. (a) Cl   (b) O   (c) Ca

52. The noble gases already have a full outer electron configuration and therefore there is no need for them to attract electrons.

53. Lithium has a $+1$ charge after the first electron is removed. It takes more energy to overcome that charge and to remove another electron than to remove a single electron from an uncharged He atom,

54. Yes. Each of these elements have an ns^1 electron and they could lose that electron in the same way elements in Group 1A do. They would then form $+1$ ions and ionic compounds such as CuCl, AgCl, and AuCl.

55. $SnBr_2$, $GeBr_2$.

56. The bond between sodium and chlorine is ionic. An electron has been transferred from a sodium atom to a chlorine atom. The substance is composed of ions not molecules. Use of the word molecule implies covalent bonding.

57. A covalent bond results from the sharing of a pair of electrons between two atoms, while an ionic bond involves the transfer of one or more electrons from one atom to another.

58. This structure shown is incorrect since the bond is ionic. It should be represented as:

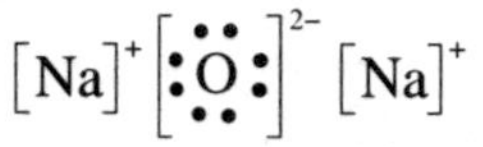

$$\left[Na\right]^+ \left[:\overset{\cdot\cdot}{\underset{\cdot\cdot}{O}}:\right]^{2-} \left[Na\right]^+$$

59. The four most electronegative elements are F, O, N, Cl.

60. highest   F, O, S, H, Mg, Cs   lowest.

61. It is possible for a molecule to be nonpolar even though it contains polar bonds. If the molecule is symmetrical, the polarities of the bonds will cancel (in a manner similar to a positive and negative number of the same size) resulting in a nonpolar molecule.
    An example is CO_2 which is linear and nonpolar.

62. Both molecules contain polar bonds. CO_2 is symmetrical about the C atom, so the polarities cancel. In CO, there is only one polar bond, therefore the molecule is polar.

63. (a)   NO < CO < NaO   (b)   GeO < SiO < CO   (c)   BSe < BS < BO

64. (a) $109.5°$ (actual angle closer to $105°$)   (b) $109.5°$ (actual angle closer to $107°$)
    (c) $109.5°$ (d) $109.5°$

65. (a)   Both use the p orbitals for bonding. B uses one s and two p orbitals while N uses one s and three p orbitals for bonding.
    (b) BF_3 is trigonal planar while NF_3 is trigonal pyramidal.
    (c) BF_3 has no lone pairs
          NF_3 has one lone pair
    (d) BF_3 has 3 very polar covalent bonds. NF_3 has 3 polar covalent bonds

66. Fluorine's electronegativity is greater than any other element. Ionic bonds form between atoms of widely different electronegativities. Therefore, Fr–F, Cs–F, Rb–F, or K–F would be ionic substances with the greatest electronegativity difference.

67. Each element in a particular column has the same number of valence electrons and therefore the same Lewis structure.

68. S $\dfrac{1.40 \text{ g}}{32.07 \frac{\text{g}}{\text{mol}}} = 0.0437 \text{ mol}$ $\dfrac{0.0437}{0.0437} = 1.00$

    O $\dfrac{2.10 \text{ g}}{16.00 \frac{\text{g}}{\text{mol}}} = 0.131 \text{ mol}$ $\dfrac{0.131}{0.0437} = 3.00$

Empirical formula is SO_3

69. We need to know the molecular formula before we can draw the Lewis structure. From the data, determine the empirical and then the molecular formula.

    C $\dfrac{14.5 \text{ g}}{12.01 \frac{\text{g}}{\text{mol}}} = 1.21 \text{ mol}$ $\dfrac{1.21}{1.21} = 1.00$

    Cl $\dfrac{85.5 \text{ g}}{35.45 \frac{\text{g}}{\text{mol}}} = 2.41 \text{ mol}$ $\dfrac{2.41}{1.21} = 2.01$

CCl_2 is the empirical formula

empirical mass $= 1(12.01 \text{ g}) + 2(35.45 \text{ g}) = 82.91 \text{ g}$

$\dfrac{166 \text{ g}}{82.91 \text{ g}} = 2.00$

Therefore, the molecular formula is $(CCl_2)_2$ or C_2Cl_4

70. (a) ionic (b) both (c) covalent (d) covalent

71. (a) ionic;   sodium phosphide
    (b) both;   ammonium iodide
    (c) covalent;   sulfur dioxide
    (d) covalent;   hydrogen sulfide
    (e) both;   copper(II) nitrate
    (f) ionic;   magnesium oxide

72. (a)

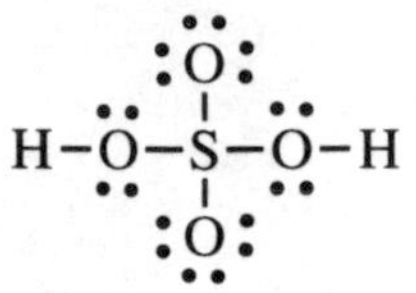

H–O–S–O–H structure with O above and O below

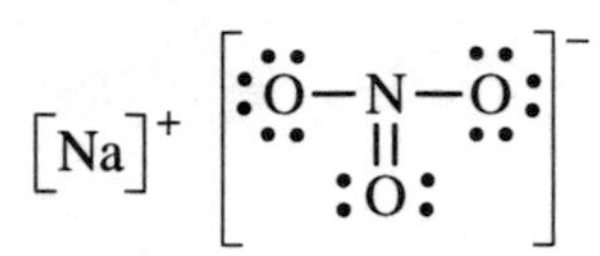

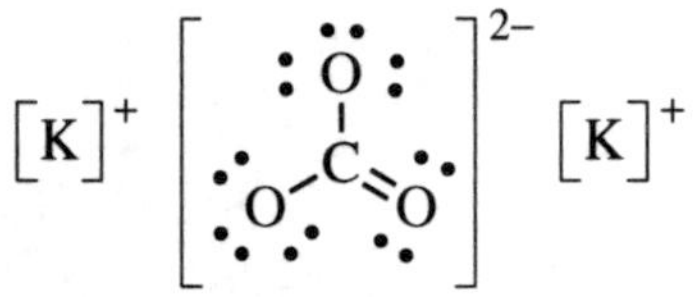

(d) H–C≡N:

(b) $[Na]^+$ $\left[:O–N–O:\right]^-$ with O below

(e) $\left[:O–S–O:\right]^{2-}$ $[Al]^{3+}$ $\left[:O–S–O:\right]^{2-}$ $[Al]^{3+}$ $\left[:O–S–O:\right]^{2-}$

(c) $[K]^+$ $\left[O–C–O\right]^{2-}$ $[K]^+$

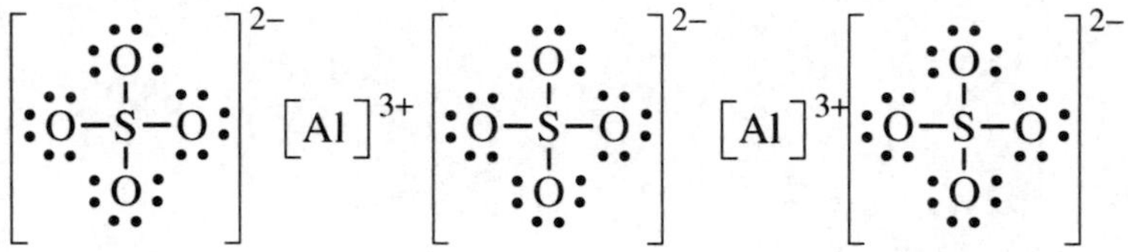

(f)

H–C–C–O–H structure with H, O, and H

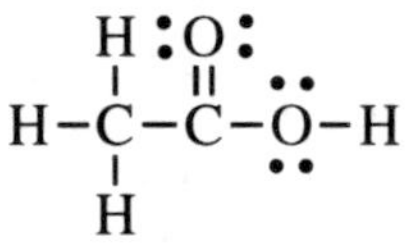

73. $(25 \text{ g Li})\left(\dfrac{1 \text{ mol}}{6.941 \text{ g}}\right)\left(\dfrac{520 \text{ kJ}}{\text{mol}}\right) = 1.9 \times 10^3 \text{ kJ}$

74. Removing the first electron from 1 mole of sodium atoms requires 496 kJ. To remove a second electron from 1 mole of sodium atoms requires 4,565 kJ. The conversions are:

$(15 \text{ mol Na})\left(\dfrac{496 \text{ KJ}}{\text{mol}}\right) = 7.4 \times 10^3 \text{ kJ}$

$(15 \text{ mol Na})\left(\dfrac{4565 \text{ KJ}}{\text{mol}}\right) = 6.8 \times 10^4 \text{ kJ}$

$(7.4 \times 10^3 \text{ kJ}) + (6.8 \times 10^4 \text{ kJ}) = 7.5 \times 10^4 \text{ kJ}$

75.

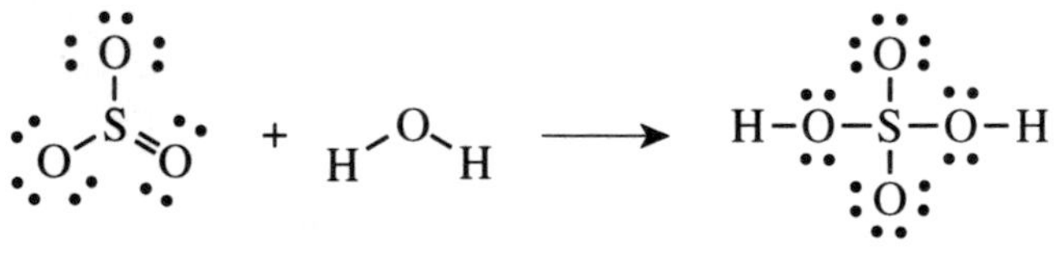

O–S–O + H–O–H → H–O–S–O–H structure

76.

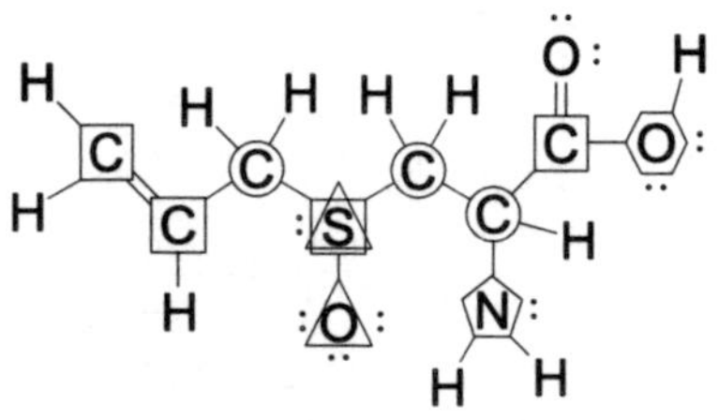

Alliin 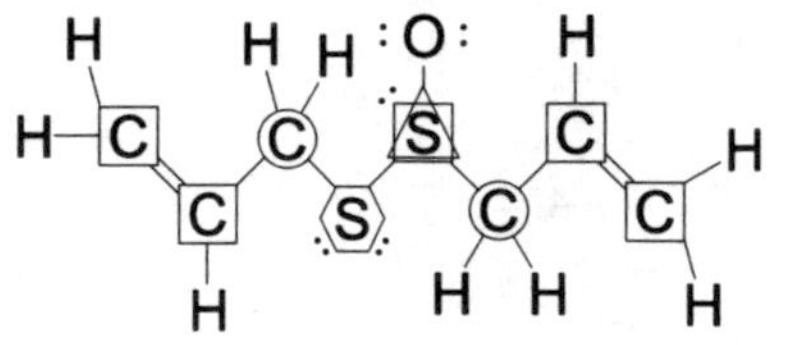   Allicin

77.

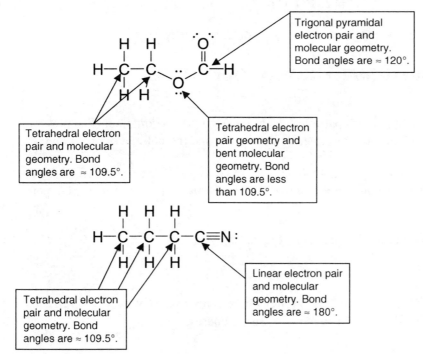

Trigonal pyramidal electron pair and molecular geometry. Bond angles are ≈ 120°.

Tetrahedral electron pair and molecular geometry. Bond angles are ≈ 109.5°.

Tetrahedral electron pair geometry and bent molecular geometry. Bond angles are less than 109.5°.

Tetrahedral electron pair and molecular geometry. Bond angles are ≈ 109.5°.

Linear electron pair and molecular geometry. Bond angles are ≈ 180°.

THE GASEOUS STATE OF MATTER

SOLUTIONS TO REVIEW QUESTIONS

1. The pressure of a gas is the force that gas particles exert on the walls of a container. The pressure depends on the temperature, the number of molecules of the gas and the volume of the container.

2. The air pressure inside the balloon is greater than the air pressure outside the balloon. The pressure inside must equal the sum of the outside air pressure plus the pressure exerted by the stretched rubber of the balloon.

3. The major components of dry air are nitrogen and oxygen.

4. 1 torr = 1 mm Hg

5. The molecules of H_2 at 100°C are moving faster. Temperature is a measure of average kinetic energy. At higher temperatures, the molecules will have more kinetic energy.

6. 1 atm corresponds to 4 L.

7. The pressure times the volume at any point on the curve is equal to the same value. This is an inverse relationship as is Boyle's law. (PV = k)

8. If $T_2 < T_1$, the volume of the cylinder would decrease (the piston would move downward).

9. $N_2(g) + O_2(g) \longrightarrow 2\,NO(g)$
   1 vol + 1 vol $\longrightarrow$ 2 vol

   According to Avogadro's law, equal volumes of nitrogen and oxygen at the same temperature and pressure contain the same number of molecules. In the reaction, nitrogen and oxygen molecules react in a 1:1 ratio. Since two volumes of nitrogen monoxide are produced, one molecule of nitrogen and one molecule of oxygen must produce two molecules of nitrogen monoxide. Therefore each nitrogen and oxygen molecule must be made up of two atoms (diatomic).

10. We refer gases to STP because some reference point is needed to relate volume to moles. A temperature and pressure must be specified to determine the moles of gas in a given volume, and 0°C and 760 torr are convenient reference points.

11. Gases are described by the following parameters:
    (a) pressure   (c) temperature
    (b) volume   (d) number of moles

12. An ideal gas is one which follows the described gas laws at all P, V, and T and whose behavior is described exactly by the Kinetic Molecular Theory.

13. Boyle's law: $P_1V_1 = P_2V_2$, ideal gas equation: $PV = nRT$
   If you have an equal number of moles of two gases at the same temperature the right side of the ideal gas equation will be the same for both gases. You can then set PV for the first gas equal to PV for the second gas (Boyle's law) because the right side of both equations will cancel.

14. Charles' law: $V_1/T_1 = V_2/T_2$, ideal gas equation: $PV = nRT$
   Rearrange the ideal gas equation to: $V/T = nR/P$
   If you have an equal number of moles of two gases at the same pressure the right side of the rearranged ideal gas equation will be the same for both. You can set V/T for the first gas equal to V/T for the second gas (Charles' law) because the right side of both equations will cancel.

15. Basic assumptions of Kinetic Molecular Theory include:
   (a) Gases consist of tiny particles.
   (b) The distance between particles is great compared to the size of the particles.
   (c) Gas particles move in straight lines. They collide with one another and with the walls of the container with no loss of energy.
   (d) Gas particles have no attraction for each other.
   (e) The average kinetic energy of all gases is the same at any given temperature. It varies directly with temperature.

16. The order of increasing molecular velocities is the order of decreasing molar masses.

   increasing molecular velocity

   $\longrightarrow$

   $Rn, F_2, N_2\ CH_4, He, H_2$

   $\longrightarrow$

   decreasing molar mass

   At the same temperature the kinetic energies of the gases are the same and equal to $\frac{1}{2} mv^2$. For the kinetic energies to be the same, the velocities must increase as the molar masses decrease.

17. Average kinetic energies of all these gases are the same, since the gases are all at the same temperature.

18. A gas is least likely to behave ideally at low temperatures. Under this condition, the velocities of the molecules decrease and attractive forces between the molecules begin to play a significant role.

19. A gas is least likely to behave ideally at high pressures. Under this condition, the molecules are forced close enough to each other so that their volume is no longer small compared to the volume of the container. Attractive forces may also occur here and sooner or later, the gas will liquefy.

20. Behavior of gases as described by the Kinetic Molecular Theory.
    (a) Boyle's law. Boyle's law states that the volume of a fixed mass of gas is inversely proportional to the pressure, at constant temperature. The Kinetic Molecular Theory assumes the volume occupied by gases is mostly empty space. Decreasing the volume of a gas by compressing it, increases the concentration of gas molecules, resulting in more collisions of the molecules and thus increased pressure upon the walls of the container.
    (b) Charles' law. Charles' law states that the volume of a fixed mass of gas is directly proportional to the absolute temperature, at constant pressure. According to Kinetic Molecular Theory, the kinetic energies of gas molecules are proportional to the absolute temperature. Increasing the temperature of a gas causes the molecules to move faster, and in order for the pressure not to increase, the volume of the gas must increase.
    (c) Dalton's law. Dalton's law states that the pressure of a mixture of gases is the sum of the pressures exerted by the individual gases. According to the Kinetic Molecular Theory, there are no attractive forces between gas molecules; therefore, in a mixture of gases, each gas acts independently and the total pressure exerted will be the sum of the pressures exerted by the individual gases.

21. Conversion of oxygen to ozone is an endothermic reaction. Evidence for this statement is that energy ($286 \text{ kJ}/3 \text{ mol } O_2$) is required to convert O_2 to O_3.

22. Oxygen atom = O   Oxygen molecule = O_2 Ozone molecule = O_3
    An oxygen molecule contains 16 electrons.

23. The pressure inside the bottle is less than atmospheric pressure. We come to this conclusion because the water inside the bottle is higher than the water in the trough (outside the bottle).

24. The density of air is 1.29 g/L. Any gas listed below air in Table 12.4 has a density greater than air. For example: O_2, H_2S, HCl, F_2, CO_2.

25. Equal volumes of H_2 and O_2 at the same T and P:
    (a) have equal number of molecules (Avogadro's law)
    (b) mass O_2 = 16 times mass of H_2
    (c) moles O_2 = moles H_2
    (d) average kinetic energies are the same (T same)
    (e) density O_2 = 16 times the density of H_2

    $$\text{density } O_2 = \left(\frac{\text{mass } O_2}{\text{volume } O_2} \right) \qquad \text{density } H_2 = \left(\frac{\text{mass } H_2}{\text{volume } H_2} \right)$$

    $$\text{volume } O_2 = \text{volume } H_2$$

    $$\left(\frac{\text{mass } O_2}{\text{density } O_2} \right) \left(\frac{\text{mass } H_2}{\text{density } H_2} \right) \qquad \text{density } O_2 = \left(\frac{\text{mass } O_2}{\text{mass } H_2} \right) (\text{density } H_2)$$

    $$\text{density } O_2 = \left(\frac{32}{2} \right) (\text{density } H_2) = 16 \, (\text{density } H_2)$$

26. Heating a mole of N_2 gas at constant pressure has the following effects:
    (a) Density will decrease. Heating the gas at constant pressure will increase its volume. The mass does not change, so the increased volume results in a lower density.
    (b) Mass does not change. Heating a substance does not change its mass.
    (c) Average kinetic energy of the molecules increases. This is a basic assumption of the Kinetic Molecular Theory.
    (d) Average velocity of the molecules will increase. Increasing the temperature increases the average kinetic energies of the molecules; hence, the average velocity of the molecules will increase also.
    (e) Number of N_2 molecules remains unchanged. Heating does not alter the number of molecules present, except if extremely high temperatures were attained. Then, the N_2 molecules might dissociate into N atoms resulting in fewer N_2 molecules.

SOLUTIONS TO EXERCISES

1. Pressure conversions:

	torr	inches Hg	kilopascals
a.	768	30.2	102
b.	752	29.6	100.
c.	745	29.3	99.3

(a) in. Hg → torr $(30.2 \text{ in. Hg})\left(\dfrac{760 \text{ torr}}{29.9 \text{ in. Hg}}\right) = 768 \text{ torr}$

  in. Hg → kPa $(30.2 \text{ in. Hg})\left(\dfrac{101.325 \text{ kPa}}{29.9 \text{ in. Hg}}\right) = 102 \text{ kPa}$

(b) torr → in. Hg $(752 \text{ torr})\left(\dfrac{29.9 \text{ in. Hg}}{760 \text{ torr}}\right) = 29.6 \text{ in. Hg}$

  torr → kPa $(752 \text{ torr})\left(\dfrac{101.325 \text{ kPa}}{760 \text{ torr}}\right) = 100. \text{ kPa}$

(c) kPa → torr $(99.3 \text{ kPa})\left(\dfrac{760 \text{ torr}}{101.325 \text{ kPa}}\right) = 745 \text{ torr}$

  kPa → in. Hg $(99.3 \text{ kPa})\left(\dfrac{29.9 \text{ in. Hg}}{101.325 \text{ kPa}}\right) = 29.3 \text{ in. Hg}$

2. Pressure conversions:

	mm Hg	lb/in.^{2}	atmospheres
a.	789	15.3	1.04
b.	1700	32	2.2
c.	1100	21	1.4

(a) mm Hg → lb/in.^{2} $(789 \text{ mm Hg})\left(\dfrac{14.7 \text{ lb/in.}^2}{760 \text{ mm Hg}}\right) = 15.3 \text{ lb/in.}^2$

  mm Hg → atm $(789 \text{ mm Hg})\left(\dfrac{1 \text{ atm}}{760 \text{ mm Hg}}\right) = 1.04 \text{ atm}$

(b) lb/in.^{2} → mm Hg $(32 \text{ lb/in.}^2)\left(\dfrac{760 \text{ mm Hg}}{14.7 \text{ lb/in.}^2}\right) = 1700 \text{ mm Hg}$

  lb/in.^{2} → atm $(32 \text{ lb/in.}^2)\left(\dfrac{1 \text{ atm}}{14.7 \text{ lb/in.}^2}\right) = 2.2 \text{ atm}$

(c) atm → mm Hg $(1.4 \text{ atm})\left(\dfrac{760 \text{ mm Hg}}{1 \text{ atm}}\right) = 1100 \text{ mm Hg}$

  atm → lb/in.^{2} $(1.4 \text{ atm})\left(\dfrac{14.7 \text{ lb/in.}^2}{1 \text{ atm}}\right) = 21 \text{ lb/in.}^2$

3. (a) torr → kPa $(953 \text{ torr})\left(\dfrac{101.3 \text{ kPa}}{760 \text{ torr}}\right) = 127 \text{ kPa}$

   (b) kPa → atm $(2.98 \text{ kPa})\left(\dfrac{1 \text{ atm}}{101.3 \text{ kPa}}\right) = 0.0294 \text{ atm}$

   (c) atm → mm Hg $(2.77 \text{ atm})\left(\dfrac{760 \text{ mm Hg}}{1 \text{ atm}}\right) = 2110 \text{ mm Hg}$

   (d) torr → atm $(372 \text{ torr})\left(\dfrac{1 \text{ atm}}{760 \text{ torr}}\right) = 0.489 \text{ atm}$

   (e) atm → cm Hg $(2.81 \text{ atm})\left(\dfrac{76.0 \text{ cm Hg}}{1 \text{ atm}}\right) = 214 \text{ cm Hg}$

4. (a) torr → kPa $(649 \text{ torr})\left(\dfrac{101.3 \text{ kPa}}{760 \text{ torr}}\right) = 86.5 \text{ kPa}$

   (b) kPa → atm $(5.07 \text{ kPa})\left(\dfrac{1 \text{ atm}}{101.3 \text{ kPa}}\right) = 0.0500 \text{ atm}$

   (c) atm → mm Hg $(3.64 \text{ atm})\left(\dfrac{760 \text{ mm Hg}}{1 \text{ atm}}\right) = 2770 \text{ mm Hg}$

   (d) torr → atm $(803 \text{ torr})\left(\dfrac{1 \text{ atm}}{760 \text{ torr}}\right) = 1.06 \text{ atm}$

   (e) atm → cm Hg $(1.08 \text{ atm})\left(\dfrac{76.0 \text{ cm Hg}}{1 \text{ atm}}\right) = 82.1 \text{ cm Hg}$

5. (a) lb/in.^{2} → atm $(1920 \text{ lb/in.}^2)\left(\dfrac{1 \text{ atm}}{14.7 \text{ lb/in.}^2}\right) = 131 \text{ atm}$

   (b) lb/in.^{2} → torr $(1920 \text{ lb/in.}^2)\left(\dfrac{760 \text{ torr}}{14.7 \text{ lb/in.}^2}\right) = 9.93 \times 10^4 \text{ torr}$

   (c) lb/in.^{2} → kPa $(1920 \text{ lb/in.}^2)\left(\dfrac{101.3 \text{ kPa}}{14.7 \text{ lb/in.}^2}\right) = 1.32 \times 10^4 \text{ kPa}$

6. (a) lb/in.^{2} → atm $(31 \text{ lb/in.}^2)\left(\dfrac{1 \text{ atm}}{14.7 \text{ lb/in.}^2}\right) = 2.1 \text{ atm}$

   (b) lb/in.^{2} → torr $(31 \text{ lb/in.}^2)\left(\dfrac{760 \text{ torr}}{14.7 \text{ lb/in.}^2}\right) = 1600 \text{ torr}$

   (c) lb/in.^{2} → kPa $(31 \text{ lb/in.}^2)\left(\dfrac{101.3 \text{ kPa}}{14.7 \text{ lb/in.}^2}\right) = 210 \text{ kPa}$

7. $P_1V_1 = P_2V_2$ or $P_2 = \dfrac{P_1V_1}{V_2}$

   (a) $\dfrac{(825 \text{ torr})(725 \text{ mL})}{(283 \text{ mL})} = 2110 \text{ torr}$

   (b) Change 2.87 L to mL $\quad \dfrac{(825 \text{ torr})(725 \text{ mL})}{(2.87 \text{ L})(1000 \text{ mL}/1 \text{ L})} = 208 \text{ torr}$

8. $P_1V_1 = P_2V_2$ or $P_2 = \dfrac{P_1V_1}{V_2}$

   (a) $\dfrac{(508 \text{ torr})(486 \text{ mL})}{(185 \text{ mL})} = 1330 \text{ torr}$

   (b) Change 6.17 L to mL $\quad \dfrac{(508 \text{ torr})(486 \text{ mL})}{(6.17 \text{ L})(1000 \text{ mL}/1 \text{ L})} = 40.0 \text{ torr}$

9. $P_1V_1 = P_2V_2$ or $V_2 = \dfrac{P_1V_1}{P_2}$

   $\dfrac{(58.2 \text{ L})(7.25 \text{ atm})}{(2.03 \text{ atm})} = 208 \text{ L}$

10. $P_1V_1 = P_2V_2$ or $V_2 = \dfrac{P_1V_1}{P_2}$

    $\dfrac{(832 \text{ L})(0.204 \text{ atm})}{(8.02 \text{ atm})} = 21.2 \text{ L}$

11. $\dfrac{V_1}{T_1} = \dfrac{V_2}{T_2}$ or $V_2 = \dfrac{V_1T_2}{T_1}$; temperatures must be in Kelvin ($^\circ$C + 273)

    (a) $\dfrac{(125 \text{ mL})(268 \text{ K})}{294 \text{ K}} = 114 \text{ mL}$

    (b) $\dfrac{(125 \text{ mL})(308 \text{ K})}{294 \text{ K}} = 131 \text{ mL}$

    (c) $\dfrac{(125 \text{ mL})(1095 \text{ K})}{294 \text{ K}} = 466 \text{ mL}$

12. $\dfrac{V_1}{T_1} = \dfrac{V_2}{T_2}$ or $V_2 = \dfrac{V_1T_2}{T_1}$; temperatures must be in Kelvin ($^\circ$C + 273)

    (a) $\dfrac{(575 \text{ mL})(298 \text{ K})}{248 \text{ K}} = 691 \text{ mL}$

    (b) $\dfrac{(575 \text{ mL})(273 \text{ K})}{248 \text{ K}} = 633 \text{ mL} \quad (32^\circ\text{F} = 0^\circ\text{C} = 273 \text{ K})$

    (c) $\dfrac{(575 \text{ mL})(318 \text{ K})}{248 \text{ K}} = 737 \text{ mL}$

13. Use the combined gas laws $\dfrac{P_1V_1}{T_1} = \dfrac{P_2V_2}{T_2}$ or $V_2 = \dfrac{P_1V_1T_2}{P_2T_1}$

$$V_2 = \frac{(0.75\ \text{atm})(1025\ \text{mL})(308\ \text{K})}{(1.25\ \text{atm})(348\ \text{K})} = 544\ \text{mL}$$

14. Use the combined gas laws $\dfrac{P_1V_1}{T_1} = \dfrac{P_2V_2}{T_2}$ or $V_2 = \dfrac{P_1V_1T_2}{P_2T_1}$

$$V_2 = \frac{(678\ \text{torr})(25.6\ \text{L})(308\ \text{K})}{(595\ \text{torr})(292\ \text{K})} = 30.8\ \text{L}$$

15. Use the combined gas law $\dfrac{P_1V_1}{T_1} = \dfrac{P_2V_2}{T_2}$ or $V_2 = \dfrac{P_1V_1T_2}{P_2T_1}$

$$V_2 = \frac{(0.950\ \text{atm})(1400.\ \text{L})(275\ \text{K})}{(4.0\ \text{torr})(1\ \text{atm}/760\ \text{torr})(291\ \text{K})} = 2.4 \times 10^5\ \text{L}$$

16. Use the combined gas law $\dfrac{P_1V_1}{T_1} = \dfrac{P_2V_2}{T_2}$ or $V_2 = \dfrac{P_1V_1T_2}{P_2T_1}$

$$V_2 = \frac{(2.50\ \text{atm})(22.4\ \text{L})(268\ \text{K})}{(1.50\ \text{atm})(300.\ \text{K})} = 33.4\ \text{L}$$

17. Use the combined gas law $\dfrac{P_1V_1}{T_1} = \dfrac{P_2V_2}{T_2}$ or $P_2 = \dfrac{P_1V_1T_2}{V_2T_1}$

$$\frac{(1.0\ \text{atm})(775\ \text{mL})(298\ \text{K})}{(615\ \text{mL})(273\ \text{K})} = 1.4\ \text{atm}$$

18. Use the combined gas law $\dfrac{P_1V_1}{T_1} = \dfrac{P_2V_2}{T_2}$ or $T_2 = \dfrac{P_2V_2T_1}{P_1V_1}$

Change 765 torr to atmospheres.

$$\frac{(765\ \text{torr})(1\ \text{atm}/760\ \text{torr})(1.5\ \text{L})(292\ \text{K})}{(1.5\ \text{atm})(2.5\ \text{L})} = 120\ \text{K} \quad (120\ \text{K} - 273),\ \text{K} = -153°\text{C}$$

19. $P_{\text{total}} = P_{O_2} + P_{H_2O\ \text{vapor}} = 772\ \text{torr}$

    $P_{H_2O\ \text{vapor}} = 21.2\ \text{torr}$

    $P_{O_2} = 772\ \text{torr} - 21.2\ \text{torr} = 751\ \text{torr}$

20. $P_{\text{total}} = P_{CH_4} + P_{H_2O\ \text{vapor}} = 749\ \text{mm Hg}$

    $P_{H_2O} = 30.0\ \text{torr} = 30.0\ \text{mm Hg}$

    $P_{CH_4} = 749\ \text{mm Hg} - 30.0\ \text{mm Hg} = 719\ \text{mm Hg}$

21. $P_{\text{total}} = P_{N_2} + P_{H_2} + P_{O_2}$

    $= 200.\ \text{torr} + 600.\ \text{torr} + 300.\ \text{torr} = 1100.\ \text{torr} = 1.100 \times 10^3\ \text{torr}$

22. $P_{total} = P_{H_2} + P_{N_2} + P_{O_2}$

    $= 325 \text{ torr} + 475 \text{ torr} + 650. \text{ torr} = 1450. \text{ torr} = 1.450 \times 10^3 \text{ torr}$

23. $P_{total} = P_{CH_4} + P_{H_2O \text{ vapor}}$ (Solubility of methane is being ignored.)

    $P_{H_2O \text{ vapor}} = 23.8 \text{ torr}$

    $P_{CH_4} = 720. \text{ torr} - 23.8 \text{ torr} = 696 \text{ torr}$

    To calculate the volume of dry methane, note that the temperature is constant, so $P_1 V_1 = P_2 V_2$ can be used.

    $$V_2 = \frac{P_1 V_1}{P_2} = \frac{(696 \text{ torr})(2.50 \text{ L CH}_4)}{(760. \text{ torr})} = 2.29 \text{ L CH}_4$$

24. $P_{total} = P_{C_3H_8} + P_{H_2O \text{ vapor}}$ C_3H_8 is propane

    $P_{H_2O \text{ vapor}} = 20.5 \text{ torr}$

    $P_{C_3H_8} = 745 \text{ torr} - 20.5 \text{ torr} = 725 \text{ torr}$

    To calculate the volume of dry propane, note that the temperature is constant, so $P_1 V_1 = P_2 V_2$ can be used.

    $$V_2 = \frac{P_1 V_1}{P_2} = \frac{(725 \text{ torr})(1.25 \text{ L C}_3\text{H}_8)}{(760. \text{ torr})} = 1.19 \text{ L C}_3\text{H}_8$$

25. 1 mol of a gas occupies 22.4 L at STP

    $$(6.26 \text{ mol N}_2)\left(\frac{22.4 \text{ L}}{1 \text{ mol}}\right) = 140. \text{ L N}_2$$

26. 1 mol of a gas occupies 22.4 L at STP

    $$(5.89 \text{ mol CO}_2)\left(\frac{22.4 \text{ L}}{1 \text{ mol}}\right) = 132 \text{ L CO}_2$$

27. (a) $(6.022 \times 10^{23} \text{ molecules CO}_2)\left(\dfrac{22.4 \text{ L}}{6.022 \times 10^{23} \text{ molecules}}\right) = 22.4 \text{ L CO}_2$

    (b) $(2.5 \text{ mol CH}_4)\left(\dfrac{22.4 \text{ L}}{\text{mol}}\right) = 56 \text{ L CH}_4$

    (c) $(12.5 \text{ g O}_2)\left(\dfrac{22.4 \text{ L}}{32.00 \text{ g}}\right) = 8.75 \text{ L O}_2$

28. (a) $(1.80 \times 10^{24} \text{ molecules SO}_3)\left(\dfrac{22.4 \text{ L}}{6.022 \times 10^{23} \text{ molecules}}\right) = 67.0 \text{ L SO}_3$

    (b) $(7.5 \text{ mol C}_2\text{H}_6)\left(\dfrac{22.4 \text{ L}}{\text{mol}}\right) = 170 \text{ L C}_2\text{H}_6$

    (c) $(25.2 \text{ g Cl}_2)\left(\dfrac{22.4 \text{ L}}{70.90 \text{ g}}\right) = 7.96 \text{ L Cl}_2$

29. $(725 \text{ mL NH}_3)\left(\dfrac{1 \text{ L}}{1000 \text{ mL}}\right)\left(\dfrac{1 \text{ mol}}{22.4 \text{ L}}\right)\left(\dfrac{17.03 \text{ g}}{\text{mol}}\right) = 0.551 \text{ g NH}_3$

30. $(945 \text{ mL C}_3\text{H}_6)\left(\dfrac{1 \text{ L}}{1000 \text{ mL}}\right)\left(\dfrac{1 \text{ mol}}{22.4 \text{ L}}\right)\left(\dfrac{42.08 \text{ g}}{\text{mol}}\right) = 1.78 \text{ g C}_3\text{H}_6$

31. $(1025 \text{ molecules CO}_2)\left(\dfrac{1 \text{ mol}}{6.022 \times 10^{23} \text{ molecules}}\right)\left(\dfrac{22.4 \text{ L}}{\text{mol}}\right) = 3.813 \times 10^{-20} \text{ L CO}_2$

32. $(10.5 \text{ L CO}_2)\left(\dfrac{1 \text{ mol}}{22.4 \text{ L}}\right)\left(\dfrac{6.022 \times 10^{23} \text{ molecules}}{\text{mol}}\right) = 2.82 \times 10^{23} \text{ molecules CO}_2$

33. density of Cl_2 gas $= 3.17 \text{ g/L}$ (from table 12.4)

    $(10.0 \text{ g Cl}_2)\left(\dfrac{1 \text{ L}}{3.17 \text{ g}}\right) = 3.15 \text{ L Cl}_2$

34. density of CH_4 gas $= 0.716 \text{ g/L}$ (from table 12.4)

    $(3.0 \text{ L CH}_4)(0.716 \text{ g/L}) = 2.1 \text{ g CH}_4$

35. $PV = nRT \quad V = \dfrac{nRT}{P} \quad V = \dfrac{(75 \text{ mol NH}_3)(0.0821 \text{ L atm/mol K})(295 \text{ K})}{\dfrac{729 \text{ torr}}{760 \dfrac{\text{torr}}{\text{atm}}}} = 1.9 \times 10^3 \text{ L NH}_3$

36. $PV = nRT \quad V = \dfrac{nRT}{P} \quad V = \dfrac{(105 \text{ mol CH}_4)(0.0821 \text{ L atm/mol K})(312 \text{ K})}{1.5 \text{ atm}} = 1.8 \times 10^3 \text{ L CH}_4$

37. $PV = nRT \quad n = \dfrac{PV}{RT} \quad n = \dfrac{(1.2 \text{ atm})(5.25 \text{ L O}_2)}{(0.0821 \text{ L atm/mol K})(299 \text{ K})} = 0.26 \text{ mol O}_2$

38. $PV = nRT \quad n = \dfrac{PV}{RT} \quad n = \dfrac{\left(\dfrac{752 \text{ torr}}{760 \dfrac{\text{torr}}{\text{atm}}}\right)(9.55 \text{ L CO}_2)}{(0.0821 \text{ L atm/mol K})(318 \text{ K})} = 0.362 \text{ mol CO}_2$

39. $PV = nRT \quad T = \dfrac{PV}{nR} \quad T = \dfrac{\left(\dfrac{732 \text{ torr}}{760 \dfrac{\text{torr}}{\text{atm}}}\right)(645 \text{ L Xe})}{(25.2 \text{ mol Xe})(0.0821 \text{ L atm/mol K})} = 300. \text{ K}$

40. $PV = nRT \quad T = \dfrac{PV}{nR} \quad T = \dfrac{\left(\dfrac{675 \text{ torr}}{760 \dfrac{\text{torr}}{\text{atm}}}\right)(725 \text{ L Ar})}{(37.5 \text{ mol Ar})(0.0821 \text{ L atm/mol K})} = 209 \text{ K}$

41. (a) Density of Gases

$$d = \left(\frac{4.003 \text{ g He}}{\text{mol}}\right)\left(\frac{1 \text{ mol}}{22.4 \text{ L}}\right) = 0.179 \text{ g/L He}$$

(b) $$d = \left(\frac{20.01 \text{ g HF}}{\text{mol}}\right)\left(\frac{1 \text{ mol}}{22.4 \text{ L}}\right) = 0.893 \text{ g/L HF}$$

(c) $$d = \left(\frac{42.08 \text{ g C}_3\text{H}_6}{\text{mol}}\right)\left(\frac{1 \text{ mol}}{22.4 \text{ L}}\right) = 1.89 \text{ g/L C}_3\text{H}_6$$

(d) $$d = \left(\frac{120.9 \text{ g CCl}_2\text{F}_2}{\text{mol}}\right)\left(\frac{1 \text{ mol}}{22.4 \text{ L}}\right) = 5.40 \text{ g/L CCl}_2\text{F}_2$$

42. (a) $$d = \left(\frac{222 \text{ g Rn}}{\text{mol}}\right)\left(\frac{1 \text{ mol}}{22.4 \text{ L}}\right) = 9.91 \text{ g/L Rn}$$

(b) $$d = \left(\frac{46.01 \text{ g NO}_2}{\text{mol}}\right)\left(\frac{1 \text{ mol}}{22.4 \text{ L}}\right) = 2.05 \text{ g/L NO}_2$$

(c) $$d = \left(\frac{80.07 \text{ g SO}_3}{\text{mol}}\right)\left(\frac{1 \text{ mol}}{22.4 \text{ L}}\right) = 3.57 \text{ g/L SO}_3$$

(d) $$d = \left(\frac{28.05 \text{ g C}_2\text{H}_4}{\text{mol}}\right)\left(\frac{1 \text{ mol}}{22.4 \text{ L}}\right) = 1.25 \text{ g/L C}_2\text{H}_4$$

43. (a) Assume 1.00 mol of NH_3 and determine the volume using the ideal gas equation, $PV = nRT$.

$$V = \frac{nRT}{P} = \frac{(1.00 \text{ mol NH}_3)(0.0821 \text{ L atm/mol K})(298 \text{ K})}{1.2 \text{ atm}} = 20. \text{ L NH}_3 \text{ at } 25°\text{C and 1.2 atm}$$

$$d = \frac{17.03 \text{ g}}{20. \text{ L NH}_3} = 0.85 \text{ g/L NH}_3$$

(b) Assume 1.00 mol of Ar and determine the volume using the ideal gas equation, $PV = nRT$.

$$V = \frac{nRT}{P} = \frac{(1.00 \text{ mol Ar})(0.0821 \text{ L atm/mol K})(348 \text{ K})}{\dfrac{745 \text{ torr}}{760 \dfrac{\text{torr}}{\text{atm}}}} = 29.1 \text{ L Ar at } 75°\text{C and 745 torr}$$

$$d = \frac{39.95 \text{ g}}{29.1 \text{ L Ar}} = 1.37 \text{ g/L Ar}$$

44. (a) Assume 1.00 mol C_2H_4 and determine the volume using the ideal gas equation, $PV = nRT$.

$$V = \frac{nRT}{P} = \frac{(1.00 \text{ mol C}_2\text{H}_4)(0.0821 \text{ L atm/mol K})(305 \text{ K})}{0.75 \text{ atm}} = 33 \text{ L C}_2\text{H}_4 \text{ at } 32°\text{C and 0.75 atm}$$

$$d = \frac{28.05 \text{ g}}{33 \text{ L C}_2\text{H}_4} = 0.85 \text{ g/L C}_2\text{H}_4$$

(b) Assume 1.00 mol of He and determine the volume using the ideal gas equation, $PV = nRT$.

$$V = \frac{nRT}{P} = \frac{(1.00 \text{ mol He})(0.0821 \text{ L atm/mol K})(330. \text{ K})}{\dfrac{791 \text{ torr}}{760 \dfrac{\text{torr}}{\text{atm}}}} = 26.0 \text{ L He at } 57°\text{C and 791 torr}$$

$$d = \frac{4.003 \text{ g}}{26.0 \text{ L He}} = 0.154 \text{ g/L He}$$

45. The balanced equation is $CaCO_3(s) \rightarrow CaO(s) + CO_2(g)$; 1 mol of a gas occupies 22.4 L at STP.

    (a) $(6.24 \text{ g CaCO}_3)\left(\dfrac{1 \text{ mol}}{100.1 \text{ g}}\right)\left(\dfrac{1 \text{ mol CO}_2}{1 \text{ mol CaCO}_3}\right)\left(\dfrac{22.4 \text{ L CO}_2}{1 \text{ mol CO}_2}\right) = 1.40 \text{ L CO}_2$

    (b) $(52.6 \text{ L CO}_2)\left(\dfrac{1 \text{ mol CO}_2}{22.4 \text{ L CO}_2}\right)\left(\dfrac{1 \text{ mol CaCO}_3}{1 \text{ mol CO}_2}\right)\left(\dfrac{100.1 \text{ g}}{1 \text{ mol}}\right) = 235 \text{ g CaCO}_3$

46. The balanced equation is $Mg(s) + 2 HCl(aq) \rightarrow MgCl_2(aq) + H_2(g)$; 1 mol of a gas occupies 22.4 L at STP.

    (a) $(42.9 \text{ g Mg})\left(\dfrac{1 \text{ mol}}{24.31 \text{ g}}\right)\left(\dfrac{1 \text{ mol H}_2}{1 \text{ mol Mg}}\right)\left(\dfrac{22.4 \text{ L}}{1 \text{ mol}}\right)\left(\dfrac{1000 \text{ mL}}{1 \text{ L}}\right) = 3.95 \times 10^4 \text{ mL H}_2$

    (b) $(825 \text{ mL H}_2)\left(\dfrac{1 \text{ L}}{1000 \text{ mL}}\right)\left(\dfrac{1 \text{ mol H}_2}{22.4 \text{ L H}_2}\right)\left(\dfrac{2 \text{ mol HCl}}{1 \text{ mol H}_2}\right) = 0.0737 \text{ mol HCl}$

47. The balanced equation is $4 NH_3(g) + 5 O_2(g) \rightarrow 4 NO(g) + 6 H_2O(g)$
    Remember that volume–volume relationships are the same as mole–mole relationships when dealing with gases at the same T and P.

    (a) $(2.5 \text{ L NH}_3)\left(\dfrac{5 \text{ L O}_2}{4 \text{ L NH}_3}\right) = 3.1 \text{ L O}_2$

    (b) $(25 \text{ L NH}_3)\left(\dfrac{6 \text{ L H}_2\text{O}}{4 \text{ L NH}_3}\right)\left(\dfrac{1 \text{ mol}}{22.4 \text{ L}}\right)\left(\dfrac{18.02 \text{ g}}{1 \text{ mol}}\right) = 30. \text{ g H}_2\text{O}$

    (c) Limiting reactant problem.

    $(25 \text{ L O}_2)\left(\dfrac{4 \text{ L NO}}{5 \text{ L O}_2}\right) = 20. \text{ L NO}$

    $(25 \text{ L NH}_3)\left(\dfrac{4 \text{ L NO}}{4 \text{ L NH}_3}\right) = 25 \text{ L NO}$

    Oxygen is the limiting reactant. 20. L NO is formed.

48. The balanced equation is $C_3H_8(g) + 5 O_2(g) \rightarrow 3 CO_2(g) + 4 H_2O(g)$
    Remember that volume–volume relationships are the same as mole–mole relationships when dealing with gases at the same T and P.

    (a) $(7.2 \text{ L C}_3\text{H}_8)\left(\dfrac{5 \text{ L O}_2}{1 \text{ L C}_3\text{H}_8}\right) = 36 \text{ L O}_2$

    (b) $(35 \text{ L C}_3\text{H}_8)\left(\dfrac{3 \text{ L CO}_2}{1 \text{ L C}_3\text{H}_8}\right)\left(\dfrac{1 \text{ mol}}{22.4 \text{ L}}\right)\left(\dfrac{44.01 \text{ g}}{1 \text{ mol}}\right) = 210 \text{ g CO}_2$

    (c) Limiting reactant problem.

    $(15 \text{ L C}_3\text{H}_8)\left(\dfrac{4 \text{ L H}_2\text{O}}{1 \text{ L C}_3\text{H}_8}\right) = 60. \text{ L H}_2\text{O}$

    $(15 \text{ L O}_2)\left(\dfrac{4 \text{ L H}_2\text{O}}{5 \text{ L O}_2}\right) = 12 \text{ L H}_2\text{O}$

    Oxygen is the limiting reactant. 12 L H₂O is formed.

49. The balanced equation is $2\,KClO_3(s) \rightarrow 2\,KCl(s) + 3\,O_2(g)$

$$(0.525\,\text{kg KCl})\left(\frac{1000\,\text{g}}{1\,\text{kg}}\right)\left(\frac{1\,\text{mol}}{74.55\,\text{g}}\right)\left(\frac{3\,\text{mol O}_2}{2\,\text{mol KCl}}\right)\left(\frac{22.4\,\text{L}}{1\,\text{mol}}\right) = 237\,\text{L O}_2$$

50. The balanced equation is $C_6H_{12}O_6(s) + 6\,O_2(g) \rightarrow 6\,CO_2(g) + 6\,H_2O(l)$

$$(1.50\,\text{kg C}_6\text{H}_{12}\text{O}_6)\left(\frac{1000\,\text{g}}{1\,\text{kg}}\right)\left(\frac{1\,\text{mol}}{180.2\,\text{g}}\right)\left(\frac{6\,\text{mol CO}_2}{1\,\text{mol C}_6\text{H}_{12}\text{O}_6}\right)\left(\frac{22.4\,\text{L}}{1\,\text{mol}}\right) = 1.12 \times 10^3\,\text{L CO}_2$$

51. Like any other gas, water in the gaseous state occupies a much larger volume than in the liquid state.

52. During the winter the air in a car's tires is colder, the molecules move slower and the pressure decreases. In order to keep the pressure at the manufacturer's recommended psi air needs to be added to the tire. The opposite is true during the summer.

53. Sample (b) has the greatest pressure because it has more molecules. Pressure is proportional to number of moles when temperature and volume remain constant.

54. Image (a) best represents the balloon because lowering the temperature of the gas will decrease its volume. The size of the balloon will also decrease, but the gas molecules will still be distributed throughout the volume of the balloon.

55. (a) the pressure will be cut in half
    (b) the pressure will double
    (c) the pressure will be cut in half
    (d) the pressure will increase to 3.7 atm or 2836 torr

$$PV = nRT \quad P = \frac{nRT}{V}$$

$$P = \frac{(1.5\,\text{mol})(0.0821\,\text{L atm/mol K})(303\,\text{K})}{10.\,\text{L}} = 3.7\,\text{atm}$$

$$P = 3.7\,\text{atm}\left(\frac{760\,\text{torr}}{1\,\text{atm}}\right) = 2.8 \times 10^3\,\text{torr}$$

56. (a)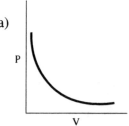

(c)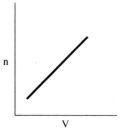

(b)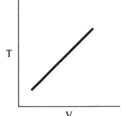

57.

Law	Factors that are constant	Factors that are variable	Graph showing the relationship of variable factors
Boyle's law	Temperature and number of moles	Pressure and volume	(c)
Charles' law	Pressure and number of moles	Volume and temperature	(b)
Avogadro's law	Pressure and temperature	Volume and number of moles	(b)

58. The can is a sealed unit and very likely still contains some of the aerosol. As the can is heated, pressure builds up in it eventually causing the can to explode and rupture with possible harm from flying debris.

59. One mole of an ideal gas occupies 22.4 liters at standard conditions. (0°C and 1 atm pressure)

$PV = nRT$

$(1.00\,\text{atm})(V) = (1.00\,\text{mol})(0.0821\,\text{L atm/mol K})(273\,\text{K})$

$V = 22.4\,\text{L}$

60. Solve for volume using $PV = nRT$

(a) $V = \dfrac{(0.2\,\text{mol Cl}_2)(0.0821\,\text{L atm/mol K})(321\,\text{K})}{(80\,\text{cm}/76\,\text{cm})\,\text{atm}} = 5\,\text{L Cl}_2$

(b) $V = \dfrac{(4.2\,\text{g NH}_3)\left(\dfrac{1\,\text{mol}}{17.03\,\text{g}}\right)\left(\dfrac{0.0821\,\text{L atm}}{\text{mol K}}\right)(262\,\text{K})}{0.65\,\text{atm}} = 8.2\,\text{L NH}_3$

(c) $V = \dfrac{(21\,\text{g SO}_3)\left(\dfrac{1\,\text{mol}}{80.07\,\text{g}}\right)\left(\dfrac{0.0821\,\text{L atm}}{\text{mol K}}\right)(328\,\text{K})}{\dfrac{110\,\text{kPa}}{101.3\dfrac{\text{kPa}}{\text{atm}}}} = 6.5\,\text{L SO}_3$

4.2 g NH_3 has the greatest volume

61. (a) 1 mol of a gas occupies 22.4 L at STP

$(1\,\text{L CH}_4)\left(\dfrac{1\,\text{mol CH}_4}{22.4\,\text{L CH}_4}\right)\left(\dfrac{6.022 \times 10^{23}\,\text{molecule CH}_4}{1\,\text{mol CH}_4}\right) = 2.69 \times 10^{22}\,\text{molecules CH}_4$

(b) Temperature must be in K; must convert torr to atm and mol to molecules

$(235°\text{F} - 32°\text{F})\dfrac{100°\text{C}}{180°\text{F}} = 113°\text{C};\quad 113 + 273 = 386\,\text{K};\quad PV = nRT \text{ or } n = \dfrac{PV}{RT}$

$\dfrac{(952\,\text{torr})(1\,\text{atm}/760\,\text{torr})(3.29\,\text{L N}_2)}{(0.0821\,\text{L atm/mol K})(386\,\text{K})} = 0.130\,\text{mol N}_2$

$(0.130\,\text{mol N}_2)\left(\dfrac{6.022 \times 10^{23}\,\text{molecule N}_2}{1\,\text{mol N}_2}\right) = 7.83 \times 10^{22}\,\text{molecule N}_2$

(c)   Temperature must be in K

$$0 + 273 = 273\ K; \quad PV = nRT \text{ or } n = \frac{PV}{RT}$$

$$\frac{(0.624\ atm)(5.05\ L\ Cl_2)}{(0.0821\ L\ atm/mol\ K)(273\ K)} = 0.141\ mol\ Cl_2$$

$$(0.141\ mol\ Cl_2)\left(\frac{6.022 \times 10^{23}\ molecule\ Cl_2}{1\ mol\ Cl_2}\right) = 8.49 \times 10^{22}\ molecule\ Cl_2$$

The container with chlorine gas contains the largest number of molecules.

62.   Assume 1 mol of each gas

(a) $SF_6 = 146.1\ g/mol\ SF_6$

$$d = \left(\frac{146.1\ g}{mol\ SF_6}\right)\left(\frac{1\ mol}{22.4\ L}\right) = 6.52\ g/L\ SF_6$$

(b)   Assume 25 °C and 1 atm pressure

$$V(at\ 25°C) = (22.4\ L\ C_2H_6)\left(\frac{298\ K}{273\ K}\right) = 24.5\ L\ C_2H_6$$

$$C_2H_6 = 30.07\ g/mol$$

$$d = \left(\frac{30.07\ g}{mol}\right)\left(\frac{1\ mol}{24.5\ L\ C_2H_6}\right) = 1.23\ g/L\ C_2H_6$$

(c)   He at −80°C and 2.15 atm

$$V = \frac{(1\ mol\ He)(0.0821\ L\ atm/mol\ K)(193\ K)}{2.15\ atm} = 7.37\ L\ He$$

$$d = \left(\frac{4.003\ g}{mol}\right)\left(\frac{1\ mol}{7.37\ L\ He}\right) = 0.543\ g/L\ He$$

SF_6 has the greatest density

63.   (a)   Sample 2 has the higher density because it has more molecules in the same volume.
       (b)   Sample 2 has the higher density because the molar mass of the individual molecules is larger.

64.   (a)   Empirical formula. Assume 100. g starting material

$$\frac{80.0\ g\ C}{12.01\ g/mol} = 6.66\ mol\ C \qquad \frac{6.66}{6.66} = 1.00$$

$$\frac{20.0\ g\ H}{1.008\ g/mol} = 19.8\ mol\ H \qquad \frac{19.8}{6.66} = 2.97$$

Empirical formula $= CH_3$
Empirical mass $= 12.01\ g + 3.024\ g = 15.03\ g$

(b) Molecular formula. $\left(\dfrac{2.01\,g}{1.5\,L}\right)\left(\dfrac{22.4\,L}{mol}\right) = 30.\ g/mol\,(molar\ mass)$

$\dfrac{30.\ g/mol}{15.03\ g/mol} = 2;\ $ Molecular formula is C_2H_6

(c) Valence electrons $= 2(4) + 6 = 14$

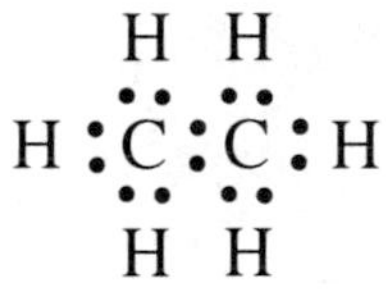

$$\begin{array}{c} \text{H}\quad\text{H} \\ \text{H}:\!\overset{..}{\underset{..}{\text{C}}}\!:\!\overset{..}{\underset{..}{\text{C}}}\!:\!\text{H} \\ \text{H}\quad\text{H} \end{array}$$

65. $PV = nRT$

(a) $\left(\dfrac{(790\ torr)(1\ atm)}{760\ torr}\right)(2.0\,L) = (n)(0.0821\ L\ atm/mol\ K)(298\ K)$

$n = 0.085\ mol\,(total\ moles)$

(b) $mol\ N_2 = total\ moles - mol\ O_2 - mol\ CO_2$

$= 0.085\ mol - \dfrac{0.65\ g\ O_2}{32.00\ g/mol} - \dfrac{0.58\ g\ CO_2}{44.01\ g/mol}$

$mol\ N_2 = 0.085\ mol - 0.020\ mol\ O_2 - 0.013\ mol\ CO_2 = 0.052\ mol\ N_2$

$(0.052\ mol\ N_2)\left(\dfrac{28.02\ g}{mol}\right) = 1.5\ g\ N_2$

(c) $P_{O_2} = (790\ torr)\left(\dfrac{0.020\ mol\ O_2}{0.085\ mol}\right) = 1.9 \times 10^2\ torr\ (O_2)$

$P_{CO_2} = (790\ torr)\left(\dfrac{0.013\ mol\ CO_2}{0.085\ mol}\right) = 1.2 \times 10^2\ torr\ (CO_2)$

$P_{N_2} = (790\ torr)\left(\dfrac{0.051\ mol\ N_2}{0.085\ mol}\right) = 4.7 \times 10^2\ torr\ (N_2)$

66. $2\,CO + O_2 \longrightarrow 2\,CO_2$
Calculate the moles of O_2 and CO to find the limiting reactant.
$PV = nRT$

O_2: $(1.8\ atm)(0.500\ L\ O_2) = (n)(0.0821\ L\ atm/mol\ K)(288\ K)$

$mol\ O_2 = 0.038\ mol$

CO: $\left(\dfrac{800\ mm\ Hg \times 1\ atm}{760\ mm\ Hg}\right)(0.500\,L) = (n)(0.0821\ L\ atm/mol\ K)(333\ K)$

$mol\ CO = 0.019\ mol$ \qquad Limiting reactant is CO

$0.0095\ mol\ O_2$ will react with $0.019\ mol\ CO$.

$(0.019\ mol\ CO)\left(\dfrac{2\ mol\ CO_2}{2\ mol\ CO}\right)\left(\dfrac{22.4\ L}{mol}\right) = 0.43\ L\ CO_2 = 430\ mL\ CO_2$

67. $PV = nRT$ or $PV = \left(\dfrac{g}{\text{molar mass}}\right)RT$

$$\left(\dfrac{1.4\,g}{cm^3}\right)\left(\dfrac{1000\,cm^3}{L}\right) = 1.4 \times 10^3\,g/L$$

$$(1.3 \times 10^9\,atm)(1.0\,L) = \left(\dfrac{1.4 \times 10^3\,g}{2.0\,g/mol}\right)(0.0821\,L\,atm/mol\,K)(T)$$

$$T = \dfrac{(1.3 \times 10^9\,atm)(1.0\,L)(2.0\,g/mol)}{(0.0821\,L\,atm/mol\,K)(1.4 \times 10^3\,g)} = 2.3 \times 10^7\,K$$

68. (a) Assume atmospheric pressure of 14.7 lb/in.² to begin with.
Total pressure in the ball $= 14.7\,lb/in.^2 + 13\,lb/in.^2 = 28\,lb/in.^2$
$PV = nRT$

$$(28\,lb/in.^2)\left(\dfrac{1\,atm}{14.7\,lb/in.^2}\right)(2.24\,L) = (n)(0.0821\,L\,atm/mol\,K)(293\,K)$$

$n = 0.18$ mol air

(b) mass of air in the ball   molar mass of air is about 29 g/mol

$$m = (0.18\,mol)\left(\dfrac{29\,g}{mol}\right) = 5.2\,g\,air$$

(c) Actually the pressure changes when the temperature changes. Since pressure is directly proportional to moles we can calculate the change in moles required to keep the pressure the same at 30°C as it was at 20°C.
$PV = nRT$

$$(28\,lb/in.^2)\left(\dfrac{1\,atm}{14.7\,lb/in.^2}\right)(2.24\,L) = (n)(0.0821\,L\,atm/mol\,K)(303\,K)$$

$n = 0.17$ mol of air required to keep the pressure the same at 30°C.
0.01 mol air (0.18 − 0.17) must be allowed to escape from the ball.

$$(0.01\,mol\,air)\left(\dfrac{29\,g}{mol}\right) = 0.29\,g \quad or \quad 0.3\,g\,air\,must\,be\,allowed\,to\,escape.$$

69. Use the combined gas laws to calculate the bursting temperature (T_2).

$$\dfrac{P_1V_1}{T_1} = \dfrac{P_2V_2}{T_2} \qquad P_1 = 65\,cm \qquad P_2 = 1.00\,atm\left(\dfrac{76\,cm}{1\,atm}\right) = 76\,cm$$

$$V_1 = 1.75\,L \qquad V_2 = 2.00\,L$$
$$T_1 = 20°C(293\,K) \qquad T_2 = T_2$$

$$T_2 = \dfrac{P_2V_2T_1}{P_1V_1} = \dfrac{(76\,cm)(2.00\,L)(293\,K)}{(65\,cm)(1.75\,L)} = 392\,K(119°C)$$

70. To double the volume of a gas, at constant pressure, the temperature (K) must be doubled.

$$\frac{V_1}{T_1} = \frac{V_2}{T_2} \qquad V_2 = 2\,V_1$$

$$\frac{V_1}{T_1} = \frac{2\,V_1}{T_2} \qquad T_2 = \frac{2\,V_1 T_1}{V_1} \qquad T_2 = 2\,T_1$$

$$T_2 = 2(300.\text{ K}) = 600.\text{ K} = 327°C$$

71. V = volume at 22°C and 740 torr

   2 V = volume after change in temperature(P constant)

   V = volume after change in pressure(T constant)

   Since temperature is constant, $P_1V_1 = P_2V_2$ or $P_2 = \dfrac{P_1V_1}{V_2}$

   $$P_2 = (740\text{ torr})\left(\frac{2\,V}{V}\right) = 1.5 \times 10^3 \text{ torr (pressure to change 2 V to V)}$$

72. Use the combined gas laws

   $$\frac{(P_1)(V_1)}{(T_1)} = \frac{(P_2)(V_2)}{(T_2)} \text{ or } T_2 = \frac{(T_1)(P_2)(V_2)}{(P_1)(V_1)}$$

   Since the volume stays constant, $V_1 = V_2$ and the equation reduces to $T_2 = \dfrac{(T_1)(P_2)}{(P_1)}$

   $$T_2 = \frac{(500.\text{ torr})(295\text{ K})}{700.\text{ torr}} = 211\text{ K} = -62°C$$

73. Use the combined gas laws

   $$\frac{(P_1)(V_1)}{(T_1)} = \frac{(P_2)(V_2)}{(T_2)} \quad \text{or} \quad T_2 = \frac{(T_1)(P_2)(V_2)}{(P_1)(V_1)}$$

   The volume of the tires remains constant (until they burst) so $V_1 = V_2$ and the equation reduces to
   $T_2 = \dfrac{(T_1)(P_2)}{(P_1)}$

   71.0°F = 21.7°C = 295 K

   $$T_2 = \frac{(44\text{ psi})(295\text{ K})}{30.\text{ psi}} = 433\text{ K} = 160°C = 320°F$$

74. Use the combined gas laws.

   $$\frac{P_1V_1}{T_1} = \frac{P_2V_2}{T_2} \quad \text{or} \quad P_2 = \frac{P_1V_1T_2}{V_2T_1} \qquad P_1 \text{ and } T_1 \text{ are at STP}$$

   $$P_2 = \frac{(1.00\text{ atm})(800.\text{ mL})(303\text{ K})}{(250.\text{ mL})(273\text{ K})} = 3.55\text{ atm}$$

75. 1 mol of a gas occupies 22.4 L at STP

$$(9.14 \text{ g SO}_2)\left(\frac{1 \text{ mol}}{64.07 \text{ g}}\right)\left(\frac{22.4 \text{ SO}_2}{1 \text{ mol SO}_2}\right) = 3.20 \text{ L SO}_2$$

76. Use the combined gas law $\dfrac{P_1 V_1}{T_1} = \dfrac{P_2 V_2}{T_2}$ or $V_2 = \dfrac{P_1 V_1 T_2}{P_2 T_1}$

First calculate the volume at STP.

$$V_2 = \frac{(400. \text{ torr})(600. \text{ mL N}_2\text{O})(273 \text{ K})}{(760. \text{ torr})(313 \text{ K})} = 275 \text{ mL N}_2\text{O} = 0.275 \text{ L N}_2\text{O}$$

At STP, a mole of any gas has a volume of 22.4 L

$$(0.275 \text{ L N}_2\text{O})\left(\frac{1 \text{ mol}}{22.4 \text{ L}}\right)\left(\frac{6.022 \times 10^{23} \text{ molecules}}{1 \text{ mol}}\right) = 7.39 \times 10^{21} \text{ molecules N}_2\text{O}$$

Each molecule of N_2O contains 3 atoms, so:

$$(7.39 \times 10^{21} \text{ molecules N}_2\text{O})\left(\frac{3 \text{ atoms}}{1 \text{ molecule N}_2\text{O}}\right) = 2.22 \times 10^{22} \text{ atoms}$$

77. Use the combined gas laws

$$\frac{(P_1)(V_1)}{(T_1)} = \frac{(P_2)(V_2)}{(T_2)} \quad \text{or} \quad P_2 = \frac{(P_1)(T_2)(V_1)}{(T_1)(V_2)}$$

Since the volume stays constant (unless it does burst), $V_1 = V_2$ and the equation reduces to

$$P_2 = \frac{(P_1)(T_2)}{(T_1)}$$

$T_1 = 25°C + 273 = 298 \text{ K}$

$T_2 = 212°F = 100°C = 373 \text{ K}$

$$P_2 = \frac{(32 \text{ lb/in.}^2)(373 \text{ K})}{298 \text{ K}} = 40. \text{ lb/in.}^2$$

At 212°F the tire pressure is 40. lb/in.^{2}
The tire will not burst.

78. A column of mercury at 1 atm pressure is 760 mm Hg high. The density of mercury is 13.6 times that of water, so a column of water at 1 atm pressure should be 13.6 times as high as that for mercury.
$(760 \text{ mm})(13.6) = 1.03 \times 10^4 \text{ mm}(33.8 \text{ ft})$

79. Use the ideal gas equation

$$PV = nRT \qquad n = \frac{RT}{PV}$$

Change 2.20×10^3 lb/in.² to atmosphere

$$\left(2.20 \times 10^3 \text{ lb/in.}^2\right)\left(\frac{1 \text{ atm}}{14.7 \text{ lb/in.}^2}\right) = 150. \text{ atm}$$

$$n = \frac{(150. \text{ atm})(55 \text{ L } O_2)}{\left(\dfrac{0.0821 \text{ L} \cdot \text{atm}}{\text{mol} \cdot \text{K}}\right)(300. \text{ K})} = 3.3 \times 10^2 \text{ mol } O_2$$

80. The conversion is: $m^3 \longrightarrow cm^3 \longrightarrow mL \longrightarrow L \longrightarrow mol$

$$\left(1.00 \text{ m}^3 \text{ Cl}_2\right)\left(\frac{100 \text{ cm}}{1 \text{ m}}\right)^3\left(\frac{1 \text{ mL}}{1 \text{ cm}^3}\right)\left(\frac{1 \text{ L}}{1000 \text{ mL}}\right)\left(\frac{1 \text{ mol}}{22.4 \text{ L}}\right) = 44.6 \text{ mol } Cl_2$$

81. First calculate the moles of gas and then convert moles to molar mass.

$$\left(0.560 \text{ L}\right)\left(\frac{1 \text{ mol}}{22.4 \text{ L}}\right) = 0.0250 \text{ mol}$$

$$\frac{1.08 \text{ g}}{0.0250 \text{ mol}} = 43.2 \text{ g/mol (molar mass)}$$

82. The conversion is: $g/L \longrightarrow g/mol$

$$\left(\frac{1.78 \text{ g}}{L}\right)\left(\frac{22.4 \text{ L}}{\text{mol}}\right) = 39.9 \text{ g/mol (molar mass)}$$

83. $PV = nRT$

(a) $\quad V = \dfrac{nRT}{P} = \dfrac{(0.510 \text{ mol } H_2)(0.0821 \text{ L atm/mol K})(320. \text{ K})}{1.6 \text{ atm}} = 8.4 \text{ L } H_2$

(b) $\quad n = \dfrac{PV}{RT} = \dfrac{(600 \text{ torr})(1 \text{ atm}/760 \text{ torr})(16.0 \text{ L } CH_4)}{(0.0821 \text{ L atm/mol K})(300. \text{ K})} = 0.513 \text{ mol } CH_4$

The molar mass for CH_4 is 16.04 g/mol

$(16.04 \text{ g/mol})(0.513 \text{ mol}) = 8.23 \text{ g } CH_4$

(c) $\quad PV = nRT$, but $n = \dfrac{g}{M}$ where M is the molar mass and g is the grams of the gas.

Thus, $PV = \dfrac{gRT}{M}$. To determine density, $d = g/V$.

Solving $PV = \dfrac{gRT}{M}$ for $\dfrac{g}{V}$ produces $\dfrac{g}{V} = \dfrac{PM}{RT}$.

$$d = \frac{g}{V} = \frac{(4.00 \text{ atm})(44.01 \text{ g/mol } CO_2)}{(0.0821 \text{ L atm/mol K})(253 \text{ K})} = 8.48 \text{ g/L } CO_2$$

(d) Since $d = \dfrac{g}{V} = \dfrac{PM}{RT}$ from part (c), solve for M (molar mass)

$$M = \frac{dRT}{P} = \frac{(2.58 \text{ g/L})(0.0821 \text{ L atm/mol K})(300. \text{ K})}{1.00 \text{ atm}} = 63.5 \text{ g/mol (molar mass)}$$

84. $C_2H_2(g) + 2\,HF(g) \longrightarrow C_2H_4F_2(g)$

$1.0 \text{ mol } C_2H_2 \longrightarrow 1.0 \text{ mol } C_2H_4F_2$

$$(5.0 \text{ mol HF})\left(\frac{1 \text{ mol } C_2H_4F_2}{2 \text{ mol HF}}\right) = 2.5 \text{ mol } C_2H_4F_2$$

C_2H_2 is the limiting reactant. 1.0 mol $C_2H_4F_2$ forms, no moles C_2H_2 remain.
According to the equation, 2.0 mol HF yields 1.0 mol $C_2H_4F_2$. Therefore,
5.0 mol HF − 2.0 mol HF = 3.0 mol HF unreacted
The flask contains 1.0 mol $C_2H_4F_2$ and 3.0 mol HF when the reaction is complete.
The flask contains 4.0 mol of gas.

$$P = \frac{nRT}{V} = \frac{(4.0 \text{ mol})(0.0821 \text{ L atm/mol K})(273 \text{ K})}{10.0 \text{ L}} = 9.0 \text{ atm}$$

85. $(8.30 \text{ mol Al})\left(\dfrac{3 \text{ mol } H_2}{2 \text{ mol Al}}\right)\left(\dfrac{22.4 \text{ L}}{\text{mol}}\right) = 279 \text{ L } H_2 \text{ at STP}$

86. Assume 100. g of material to start with. Calculate the empirical formula.

C $(85.7 \text{ g})\left(\dfrac{1 \text{ mol}}{12.01 \text{ g}}\right) = 7.14 \text{ mol}$ $\dfrac{7.14}{7.14} = 1.00 \text{ mol}$

H $(14.3 \text{ g})\left(\dfrac{1 \text{ mol}}{1.008 \text{ g}}\right) = 14.2 \text{ mol}$ $\dfrac{14.2}{7.14} = 1.99 \text{ mol}$

The empirical formula is CH_2. To determine the molecular formula, the molar mass must be known.

$$\left(\frac{2.50 \text{ g}}{\text{L}}\right)\left(\frac{22.4 \text{ L}}{\text{mol}}\right) = 56.0 \text{ g/mol (molar mass)}$$

The empirical formula mass is 14.0 $\dfrac{56.0}{14.0} = 4$

Therefore, the molecular formula is $(CH_2)_4 = C_4H_8$

87. $2\,CO(g) + O_2(g) \rightarrow 2\,CO_2(g)$ Determine the limiting reactant

$$(10.0 \text{ mol CO})\left(\frac{2 \text{ mol } CO_2}{2 \text{ mol CO}}\right) = 10.0 \text{ mol } CO_2 \text{(from CO)}$$

$$(8.0 \text{ mol } O_2)\left(\frac{2 \text{ mol } CO_2}{1 \text{ mol } O_2}\right) = 16 \text{ mol } CO_2 \text{(from } O_2)$$

CO: the limiting reactant, O_2: in excess, 3.0 mol O_2 unreacted.

(a)   10.0 mol CO react with 5.0 mol O_2

      10.0 mol CO_2 and 3.0 mol O_2 are present, no CO will be present.

(b) $P = \dfrac{nRT}{V} = \dfrac{(13\,mol)(0.0821\,L\,atm/mol\,K)(273\,K)}{10.\,L} = 29\,atm$

88. $2\,KClO_3(s) \xrightarrow{\Delta} 2\,KCl(s) + 3\,O_2(g)$

First calculate the moles of O_2 produced. Then calculate the grams of $KClO_3$ required to produce the O_2. Then calculate the % $KClO_3$.

$(0.25\,L\,O_2)\left(\dfrac{1\,mol}{22.4\,L}\right) = 0.011\,mol\,CO_2$

$(0.011\,mol\,O_2)\left(\dfrac{2\,mol\,KClO_3}{3\,mol\,O_2}\right)\left(\dfrac{122.6\,g}{mol}\right) = 0.90\,g\,KClO_3$ in the sample

$\left(\dfrac{0.90\,g}{1.20\,g}\right)(100) = 75\%\,KClO_3$ in the mixture

89.   Assume 1.00 L of air. The mass of 1.00 L of air is 1.29 g.

$\dfrac{P_1 V_1}{T_1} = \dfrac{P_2 V_2}{T_2}$

$V_2 = \dfrac{P_1 V_1 T_2}{P_2 T_1} = \dfrac{(760\,torr)(1.00\,L)(290\,K)}{(450\,torr)(273\,K)} = 1.8\,L$

$d = \dfrac{m}{V} = \dfrac{1.29\,g}{1.8\,L} = 0.72\,g/L$

90.   Each gas behaves as though it were alone in a 4.0 L system.

(a)   After expansion: $P_1 V_1 = P_2 V_2$

    For CO_2 $P_2 = \dfrac{P_1 V_1}{V_2} = \dfrac{(150.\,torr)(3.0\,L)}{4.0\,L} = 1.1 \times 10^2\,torr$

    For H_2 $P_2 = \dfrac{P_1 V_1}{V_2} = \dfrac{(50.\,torr)(1.0\,L)}{4.0\,L} = 13\,torr$

(b) $P_{total} = P_{H_2} + P_{CO_2} = 110\,torr + 13\,torr = 120\,torr$ (2 sig. figures)

91.   Use the combined gas laws

$\dfrac{(P_1)(V_1)}{(T_1)} = \dfrac{(P_2)(V_2)}{(T_2)}$ or $P_2 = \dfrac{(P_1)(T_2)(V_1)}{(T_1)(V_2)}$

$P_1 = 40.0\,atm$ $P_2 = unknown$

$V_1 = 50.0\,L$ $V_2 = 50.0\,L$

$T_1 = 25°C = 298\,K$ $T_2 = 25°C + 152°C = 177°C = 450\,K$

$$P_2 = \frac{(40.0\text{ atm})(450\text{ K})(\cancel{50.0\text{ L}})}{(298\text{ K})(\cancel{50.0\text{ L}})} = 60.4\text{ atm}$$

(Note that the volume canceled out of the expression because it stays constant.)

92. You can identify the gas by determining its density.

mass of gas $= 1.700\text{ g} - 0.500\text{ g} = 1.200\text{ g}$

volume of gas: Charles law problem. Correct volume to 273 K

$$\frac{V_1}{T_1} = \frac{V_2}{T_2} \qquad V_2 = \frac{V_1 T_2}{T_1} = \frac{(0.4478\text{ L})(273\text{ K})}{323\text{ K}} = 0.3785\text{ L}$$

$$d = \frac{m}{V} = \frac{1.200\text{ g}}{0.3785\text{ L}} = 3.170\text{ g/L}$$

gas is chlorine (see Table 12.4)

93. **1st step:** find the volume of CO_2 that must be produced

$$\text{volume CO}_2 = (\text{volume of batter})(55.0\%) = (1.32\text{ L batter})\left(\frac{55.0\text{ L CO}_2}{100\text{ L batter}}\right) = 0.726\text{ L CO}_2$$

2nd step: find the moles of CO_2 needed to produce the necessary rise in the cupcakes

$$PV = nRT \quad \text{or} \quad n = \frac{PV}{RT}$$

$$P = (738\text{ torr})\left(\frac{1\text{ atm}}{760\text{ torr}}\right) = 0.971\text{ atm}$$
$$V = 0.726\text{ L}$$
$$T = 325°F = 163°C = 436\text{ K}$$

$$n = \frac{(0.971\text{ atm})(0.726\text{ L})}{(0.0821\text{ L atm/mol K})(436\text{ K})} = 0.0197\text{ mol CO}_2$$

3rd step: Calculate the mass of sodium bicarbonate which will produce the CO_2 needed

$$3\text{ NaHCO}_3 + \text{H}_3\text{C}_6\text{H}_5\text{O}_7 \rightarrow \text{Na}_3\text{C}_6\text{H}_5\text{O}_7 + 3\text{ H}_2\text{O} + 3\text{ CO}_2$$

$$(0.0197\text{ mol CO}_2)\left(\frac{3\text{ mol NaHCO}_3}{3\text{ mol CO}_2}\right)\left(\frac{84.02\text{ g}}{1\text{ mol}}\right) = 1.66\text{ g NaHCO}_3$$

4th step: calculate mass of sodium bicarbonate needed to add since only 63.7% will decompose to form carbon dioxide

$$(1.66\text{ g NaHCO}_3\text{ reacted})\left(\frac{100\text{ g NaHCO}_3\text{ added}}{63.7\text{ g NaHCO}_3\text{ reacted}}\right) = 2.61\text{ g NaHCO}_3\text{ added}$$

CHAPTER 13

LIQUIDS

SOLUTIONS TO REVIEW QUESTIONS

1. At 0°C, all three substances, H_2S, H_2Se, and H_2Te, are gases, because they all have boiling points below 0°C.

2. Liquids are made of particles that are close together, they are not compressible and they have definite volume. Solids also exhibit similar properties.

3. Liquids take the shape of the container they are in. Gases also exhibit this property.

4. The water in both containers would have the same vapor pressure, for it is a function of the temperature of the liquid.

5. Vapor pressure is the pressure exerted by a vapor when it is in equilibrium with its liquid. The liquid molecules on the surface and the gas molecules above the liquid are the important players in vapor pressure. One liter of water will evaporate at the same rate as 100 mL of water if they are at the same temperature and have the same surface area. The liquid and vapor in both cases will come to equilibrium in the same amount of time and they will have the same vapor pressure. Vapor pressure is a characteristic of the type of liquid, not the number of molecules of a liquid.

6. In Figure 13.5, it would be case (b) in which the atmosphere would reach saturation. The vapor pressure of water is the same in both (a) and (b), but since (a) is an open container the vapor escapes into the atmosphere and doesn't reach saturation.

7. If ethyl ether and ethyl alcohol were both placed in a closed container, (a) both substances would be present in the vapor, for both are volatile liquids; (b) ethyl ether would have more molecules in the vapor because it has a higher vapor pressure at a given temperature.

8. Rubbing alcohol feels cold when applied to the skin, because the evaporation of the alcohol absorbs heat from the skin. The alcohol has a fairly high vapor pressure (low boiling point) and evaporates quite rapidly. This produces the cooling effect.

9. (a) Order of increasing rate of evaporation: mercury, acetic acid, water, toluene, benzene, carbon tetrachloride, methyl alcohol, bromine.
    (b) Highest boiling point is mercury. Lowest boiling point is bromine.

10. As temperature increases, molecular velocities increase. At higher molecular velocities, it becomes easier for molecules to break away from the attractive forces in the liquid.

11. The pressure of the atmosphere must be 1.00 atmosphere, otherwise the water would be boiling at some other temperature.

12. The thermometer would be at about 70°C. The liquid is boiling, which means its vapor pressure equals the confining pressure. From Table 13.7, we find that ethyl alcohol has a vapor pressure of 543 torr at 70°C.

13. At 30 torr, H_2O would boil at approximately 29°C, ethyl alcohol at 14°C, and ethyl ether at some temperature below 0°C.

14. (a) At a pressure of 500 torr, water boils at 88°C.
    (b) The normal boiling point of ethyl alcohol is 78°C.
    (c) At a pressure of 0.50 atm (380 torr), ethyl ether boils at 16°C.

15. Water boils when its vapor pressure equals the prevailing atmospheric pressure over the water. In order for water to boil at 50°C, the pressure over the water would need to be reduced to a point equal to the vapor pressure of the water (92.5 torr).

16. In a pressure cooker, the temperature at which the water boils increases above its normal boiling point, because the water vapor (steam) formed by boiling cannot escape. This results in an increased pressure over water and, consequently, an increased boiling temperature.

17. Vapor pressure varies with temperature. The temperature at which the vapor pressure of a liquid equals the prevailing pressure is the boiling point of the liquid.

18. Ammonia would have a higher vapor pressure than SO_2 at −40°C because it has a lower boiling point (NH_3 is more volatile than SO_2).

19. As the temperature of a liquid increases, the kinetic energy of the molecules as well as the vapor pressure of the liquid increases. When the vapor pressure of the liquid equals the external pressure, boiling begins with many of the molecules having enough energy to escape from the liquid. Bubbles of vapor are formed throughout the liquid and these bubbles rise to the surface, escaping as boiling continues.

20. 34.6°C, the boiling point of ethyl ether. (See Table 13.1)

21. The potential energy is greater in the liquid water than in the ice. The heat necessary to melt the ice increases the potential energy of the liquid, thus allowing the molecules greater freedom of motion. The potential energy of steam (gas) is greater than that of liquid water.

22. Based on Figure 13.8:
    (a) Line BC is horizontal because the temperature remains constant during the entire process of melting. The energy input is absorbed in changing from the solid to the liquid state.
    (b) During BC, both solid and liquid phases are present.
    (c) The line DE represents the change from liquid water to steam (vapor) at the boiling temperature of water.

23. Apply heat to an ice-water mixture, the heat energy is absorbed to melt the ice (heat of fusion), rather than warm the water, so the temperature remains constant until all the ice has melted.

24. Ice at 0°C contains less heat energy than water at 0°C. Heat must be added to convert ice to water, so the water will contain that much additional heat energy.

25. The boiling liquid remains at constant temperature because the added heat energy is being used to convert the liquid to a gas, i.e., to supply the heat of vaporization for the liquid at its boiling point.

26. Intermolecular forces are the attractive forces between molecules. These forces hold molecules together to form liquids and solids. Intramolecular forces are forces between atoms in a molecule. These forces hold the atoms in a molecule together.

27. Polar bonds are covalent bonds in which atoms do not share electrons equally.

28. The ability of a molecule to form instantaneous dipoles is most dependant on its number of electrons.

29. The heat of vaporization of water would be lower if water molecules were linear instead of bent. If linear, the molecules of water would be nonpolar. The relatively high heat of vaporization of water is a result of the molecule being highly polar and having strong dipole-dipole and hydrogen bonding attraction for other water molecules.

30. Ethyl alcohol exhibits hydrogen bonding; ethyl ether does not. This is indicated by the high heat of vaporization of ethyl alcohol, even though its molar mass is much less than the molar mass of ethyl ether.

31. Although a linear water molecule would be non-polar due to its symmetry, the individual O—H bonds would still be polar meaning that the hydrogen's would still be able to form hydrogen bonds. The number of possible hydrogen bonds would not change. The overall intermolecular attractive force would be lower though due to its non-polar nature.

32. Water, at $80°C$, will have fewer hydrogen bonds than water at $40°C$. At the higher temperature, the molecules of water are moving faster than at the lower temperature. This results in less hydrogen bonding at the higher temperature.

33. $H_2NCH_2CH_2NH_2$ has two polar NH_2 groups. It should, therefore, show more hydrogen bonding and a higher boiling point ($117°C$) versus $49°C$ for $CH_3CH_2CH_2NH_2$.

34. Water has a relatively high boiling point because there is a high attraction between molecules due to hydrogen bonding.

35. HF has a higher boiling point than HCl because of the strong hydrogen bonding in HF (F is the most electronegative element). Neither F_2 nor Cl_2 will have hydrogen bonding, so the compound, F_2, with the lower molar mass, has the lower boiling point.

36. Prefixes preceding the word hydrate are used in naming hydrates, indicating the number of molecules of water present in the formulas. The prefixes used are:

mono = 1	di = 2	tri = 3	tetra = 4	penta = 5
hexa = 6	hepta = 7	octa = 8	nona = 9	deca = 10

37.

38. Melting point, boiling point, heat of fusion, heat of vaporization, density, and crystal structure in the solid state are some of the physical properties of water that would be very different, if the molecules were linear and nonpolar instead of bent and highly polar. For example, the boiling point, melting point, heat of fusion and heat of vaporization would be lower because linear molecules have no dipole moment and the attraction among molecules would be much less.

39. Physical properties of water:
    (a) melting point, 0°C
    (b) boiling point, 100°C (at 1 atm pressure)
    (c) colorless
    (d) odorless
    (e) tasteless
    (f) heat of fusion, 335 J/g (80 cal/g)
    (g) heat of vaporization, 2.26 kJ/g (540 cal/g)
    (h) density = 1.0 g/mL (at 4°C)
    (i) specific heat = 4.184 J/g°C

40. For water, to have its maximum density, the temperature must be 4°C, and the pressure sufficient to keep it liquid, $d = 1.0$ g/mL

41. Ice floats in water because it is less dense than water. The density of ice at 0°C is 0.915 g/mL. Liquid water, however, has a density of 1.00 g/mL. Ice will sink in ethyl alcohol, which has a density of 0.789 g/mL.

42. If the lake is in an area where the temperature is below freezing for part of the year, the expected temperature would be 4°C at the bottom of the lake. This is because the surface water would cool to 4°C (maximum density) and sink.

43. The formation of hydrogen and oxygen from water is an endothermic reaction, due to the following evidence:
    (a) Energy must continually be provided to the system for the reaction to proceed. The reaction will cease when the energy source is removed.
    (b) The reverse reaction, burning hydrogen in oxygen, releases energy as heat.

SOLUTIONS TO EXERCISES

1. $CH_3Br < CH_3Cl < CH_3F$

2. $H_2Se < H_2S < H_2O$

3. $CCl_4, < CBr_4 < CI_4$

4. $CO_2 < SO_2 < CS_2$

5. In which of the following substances would you expect to find hydrogen bonding?
   (a) C_3H_7OH will hydrogen bond; one of the hydrogens is bonded to oxygen.
   (b) H_2O_2 will hydrogen bond; hydrogen is bonded to oxygen.
   (c) $CHCl_3$ will not hydrogen bond; hydrogen is not bonded to fluorine, oxygen, or nitrogen.
   (d) PH_3 will not hydrogen bond; hydrogen is not bonded to fluorine, oxygen, or nitrogen.
   (e) HF will hydrogen bond; hydrogen is bonded to fluorine.

6. (a) HI will not hydrogen bond; hydrogen is not bonded to fluorine, oxygen, or nitrogen.
   (b) NH_3 will hydrogen bond; hydrogen is bonded to nitrogen.
   (c) CH_2F_2 will not hydrogen bond; hydrogen is not bonded to fluorine, oxygen, or nitrogen.
   (d) C_2H_5OH will hydrogen bond; one of the hydrogens is bonded to oxygen.
   (e) H_2O will hydrogen bond; hydrogen is bonded to oxygen.

7. (a) 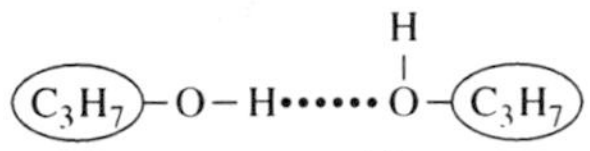 C_3H_7OH

   (b) 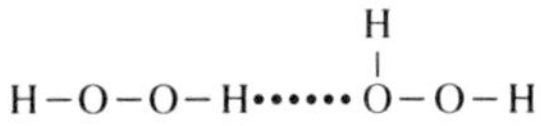 H_2O_2

   (c) HF

8. (a) 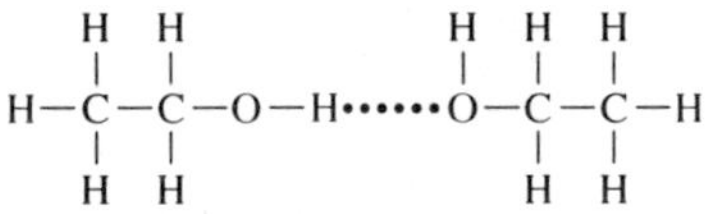 C_2H_5OH

   (b) 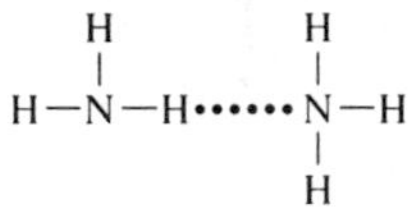 NH_3

   (c) 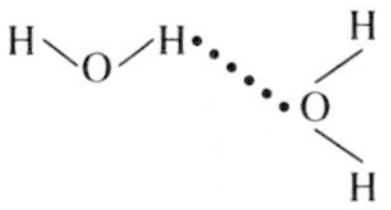 H_2O

9. The adhesive forces between the cotton fabric of the T-shirt and the water are stronger than the cohesive forces between water molecules causing the water to absorb into the fabric. The cohesive forces between water molecules is stronger than the adhesive forces between the raincoat fabric and water causing the water to bead up on the raincoat.

10. Water forming beaded droplets is an example of cohesive forces. The water molecules have stronger attractive forces for other water molecules than they do for the surface.

11. (a) barium bromide dihydrate
    (b) aluminum chloride hexahydrate
    (c) iron(III) phosphate tetrahydrate

12. (a) magnesium ammonium phosphate hexahydrate
    (b) iron(II) sulfate heptahydrate
    (c) tin(IV) chloride pentahydrate

13. $(25.0 \text{ g Na}_2\text{CO}_3 \cdot 10 \text{ H}_2\text{O}) \left(\dfrac{1 \text{ mol}}{286.2 \text{ g}} \right) = 0.0874 \text{ mol Na}_2\text{CO}_3 \cdot 10 \text{ H}_2\text{O}$

14. $(25.0 \text{ g Na}_2\text{B}_4\text{O}_7 \cdot 10 \text{ H}_2\text{O}) \left(\dfrac{1 \text{ mol}}{381.4 \text{ g}} \right) = 0.0655 \text{ mol Na}_2\text{B}_4\text{O}_7 \cdot 10 \text{ H}_2\text{O}$

15. (a) $(125 \text{ g MgSO}_4 \cdot 7 \text{ H}_2\text{O}) \left(\dfrac{1 \text{ mol}}{246.5 \text{ g}} \right) \left(\dfrac{7 \text{ mol H}_2\text{O}}{1 \text{ mol MgSO}_4 \cdot 7 \text{ H}_2\text{O}} \right) \left(\dfrac{18.02 \text{ g}}{\text{mol}} \right) = 64.0 \text{ g H}_2\text{O}$

    (b) $(125 \text{ g MgSO}_4 \cdot 7 \text{ H}_2\text{O}) \left(\dfrac{1 \text{ mol}}{246.5 \text{ g}} \right) \left(\dfrac{1 \text{ mol MgSO}_4}{1 \text{ mol MgSO}_4 \cdot 7 \text{ H}_2\text{O}} \right) \left(\dfrac{120.4 \text{ g}}{\text{mol}} \right) = 61.1 \text{ g MgSO}_4$

16. (a) $(125 \text{ g AlCl}_3 \cdot 6 \text{ H}_2\text{O}) \left(\dfrac{1 \text{ mol}}{241.4 \text{ g}} \right) \left(\dfrac{6 \text{ mol H}_2\text{O}}{1 \text{ mol AlCl}_3 \cdot 6 \text{ H}_2\text{O}} \right) \left(\dfrac{18.02 \text{ g}}{\text{mol}} \right) = 56.0 \text{ g H}_2\text{O}$

    (b) $(125 \text{ g AlCl}_3 \cdot 6 \text{ H}_2\text{O}) \left(\dfrac{1 \text{ mol}}{241.4 \text{ g}} \right) \left(\dfrac{1 \text{ mol AlCl}_3}{1 \text{ mol AlCl}_3 \cdot 6 \text{ H}_2\text{O}} \right) \left(\dfrac{133.3 \text{ g}}{\text{mol}} \right) = 69.0 \text{ g AlCl}_3$

17. $\% \text{ H}_2\text{O} = \left(\dfrac{6 \text{ H}_2\text{O}}{1 \text{ CoCl}_2 \cdot 6 \text{ H}_2\text{O}} \right) = \left(\dfrac{(6)(18.02 \text{ g})}{(238.0 \text{ g})} \right) (100) = 45.43\% \text{ H}_2\text{O}$

18. Assume 1 mol of the compound which contains 2 mol of water.

    $\% \text{ H}_2\text{O} = \left(\dfrac{\text{g H}_2\text{O}}{\text{g CaSO}_4 \cdot 2 \text{ H}_2\text{O}} \right) (100) = \left(\dfrac{(2)(18.02 \text{ g})}{(172.2 \text{ g})} \right) (100) = 20.93\% \text{ H}_2\text{O}$

19. Assume 100. g of the compound.

    $(0.2066)(100. \text{ g}) = 20.66 \text{ g Fe}$ $(0.3935)(100. \text{ g}) = 39.35 \text{ g Cl}$

    $(0.3999)(100. \text{ g}) = 39.99 \text{ g H}_2\text{O}$

    $(20.66 \text{ g Fe}) \left(\dfrac{1 \text{ mol}}{55.85 \text{ g}} \right) = 0.3699 \text{ mol Fe}$ $\dfrac{0.3699}{0.3699} = 1.000$

    $(39.35 \text{ g Cl}) \left(\dfrac{1 \text{ mol}}{35.45 \text{ g}} \right) = 1.110 \text{ mol Cl}$ $\dfrac{1.110}{0.3699} = 3.001$

    $(39.99 \text{ g H}_2\text{O}) \left(\dfrac{1 \text{ mol}}{18.02 \text{ g}} \right) = 2.219 \text{ mol H}_2\text{O}$ $\dfrac{2.219}{0.3699} = 5.999$

    The empirical formula is $\text{FeCl}_3 \cdot 6 \text{ H}_2\text{O}$

20. Assume 100. g of the compound.

    $(0.2469)(100.\text{ g}) = 24.69$ g Ni $(0.2983)(100.\text{ g}) = 29.83$ g Cl

    $(0.4548)(100.\text{ g}) = 45.48$ g H_2O

    $(24.69 \text{ g Ni})\left(\dfrac{1 \text{ mol}}{58.69 \text{ g}}\right) = 0.4207$ mol Ni $\qquad \dfrac{0.4207}{0.4207} = 1.000$

    $(29.83 \text{ g Cl})\left(\dfrac{1 \text{ mol}}{35.45 \text{ g}}\right) = 0.8415$ mol Cl $\qquad \dfrac{0.8415}{0.4207} = 2.000$

    $(45.48 \text{ g } H_2O)\left(\dfrac{1 \text{ mol}}{18.02 \text{ g}}\right) = 2.524$ mol H_2O $\qquad \dfrac{2.524}{0.4207} = 6.000$

    The formula is $NiCl_2 \cdot 6\,H_2O$

21. Energy (E_a) to heat the water to steam from $15°C \rightarrow 100.°C$

    $E_a = (m)(\text{specific heat})(\Delta t) = (275 \text{ g})\left(\dfrac{4.184 \text{ J}}{\text{g}°C}\right)(85°C) = 9.8 \times 10^4$ J

    Energy (E_b) to convert water at $100°C$ to steam: heat of vaporization $= 2.26 \times 10^3$ J/g
    $E_b = (m)(\text{heat of vaporization}) = (275 \text{ g})(2.26 \times 10^3 \text{ J/g}) = 6.22 \times 10^5$ J
    $E_{total} = E_a + E_b = (9.8 \times 10^4 \text{ J}) + (6.22 \times 10^5 \text{ J}) = 7.20 \times 10^5$ J

22. Energy (E_a) to cool the water from $35°C \rightarrow 0°C$

    $E_a = (m)(\text{specific heat})(\Delta t) = (325 \text{ g})\left(\dfrac{4.184 \text{ J}}{\text{g}°C}\right)(35°C) = 4.8 \times 10^4$ J

    Energy (E_b) to convert water to ice: heat of fusion $= 335$ J/g
    $E_b = (m)(\text{heat of fusion}) = (325 \text{ g})(335 \text{ J/g}) = 1.09 \times 10^5$ J
    $E_{total} = E_a + E_b = (4.8 \times 10^4 \text{ J}) + (1.09 \times 10^5 \text{ J}) = 1.57 \times 10^5$ J

23. Energy released in cooling the water: $25°C$ to $0°C$

    $E = (m)(\text{specific heat})(\Delta t) = (300. \text{ g})\left(\dfrac{4.184 \text{ J}}{\text{g}°C}\right)(25°C) = 3.1 \times 10^4$ J

    Energy required to melt the ice
    $E = (m)(\text{heat of fusion}) = (100. \text{ g})(335 \text{ J/g}) = 3.35 \times 10^4$ J

    Less energy is released in cooling the water than is required to melt the ice. Ice will remain and the water will be at $0°C$.

24. Energy to heat the water = energy to condense the steam

    $(300. \text{ g})\left(\dfrac{4.184 \text{ J}}{\text{g}°C}\right)(100.°C - 25°C) = (m)(2259 \text{ J/g})$

    $m = 42$ g (grams of steam required to heat the water to $100.°C$) 42 g of steam are required to heat 300. g of water to $100.°C$. Since only 35 g of steam are added to the system, the final temperature will be less than $100.°C$. Not sufficient steam.

25. Energy lost by warm water = energy gained by the ice

x = final temperature

$$\text{mass}(H_2O) = (1.5 \text{ L } H_2O)\left(\frac{1000 \text{ mL}}{L}\right)\left(\frac{1.0 \text{ g}}{mL}\right) = 1500 \text{ g}$$

$$(1500 \text{ g})\left(\frac{4.184 \text{ J}}{g\,^\circ C}\right)(75^\circ C - x) = (75 \text{ g})\left(335\frac{J}{g}\right) + (75 \text{ g})\left(\frac{4.184 \text{ J}}{g\,^\circ C}\right)(x - 0^\circ C)$$

$$\left(4.707 \times 10^5 \text{ J}\right) - (6276x \text{ J}/^\circ C) = 2.51 \times 10^4 \text{ J} + 313.8x \text{ J}/^\circ C$$

$$4.46 \times 10^5 \text{ J} = 6.5898 \times 10^3 x \text{ J}/^\circ C$$

$$x = 68^\circ C$$

26. $E = (m)$(heat of fusion)

(500. g)(335 J/g) = 167,000 J needed to melt the ice
9560 J < 167,500 J

Since 167,500 J are required to melt all the ice, and only 9560 J are available, the system will be at 0°C. It will be a mixture of ice and water.

27. Water forms droplets because of surface tension, or the desire for a droplet of water to minimize its ratio of surface area to volume. The molecules of water inside the drop are attracted to other water molecules all around them, but on the surface of the droplet the water molecules feel an inward attraction only. This inward attraction of surface water molecules for internal water molecules is what holds the droplets together and minimizes their surface area.

28. Steam molecules will cause a more severe burn. Steam molecules contain more energy at 100°C than water molecules at 100°C due to the energy absorbed during the vaporization stage (heat of vaporization).

29. The alcohol has a higher vapor pressure than water and thus evaporates faster than water. When the alcohol evaporates it absorbs energy from the water, cooling the water. Eventually the water will lose enough energy to change from a liquid to a solid (freeze).

30. When one leaves the swimming pool, water starts to evaporate from the skin of the body. Part of the energy needed for evaporation is absorbed from the skin, resulting in the cool feeling.

31.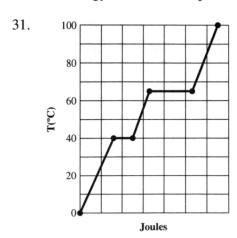

(a) From 0°C to 40.°C solid X warms until at 40.°C it begins to melt. The temperature remains at 40.0°C until all of X is melted. After that, liquid X will warm steadily to 65°C where it will boil and remain at 65°C until all of the liquid becomes vapor. Beyond 65°C, the vapor will warm steadily until 100°C.

(b)

Joules needed (0°C to 40°C)	=	(60. g)(3.5 J/g°C)(40.°C)	=	8400 J
Joules needed at 40°C	=	(60. g)(80. J/g)	=	4800 J
Joules needed (40°C to 65°C)	=	(60. g)(3.5 J/g°C)(25°C)	=	5300 J
Joules needed at 65°C	=	(60. g)(190 J/g)	=	11,000 J
Joules needed (65°C to 100°C)	=	(60. g)(3.5 J/g°C)(35°C)	=	7400 L
Total Joules needed				37,000 J

(each step rounded to two significant figures)

32. During phase changes (ice melting to liquid water or liquid water evaporating to steam), all the heat energy is used to cause the phase change. Once the phase change is complete the heat energy is once again used to increase the temperature of the substance.

33. As the temperature of a liquid increases, the molecules gain kinetic energy thereby increasing their escaping tendency (vapor pressure).

34. Since boiling occurs when vapor pressure equals atmospheric pressure, the graph in Figure 13.7 indicates that water will boil at about 78°C or 172°F at 330 torr pressure.

35. $CuSO_4$ (anhydrous) is greenish white. When exposed to moisture, it turns bright blue forming $CuSO_4 \cdot 5\,H_2O$. The color change is an indicator of moisture in the environment.

36. $MgSO_4 \cdot 7\,H_2O$ $Na_2HPO_4 \cdot 12\,H_2O$

37. For the noble gases the boiling point increases as the molar mass increases. This suggests that boiling points are directly related to molar masses. This is consistent with an increase in dispersion forces with an increase in molar mass which would result in an increase in boiling point.

38. The elevation in Santa Fe is much higher than in Santa Barbara. This means that the air pressure is also lower. Since the boiling temperature of water is dependant on the atmospheric pressure, it will be lower in Santa Fe than in Santa Barbara. Boiling the eggs for 8 minutes at a lower pressure will result in undercooked eggs.

39. Water is a very polar molecule. Glass, an oxide of silicon, is composed of polar silicon-oxygen bonds. Because both water and glass are composed of polar molecules there are strong intermolecular forces between them which create a strong adhesive force. The water climbs up the walls of the graduated cylinder in order to maximize the water-glass interactions resulting in a downward curving meniscus. When water is poured into a plastic cylinder the intermolecular forces between water and the non-polar hydrocarbons composing the plastic are very weak. The cohesive forces between water molecules are stronger than the adhesive forces with the plastic resulting in an almost flat meniscus which minimizes interactions between the water and the plastic.

40.

T (°C)	drops/min
10	3
20	17
30	52
50	87
70	104

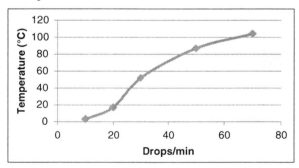

As the temperature increases the rate of flow increases. As the amount of energy in the honey increases, the individual molecules have enough energy to overcome the strong intermolecular forces holding them together allowing the liquid to flow more easily.

41. The wax on the floors is composed of long chain hydrocarbons which are very non-polar. When polar water is spilled on the floor it will bead up to minimize contact with the non-polar wax. When the non-polar hexane is spilled on the floor it will spread out so that it has more contact with the non-polar wax.

42. (a)   The boiling point of acetic acid is approximately 119°C and the melting point is approximately 18°C according to this diagram.

   (b) and (c)

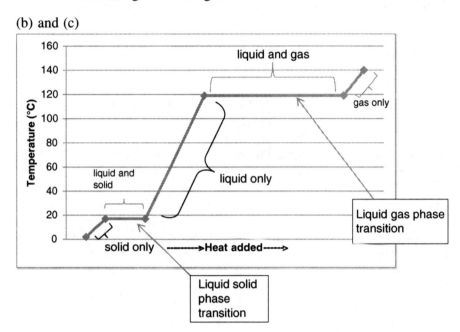

43. (a)   Melt ice: $E_a = (m)(\text{heat of fusion}) = (225\text{ g})(335\text{ J/g}) = 7.54 \times 10^4$ J
     (b)   Warm the water: $E_b = (m)(\text{specific heat})(\Delta t)$

$$= (225\text{ g})\left(\frac{4.184\text{ J}}{\text{g}^\circ\text{C}}\right)(100.^\circ\text{C}) = 9.41 \times 10^4 \text{ J}$$

   (c)   Vaporize the water:

   $E_c = (m)(\text{heat of vaporization}) = (225\text{ g})(2259\text{ J/g}) = 5.08 \times 10^5$ J

   $E_{total} = E_a + E_b + E_c = 6.78 \times 10^5$ J

$$\left(6.78 \times 10^5 \text{ J}\right)\left(\frac{1\text{ cal}}{4.184\text{ J}}\right) = 1.62 \times 10^5 \text{ cal}$$

44. The heat of vaporization of water is 2.26 kJ/g.

$$(2.26\text{ kJ/g})\left(\frac{18.02\text{ g}}{\text{mol}}\right) = 40.7 \text{ kJ/mol}$$

45. $E = (m)(\text{specific heat})(\Delta t) = (250.\text{ g})\left(\frac{0.096\text{ cal}}{\text{g}^\circ\text{C}}\right)(150. - 20.0^\circ\text{C})$

$$= 3.1 \times 10^3 \text{ cal}(3.1 \text{ kcal})$$

46. Energy liberated when steam at 100.0°C condenses to water at 100.0°C

$$(50.0 \text{ mol steam})\left(\frac{18.02 \text{ g}}{\text{mol}}\right)\left(\frac{2.26 \text{ kJ}}{\text{g}}\right)\left(\frac{1000 \text{ J}}{\text{kJ}}\right) = 2.04 \times 10^6 \text{ J}$$

Energy liberated in cooling water from 100.0°C to 30.0°C

$$(50.0 \text{ mol H}_2\text{O})\left(\frac{18.02 \text{ g}}{\text{mol}}\right)\left(\frac{4.184 \text{ J}}{\text{g°C}}\right)(100.0°\text{C} - 30.0°\text{C}) = 2.64 \times 10^5 \text{ J}$$

Total energy liberated $= 2.04 \times 10^6 \text{ J} + 2.64 \times 10^5 \text{ J} = 2.30 \times 10^6 \text{ J}$

47. Energy to warm the ice from −10.0°C to 0°C

$$(100. \text{ g})\left(\frac{2.01 \text{ J}}{\text{g°C}}\right)(10.0°\text{C}) = 2010 \text{ J}$$

Energy to melt the ice at 0°C

$$(100. \text{ g})(335 \text{ J/g}) = 33,500 \text{ J}$$

Energy to heat the water from 0°C to 20.0°C

$$(100. \text{ g})\left(\frac{4.184 \text{ J}}{\text{g°C}}\right)(20.0°\text{C}) = 8370 \text{ J}$$

$E_{\text{total}} = 2010 \text{ J} + 33,500 \text{ J} + 8370 \text{ J} = 4.39 \times 10^4 \text{ J} = 43.9 \text{ kJ}$

48. $2 \text{ H}_2\text{O}(l) \rightarrow 2 \text{ H}_2(g) + \text{O}_2(g)$

The conversion is L $\text{O}_2 \rightarrow$ mol $\text{O}_2 \rightarrow$ mol $\text{H}_2\text{O} \rightarrow$ g H_2O

$$(25.0 \text{ L O}_2)\left(\frac{1 \text{ mol}}{22.4 \text{ L}}\right)\left(\frac{2 \text{ mol H}_2\text{O}}{1 \text{ mol O}_2}\right)\left(\frac{18.02 \text{ g}}{\text{mol}}\right) = 40.2 \text{ g H}_2\text{O}$$

49. The conversion is

$$\frac{\text{mol}}{\text{day}} \rightarrow \frac{\text{molecules}}{\text{day}} \rightarrow \frac{\text{molecules}}{\text{hr}} \rightarrow \frac{\text{molecules}}{\text{min}} \rightarrow \frac{\text{molecules}}{\text{s}}$$

$$\left(\frac{1.00 \text{ mol H}_2\text{O}}{\text{day}}\right)\left(\frac{6.022 \times 10^{23} \text{ molecules}}{\text{mol}}\right)\left(\frac{1.00 \text{ day}}{24 \text{ hr}}\right)\left(\frac{1 \text{ hr}}{60 \text{ min}}\right)\left(\frac{1 \text{ min}}{60 \text{ s}}\right) = 6.97 \times 10^{18} \text{ molecules H}_2\text{O/s}$$

50. Liquid water has a density of 1.00 g/mL.

$$d = \frac{m}{V} \quad V = \frac{m}{d} = \frac{18.02 \text{ g}}{1.00 \text{ g/mL}} = 18.0 \text{ mL} \quad (\text{volume of 1 mole})$$

1.00 mole of water vapor at STP has a volume of 22.4 L (gas)

51. $2\,H_2(g) + O_2(g) \rightarrow 2\,H_2O(g)$

(a) $(80.0\text{ mL }H_2)\left(\dfrac{1\text{ mL }O_2}{2\text{ mL }H_2}\right) = 40.0\text{ mL }O_2$ react with 80.0 mL of H_2

Since 60.0 mL of O_2 are available, some oxygen remains unreacted.

(b) $60.0\text{ mL} - 40.0\text{ mL} = 20.0\text{ mL }O_2$ unreacted.

52. Energy absorbed by the student when steam at 100.°C changes to water at 100.°C

$(1.5\text{ g steam})\left(\dfrac{2.26\text{ kJ}}{g}\right) = 3.4\text{ kJ}\quad(3.4\times10^3\text{ J})$

Energy absorbed when water cools from 100.°C to 20.0°C

$E = (m)(\text{specific heat})(\Delta t)$

$E = (1.5\text{ g})\left(\dfrac{4.184\text{ J}}{g°C}\right)(100.°C - 20.0°C) = 5.0\times10^2\text{ J}$

$E_{total} = 3.4\times10^3\text{ J} + 5.0\times10^2\text{ J} = 3.9\times10^3\text{ J}$

53. (a) won't hydrogen bond – no hydrogen in the molecule
(b) will hydrogen bond

$CH_3-O-H\cdots\cdots O-CH_3$
                              $|$
                              H

(c) won't hydrogen bond; no hydrogen covalently bonded to a strongly electronegative element
(d) will hydrogen bond

$H-O\cdots\cdot H$
      $|\quad\;|$
      $H\;\;H$

(e) won't hydrogen bond; hydrogen must be attached to N, O, or F

54. During the fusion (melting) of a substance the temperature remains constant so a temperature factor is not needed.

55. Energy needed to heat Cu to its melting point:

$E = (50.0\text{ g})(0.385\text{ J/g°C})(1083°C - 25.0°C) = 2.04\times10^4\text{ J}$

Energy needed to melt the Cu.

$E = (50.0\text{ g})(134\text{ J/g}) = 6.70\times10^3\text{ J}$

$E_{total} = 2.04\times10^4\text{ J} + 6.70\times10^3\text{ J} = 2.71\times10^4\text{ J}$

56. Energy released by the soup to convert ice at 0°C to water at 0°C.

$(75\text{ g ice})\left(\dfrac{335\text{ J}}{g}\right) = 2.5\times10^4\text{J}$

Energy released by soup to increase temperature of water from 0°C to 87°C.

$$E = (m)(\text{specific heat})(\Delta t)$$

$$E = (75\text{ g})\left(\frac{4.184\text{ J}}{\text{g}°\text{C}}\right)(87°\text{C}) = 2.7 \times 10^4\text{J}$$

$$E_{\text{total}} = 2.5 \times 10^4\text{J} + 2.7 \times 10^4\text{J} = 5.2 \times 10^4\text{J}$$

57. $2\text{ H}_2\text{O}(l) \rightarrow 2\text{ H}_2(g) + \text{O}_2(g)$

58. At 500 torr, water will boil at approximately 88°C or 190°F

    At 300 torr, water will boil at approximately 76°C or 169°F

    At 100 torr, water will boil at approximately 51°C or 124°F

59. The mass of borax in the box is 2467.4 g – 492.5 g = 1974.9 g borax

$$1974.9\text{ g} \times \frac{1\text{ lb}}{453.6\text{ g}} = 4.35\text{ lb borax}$$

The borax in the box is a hydrate. If it is left to sit out on a hot summer day in Phoenix, Arizona some of the water of hydration may be lost resulting in a reduction in mass. You may have started with the correct amount of borax, but some of the water has evaporated leaving you with less mass.

If 5 lb of borax were heated to drive off all of the waters of hydration, the remaining solid would have a mass of

$$(5.0\text{ lb})\left(\frac{453.6\text{ g}}{1\text{ lb}}\right)\left(\frac{1\text{ mol}}{381.4\text{ g}}\right)\left(\frac{1\text{ mol Na}_2\text{B}_4\text{O}_7}{1\text{ mol Na}_2\text{B}_4\text{O}_7 \cdot 10\text{ H}_2\text{O}}\right)\left(\frac{201.2\text{ g}}{1\text{ mol}}\right)\left(\frac{1\text{ lb}}{453.6\text{ g}}\right) = 2.63\text{ lb Na}_2\text{B}_4\text{O}_7$$

The mass lost was less than the mass of water in the borax.

60. A lake freezes from the top down because the density of water as a solid is lower than that of the liquid, causing the ice to float on the top of the liquid. Since the liquid remains below the solid, marine, and aquatic life continues.

61. Heat lost by warm water = heat gained by ice
    m = grams of ice to lower temperature of water to 0.0°C.

$$(120.\text{ g})\left(\frac{4.184\text{ J}}{\text{g}°\text{C}}\right)(45°\text{C} - 0.0°\text{C}) = (m)(335\text{ J/g})$$

68 g = m (grams of ice melted)
68 g of ice melted. Therefore, 150. g – 68 g = 82 g ice remain.

CHAPTER 14

SOLUTIONS

SOLUTIONS TO REVIEW QUESTIONS

1. A true solution is one in which the size of the particles of solute are between 0.1 – 1 nm. True solutions are homogeneous and the ratio of solute to solvent can be varied. They can be colored or colorless but are transparent. The solute remains distributed evenly in the solution, it will not settle out.

2. The two components of a solution are the solute and the solvent. The solute is dissolved into the solvent or is the least abundant component. The solvent is the dissolving agent or the most abundant component.

3. It is not always apparent which component in a solution is the solute. For example, in a solution composed of equal volumes of two liquids, the designation of solute and solvent would be simply a matter of preference on the part of the person making the designation.

4. The ions or molecules of a dissolved solute do not settle out because the individual particles are so small that the force of molecular collisions is large compared to the force of gravity.

5. Yes. It is possible to have one solid dissolved in another solid. Metal alloys are of this type. Atoms of one metal are dissolved among atoms of another metal.

6. Orange. The three reference solutions are KCl, $KMnO_4$ and $K_2Cr_2O_7$. They all contain K^+ ions in solution. The different colors must result from the different anions dissolved in the solutions: MnO_4^- (purple) and $Cr_2O_7^{2-}$ (orange). Therefore, it is predictable that the $Cr_2O_7^{2-}$ ion present in an aqueous solution of $Na_2Cr_2O_7$ will impart an orange color to the solution.

7. Air is considered to be a solution because it is a homogeneous mixture of several gaseous substances and does not have a fixed composition.

8.

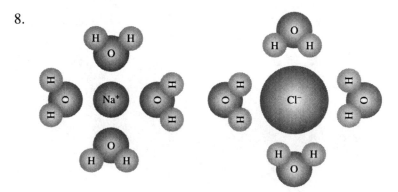

These diagrams are intended to illustrate the orientation of the water molecules about the ions, not the number of water molecules.

9. From Table 14.2, approximately 26.8 g of KBr would dissolve in 50 g water at 0°C.

10. From Figure 14.4, solubilities in water at 25°C:
    (a) NH$_4$Cl   39 g/100 g H$_2$O
    (b) CuSO$_4$ 22 g/100 g H$_2$O
    (c) NaNO$_3$ 91 g/100 g H$_2$O

11. Going down Group 1A, the solubilities of both the chlorides and bromides decrease.

12. From Fig. 14.4, the solubility, in grams of solute per 100 g of H$_2$O:
    (a) NH$_4$Cl at 35°C, 43 g   (c) SO$_2$ gas at 30°C, 8 g at 1 atm
    (b) CuSO$_4$ at 60°C, 39 g   (d) NaNO$_3$ at 15°C, 82 g

13. Li$_2$SO$_4$

14. $\dfrac{6.5\,\text{g}}{21.5\,\text{g}} \times 100 = 30$ mass percent. The solution will be unsaturated.

15. $\dfrac{40.\,\text{g}}{115\,\text{g}} \times 100 = 35$ mass percent. The solution will be saturated.

16. A supersaturated solution of NaC$_2$H$_3$O$_2$ may be prepared in the following sequence:
    (a) Determine the mass of NaC$_2$H$_3$O$_2$ necessary to saturate a specific amount of water at room temperature.
    (b) Place a bit more NaC$_2$H$_3$O$_2$ in the water than the amount needed to saturate the solution.
    (c) Heat and stir the solution until all the solid dissolves.
    (d) Cover the container and allow it to cool undisturbed. The cooled solution, which should contain no solid NaC$_2$H$_3$O$_2$, is supersaturated.

    To test for supersaturation, add one small crystal of NaC$_2$H$_3$O$_2$ to the solution. Immediate crystallization is an indication that the solution was supersaturated.

17. Hexane and benzene are both nonpolar molecules. There are no strong intermolecular forces between molecules of either substance or with each other, so they are miscible. Sodium chloride consists of ions strongly attracted to each other by electrical attractions. The hexane molecules, being nonpolar, have no strong forces to pull the ions apart, so sodium chloride is insoluble in hexane.

18. Coca Cola has two main characteristics, taste and fizz (carbonation). The carbonation is due to a dissolved gas, carbon dioxide. Since dissolved gases become less soluble as temperature increases, warm Coca Cola would be flat, with little to no carbonation. It is, therefore, unappealing to most people.

19. The solubility of gases in liquids is greatly affected by the pressure of a gas above the liquid. The greater the pressure, the more soluble the gas. There is very little effect of pressure regarding the dissolution of solids in liquids.

20. In a saturated solution, the net rate of dissolution is zero. There is no further increase in the amount of dissolved solute, even though undissolved solute is continuously dissolving, because dissolved solute is continuously coming out of solution, crystallizing at a rate equal to the rate of dissolving.

21. The champagne would spray out of the bottle all over the place. The rise in temperature and the increase in kinetic energy of the molecules by shaking both act to decrease the solubility of gas within the liquid. The pressure inside the bottle would be great. As the cork is popped, much of the gas would escape from the liquid very rapidly, causing the champagne to spray.

22.

Cube	1 cm	0.01 cm
Volume	1 cm^3	1×10^{-6} cm^3
Number/1 cm cube	1	$10^6 \left[(1\text{ cm}^3)/(1 \times 10^{-6}\text{cm}^3) = 10^6\text{cubes} \right]$
Area of face	1 cm^2	1×10^{-4} cm^2
Total surface area	6 cm^2	6×10^2 cm^2

$$\left(1 \times 10^6 \text{ cubes}\right)\left(6 \text{ faces/cube}\right)\left(1 \times 10^{-4}\text{cm}^2/\text{face}\right) = 6 \times 10^2 \text{ cm}^2$$

23. The rate of dissolving decreases. The rate of dissolving is at its maximum when the solute and solvent are first mixed.

24. A teaspoon of sugar would definitely dissolve more rapidly in 200 mL of hot coffee than in 200 mL of iced tea. The much greater thermal agitation of the hot coffee will help break the sugar molecules away from the undissolved solid and disperse them throughout the solution. Other solutes in coffee and tea would have no significant effect. The temperature difference is the critical factor.

25. For a given mass of solute, the smaller the particles, the faster the dissolution of the solute. This is due to the smaller particles having a greater surface area exposed to the dissolving action of the solvent.

26. When crystals of $AgNO_3$ and $NaCl$ are mixed, the contact between the individual ions is not intimate enough for the double displacement reaction to occur. When solutions of the two chemicals are mixed, the ions are free to move and come into intimate contact with each other, allowing the reaction to occur easily. The $AgCl$ formed is insoluble.

27. A 16 molar solution of nitric acid is a solution that contains 16 moles HNO_3 per liter of solution.

28. The two solutions contain the same number of chloride ions. One liter of 1 M NaCl contains 1 mole of NaCl, therefore 1 mole of chloride ions. 0.5 liter of 1 M $MgCl_2$ contains 0.5 mol of $MgCl_2$ and 1 mole of chloride ions.

$$(0.5\text{ L})\left(\frac{1 \text{ mol MgCl}_2}{\text{L}}\right)\left(\frac{2 \text{ mol Cl}^-}{1 \text{ mol MgCl}_2}\right) = 1 \text{ mol Cl}^-$$

29. The number of grams of NaCl in 750 mL of 5.0 molar solution is

$$(0.75\text{ L})\left(\frac{5.0 \text{ mol NaCl}}{\text{L}}\right)\left(\frac{58.44 \text{ g}}{1 \text{ mol}}\right) = 2.2 \times 10^2 \text{ g NaCl}$$

Dissolve the 220 g of NaCl in a minimum amount of water, then dilute the resulting solution to a final volume of 750 mL (0.75 L).

30. Ranking of the specified bases in descending order of the volume of each required to react with 1 liter of 1 M HCl. The volume of each required to yield 1 mole of OH^- ion is shown.

    (a)   1 M NaOH   1 liter
    (b)   0.6 M $Ba(OH)_2$ 0.83 liter
    (c)   2 M KOH   0.50 liter
    (d)   1.5 M $Ca(OH)_2$ 0.33 liter

31. The boiling point of a liquid or solution is the temperature at which the vapor pressure of the liquid equals the pressure of the atmosphere. Since a solution containing a nonvolatile solute has a lower vapor pressure than the pure solvent, the boiling point of the solution must be at a higher temperature than for the pure solvent. At the higher boiling temperature the vapor pressure of the solution equals the atmospheric pressure.

32. The freezing point is the temperature at which a liquid changes to a solid. The vapor pressure of a solution is lower than that of a pure solvent. Therefore, the vapor pressure curve of the solution intersects the vapor pressure curve of the pure solvent, at a temperature lower than the freezing point of the pure solvent. (See Figure 14.8b) At this point of intersection, the vapor pressure of the solution equals the vapor pressure of the pure solvent.

33. Water and ice are different phases of the same substance in equilibrium at the freezing point of water, 0°C. The presence of the methanol lowers the vapor pressure and hence the freezing point of water. If the ratio of alcohol to water is high, the freezing point can be lowered as much as 10°C or more.

34. Effectiveness in lowering the freezing point of 500. g water:

    (a)   100. g (2.17 mol) of ethyl alcohol is more effective than 100. g (0.292 mol) of sucrose.
    (b)   20.0 g (0.435 mol) of ethyl alcohol is more effective than 100. g (0.292 mol) of sucrose.
    (c)   20.0 g (0.625 mol) of methyl alcohol is more effective than 20.0 g (0.435 mol) of ethyl alcohol.

35. Both molarity and molality describe the concentration of a solution. However, molarity is the ratio of moles of solute per liter of solution, and molality is the ratio of moles of solute per kilogram of solvent.

36. 5 molal NaCl = 5 mol $NaCl/kg\ H_2O$; 5 molar NaCl = 5 mol NaCl/L of solution. The volume of the 5 molal solution will be larger than 1 liter (1 L H_2O + 5 mol NaCl). The volume of the 5 molar solution is exactly 1 L (5 mol NaCl + sufficient H_2O to produce 1 L of solution). The molarity of a 5 molal solution is therefore, less than 5 molar.

37. A nonvolatile solute (such as salt) lowers the freezing point of water. Adding salt to icy roads in winter melts the ice because the salt lowers the freezing point of water.

38. Because the concentration of water is greater in the thistle tube, the water will flow through the membrane from the thistle tube to the urea solution in the beaker. The solution level in the thistle tube will fall.

39. A semipermeable membrane will allow water molecules to pass through in both directions. If it has pure water on one side and 10% sugar solutions on the other side of the membrane, there is a higher concentration of water molecules on the pure water side. Therefore, there are more water molecule impacts per second on the pure water side of the membrane. The net result is more water molecules pass from the pure water to the sugar solution. Osmotic pressure effect.

40. The urea solution will have the greater osmotic pressure because it has 1.67 mol solute/kg H_2O, while the glucose solution has only 0.83 mol solute/kg H_2O.

41. A lettuce leaf immersed in salad dressing containing salt and vinegar will become limp and wilted as a result of osmosis. As the water inside the leaf flows into the dressing where the solute concentration is higher the leaf becomes limp from fluid loss. In water, osmosis proceeds in the opposite direction flowing into the lettuce leaf maintaining a high fluid content and crisp leaf.

42. The concentration of solutes (such as salts) is higher in seawater than in body fluids. The survivors who drank seawater suffered further dehydration from the transfer of water by osmosis from body tissues to the intestinal tract.

SOLUTIONS TO EXERCISES

1. Reasonably soluble: (b) K_2SO_4 (c) Na_3PO_4 (d) NaOH

    Insoluble: (a) AgCl   (e) PbI_2 (f) $SnCO_3$

2. Reasonably soluble: (b) $Cu(NO_3)_2$ (d) $NH_4C_2H_3O_2$ (f) $AgNO_3$

    Insoluble: (a) $Ba_3(PO_4)_2$ (c) $Fe(OH)_3$ (e) MgO

3. Mass percent calculations

    (a)   15.0 g KCl + 100.0 g H_2O = 115.0 g solution

$$\left(\frac{15.0\ g}{115.0\ g}\right)(100) = 13.0\%\ KCl$$

    (b)   2.50 g Na_3PO_4 + 10.0 g H_2O = 12.5 g solution

$$\left(\frac{2.50\ g}{12.5\ g}\right)(100) = 20.0\%\ Na_3PO_4$$

    (c)   (0.20 mol $NH_4C_2H_3O_2$) $\left(\frac{77.09\ g}{1\ mol}\right) = 15\ g\ NH_4C_2H_3O_2$

15 g $NH_4C_2H_3O_2$ + 125 g H_2O = 140. g solution

$$\left(\frac{15\ g}{140\ g}\right)(100) = 11\%\ NH_4C_2H_3O_2$$

    (d)   (1.50 mol NaOH) $\left(\frac{40.00\ g}{1\ mol}\right) = 60.0\ g\ NaOH$

(33.0 mol H_2O) $\left(\frac{18.02\ g}{1\ mol}\right) = 595\ g\ H_2O$

60.0 g NaOH + 595 g H_2O = 655 g solution

$$\left(\frac{60.0\ g}{655\ g}\right)(100) = 9.16\%\ NaOH$$

4. Mass percent calculations

    (a)   25.0 g $NaNO_3$ in 125.0 g H_2O = 150.0 g solution

$$\left(\frac{25.0\ g}{150.0\ g}\right)(100) = 16.7\%\ NaNO_3$$

    (b)   1.25 g $CaCl_2$ in 35.0 g H_2O = 36.3 g solution

$$\left(\frac{1.25\ g}{36.3\ g}\right)(100) = 3.44\%\ CaCl_2$$

(c) $(0.75 \text{ mol } K_2CrO_4) + \left(\dfrac{194.2 \text{ g}}{1 \text{ mol}}\right) = 150 \text{ g } K_2CrO_4$

$150 \text{ g } K_2CrO_4 + 225 \text{ g } H_2O = 380 \text{ g solution}$

$\left(\dfrac{150 \text{ g}}{380 \text{ g}}\right)(100) = 39\% \text{ } K_2CrO_4$

(d) $(1.20 \text{ mol } H_2SO_4) \left(\dfrac{98.09 \text{ g}}{1 \text{ mol}}\right) = 118 \text{ g } H_2SO_4$

$(72.5 \text{ mol } H_2O) \left(\dfrac{18.02 \text{ g}}{1 \text{ mol}}\right) = 1.31 \times 10^3 \text{ g } H_2O$

$118 \text{ g } H_2SO_4 + 1.31 \times 10^3 \text{ g } H_2O = 1.43 \times 10^3 \text{ g solution}$

$\left(\dfrac{118 \text{ g}}{1.43 \times 10^3 \text{ g}}\right)(100) = 8.25\% \text{ } H_2SO_4$

5. $5.23 \text{ g NaClO} \left(\dfrac{100 \text{ g solution}}{21.5 \text{ g NaClO}}\right) = 24.3 \text{ g solution}$

6. $42.8 \text{ g Fe}_2O_3 \left(\dfrac{100 \text{ g solution}}{30.0 \text{ g Fe}_2O_3}\right) = 143 \text{ g solution}$

7. (a) $(25 \text{ g solution}) \left(\dfrac{7.5 \text{ g CaSO}_4}{100. \text{ g solution}}\right) = 1.9 \text{ g CaSO}_4$

   (b) $25 \text{ g solution} - 1.9 \text{ g solute} = 23 \text{ g solvent}$

8. (a) $(75 \text{ g solution}) \left(\dfrac{12.0 \text{ g BaCl}_2}{100. \text{ g solution}}\right) = 9.0 \text{ g BaCl}_2$

   (b) $75 \text{ g solution} - 9.0 \text{ g solute} = 66 \text{ g solvent}$

9. Mass/volume percent.

   (a) $\left(\dfrac{15.0 \text{ g } C_2H_5OH}{150.0 \text{ mL solution}}\right)(100) = 10.0\% \text{ } C_2H_5OH$

   (b) $\left(\dfrac{25.2 \text{ g NaCl}}{125.5 \text{ mL solution}}\right)(100) = 20.1\% \text{ NaCl}$

10. Mass/volume percent.

   (a) $\left(\dfrac{175.2 \text{ g } C_{12}H_{22}O_{11}}{275.5 \text{ mL solution}}\right)(100) = 63.59\% \text{ } C_{12}H_{22}O_{11}$

   (b) $\left(\dfrac{35.5 \text{ g of } CH_3OH}{75.0 \text{ mL solution}}\right)(100) = 47.3\% \text{ } CH_3OH$

11. Volume percent.

   (a) $\left(\dfrac{50.0 \text{ mL hexanol}}{125 \text{ mL solution}}\right)(100) = 40.0\%$ hexanol

   (b) $\left(\dfrac{2.0 \text{ mL ethanol}}{15.0 \text{ mL solution}}\right)(100) = 13\%$ ethanol

   (c) $\left(\dfrac{15.0 \text{ mL acetone}}{325 \text{ mL solution}}\right)(100) = 4.62\%$ acetone

12. Volume percent.

   (a) $\left(\dfrac{37.5 \text{ mL butanol}}{275 \text{ mL solution}}\right)(100) = 13.6\%$ butanol

   (b) $\left(\dfrac{4.0 \text{ mL methanol}}{25.0 \text{ mL solution}}\right)(100) = 16\%$ methanol

   (c) $\left(\dfrac{45.0 \text{ mL isoamyl alcohol}}{750. \text{ mL solution}}\right)(100) = 6.00\%$ isoamyl alcohol

13. Molarity problems $\left(M = \dfrac{\text{mol}}{\text{L}}\right)$

   (a) $\left(\dfrac{0.25 \text{ mol}}{75.0 \text{ mL}}\right)\left(\dfrac{1000 \text{ mL}}{1 \text{ L}}\right) = 3.3\ M$

   (b) $\left(\dfrac{1.75 \text{ mol KBr}}{0.75 \text{ L}}\right) = 2.3\ M$ KBr

   (c) $\left(\dfrac{35.0 \text{ g NaC}_2\text{H}_3\text{O}_2}{1.25 \text{ L}}\right)\left(\dfrac{1 \text{ mol}}{82.03 \text{ g}}\right) = 0.341\ M$ NaC$_2$H$_3$O$_2$

   (d) $\left(\dfrac{77 \text{ g CuSO}_4\cdot 5\text{H}_2\text{O}}{1.0 \text{ L}}\right)\left(\dfrac{1 \text{ mol}}{249.7 \text{ g}}\right) = 0.30\ M$ CuSO$_4$

14. Molarity problems $\left(M = \dfrac{\text{mol}}{\text{L}}\right)$

   (a) $\left(\dfrac{0.50 \text{ mol}}{125 \text{ mL}}\right)\left(\dfrac{1000 \text{ mL}}{1 \text{ L}}\right) = 4.0\ M$

   (b) $\left(\dfrac{2.25 \text{ mol}}{1.50 \text{ L}}\right) = 1.50\ M$ CaCl$_2$

   (c) $\left(\dfrac{275 \text{ g}}{775 \text{ mL}}\right)\left(\dfrac{1 \text{ mol}}{180.2 \text{ g}}\right)\left(\dfrac{1000 \text{ mL}}{1 \text{ L}}\right) = 1.97\ M$ C$_6$H$_{12}$O$_6$

   (d) $\left(\dfrac{125 \text{ g MgSO}_4\cdot 7\text{ H}_2\text{O}}{2.50 \text{ L}}\right)\left(\dfrac{1 \text{ mol}}{246.5 \text{ g}}\right) = 0.203\ M$ MgSO$_4$

15. Molarity $= \dfrac{\text{mol solute}}{\text{L solution}}$ or mol solute $= $ (L solution)(Molarity)

    (a) $(1.5\,\text{L})\left(\dfrac{1.20\,\text{mol}\,H_2SO_4}{L}\right) = 1.8\,\text{mol}\,H_2SO_4$

    (b) $(25.0\,\text{mL})\left(\dfrac{1\,L}{1000\,\text{mL}}\right)\left(\dfrac{0.0015\,\text{mol}\,BaCl_2}{L}\right) = 3.8 \times 10^{-5}\,\text{mol}\,BaCl_2$

    (c) $(125\,\text{mL})\left(\dfrac{1\,L}{1000\,\text{mL}}\right)\left(\dfrac{0.35\,\text{mol}\,K_3PO_4}{L}\right) = 0.044\,\text{mol}\,K_3PO_4$

16. Molarity $= \dfrac{\text{mol solute}}{\text{L solution}}$ or mol solute $= $ (L solution) (Molarity)

    (a) $(0.75\,\text{L})\left(\dfrac{1.50\,\text{mol}\,HNO_3}{L}\right) = 1.1\,\text{mol}\,HNO_3$

    (b) $(10.0\,\text{mL})\left(\dfrac{1\,L}{1000\,\text{mL}}\right)\left(\dfrac{0.75\,\text{mol}\,NaClO_3}{L}\right) = 7.5 \times 10^{-3}\,\text{mol}\,NaClO_3$

    (c) $(175\,\text{mL})\left(\dfrac{1\,L}{1000\,\text{mL}}\right)\left(\dfrac{0.50\,\text{mol}\,LiBr}{L}\right) = 0.088\,\text{mol}\,LiBr$

17. (a) $(2.5\,\text{L})\left(\dfrac{0.75\,\text{mol}\,K_2CrO_4}{L}\right)\left(\dfrac{194.2\,g}{1\,\text{mol}}\right) = 360\,g\,K_2CrO_4$

    (b) $(75.2\,\text{mL})\left(\dfrac{1\,L}{1000\,\text{mL}}\right)\left(\dfrac{0.050\,\text{mol}\,HC_2H_3O_2}{L}\right)\left(\dfrac{60.05\,g}{1\,\text{mol}}\right) = 0.226\,g\,HC_2H_3O_2$

    (c) $(250\,\text{mL})\left(\dfrac{1\,L}{1000\,\text{mL}}\right)\left(\dfrac{16\,\text{mol}\,HNO_3}{L}\right)\left(\dfrac{63.02\,g}{1\,\text{mol}}\right) = 250\,g\,HNO_3$

18. (a) $(1.20\,\text{L})\left(\dfrac{18\,\text{mol}\,H_2SO_4}{L}\right)\left(\dfrac{98.09\,g}{1\,\text{mol}}\right) = 2.1 \times 10^3\,g\,H_2SO_4$

    (b) $(27.5\,\text{mL})\left(\dfrac{1\,L}{1000\,\text{mL}}\right)\left(\dfrac{1.5\,\text{mol}\,KMnO_4}{L}\right)\left(\dfrac{158.0\,g}{1\,\text{mol}}\right) = 6.52\,g\,KMnO_4$

    (c) $(120\,\text{mL})\left(\dfrac{1\,L}{1000\,\text{mL}}\right)\left(\dfrac{0.025\,\text{mol}\,Fe_2(SO_4)_3}{L}\right)\left(\dfrac{399.9\,g}{1\,\text{mol}}\right) = 1.2\,g\,Fe_2(SO_4)_3$

19. (a) $(0.15\,\text{mol}\,H_3PO_4)\left(\dfrac{1\,L}{0.750\,\text{mol}\,H_3PO_4}\right)\left(\dfrac{1000\,\text{mL}}{1\,L}\right) = 2.0 \times 10^2\,\text{mL}$

    (b) $(35.5\,g\,H_3PO_4)\left(\dfrac{1\,\text{mol}}{97.99\,g}\right)\left(\dfrac{1\,L}{0.750\,\text{mol}\,H_3PO_4}\right)\left(\dfrac{1000\,\text{mL}}{1\,L}\right) = 483\,\text{mL}$

    (c) $7.34 \times 10^{22}\,\text{molecules}\,H_3PO_4\left(\dfrac{1\,\text{mol}}{6.022 \times 10^{23}\,\text{molecules}}\right)\left(\dfrac{1\,L}{0.750\,\text{mol}\,H_3PO_4}\right)\left(\dfrac{1000\,\text{mL}}{1\,L}\right)$

       $= 163\,\text{mL}$

20. (a) $(0.85 \text{ mol NH}_4\text{Cl})\left(\dfrac{1 \text{ L}}{0.250 \text{ mol NH}_4\text{Cl}}\right)\left(\dfrac{1000 \text{ mL}}{1 \text{ L}}\right) = 3.4 \times 10^3 \text{ mL}$

(b) $(25.2 \text{ g NH}_4\text{Cl})\left(\dfrac{1 \text{ mol}}{53.49 \text{ g}}\right)\left(\dfrac{1 \text{ L}}{0.250 \text{ mol NH}_4\text{Cl}}\right)\left(\dfrac{1000 \text{ mL}}{\text{L}}\right) = 1.88 \times 10^3 \text{ mL}$

(c) $2.06 \times 10^{20} \text{ formula units NH}_4\text{Cl}\left(\dfrac{1 \text{ mol}}{6.022 \times 10^{23} \text{ formula units}}\right)\left(\dfrac{1 \text{ L}}{0.250 \text{ mol NH}_4\text{Cl}}\right)\left(\dfrac{1000 \text{ mL}}{1 \text{ L}}\right)$

$= 1.37 \text{ mL}$

21. Dilution problem $V_1 M_1 = V_2 M_2$

(a) $V_1 = 125 \text{ mL}$ $V_2 = (125 \text{ mL} + 775 \text{ mL}) = 900. \text{ mL}$
$M_1 = 5.0 \, M$ $M_2 = M_2$
$(125 \text{ mL})(5.0 \, M) = (900. \text{ mL})(M_2)$
$M_2 = \dfrac{(125 \text{ mL})(5.0 \, M)}{900. \text{ mL}} = 0.694 \, M$

(b) $V_1 = 250 \text{ mL}$ $V_2 = (250 \text{ mL} + 750 \text{ mL}) = 1.00 \times 10^3 \text{ mL}$
$M_1 = 0.25 \, M$ $M_2 = M_2$
$(250 \text{ mL})(0.25 \, M) = (1.00 \times 10^3 \text{ mL})(M_2)$
$M_2 = \dfrac{(250 \text{ mL})(0.25 \, M)}{1.00 \times 10^3 \text{ mL}} = 0.063 \, M$

(c) First calculate the moles of HNO_3 in each solution. Then calculate the molarity.

$(75 \text{ mL})\left(\dfrac{1 \text{ L}}{1000 \text{ mL}}\right)\left(\dfrac{0.50 \text{ mol}}{1 \text{ L}}\right) = 0.038 \text{ mol HNO}_3$

$(75 \text{ mL})\left(\dfrac{1 \text{ L}}{1000 \text{ mL}}\right)\left(\dfrac{1.5 \text{ mol}}{1 \text{ L}}\right) = 0.11 \text{ mol HNO}_3$

Total mol $= 0.15$ mol

Total volume $= 75 \text{ mL} + 75 \text{ mL} = 150. \text{ mL} = 0.150 \text{ L}$

$\dfrac{0.150 \text{ mol}}{0.150 \text{ L}} = 1.00 \, M$

22. Dilution problem $V_1 M_1 = V_2 M_2$

(a) $V_1 = 175 \text{ mL}$ $V_2 = (175 \text{ mL} + 275 \text{ mL}) = 450. \text{ mL}$
$M_1 = 3.0 \, M$ $M_2 = M_2$
$(175 \text{ mL})(3.0 \, M) = (450. \text{ mL})(M_2)$
$M_2 = \dfrac{(175 \text{ mL})(3.0 \, M)}{450. \text{ mL}} = 1.2 \, M$

(b) $V_1 = 350\,\text{mL}$ $V_2 = (350\,\text{mL} + 150\,\text{mL}) = 5.0 \times 10^2\,\text{mL}$

$M_1 = 0.10\,M$ $M_2 = M_2$

$(350\,\text{mL})(0.10\,M) = (5.0 \times 10^2\,\text{mL})(M_2)$

$M_2 = \dfrac{(350\,\text{mL})(0.10\,M)}{5.0 \times 10^2\,\text{mL}} = 0.070\,M$

(c) First calculate the moles of HCl in each solution. Then calculate the molarity.

$$(50.0\,\text{mL})\left(\frac{1\,\text{L}}{1000\,\text{mL}}\right)\left(\frac{0.250\,\text{mol HCl}}{1\,\text{L}}\right) = 0.0125\,\text{mol HCl}$$

$$(25.0\,\text{mL})\left(\frac{1\,\text{L}}{1000\,\text{mL}}\right)\left(\frac{0.500\,\text{mol HCl}}{1\,\text{L}}\right) = 0.0125\,\text{mol HCl}$$

Total mol $= 0.0250\,\text{mol HCl}$

Total volume $= 50.0\,\text{mL} + 25.0\,\text{mL} = 75.0\,\text{mL} = 0.0750\,\text{L}$

$\dfrac{0.250\,\text{mol}}{0.0750\,\text{L}} = 0.333\,M$

23. $V_1 M_1 = V_2 M_2$

(a) $(V_1)(15\,M) = (750\,\text{mL})(3.0\,M)$

$V_1 = \dfrac{(750\,\text{mL})(3.0\,M)}{15\,M} = 150\,\text{mL}\ 15\,M\ H_3PO_4$

(b) $(V_1)(16\,M) = (250\,\text{mL})(0.50\,M)$

$V_1 = \dfrac{(250\,\text{mL})(0.50\,M)}{16\,M} = 7.8\,\text{mL}\ 16\,M\ HNO_3$

24. $V_1 M_1 = V_2 M_2$

(a) $(V_1)(18\,M) = (225\,\text{mL})(2.0\,M)$

$V_1 = \dfrac{(225\,\text{mL})(2.0\,M)}{18\,M} = 25\,\text{mL}\ 18\,M\ H_2SO_4$

(b) $(V_1)(15\,M) = (75\,\text{mL})(1.0\,M)$

$V_1 = \dfrac{(75\,\text{mL})(1.0\,M)}{15\,M} = 5.0\,\text{mL}\ 15\,M\ NH_3$

25. $(0.125\,\text{L})\left(\dfrac{6.0\,\text{mol HC}_2\text{H}_3\text{O}_2}{\text{L}}\right) = 0.75\,\text{mol HC}_2\text{H}_3\text{O}_2$

(a) Final volume after mixing
$125\,\text{mL} + 525\,\text{mL} = 650.\,\text{mL} = 0.650\,\text{L}$

$\dfrac{0.75\,\text{mol HC}_2\text{H}_3\text{O}_2}{0.650\,\text{L}} = 1.2\,M\ HC_2H_3O_2$

(b) $(175 \text{ mL})\left(\dfrac{1 \text{ L}}{1000 \text{ mL}}\right)\left(\dfrac{1.5 \text{ mol HC}_2\text{H}_3\text{O}_2}{\text{L}}\right) = 0.26 \text{ mol HC}_2\text{H}_3\text{O}_2$

Total moles $= 0.75 \text{ mol} + 0.26 \text{ mol} = 1.01 \text{ mol HC}_2\text{H}_3\text{O}_2$

Final volume $= 125 \text{ mL} + 175 \text{ mL} = 300. \text{ mL} = 0.300 \text{ L}$

$\dfrac{1.01 \text{ mol HC}_2\text{H}_3\text{O}_2}{0.300 \text{ L}} = 3.37 \, M \text{ HC}_2\text{H}_3\text{O}_2$

26. $(0.175 \text{ L})\left(\dfrac{3.0 \text{ mol HCl}}{\text{L}}\right) = 0.53 \text{ mol HCl}$

(a)   Final volume after mixing

$175 \text{ mL} + 250 \text{ mL} = 425 \text{ mL} = 0.425 \text{ L}$

$\dfrac{0.53 \text{ mol HCl}}{0.425 \text{ L}} = 1.2 \, M \text{ HCl}$

(b) $(115 \text{ mL})\left(\dfrac{1 \text{ L}}{1000 \text{ mL}}\right)\left(\dfrac{6.0 \text{ mol HCl}}{\text{L}}\right) = 0.69 \text{ mol HCl}$

Total moles $= 0.53 \text{ mol} + 0.69 \text{ mol} = 1.22 \text{ mol HCl}$

Final volume $= 175 \text{ mL} + 115 \text{ mL} = 290. \text{ mL} = 0.290 \text{ L}$

$\dfrac{1.22 \text{ mol HCl}}{0.290 \text{ L}} = 4.21 \, M \text{ HCl}$

27. $3 \text{ Ca(NO}_3)_2(aq) + 2 \text{ Na}_3\text{PO}_4(aq) \longrightarrow \text{Ca}_3(\text{PO}_4)_2(s) + 6 \text{ NaNO}_3(aq),$

(a) $(2.7 \text{ mol Na}_3\text{PO}_4)\left(\dfrac{1 \text{ mol Ca}_3(\text{PO}_4)_2}{2 \text{ mol Na}_3\text{PO}_3}\right) = 1.4 \text{ mol Ca}_3(\text{PO}_4)_2$

(b) $\left(0.75 \text{ mol Ca(NO}_3)_2\right)\left(\dfrac{6 \text{ mol NaNO}_3}{3 \text{ mol Ca(NO}_3)_2}\right) = 1.5 \text{ mol NaNO}_3$

(c) $\text{L Ca(NO}_3)_2 \longrightarrow \text{mol Ca(NO}_3)_2 \longrightarrow \text{mol Na}_3\text{PO}_4$

$\left(1.45 \text{ L Ca(NO}_3)_2\right)\left(\dfrac{0.225 \text{ mol}}{\text{L}}\right)\left(\dfrac{2 \text{ mol Na}_3\text{PO}_4}{3 \text{ mol Ca(NO}_3)_2}\right) = 0.218 \text{ mol Na}_3\text{PO}_4$

(d) $\text{mL Ca(NO}_3)_2 \longrightarrow \text{mol Ca(NO}_3)_2 \longrightarrow \text{mol Ca}_3(\text{PO}_4)_2 \longrightarrow \text{g Ca}_3(\text{PO}_4)_2$

$\left(125 \text{ mL Ca(NO}_3)_2\right)\left(\dfrac{0.500 \text{ mol}}{1000 \text{ mL}}\right)\left(\dfrac{1 \text{ mol Ca}_3(\text{PO}_4)_2}{3 \text{ mol Ca(NO}_3)_2}\right)\left(\dfrac{310.18 \text{ g}}{\text{mol}}\right) = 6.46 \text{ g Ca}_3(\text{PO}_4)_2$

(e) $\text{mL Ca(NO}_3)_2 \longrightarrow \text{mol Ca(NO}_3)_2 \longrightarrow \text{mol Na}_3\text{PO}_4 \longrightarrow \text{mL Na}_3\text{PO}_4$

$\left(15.0 \text{ mL Ca(NO}_3)_2\right)\left(\dfrac{0.50 \text{ mol}}{1000 \text{ mL}}\right)\left(\dfrac{2 \text{ mol Na}_3\text{PO}_4}{3 \text{ mol Ca(NO}_3)_2}\right)\left(\dfrac{1000 \text{ mL}}{0.25 \text{ mol}}\right) = 20. \text{ mL Na}_3\text{PO}_4$

(f)   Find mol Ca(NO$_3$)$_2$ $\text{mL Na}_3\text{PO}_4 \longrightarrow \text{mol Na}_3\text{PO}_4 \longrightarrow \text{mol Ca(NO}_3)_2$

$\left(50.0 \text{ mL Na}_3\text{PO}_4\right)\left(\dfrac{2.0 \text{ mol}}{1000 \text{ mL}}\right)\left(\dfrac{3 \text{ mol Ca(NO}_3)_2}{2 \text{ mol Na}_3\text{PO}_4}\right) = 0.15 \text{ mol Ca(NO}_3)_2$

$M = \dfrac{\text{mol}}{\text{L}} \quad M = \left(\dfrac{0.15 \text{ mol Ca(NO}_3)_2}{0.0500 \text{ L}}\right) = 3.0 \, M \text{ Ca(NO}_3)_2$

28. $2\,NaOH(aq) + H_2SO_4(aq) \longrightarrow Na_2SO_4(aq) + 2\,H_2O(l)$,

(a) $(3.6\,mol\,H_2SO_4)\left(\dfrac{1\,mol\,Na_2SO_4}{1\,mol\,H_2SO_4}\right) = 3.6\,mol\,Na_2SO_4$

(b) $(0.025\,mol\,NaOH)\left(\dfrac{2\,mol\,H_2O}{2\,mol\,NaOH}\right) = 0.025\,mol\,H_2O$

(c) $L\,H_2SO_4 \longrightarrow mol\,H_2SO_4 \longrightarrow mol\,NaOH$

$(2.50\,L\,H_2SO_4)\left(\dfrac{0.125\,mol}{1\,L}\right)\left(\dfrac{2\,mol\,NaOH}{1\,mol\,H_2SO_4}\right) = 0.625\,mol\,NaOH$

(d) $mL\,NaOH \longrightarrow mol\,NaOH \longrightarrow mol\,Na_2SO_4 \longrightarrow g\,Na_2SO_4$

$(25\,mL\,NaOH)\left(\dfrac{0.050\,mol}{1000\,mL}\right)\left(\dfrac{1\,mol\,Na_2SO_4}{2\,mol\,NaOH}\right)\left(\dfrac{142.1\,g}{mol}\right) = 0.089\,g\,Na_2SO_4$

(e) $mL\,NaOH \longrightarrow mol\,NaOH \longrightarrow mol\,H_2SO_4 \longrightarrow mL\,H_2SO_4$

$(25.5\,mL\,NaOH)\left(\dfrac{0.750\,mol}{1000\,mL}\right)\left(\dfrac{1\,mol\,H_2SO_4}{2\,mol\,NaOH}\right)\left(\dfrac{1000\,mL}{0.250\,mol}\right) = 38.25\,mL\,H_2SO_4$

(f) Find mol NaOH $mL\,H_2SO_4 \longrightarrow mol\,H_2SO_4 \longrightarrow mol\,NaOH$

$(35.72\,mL\,H_2SO_4)\left(\dfrac{0.125\,mol}{1000\,mL}\right)\left(\dfrac{2\,mol\,NaOH}{1\,mol\,H_2SO_4}\right) = 8.93 \times 10^{-3}\,mol\,NaOH$

$M = \dfrac{mol}{L} \quad M = \left(\dfrac{8.93 \times 10^{-3}\,mol\,NaOH}{0.04820\,L}\right) = 0.185\,M\,NaOH$

29. $2\,KMnO_4(aq) + 16\,HCl(aq) \longrightarrow 2\,MnCl_2(aq) + 5\,Cl_2(g) + 8\,H_2O(l) + 2\,KCl(aq)$

(a) $(15.0\,mL\,HCl)\left(\dfrac{0.250\,mol}{1000\,mL}\right)\left(\dfrac{8\,mol\,H_2O}{16\,mol\,HCl}\right) = 1.88 \times 10^{-3}\,mol\,H_2O$

(b) $(1.85\,mol\,MnCl_2)\left(\dfrac{2\,mol\,KMnO_4}{2\,mol\,MnCl_2}\right)\left(\dfrac{1\,L\,KMnO_4}{0.150\,mol\,KMnO_4}\right) = 12.3\,L\,KMnO_4$

(c) $(125\,mL\,KCl)\left(\dfrac{0.525\,mol}{1000\,mL}\right)\left(\dfrac{16\,mol\,HCl}{2\,mol\,KCl}\right)\left(\dfrac{1000\,mL}{2.50\,mol}\right) = 210.\,mL\,HCl$

(d) $(15.60\,mL\,KMnO_4)\left(\dfrac{0.250\,mol}{1000\,mL}\right)\left(\dfrac{16\,mol\,HCl}{2\,mol\,KMnO_4}\right) = 0.0312\,mol\,HCl$

$M = \left(\dfrac{0.0312\,mol\,HCl}{0.02220\,L}\right) = 1.41\,M\,HCl$

(e) $mL\,HCl \longrightarrow mol\,HCl \longrightarrow mol\,Cl_2 \longrightarrow L\,Cl_2$ (gas at STP)

$(125\,mL\,HCl)\left(\dfrac{2.5\,mol}{1000\,mL}\right)\left(\dfrac{5\,mol\,Cl_2}{16\,mol\,HCl}\right)\left(\dfrac{22.4\,L}{mol}\right) = 2.2\,L\,Cl_2$

(f) Limiting reactant problem. Convert volume of both reactants to liters of Cl_2 gas.

$$(15.0 \text{ mL HCl})\left(\frac{0.750 \text{ mol}}{1000 \text{ mL}}\right)\left(\frac{5 \text{ mol Cl}_2}{16 \text{ mol HCl}}\right)\left(\frac{22.4 \text{ L}}{\text{mol}}\right) = 0.0788 \text{ L Cl}_2$$

$$(12.0 \text{ mL KMnO}_4)\left(\frac{0.550 \text{ mol}}{1000 \text{ mL}}\right)\left(\frac{5 \text{ mol Cl}_2}{2 \text{ mol KMnO}_4}\right)\left(\frac{22.4 \text{ L}}{\text{mol}}\right) = 0.373 \text{ L Cl}_2$$

HCl is limiting reactant. 0.0788 L of Cl_2 are produced.

30. $K_2CO_3(aq) + 2\,HC_2H_3O_2(aq) \longrightarrow 2\,KC_2H_3O_2(aq) + H_2O(l) + CO_2(g)$

(a) $(25.0 \text{ mL HC}_2\text{H}_3\text{O}_2)\left(\dfrac{0.150 \text{ mol}}{1000 \text{ mL}}\right)\left(\dfrac{1 \text{ mol H}_2\text{O}}{2 \text{ mol HC}_2\text{H}_3\text{O}_2}\right) = 1.88 \times 10^{-3} \text{ mol H}_2\text{O}$

(b) $(17.5 \text{ mol KC}_2\text{H}_3\text{O}_2)\left(\dfrac{1 \text{ mol K}_2\text{CO}_3}{2 \text{ mol KC}_2\text{H}_3\text{O}_2}\right)\left(\dfrac{1 \text{ L K}_2\text{CO}_3}{0.210 \text{ mol K}_2\text{CO}_3}\right) = 41.7 \text{ L K}_2\text{CO}_3$

(c) $(75.2 \text{ mL K}_2\text{CO}_3)\left(\dfrac{0.750 \text{ mol}}{1000 \text{ mL}}\right)\left(\dfrac{2 \text{ mol HC}_2\text{H}_3\text{O}_2}{1 \text{ mol K}_2\text{CO}_3}\right)\left(\dfrac{1000 \text{ mL HC}_2\text{H}_3\text{O}_2}{1.25 \text{ mol HC}_2\text{H}_3\text{O}_2}\right) = 90.2 \text{ mL HC}_2\text{H}_3\text{O}_2$

(d) $(18.50 \text{ mL K}_2\text{CO}_3)\left(\dfrac{0.250 \text{ mol}}{1000 \text{ mL}}\right)\left(\dfrac{2 \text{ mol HC}_2\text{H}_3\text{O}_2}{1 \text{ mol K}_2\text{CO}_3}\right) = 9.25 \times 10^{-3} \text{ mol HC}_2\text{H}_3\text{O}_2$

$$M = \left(\frac{9.25 \times 10^{-3} \text{ mol HC}_2\text{H}_3\text{O}_2}{0.01015 \text{ L}}\right) = 0.911 \, M \text{ HC}_2\text{H}_3\text{O}_2$$

(e) $\text{mL HC}_2\text{H}_3\text{O}_2 \longrightarrow \text{mol HC}_2\text{H}_3\text{O}_2 \longrightarrow \text{mol CO}_2 \longrightarrow \text{L CO}_2 \text{ (gas at STP)}$

$$(105 \text{ mL of HC}_2\text{H}_3\text{O}_2)\left(\frac{1.5 \text{ mol}}{1000 \text{ mL}}\right)\left(\frac{1 \text{ mol CO}_2}{2 \text{ mol HC}_2\text{H}_3\text{O}_2}\right)\left(\frac{22.4 \text{ L}}{\text{mol}}\right) = 1.8 \, \text{L CO}_2$$

(f) Limiting reactant problem. Convert volume of both reactants to liters of CO_2 gas.

$$(25.0 \text{ mL K}_2\text{CO}_3)\left(\frac{0.350 \text{ mol}}{1000 \text{ mL}}\right)\left(\frac{1 \text{ mol CO}_2}{1 \text{ mol K}_2\text{CO}_3}\right)\left(\frac{22.4 \text{ L}}{\text{mol}}\right) = 0.196 \text{ L CO}_2$$

$$(25.0 \text{ mL HC}_2\text{H}_3\text{O}_2)\left(\frac{0.250 \text{ mol}}{1000 \text{ mL}}\right)\left(\frac{1 \text{ mol CO}_2}{2 \text{ mol HC}_2\text{H}_3\text{O}_2}\right)\left(\frac{22.4 \text{ L}}{\text{mol}}\right) = 0.0700 \text{ L CO}_2$$

$HC_2H_3O_2$ is the limiting reactant. 0.0700 L of CO_2 are produced.

31. Molality $= m = \dfrac{\text{mol solute}}{\text{kg solvent}}$

(a) $\left(\dfrac{2.0 \text{ mol HCl}}{175 \text{ g H}_2\text{O}}\right)\left(\dfrac{1000 \text{ g}}{\text{kg}}\right) = 11 \, m \text{ HCl}$

(b) $\left(\dfrac{14.5 \text{ g C}_{12}\text{H}_{22}\text{O}_{11}}{550.0 \text{ g H}_2\text{O}}\right)\left(\dfrac{1000 \text{ g}}{\text{kg}}\right)\left(\dfrac{1 \text{ mol}}{342.3 \text{ g}}\right) = 0.0770 \, m \text{ C}_{12}\text{H}_{22}\text{O}_{11}$

(c) $\left(\dfrac{25.2 \text{ mL CH}_3\text{OH}}{595 \text{ g CH}_3\text{CH}_2\text{OH}}\right)\left(\dfrac{0.791 \text{ g}}{\text{mL}}\right)\left(\dfrac{1000 \text{ g}}{\text{kg}}\right)\left(\dfrac{1 \text{ mol}}{32.04 \text{ g}}\right) = 0.985 \, m \text{ CH}_3\text{OH}$

32. Molality $= m = \dfrac{\text{mol solute}}{\text{kg solvent}}$

   (a) $\left(\dfrac{125 \text{ mol } CaCl_2}{750.0 \text{ g } H_2O}\right)\left(\dfrac{1000 \text{ g}}{\text{kg}}\right) = 1.67 \, m \, CaCl_2$

   (b) $\left(\dfrac{2.5 \text{ g } C_6H_{12}O_6}{525 \text{ g } H_2O}\right)\left(\dfrac{1000 \text{ g}}{\text{kg}}\right)\left(\dfrac{1 \text{ mol}}{180.2 \text{ g}}\right) = 0.026 \, m \, C_6H_{12}O_6$

   (c) $\left(\dfrac{17.5 \text{ mL } (CH_3)_2CHOH}{35.5 \text{ mL } H_2O}\right)\left(\dfrac{0.785 \text{ g}}{\text{mL}}\right)\left(\dfrac{1 \text{ mL}}{1.00 \text{ g}}\right)\left(\dfrac{1000 \text{ g}}{\text{kg}}\right)\left(\dfrac{1 \text{ mol}}{60.09 \text{ g}}\right) = 6.44 \, m \, (CH_3)_2CHOH$

33. (a) $\left(\dfrac{2.68 \text{ g } C_{10}H_8}{38.4 \text{ g } C_6H_6}\right)\left(\dfrac{1 \text{ mol}}{128.2 \text{ g } C_{10}H_8}\right)\left(\dfrac{1000 \text{ g}}{\text{kg}}\right) = 0.544 \, m$

   (b) $K_f \text{ (for benzene)} = \dfrac{5.1°C}{m}$ \qquad Freezing point of benzene $= 5.5°C$

   $\Delta t_f = (0.544 \, m)\left(\dfrac{5.1°C}{m}\right) = 2.8°C$

   Freezing point of solution $= 5.5°C - 2.8°C = 2.7°C$

   (c) $K_b \text{ (for benzene)} = \dfrac{2.53°C}{m}$ \qquad Boiling point of benzene $= 80.1°C$

   $\Delta t_b = (0.544 \, m)\left(\dfrac{2.53°C}{m}\right) = 1.38°C$

   Boiling point of solution $= 80.1°C + 1.38°C = 81.5°C$

34. (a) $\left(\dfrac{100.0 \text{ g } C_2H_6O_2}{150.0 \text{ g } H_2O}\right)\left(\dfrac{1 \text{ mol}}{62.07 \text{ g}}\right)\left(\dfrac{1000 \text{ g}}{\text{kg}}\right) = 10.74 \, m \, C_2H_6O_2$

   (b) $\Delta t_b = mK_b = (10.74 \, m)\left(\dfrac{0.512°C}{m}\right) = 5.50°C$ (Increase in boiling point)

   Boiling point $= 100.00°C + 5.50°C = 105.50°C$

   (c) $\Delta t_f = mK_f = (10.74 \, m)\left(\dfrac{1.86°C}{m}\right) = 20.0°C$ (Decrease in freezing point)

   Freezing point $= 0.00°C - 20.0°C = -20.0°C$

35. Freezing point of acetic acid is $16.6°C$ \qquad K_f acetic acid $= \dfrac{3.90°C}{m}$

   $\Delta t_f = 16.6°C - 13.2°C = 3.4°C$

   $\Delta t_f = mK_f$

   $m = \dfrac{3.4°C}{3.90°C/m} = 0.87 \, m$

   Convert 8.00 g unknown/60.0 g $HC_2H_3O_2$ to g/mol (molar mass)

Conversion: $\dfrac{\text{g unknown}}{\text{g HC}_2\text{H}_3\text{O}_2} \longrightarrow \dfrac{\text{g unknown}}{\text{kg HC}_2\text{H}_3\text{O}_2} \longrightarrow \dfrac{\text{g}}{\text{mol}}$

$$\left(\dfrac{8.00 \text{ g unknown}}{60.0 \text{ g HC}_2\text{H}_3\text{O}_2}\right)\left(\dfrac{1000 \text{ g}}{\text{kg}}\right)\left(\dfrac{1 \text{ kg HC}_2\text{H}_3\text{O}_2}{0.87 \text{ mol unknown}}\right) = 153 \text{ g/mol}$$

36. $\Delta t_f = 2.50°C \qquad K_f(\text{for } H_2O) = \dfrac{1.86°C}{m}$

$\Delta t_f = mK_f$

$m = \dfrac{2.50°C}{1.86°C/m} = 1.34\, m$

Convert 4.80 g unknown/22.0 g H_2O to g/mol (molar mass)

$$\left(\dfrac{4.80 \text{ g unknown}}{22.0 \text{ g } H_2O}\right)\left(\dfrac{1000 \text{ g}}{\text{kg}}\right)\left(\dfrac{1 \text{ kg } H_2O}{1.34 \text{ mol unknown}}\right) = 163 \text{ g/mol}$$

37. Jasmine tea and gasoline are true solutions.

38. Concord grape juice and stainless steel are true solutions.

39. Chromium metal is not a true solution because it is only a single substance. Muddy water is not a true solution because the dirt particles will settle out.

40. Red paint and oil and vinegar salad dressing are not true solutions because they will separate out if they are not shaken or stirred.

41. (a) The teaspoon of sugar will dissolve faster because it has a smaller particle size.
     (b) The copper(II) sulfate will dissolve in the pure water faster because solutes dissolve in a pure solvent faster than in a solvent with some solute already dissolved.
     (c) The sweetener will dissolve faster in a cup of hot tea because the rate of dissolving increases with increasing temperature.
     (d) The silver nitrate will dissolve faster in the sloshing water because agitation increases the rate of dissolution.

42. (a) The amino acid will dissolve faster in the solvent at 75°C because the rate of dissolving increases with increasing temperature.
     (b) The powdered sodium acetate will dissolve faster because it has a smaller particle size.
     (c) The carton of table salt will dissolve faster because it has both a smaller particle size and the temperature is higher.
     (d) The acetaminophen will dissolve faster in the infant pain medication because it has less acetaminophen already dissolved in it.

43. Molarity Problem $\left(M = \dfrac{\text{mol}}{\text{L}}\right)$

$$\left(\dfrac{396.1 \text{ g KI}}{750.0 \text{ mL}}\right)\left(\dfrac{1 \text{ mol}}{166.0 \text{ g}}\right)\left(\dfrac{1000 \text{ mL}}{1 \text{ L}}\right) = 3.182\, M \text{ KI}$$

44. Molarity Problem $\left(M = \dfrac{\text{mol}}{\text{L}} \right)$

$$\left(\frac{74.15 \text{ g HgCl}_2}{250.0 \text{ mL}} \right) \left(\frac{1 \text{ mol}}{271.5 \text{ g}} \right) \left(\frac{1000 \text{ mL}}{1 \text{ L}} \right) = 1.092 \, M \text{ HgCl}_2$$

45. Salt (NaCl) is an ionic compound. When it is dissolved in water the sodium and chloride ions separate or dissociate. The polar water molecules are attracted to the sodium and chloride ions and are hydrated (surrounded by water molecules). The sodium and chloride ions are separated from one another and distributed throughout the water in this way. Na^+ and Cl^- are hydrated in aqueous solution: See Question 8 in the Review Questions.

46. Sugar molecules are not ionic; therefore they do not dissociate when dissolved in water. However, sugar molecules are polar, so water molecules are attracted to them and the sugar becomes hydrated. The water molecules help separate the sugar molecules from each other and distribute them throughout the solution.

47. Sugar and salt behave differently when dissolved in water because salt is an ionic compound and sugar is a molecular compound.

48. An isotonic sodium chloride solution has the same osmotic pressure as human blood plasma. When blood cells are placed in an isotonic solution the osmotic pressure inside the cells is equal to the osmotic pressure outside the cells so there is no change in the appearance of the blood cells.

49. The $KMnO_4$ crystals give the solution its purple color. The purple streaks are formed because the solute has not been evenly distributed throughout that solvent yet. The MnO_4^- has a purple color in solution.

50. The line for KNO_3 slopes upward, because the solubility increases as the temperature increases. KNO_3 has the steepest slope of all the compounds given in the diagram. It exhibits the greatest increase in the number of grams of solute that is able to dissolve in 100 g of water than any other compound in the diagram as the temperature increases.

51. $\left(\dfrac{9.0 \text{ g NaCl}}{1 \text{ L}} \right) \left(\dfrac{1 \text{ mol}}{58.44 \text{ g}} \right) = 0.15 \, M \text{ NaCl}$

52. $1.0 \, m \text{ HCl} = \dfrac{1 \text{ mol HCl}}{1 \text{ kg H}_2\text{O}} = \dfrac{36.46 \text{ g HCl}}{1000 \text{ g H}_2\text{O}}$

Total mass of solution $= 1000 \text{ g} + 36.46 \text{ g} = 1036.46 \text{ g}$

Therefore, $1.0 \, m \text{ HCl} = \dfrac{1 \text{ mol HCl}}{1036.46 \text{ g HCl solution}}$

$NaOH + HCl \longrightarrow NaCl + H_2O$

Calculate the grams NaOH to neutralize HCl

$$(250.0 \text{ g solution}) \left(\frac{1 \text{ mol HCl}}{1036.46 \text{ g solution}} \right) \left(\frac{1 \text{ mol NaOH}}{1 \text{ mol HCl}} \right) \left(\frac{40.00 \text{ g}}{\text{mol}} \right) = 9.648 \text{ g NaOH}$$

Calculate the grams of 10.0% NaOH solution that contains 9.648 g NaOH.

$$\frac{9.648 \text{ g NaOH}}{x} = \frac{10.0 \text{ g NaOH}}{100.0 \text{ g } 10.0\% \text{ NaOH solution}}$$

$x = 96.5$ g 10% NaOH solution

53. (a) $(1.0 \text{ L syrup})\left(\frac{1000 \text{ mL}}{\text{L}}\right)\left(\frac{1.06 \text{ g}}{\text{mL}}\right)\left(\frac{15.0 \text{ g sugar}}{100. \text{ g syrup}}\right) = 1.6 \times 10^2 \text{ g sugar}$

(b) $\left(\frac{1.6 \times 10^2 \text{g } C_{12}H_{22}O_{11}}{\text{L}}\right)\left(\frac{1 \text{ mol}}{342.3 \text{ g}}\right) = 0.47 \, M \, C_{12}H_{22}O_{11}$

(c) $m = \frac{\text{mol sugar}}{\text{kg } H_2O}$ 15% sugar by mass $= 15.0 \text{ g } C_{12}H_{22}O_{11} + 85.0 \text{ g } H_2O$

$$\left(\frac{15.0 \text{ g } C_{12}H_{22}O_{11}}{85.0 \text{ g } H_2O}\right)\left(\frac{1000 \text{ g}}{1 \text{ kg}}\right)\left(\frac{1 \text{ mol}}{342.3 \text{ g}}\right) = 0.516 \, m \, C_{12}H_{22}O_{11}$$

54. $K_f = \dfrac{5.1°C}{m}$ $\Delta t_f = 0.614°C$

$$\left(\frac{3.84 \text{ g } C_4H_2N}{250.0 \text{ g } C_6H_6}\right)\left(\frac{1000 \text{ g}}{\text{kg}}\right) = \frac{15.4 \text{ g } C_4H_2N}{\text{kg } C_6H_6}$$

$\Delta t_f = mK_f$

$$m = \frac{0.614°C}{5.1°C/m} = 0.12 \, m = \frac{0.12 \text{ mol } C_4H_2N}{\text{kg } C_6H_6}$$

$$\left(\frac{15.4 \text{ g } C_4H_2N}{\text{kg } C_6H_6}\right)\left(\frac{1 \text{ kg } C_6H_6}{0.12 \text{ mol } C_4H_2N}\right) = 128 \text{ g/mol} = 1.3 \times 10^2 \text{ g/mol}$$

Empirical mass $(C_4H_2N) = 64.07$ g

$$\frac{130 \text{ g}}{64.07 \text{ g}} = 2.0 \text{ (number of empirical formulas per molecular formula)}$$

Therefore, the molecular formula is twice the empirical formula, or $C_8H_4N_2$.

55. $(12.0 \text{ mol HCl})\left(\dfrac{36.46 \text{ g}}{\text{mol}}\right) = 438 \text{ g HCl in } 1.00 \text{ L solution}$

$(1.00 \text{ L})\left(\dfrac{1.18 \text{ g solution}}{\text{mL}}\right)\left(\dfrac{1000 \text{ mL}}{\text{L}}\right) = 1180 \text{ g solution}$

1180 g solution $- 438$ g HCl $= 742$ g H_2O (0.742 kg H_2O)

Since molality $= \dfrac{\text{mol HCl}}{\text{kg } H_2O} = \dfrac{12.0 \text{ mol HCl}}{0.742 \text{ kg } H_2O} = 16.2 \, m \text{ HCl}$

56. First calculate the g KNO_3 in the solution.

The conversion is: $\dfrac{mg\ K^+}{mL} \longrightarrow \dfrac{1\ g}{1000\ mg} \longrightarrow \dfrac{g\ KNO_3}{g\ K^+}(450\ mL) \longrightarrow g\ KNO_3$

$\left(\dfrac{5.5\ mg\ K^+}{mL}\right)\left(\dfrac{1\ g}{1000\ mg}\right)\left(\dfrac{101.1\ g\ KNO_3}{39.10\ g\ K^+}\right)(450\ mL) = 6.4\ g\ KNO_3$

Now calculate the mol KNO_3 and the molarity.

$(6.4\ g\ KNO_3)\left(\dfrac{1\ mol}{101.1\ g}\right) = 0.063\ mol\ KNO_3$

$\dfrac{0.063\ mol\ KNO_3}{0.450\ L} = 0.14\ M$

57. $(16\ fl.\ oz\ witch\ hazel)\left(\dfrac{1\ qt}{32\ oz}\right)\left(\dfrac{946.1\ mL}{qt}\right)\left(\dfrac{14\ mL\ ethyl\ alcohol}{100\ mL\ witch\ hazel}\right) = 66\ mL\ ethyl\ alcohol$

58. Verification of K_b for water

$\Delta t_b = mK_b \qquad \Delta t_b = 101.62°C - 100°C = 1.62°C \qquad K_b = \dfrac{\Delta t_b}{m}$

First calculate the molality of the solution.

$m = \dfrac{16.10\ g\ C_2H_6O_2}{(62.07\ g/mol)(0.0820\ kg\ H_2O)} = \dfrac{3.16\ mol\ C_2H_6O_2}{kg\ H_2O}$

$K_b = \dfrac{\Delta t_b}{m} = \dfrac{1.62°C}{3.16\ mol/kg\ H_2O} = \dfrac{0.513°C\ kg\ H_2O}{mol}$

59. (a) $(500.0\ mL\ solution)\left(\dfrac{0.90\ g\ NaCl}{100.\ mL\ solution}\right) = 4.5\ g\ NaCl$

(b) $\left(\dfrac{4.5\ g\ NaCl}{x\ mL}\right)(100) = 9.0\% \qquad x = $ volume of 9.0% solution

$x = \dfrac{4.5\ g\ NaCl}{9.0\%}(100) = 50.\ mL(4.5\ g\ NaCl\ in\ solution)$

500. mL − 50. mL = 450. mL H_2O must evaporate

60. From Figure 14.4, the solubility of KNO_3 in H_2O at 20°C is 32 g per 100. g H_2O.

$(50.0\ g\ KNO_3)\left(\dfrac{100.\ g\ H_2O}{32.0\ g\ KNO_3}\right) = 156\ g\ H_2O$ to produce a saturated solution.

175 g H_2O − 156 g H_2O = 19 g H_2O must be evaporated.

61. 7.35 mL oil of wintergreen $\left(\dfrac{100\ mL\ solution}{0.25\ mL\ oil\ of\ wintergreen}\right)\left(\dfrac{1\ L}{1000\ mL}\right) = 2.94\ L\ solution$

62. (a) $(1.00 \text{ L solution})\left(\dfrac{1000 \text{ mL solution}}{\text{L solution}}\right)\left(\dfrac{1.21 \text{ g}}{\text{mL}}\right)\left(\dfrac{35.0 \text{ g HNO}_3}{100. \text{ g solution}}\right) = 424 \text{ g HNO}_3$

    (b) $(500. \text{ g HNO}_3)\left(\dfrac{1000 \text{ mL solution}}{424 \text{ g HNO}_3}\right)\left(\dfrac{1.00 \text{ L}}{1000 \text{ mL}}\right) = 1.18 \text{ L solution}$

63. $\text{Molarity} = M = \dfrac{\text{mol solute}}{\text{L solution}}$

$\left(\dfrac{85 \text{ g H}_3\text{PO}_4}{100 \text{ g solution}}\right)\left(\dfrac{1.7 \text{ g solution}}{\text{mL solution}}\right)\left(\dfrac{1000 \text{ mL}}{\text{L}}\right)\left(\dfrac{1 \text{ mol H}_3\text{PO}_4}{97.99 \text{ g}}\right) = 15 \ M \ \text{H}_3\text{PO}_4$

64. First calculate the molarity of the solution

$\left(\dfrac{80.0 \text{ g H}_2\text{SO}_4}{500. \text{ mL}}\right)\left(\dfrac{1000 \text{ mL}}{\text{L}}\right)\left(\dfrac{1 \text{ mol}}{98.09 \text{ g}}\right) = 1.63 \ M \ \text{H}_2\text{SO}_4$

$M_1V_1 = M_2V_2$

$(1.63 \ M)(500. \text{ mL}) = (0.10 \ M)(V_2)$

$V_2 = \dfrac{(1.63 \ M)(500. \text{ mL})}{0.10 \ M} = 8.2 \times 10^3 \text{mL} = 8.2 \text{ L}$

65. $30.0 \text{ gal}\left(\dfrac{4 \text{ qt}}{1 \text{ gal}}\right)\left(\dfrac{1 \text{ L}}{1.057 \text{ qt}}\right)\left(\dfrac{4.28 \text{ mol C}_3\text{H}_8\text{O}_3}{1 \text{ L}}\right)\left(\dfrac{92.09 \text{ g}}{1 \text{ mol}}\right)\left(\dfrac{1 \text{ lb}}{453.6 \text{ g}}\right) = 98.6 \text{ lb C}_3\text{H}_8\text{O}_3$

66. $\text{Mg} + 2 \text{ HCl} \longrightarrow \text{MgCl}_2 + \text{H}_2(g)$

    (a) $\text{mL HCl} \longrightarrow \text{mol HCl} \longrightarrow \text{mol H}_2$

$(200.0 \text{ mL HCl})\left(\dfrac{3.00 \text{ mol}}{1000 \text{ mL}}\right)\left(\dfrac{1 \text{ mol H}_2}{2 \text{ mol HCl}}\right) = 0.300 \text{ mol H}_2$

    (b) $PV = nRT$

$P = (720 \text{ torr})\left(\dfrac{1 \text{ atm}}{760 \text{ torr}}\right) = 0.95 \text{ atm}$

$T = 27°C = 300. \text{ K}$

$n = 0.300 \text{ mol}$

$V = \dfrac{nRT}{P} = \dfrac{(0.300 \text{ mol})(0.0821 \text{ L atm/mol K})(300. \text{ K})}{0.95 \text{ atm}} = 7.8 \text{ L H}_2$

67. $\text{Mg(OH)}_2 + 2 \text{ HCl} \longrightarrow \text{MgCl}_2 + 2 \text{ H}_2\text{O}$

$\text{Al(OH)}_3 + 3 \text{ HCl} \longrightarrow \text{AlCl}_3 + 3 \text{ H}_2\text{O}$

Calculate the moles of HCl neutralized by each base.

$$\left(1.20 \text{ g Mg(OH)}_2\right)\left(\frac{1 \text{ mol}}{58.33 \text{ g}}\right)\left(\frac{2 \text{ mol HCl}}{1 \text{ mol Mg(OH)}_2}\right) = 0.0411 \text{ mol HCl}$$

$$\left(1.00 \text{ g Al(OH)}_3\right)\left(\frac{1 \text{ mol}}{78.00 \text{ g}}\right)\left(\frac{3 \text{ mol HCl}}{1 \text{ mol Al(OH)}_3}\right) = 0.0385 \text{ mol HCl}$$

1.20 g $Mg(OH)_2$ reacts with more HCl than 1.00 g $Al(OH)_3$. Therefore, $Mg(OH)_2$ is more effective in neutralizing stomach acid.

68. (a) With equal masses of CH_3OH and C_2H_5OH, the substance with the lower molar mass will represent more moles of solute in solution. Therefore, the CH_3OH will be more effective than C_2H_5OH as an antifreeze.

    (b) Equal molal solutions will lower the freezing point of the solution by the same amount.

69. Calculate molarity and molality. Assume 1000 mL of solution to calculate the amounts of H_2SO_4 and H_2O in the solution.

$$\left(1000 \text{ mL solution}\right)\left(\frac{1.29 \text{ g}}{\text{mL}}\right) = 1.29 \times 10^3 \text{ g solution}$$

$$\left(1.29 \times 10^3 \text{ g solution}\right)\left(\frac{38 \text{ g H}_2\text{SO}_4}{100 \text{ g solution}}\right) = 4.9 \times 10^2 \text{ g H}_2\text{SO}_4$$

1.29×10^3 g solution $- 4.9 \times 10^2$ g $H_2SO_4 = 8.0 \times 10^2$ g H_2O in the solution

$$m = \left(\frac{490 \text{ g H}_2\text{SO}_4}{8.0 \times 10^2 \text{ g H}_2\text{O}}\right)\left(\frac{1000 \text{ g}}{\text{kg}}\right)\left(\frac{1 \text{ mol}}{98.09 \text{ g}}\right) = 6.2 \, m \text{ H}_2\text{SO}_4$$

$$M = \left(\frac{4.9 \times 10^2 \text{ g H}_2\text{SO}_4}{\text{L}}\right)\left(\frac{1 \text{ mol}}{98.09 \text{ g}}\right) = 5.0 \, M \text{ H}_2\text{SO}_4$$

70. Freezing point depression is 5.4°C

    (a) $\Delta t_f = mK_f$

    $$m = \frac{\Delta t_f}{K_f} = \frac{5.4°\text{C}}{1.86°\text{C kg solvent/mol solute}} = 2.9 \, m$$

    (b) K_b(for H_2O) $= \dfrac{0.512°\text{C kg solvent}}{\text{mol solute}} = \dfrac{0.512°\text{C}}{m}$

    $$\Delta t_b = mK_b = (2.9 \, m)\left(\frac{0.512°\text{C}}{m}\right) = 1.5°\text{C}$$

    Boiling point $= 100°\text{C} + 1.5°\text{C} = 101.5°\text{C}$

71. Freezing point depression $= 0.372°\text{C}$ $K_f = \dfrac{1.86°\text{C}}{m}$

    $\Delta t_f = mK_f$

    $$m = \frac{0.372°\text{C}}{1.86°\text{C}/m} = 0.200 \, m$$

$$(6.20 \text{ g C}_2\text{H}_6\text{O}_2)\left(\frac{1 \text{ mol}}{62.07 \text{ g}}\right) = 0.100 \text{ mol C}_2\text{H}_6\text{O}_2$$

$$(0.100 \text{ mol C}_2\text{H}_6\text{O}_2)\left(\frac{1 \text{ kg H}_2\text{O}}{0.200 \text{ mol C}_2\text{H}_6\text{O}_2}\right)\left(\frac{1000 \text{ g H}_2\text{O}}{\text{kg H}_2\text{O}}\right) = 500. \text{ g H}_2\text{O}$$

72. (a) Freezing point depression $= 20.0°\text{C}$

$$12.0 \text{ L H}_2\text{O}\left(\frac{1000 \text{ mL}}{\text{L}}\right)\left(\frac{1.00 \text{ g}}{\text{mL}}\right) = 1.20 \times 10^4 \text{ g H}_2\text{O}$$

$$\Delta t_f = mK_f$$

$$m = \frac{20.0°\text{C}}{1.86°\text{C}/m} = 10.8 \, m$$

$$(1.20 \times 10^4 \text{ g H}_2\text{O})\left(\frac{10.8 \text{ mol C}_2\text{H}_6\text{O}_2}{1000 \text{ g H}_2\text{O}}\right)\left(\frac{62.07 \text{ g}}{\text{mol}}\right) = 8.04 \times 10^3 \text{ g C}_2\text{H}_6\text{O}_2$$

(b) $$(8.04 \times 10^3 \text{ g C}_2\text{H}_6\text{O}_2)\left(\frac{1.00 \text{ mL}}{1.11 \text{ g}}\right) = 7.24 \times 10^3 \text{ mL C}_2\text{H}_6\text{O}_2$$

(c) $°\text{F} = 1.8(°\text{C}) + 32 = 1.8(-20.0) + 32 = -4.0°\text{F}$

73. $\text{HNO}_3 + \text{NaHCO}_3 \longrightarrow \text{NaNO}_3 + \text{H}_2\text{O} + \text{CO}_2$

First calculate the grams of NaHCO_3 in the sample.

mL $\text{HNO}_3 \longrightarrow$ L $\text{HNO}_3 \longrightarrow$ mol $\text{HNO}_3 \longrightarrow$ mol $\text{NaHCO}_3 \longrightarrow$ g NaHCO_3

$$(150 \text{ mL HNO}_3)\left(\frac{1 \text{ L}}{1000 \text{ mL}}\right)\left(\frac{0.055 \text{ mol}}{\text{L}}\right)\left(\frac{1 \text{ mol NaHCO}_3}{1 \text{ mol HNO}_3}\right)\left(\frac{84.01 \text{ g}}{\text{mol}}\right)$$

$$= 0.69 \text{ g NaHCO}_3 \text{ in the sample}$$

$$\left(\frac{0.69 \text{ g NaHCO}_3}{1.48 \text{ g sample}}\right)(100) = 47\% \text{ NaHCO}_3$$

74. (a) Dilution problem: $M_1V_1 = M_2V_2$

$$(1.5 \, M)(8.4 \text{ L}) = (17.8 \, M)(V_2)$$

$$V_2 = \frac{(1.5 \, M)(8.4 \text{ L})}{17.8 \, M} = 0.71 \text{ L H}_2\text{SO}_4$$

If 0.71 L of 17.8 M H_2SO_4 are required, then 7.7 L of water must also be added to bring the total volume to 8.4 L.

(b) $$\left(\frac{17.8 \text{ mol H}_2\text{SO}_4}{1000. \text{ mL}}\right)(1.00 \text{ mL}) = 0.0178 \text{ mol H}_2\text{SO}_4$$

(c) $$\left(\frac{1.5 \text{ mol H}_2\text{SO}_4}{1000. \text{ mL}}\right)(1.00 \text{ mL}) = 0.0015 \text{ mol H}_2\text{SO}_4 \text{ in each mL}$$

75. moles HNO_3 total $=$ moles HNO_3 from 3.00 M $+$ moles HNO_3 from 12.0 M

$$M_T V_T = M_{3.00 \, M} V_{3.00 \, M} + M_{12.0 \, M} V_{12.0 \, M}$$

Assume preparation of 1000. mL of 6 M solution

Let y = volume of 3.00 M solution; volume of 12.0 M = 1000. mL − y

$(6.00\ M)(1000.\ \text{mL}) = (3.00\ M)(y) + (12.0\ M)(1000.\ \text{mL} - y)$

$6000.\ \text{mL} = 3.00\ y\ \text{mL} + 12{,}000\ \text{mL} - 12.0\ y$

$6000.\ \text{mL} = 9.00\ y \qquad y = \dfrac{6000.\ \text{mL}}{9.00} = 667\ \text{mL}\ 3\ M$

$1000.\ \text{mL} - 667\ \text{mL} = 333\ \text{mL}\ 12\ M$

Mix together 667 mL 3.00 M HNO$_3$ and 333 mL of 12.0 M HNO$_3$ to get 1000. mL of 6,00 M HNO$_3$.

76. $HBr + NaOH \longrightarrow NaBr + H_2O$

First calculate the molarity of the diluted HBr solution.

The reaction is 1 mol HBr to 1 mol NaOH, so

$M_A V_A = M_B V_B$

$(M_A)(100.0\ \text{mL}) = (0.37\ M)(88.4\ \text{mL})$

$M_A = \dfrac{(0.37\ M)(88.4\ \text{mL})}{100.00\ \text{mL}} = 0.33\ M\ \text{HBr (diluted solution)}$

Now calculate the molarity of the HBr before dilution.

$M_1 V_1 = M_2 V_2$

$(M_1)(20.0\ \text{mL}) = (0.33\ M)(240.\ \text{mL})$

$M_1 = \dfrac{(0.33\ M)(240.\ \text{mL})}{20.0\ \text{mL}} = 4.0\ M\ \text{HBr (original solution)}$

77. $Ba(NO_3)_2 + 2\,KOH \longrightarrow Ba(OH)_2 + 2\,KNO_3$

This is a limiting reactant problem. First calculate the moles of each reactant and determine the limiting reactant.

$M \times L = \left(\dfrac{\text{moles}}{L}\right)(L) = \text{moles}$

$\left(\dfrac{0.642\ \text{mol}}{L}\right)(0.0805\ L) = 0.0517\ \text{mol Ba}(NO_3)_2$

$\left(\dfrac{0.743\ \text{mol}}{L}\right)(0.0445\ L) = 0.0331\ \text{mol KOH}$

According to the equation, twice as many moles of KOH as Ba(NO$_3$)$_2$ are needed, so KOH is the limiting reactant. The mass of Ba(OH)$_2$ formed will be 2.84 g.

$0.0331\ \text{mol KOH} \left(\dfrac{1\,\text{mol Ba(OH)}_2}{2\,\text{mol KOH}}\right)\left(\dfrac{171.3\ \text{g Ba(OH)}_2}{1\ \text{mol Ba(OH)}_2}\right) = 2.84\ \text{g Ba(OH)}_2$

78.   (a) $\left(\dfrac{0.25\ \text{mol}}{L}\right)(0.0458\ L) = 0.011\ \text{mol Li}_2CO_3$

(b) $\left(\dfrac{0.25 \text{ mol}}{L}\right)(0.75 \text{ L})\left(\dfrac{73.89 \text{ g}}{\text{mol}}\right) = 14 \text{ g Li}_2\text{CO}_3$

(c) $(6.0 \text{ g Li}_2\text{CO}_3)\left(\dfrac{1 \text{ mol}}{73.89 \text{ g}}\right)\left(\dfrac{1000. \text{ mL}}{0.25 \text{ mol}}\right) = 3.2 \times 10^2 \text{ mL solution}$

(d) Assume 1000. mL solution

$\left(\dfrac{1.22 \text{ g}}{\text{mL}}\right)(1000. \text{ mL}) = 1220 \text{ g solution}$

$\left(\dfrac{0.25 \text{ mol}}{L}\right)\left(\dfrac{73.89 \text{ g Li}_2\text{CO}_3}{\text{mol}}\right) = 18 \text{ g Li}_2\text{CO}_3 \text{ per L solution}$

$\% = \left(\dfrac{\text{g solute}}{\text{g solution}}\right)(100) = \left(\dfrac{18 \text{ g}}{1220 \text{ g}}\right)(100) = 1.5\% \text{ (mass percent)}$

79. Molarity Problem $\left(M = \dfrac{\text{mol}}{L}\right)$

$\left(\dfrac{0.625 \text{ g EuCl}_3\cdot 6 \text{ H}_2\text{O}}{500.0 \text{ mL}}\right)\left(\dfrac{1 \text{ mol}}{366.5 \text{ g}}\right)\left(\dfrac{1000 \text{ mL}}{1 \text{ L}}\right) = 3.40 \times 10^{-3} M \text{ EuCl}_3\cdot 6 \text{ H}_2\text{O}$

80. Picture A represents potassium chloride, picture B represents sucrose, and picture C represents sodium phosphate.

81. Freezing Point Depression $\left(\text{freezing point of water is } 0°C, \quad K_f \text{ water} = \dfrac{1.86°C}{m}\right)$

$\Delta t_f = 0°C - (-12.7°C) = 12.7°C$

$\Delta t_f = mK_f$

$m_{\text{ions}} = \dfrac{12.7°C}{1.86°C/m} = 6.83 \, m \text{ ions}$

To determine molality of sodium chloride remember there are 2 ions formed when sodium chloride dissociates in water.

$m_{\text{NaCl}} = \left(\dfrac{6.83 \text{ mol ions}}{\text{kg water}}\right)\left(\dfrac{1 \text{ mol NaCl}}{2 \text{ mol ions}}\right) = 3.42 \, m \text{ NaCl}$

82. Salt will cause the loss of water from a snail due to osmosis. When salt is sprinkled on a snail it dissolves in the watery coating of the snail. Because the concentration of salt outside the snail is now greater than the concentration of salt on the inside, water will flow out of the snail to dilute the more concentrated solution on its surface.

83. A slice of eggplant coated with a layer of salt will release water as a result of osmosis. When salt is sprinkled on the surface of a slice of eggplant, the water inside the slice will flow out onto the surface to dilute the salt. As the water flows out of the slice of eggplant it will also draw out some of the compounds which are responsible for the bitter flavor of the eggplant.

84. Boiling Point Elevation $\left(\text{boiling point of water is } 100°C, \quad K_b \text{ water} = \dfrac{0.512°C}{m}\right)$

$\Delta t_b = 127°C - 100°C = 27°C$

$\Delta t_b = mK_b$

$m = \dfrac{27°C}{0.512°C/m} = 53 \, m \text{ suger}$

Taffy is made from a very concentrated sugar solution!

85. The balanced equation is

$2\,HCl(aq) + Na_2SO_3(aq) \longrightarrow 2\,NaCl(aq) + H_2O(l) + SO_2(g)$

$(125 \text{ mL HCl})\left(\dfrac{2.50 \text{ mol}}{1000 \text{ mL}}\right)\left(\dfrac{1 \text{ mol SO}_2}{2 \text{ mol HCl}}\right) = 0.156 \text{ mol SO}_2$

$(75 \text{ mL Na}_2\text{SO}_3)\left(\dfrac{1.75 \text{ mol}}{1000 \text{ mL}}\right)\left(\dfrac{1 \text{ mol SO}_2}{1 \text{ mol Na}_2\text{SO}_3}\right) = 0.131 \text{ mol SO}_2$

Na_2SO_3 is the limiting reactant; 0.131 mol of SO_2 gas will be produced.

The gas is at non-standard conditions, so use $PV = nRT$ to find the liters of SO_2.

$V = \dfrac{nRT}{P} \quad V = \dfrac{(0.131 \text{ mol})(0.0821 \text{ L atm/mol K})(295 \text{ K})}{(775 \text{ torr})(1 \text{ atm}/760 \text{ torr})} = 3.11 \text{ L SO}_2$

86. mass of solute = mass of container and solute − mass of container − mass of water
mass of water = (5.549 moles) (18.02 g/mol) = 100.0 g
mass of solute = 563 g − 375 g − 100.0 g = 88 g
solubility in water = g solute/100 g H_2O at 20° C
Using this data, solubility is 88 g solute/100.0 g water = $NaNO_3$ (see Table 14.3).

ACIDS, BASES, AND SALTS

SOLUTIONS TO REVIEW QUESTIONS

1. The Arrhenius definition is restricted to aqueous solutions, while the Brønsted-Lowry definition is not.

2. By the Arrhenius theory, an acid is a substance that produces hydrogen ions in aqueous solution. A base is a substance that produces hydroxide ions in aqueous solution.

   By the Brønsted-Lowry theory, an acid is a proton donor, while a base accepts protons. Since a proton is a hydrogen ion, then the two theories are very similar for acids, but not bases. A chloride ion can accept a proton (producing HCl), so it is a Brønsted-Lowry base, but would not be a base by the Arrhenius theory, since it does not produce hydroxide ions.

   By the Lewis theory, an acid is an electron pair acceptor, and a base is an electron pair donor. Many individual substances would be similarly classified as bases by Brønsted-Lowry or Lewis theories, since a substance with an electron pair to donate, can accept a proton. But, the Lewis definition is almost exclusively applied to reactions where the acid and base combine into a single molecule. The Brønsted-Lowry definition is usually applied to reactions that involve a transfer of a proton from the acid to the base. The Arrhenius definition is most often applied to individual substances, not to reactions. According to the Arrhenius theory, neutralization involves the reaction between a hydrogen ion and a hydroxide ion to form water.

   Neutralization, according to the Brønsted-Lowry theory, involves the transfer of a proton to a negative ion. The formation of a covalent bond constitutes a Lewis neutralization.

3. Neutralization reactions:
   Arrhenius: $HCl + NaOH \longrightarrow NaCl + H_2O$ $(H^+ + OH^- \longrightarrow H_2O)$

   Brønsted-Lowry: $HCl + KCN \longrightarrow HCN + KCl$ $(H^+ + CN^- \longrightarrow HCN)$

   Lewis: $AlCl_3 + NaCl \longrightarrow AlCl_4^- + Na^+$

4. (a) $\left[:\!\ddot{Br}\!:\right]^-$ (b) $\left[:\!\ddot{O}\!:H\right]^-$ (c) $\left[:C:::N:\right]^-$

   These ions are considered to be bases according to the Brønsted-Lowry theory, because they can accept a proton at any of their unshared pairs of electrons. They are considered to be bases according to the Lewis acid-base theory, because they can donate an electron pair.

5. Metals that lie above hydrogen in the activity series will form hydrogen gas when they react with an acid.

6. Carbonates will form carbon dioxide when they react with an acid.

7. $NaNO_3$, sodium nitrate; $Ca(NO_3)_2$, calcium nitrate; $Al(NO_3)_3$, aluminium nitrate. These are three of the many possible salts which can be formed from nitric acid.

8. LiCl, lithium chloride; Li_2SO_4, lithium sulfate; Li_3PO_4, lithium phosphate. These are three of the many possible salts which can be formed from lithium hydroxide.

9. An electrolyte must be present in the solution for the bulb to glow.

10. Electrolytes include acids, bases, and salts. (Electrolytes are any compound that conducts electricity in solution.)

11. First, the orientation of the polar water molecules about the Na^+ and Cl^- is different. The positive end (hydrogen) of the water molecule is directed towards Cl^-, while the negative end (oxygen) of the water molecule is directed towards the Na^+. Second, more water molecules will fit around Cl^-, since it is larger than the Na^+ ion.

12. The electrolytic compounds are acids, bases, and salts.

13. Names of the compounds in Table 15.3

H_2SO_4	sulfuric acid	$HC_2H_3O_2$	acetic acid
HNO_3	nitric acid	H_2CO_3	carbonic acid
HCl	hydrochloric acid	HNO_2	nitrous acid
HBr	hydrobromic acid	H_2SO_3	sulfurous acid
$HClO_4$	perchloric acid	H_2S	hydrosulfuric acid
NaOH	sodium hydroxide	$H_2C_2O_4$	oxalic acid
KOH	potassium hydroxide	H_3BO_3	boric acid
$Ca(OH)_2$	calcium hydroxide	HClO	hypochlorous acid
$Ba(OH)_2$	barium hydroxide	NH_3	ammonia
		HF	hydrofluoric acid

14. Hydrogen chloride dissolved in water conducts an electric current. HCl reacts with polar water molecules to produce H_3O^+ and Cl^- ions, which conduct an electric current. Hexane is a nonpolar solvent, so it cannot pull the HCl molecules apart. Since there are no ions in the hexane solution, it does not conduct an electric current. HCl does not ionize in hexane.

15. In their crystalline structure, salts exist as positive and negative ions in definite geometric arrangement to each other, held together by the attraction of the opposite charges. When dissolved in water, the salt dissociates as the ions are pulled away from each other by the polar water molecules.

16. Testing the electrical conductivity of the solutions shows that CH_3OH is a nonelectrolyte, while NaOH is an electrolyte. This indicates that the OH group in CH_3OH must be covalently bonded to the CH_3 group.

17. Molten NaCl conducts electricity because the ions are free to move. In the solid state, however, the ions are immobile and do not conduct electricity.

18. Dissociation is the separation of already existing ions in an ionic compound. Ionization is the formation of ions from molecules. The dissolving of NaCl is a dissociation, since the ions already exist in the crystalline compound. The dissolving of HCl in water is an ionization process, because ions are formed from HCl molecules and H_2O.

19. Strong electrolytes are those which are essentially 100% ionized or dissociated in water. Weak electrolytes are those which are only slightly ionized in water.

20. Ions are hydrated in solution because there is an electrical attraction between the charged ions and the polar water molecules.

21. The main distinction between water solutions of strong and weak electrolytes is the degree of ionization of the electrolyte. A solution of an electrolyte contains many more ions than does a solution of a nonelectrolyte. Strong electrolytes are essentially 100% ionized. Weak electrolytes are only slightly ionized in water.

22. The HCl molecule is polar and, consequently, is much more soluble in the polar solvent, water, than in the nonpolar solvent, hexane. There is also a chemical reaction between HCl and H_2O molecules.
$$HCl + H_2O \longrightarrow H_3O^+ + Cl^-$$

23. The pH for a solution with a hydrogen ion concentration of 0.003 M will be between 2 and 3.

24. Tomato juice is more acidic than blood, since its pH is lower.

25. (a) In a neutral solution, the concentration of H^+ and OH^- are equal.
    (b) In an acid solution, the concentration of H^+ is greater than the concentration of OH^-.
    (c) In a basic solution, the concentration of OH^- is greater than the concentration of H^+.

26. Pure water is neutral because when it ionizes it produces equal molar concentrations of acid $[H^+]$ and base $[OH^-]$ ions.

27. A neutral solution is one in which the concentration of acid is equal to the concentration of base $[H^+] = [OH^-]$. An acidic solution is one in which the concentration of acid is greater than the concentration of base $[H^+] > [OH^-]$. A basic solution is one in which the concentration of base is greater than the concentration of acid $[H^+] < [OH^-]$.

28. A titration is used to determine the concentration of a specific substance (often an acid or a base) in a sample. A titration determines the volume of a reagent of known concentration that is required to completely react with a volume of a sample of unknown concentration. An indicator is used to help visualize the endpoint of a titration. The endpoint is the point at which enough of the reagent of known concentration has been added to the sample of unknown concentration to completely react with the unknown solution. An indicator color change is visible when the endpoint has been reached.

29. The net ionic equation for an acid-base reaction in aqueous solutions is:

$$H^+ + OH^- \longrightarrow H_2O$$

30. Acid rain is caused by the release of nitrogen and sulfur oxides into the air. When these oxides are carried through the atmosphere they react with water and form sulfuric acid (H_2SO_4) and nitric acid (HNO_3). Precipitation (rain or snow) carries the acids to the ground.

SOLUTIONS TO EXERCISES

1. Conjugate acid – base pairs:
    (a) $NH_3 - NH_4^+$; $H_2O - OH^-$
    (b) $HC_2H_3O_2 - C_2H_3O_2^-$; $H_2O - H_3O^+$
    (c) $H_2PO_4^- - HPO_4^{2-}$; $OH^- - H_2O$
    (d) $HCl - Cl^-$; $H_2O - H_3O^+$

2. Conjugate acid – base pairs:
    (a) $H_2S - HS^-$; $NH_3 - NH_4^+$
    (b) $HSO_4^- - SO_4^{2-}$; $NH_3 - NH_4^+$
    (c) $HBr - Br^-$; $CH_3O^- - CH_3OH$
    (d) $HNO_3 - NO_3^-$; $H_2O - H_3O^+$

3. (a) $Zn(s) + 2\,HCl(aq) \rightarrow ZnCl_2(aq) + H_2(g)$
    (b) $Al(OH)_3(s) + 3\,H_2SO_4(aq) \rightarrow Al_2(SO_4)_3(aq) + 6\,H_2O(l)$
    (c) $Na_2CO_3(aq) + 2\,HC_2H_3O_2(aq) \rightarrow 2\,NaC_2H_3O_2(aq) + H_2O(l) + CO_2(g)$
    (d) $MgO(s) + 2\,HI(aq) \rightarrow MgI_2(aq) + H_2O(l)$
    (e) $Ca(HCO_3)_2(s) + 2\,HBr(aq) \rightarrow CaBr_2(aq) + 2\,H_2O(l) + 2\,CO_2(g)$
    (f) $3\,KOH(aq) + H_3PO_4(aq) \rightarrow K_3PO_4(aq) + 3\,H_2O(l)$

4. Complete and balance these equations:
    (a) $Fe_2O_3(s) + 6\,HBr(aq) \rightarrow 2\,FeBr_3(aq) + 3\,H_2O(l)$
    (b) $2\,Al(s) + 3\,H_2SO_4(aq) \rightarrow Al_2(SO_4)_3(aq) + 3\,H_2(g)$
    (c) $2\,NaOH(aq) + H_2CO_3(aq) \rightarrow Na_2CO_3(aq) + 2\,H_2O(l)$
    (d) $Ba(OH)_2(s) + 2\,HClO_4(aq) \rightarrow Ba(ClO_4)_2(aq) + 2\,H_2O(l)$
    (e) $Mg(s) + 2\,HClO_4(aq) \rightarrow Mg(ClO_4)_2(aq) + H_2(g)$
    (f) $K_2O(s) + 2\,HI(aq) \rightarrow 2\,KI(aq) + H_2O(l)$

5. (a) $Zn + (2\,H^+ + 2\,Cl^-) \rightarrow (Zn^{2+} + 2\,Cl^-) + H_2$
        $Zn + 2\,H^+ \rightarrow Zn^{2+} + H_2$

    (b) $2\,Al(OH)_3 + (6\,H^+ + 3\,SO_4^{2-}) \rightarrow (2\,Al^{3+} + 3\,SO_4^{2-}) + 6\,H_2O$
        $Al(OH)_3 + 3\,H^+ \rightarrow Al^{3+} + 3\,H_2O$

    (c) $(2\,Na^+ + CO_3^{2-}) + 2\,HC_2H_3O_2 \rightarrow (2\,Na^+ + 2\,C_2H_3O_2^-) + H_2O + CO_2$
        $CO_3^{2-} + 2\,HC_2H_3O_2 \rightarrow 2\,C_2H_3O_2^- + H_2O + CO_2$

    (d) $MgO + (2\,H^+ + 2\,I^-) \rightarrow (Mg^{2+} + 2\,I^-) + H_2O$
        $MgO + 2\,H^+ \rightarrow Mg^{2+} + H_2O$

    (e) $Ca(HCO_3)_2 + (2\,H^+ + 2\,Br^-) \rightarrow (Ca^{2+} + 2\,Br^-) + 2\,H_2O + 2\,CO_2$
        $Ca(HCO_3)_2 + 2\,H^+ \rightarrow Ca^{2+} + H_2O + CO_2$

    (f) $(3\,K^+ + 3\,OH^-) + H_3PO_4 \rightarrow (3\,K^+ + PO_4^{3-}) + 3\,H_2O$
        $3\,OH^- + H_3PO_4 \rightarrow PO_4^{3-} + 3\,H_2O$

6. (a) $Fe_2O_3 + (6\,H^+ + 6\,Br^-) \rightarrow (2\,Fe^{3+} + 6\,Br^-) + 3\,H_2O$

   $Fe_2O_3 + 6\,H^+ \rightarrow 2\,Fe^{3+} + 3\,H_2O$

   (b) $2\,Al + (6H^+ \rightarrow 3\,SO_4^{2-}) \rightarrow (2\,Al^{3+} + 3\,SO_4^{2-}) + 3\,H_2$

   $2\,Al + 6\,H^+ \rightarrow 2\,Al^{3+} + 3\,H_2$

   (c) $(2\,Na^+ + 2\,OH^-) + H_2CO_3 \rightarrow (2\,Na^+ + CO_3^{2-}) + 2\,H_2O$

   $2\,OH^- + H_2CO_3 \rightarrow CO_3^{2-} + 2\,H_2O$

   (d) $Ba(OH)_2 + (2\,H^+ + 2\,ClO_4^-) \rightarrow (Ba^{2+} + 2\,ClO_4^-) + 2\,H_2O$

   $Ba(OH)_2 + 2\,H^+ \rightarrow Ba^{2+} + 2\,H_2O$

   (e) $Mg + (2\,H^+ + 2\,ClO_4^-) \rightarrow (Mg^{2+} + 2\,ClO_4^-) + H_2$

   $Mg + 2\,H^+ \rightarrow Mg^{2+} + H_2$

   (f) $K_2O + (2\,H^+ + 2\,I^-) \rightarrow (2\,K^+ + 2\,I^-) + H_2O$

   $K_2O + 2\,H^+ \rightarrow 2\,K^+ + H_2O$

7. (a) $HNO_3 + NaOH \rightarrow H_2O + NaNO_3$
   (b) $2\,HC_2H_3O_2 + Ba(OH)_2 \rightarrow 2\,H_2O + Ba(C_2H_3O_2)_2$
   (c) $HClO_4 + NH_4OH \rightarrow H_2O + NH_4ClO_4$

8. (a) $2\,HBr + Mg(OH)_2 \rightarrow 2\,H_2O + MgBr_2$
   (b) $H_3PO_4 + 3\,KOH \rightarrow 3\,H_2O + K_3PO_4$
   (c) $H_2SO_4 + 2\,NH_4OH \rightarrow 2\,H_2O + (NH_4)_2SO_4$

9. LiOH and H_2S must be reacted

   $2\,LiOH + H_2S \rightarrow 2\,H_2O + Li_2S$

10. $Ca(OH)_2$ and H_2CO_3 must be reacted

    $Ca(OH)_2 + H_2CO_3 \rightarrow 2\,H_2O + CaCO_3$

11. The following compounds are electrolytes:

    (a) SO_3, acid in water   (e) $CuBr_2$, salt
    (b) K_2CO_3, salt   (f) HI, acid in water

12. The following compounds are electrolytes:

    (b) P_2O_5, acid in water   (d) $LiOH$, base
    (c) $NaClO$, salt   (f) $KMnO_4$, salt

13. Molarity of ions.

    (a) $(1.25\,M\ CuBr_2)\left(\dfrac{1\ mol\ Cu^{2+}}{1\ mol\ CuBr_2}\right) = 1.25\,M\ Cu^{2+}$

    $(1.25\,M\ CuBr_2)\left(\dfrac{2\ mol\ Br^-}{1\ mol\ CuBr_2}\right) = 2.50\,M\ Br^-$

(b) $(0.75\,M\,\text{NaHCO}_3)\left(\dfrac{1\,\text{mol Na}^+}{1\,\text{mol NaHCO}_3}\right) = 0.75\,M\,\text{Na}^+$

$(0.75\,M\,\text{NaHCO}_3)\left(\dfrac{1\,\text{mol HCO}_3^-}{1\,\text{mol NaHCO}_3}\right) = 0.75\,M\,\text{HCO}_3^-$

(c) $(3.50\,M\,\text{K}_3\text{AsO}_4)\left(\dfrac{3\,\text{mol K}^+}{1\,\text{mol K}_3\text{AsO}_4}\right) = 10.5\,M\,\text{K}^+$

$(3.50\,M\,\text{K}_3\text{AsO}_4)\left(\dfrac{1\,\text{mol AsO}_4^{3-}}{1\,\text{mol K}_3\text{AsO}_4}\right) = 3.50\,M\,\text{AsO}_4^{3-}$

(d) $(0.65\,M\,(\text{NH}_4)_2\text{SO}_4)\left(\dfrac{2\,\text{mol NH}_4^+}{1\,\text{mol }(\text{NH}_4)_2\text{SO}_4}\right) = 1.3\,M\,\text{NH}_4^+$

$(0.65\,M\,(\text{NH}_4)_2\text{SO}_4)\left(\dfrac{1\,\text{mol SO}_4^{2-}}{1\,\text{mol }(\text{NH}_4)_2\text{SO}_4}\right) = 0.65\,M\,\text{SO}_4^{2-}$

14. Molarity of ions.

(a) $(2.25\,M\,\text{FeCl}_3)\left(\dfrac{1\,\text{mol Fe}^{3+}}{1\,\text{mol FeCl}_3}\right) = 2.25\,M\,\text{Fe}^{3+}$

$(2.25\,M\,\text{FeCl}_3)\left(\dfrac{3\,\text{mol Cl}^-}{1\,\text{mol FeCl}_3}\right) = 6.75\,M\,\text{Cl}^-$

(b) $(1.20\,M\,\text{MgSO}_4)\left(\dfrac{1\,\text{mol }Mg^{2+}}{1\,\text{mol MgSO}_4}\right) = 1.20\,M\,Mg^{2+}$

$(1.20\,M\,\text{MgSO}_4)\left(\dfrac{1\,\text{mol SO}_4^{2-}}{1\,\text{mol MgSO}_4}\right) = 1.20\,M\,\text{SO}_4^{2-}$

(c) $(0.75\,M\,\text{NaH}_2\text{PO}_4)\left(\dfrac{1\,\text{mol Na}^+}{1\,\text{mol NaH}_2\text{PO}_4}\right) = 0.75\,M\,\text{Na}^+$

$(0.75\,M\,\text{NaH}_2\text{PO}_4)\left(\dfrac{1\,\text{mol H}_2\text{PO}_4^-}{1\,\text{mol NaH}_2\text{PO}_4}\right) = 0.75\,M\,\text{H}_2\text{PO}_4^-$

(d) $(0.35\,M\,\text{Ca}(\text{ClO}_3)_2)\left(\dfrac{1\,\text{mol Ca}^{2+}}{1\,\text{mol Ca}(\text{ClO}_3)_2}\right) = 0.35\,M\,\text{Ca}^{2+}$

$(0.35\,M\,\text{Ca}(\text{ClO}_3)_2)\left(\dfrac{2\,\text{mol ClO}_3^-}{1\,\text{mol Ca}(\text{ClO}_3)_2}\right) = 0.70\,M\,\text{ClO}_3^-$

15. We will use the data from No. 13 to solve these problems. 100 mL = 0.100 L

(a) $(0.100\,\text{L})\left(\dfrac{1.25\,\text{mol Cu}^{2+}}{\text{L}}\right)\left(\dfrac{63.55\,\text{g}}{\text{mol}}\right) = 7.94\,\text{g Cu}^{2+}$

$(0.100\,\text{L})\left(\dfrac{2.50\,\text{mol Br}^-}{\text{L}}\right)\left(\dfrac{79.90\,\text{g}}{\text{mol}}\right) = 20.0\,\text{g Br}^-$

(b) $(0.100 \, \text{L}) \left(\dfrac{0.75 \, \text{mol Na}^+}{\text{L}} \right) \left(\dfrac{22.99 \, \text{g}}{\text{mol}} \right) = 1.7 \, \text{g Na}^+$

$(0.100 \, \text{L}) \left(\dfrac{0.75 \, \text{mol HCO}_3^-}{\text{L}} \right) \left(\dfrac{61.02 \, \text{g}}{\text{mol}} \right) = 4.6 \, \text{g HCO}_3^-$

(c) $(0.100 \, \text{L}) \left(\dfrac{10.5 \, \text{mol K}^+}{\text{L}} \right) \left(\dfrac{39.10 \, \text{g}}{\text{mol}} \right) = 41.1 \, \text{g K}^+$

$(0.100 \, \text{L}) \left(\dfrac{3.50 \, \text{mol AsO}_4^{3-}}{\text{L}} \right) \left(\dfrac{138.9 \, \text{g}}{\text{mol}} \right) = 48.6 \, \text{g AsO}_4^{3-}$

(d) $(0.100 \, \text{L}) \left(\dfrac{1.3 \, \text{mol NH}_4^+}{\text{L}} \right) \left(\dfrac{18.04 \, \text{g}}{\text{mol}} \right) = 2.3 \, \text{g NH}_4^+$

$(0.100 \, \text{L}) \left(\dfrac{0.65 \, \text{mol SO}_4^{2-}}{1 \, \text{L}} \right) \left(\dfrac{96.07 \, \text{g}}{1 \, \text{mol}} \right) = 6.2 \, \text{g SO}_4^{2-}$

16.   We will use the data from No. 14 to solve these problems 100 mL = 0.100 L

(a) $(0.100 \, \text{L}) \left(\dfrac{2.25 \, \text{mol Fe}^{3+}}{\text{L}} \right) \left(\dfrac{55.85 \, \text{g}}{\text{mol}} \right) = 12.6 \, \text{g Fe}^{3+}$

$(0.100 \, \text{L}) \left(\dfrac{6.75 \, \text{mol Cl}^-}{\text{L}} \right) \left(\dfrac{35.45 \, \text{g}}{\text{mol}} \right) = 23.9 \, \text{g Cl}^-$

(b) $(0.100 \, \text{L}) \left(\dfrac{1.20 \, \text{mol Mg}^{2+}}{\text{L}} \right) \left(\dfrac{24.31 \, \text{g}}{\text{mol}} \right) = 2.92 \, \text{g Mg}^{2+}$

$(0.100 \, \text{L}) \left(\dfrac{1.20 \, \text{mol SO}_4^{2-}}{\text{L}} \right) \left(\dfrac{96.07 \, \text{g}}{\text{mol}} \right) = 11.5 \, \text{g SO}_4^{2-}$

(c) $(0.100 \, \text{L}) \left(\dfrac{0.75 \, \text{mol Na}^+}{\text{L}} \right) \left(\dfrac{22.99 \, \text{g}}{\text{mol}} \right) = 1.7 \, \text{g Na}^+$

$(0.100 \, \text{L}) \left(\dfrac{0.75 \, \text{mol H}_2\text{PO}_4^-}{\text{L}} \right) \left(\dfrac{96.99 \, \text{g H}_2\text{PO}_4^-}{\text{mol}} \right) = 7.3 \, \text{g H}_2\text{PO}_4^-$

(d) $(0.100 \, \text{L}) \left(\dfrac{0.35 \, \text{mol Ca}^{2+}}{\text{L}} \right) \left(\dfrac{40.08 \, \text{g}}{\text{mol}} \right) = 1.4 \, \text{g Ca}^{2+}$

$(0.100 \, \text{L}) \left(\dfrac{0.70 \, \text{mol ClO}_3^-}{\text{L}} \right) \left(\dfrac{83.45 \, \text{g}}{\text{mol}} \right) = 5.8 \, \text{g ClO}_3^-$

17. $\text{pH} = -\log[\text{H}^+] \quad [\text{H}^+] = 10^{-\text{pH}}$
   (a) $[\text{H}^+] = 1 \times 10^{-5}$
   (b) $[\text{H}^+] = 2 \times 10^{-7}$
   (c) $[\text{H}^+] = 1 \times 10^{-8}$
   (d) $[\text{H}^+] = 2 \times 10^{-10}$

18. $pH = -\log[H^+]$ $[H^+] = 10^{-pH}$
    (a) $[H^+] = 1 \times 10^{-7}$
    (b) $[H^+] = 5 \times 10^{-5}$
    (c) $[H^+] = 2 \times 10^{-6}$
    (d) $[H^+] = 5 \times 10^{-11}$

19. (a) $(55.5 \text{ mL})\left(\dfrac{0.50 \text{ mol HCl}}{1000 \text{ mL}}\right) = 0.028 \text{ mol HCl}$

    $(75.0 \text{ mL})\left(\dfrac{1.25 \text{ mol HCl}}{1000 \text{ mL}}\right) = 0.0938 \text{ mol HCl}$

    Total mol HCl = 0.028 mol + 0.0938 mol = 0.122 mol HCl

    Total volume = 0.0555 L + 0.0750 L = 0.1305 L

    $\dfrac{0.122 \text{ mol HCl}}{0.1305 \text{ L}} = 0.935 \, M \text{ HCl}$

    $(0.935 \, M \text{ HCl})\left(\dfrac{1 \text{ mol H}^+}{1 \text{ mol HCl}}\right) = 0.935 \, M \text{ H}^+$

    $(0.935 \, M \text{ HCl})\left(\dfrac{1 \text{ mol Cl}^-}{1 \text{ mol HCl}}\right) = 0.935 \, M \text{ Cl}^-$

    (b) $(125 \text{ mL})\left(\dfrac{0.75 \text{ mol CaCl}_2}{1000 \text{ mL}}\right) = 0.094 \text{ mol CaCl}_2$

    $(125 \text{ mL})\left(\dfrac{0.25 \text{ mol CaCl}_2}{1000 \text{ mL}}\right) = 0.031 \text{ mol CaCl}_2$

    Total mol CaCl$_2$ = 0.094 mol + 0.031 mol = 0.125 mol CaCl$_2$

    Total volume = 0.125 L + 0.125 L = 0.250 L

    $\dfrac{0.125 \text{ mol CaCl}_2}{0.250 \text{ L}} = 0.500 \, M \text{ CaCl}_2$

    $(0.500 \, M \text{ CaCl}_2)\left(\dfrac{1 \text{ mol Ca}^{2+}}{1 \text{ mol CaCl}_2}\right) = 0.500 \, M \text{ Ca}^{2+}$

    $(0.500 \, M \text{ CaCl}_2)\left(\dfrac{2 \text{ mol Cl}^-}{1 \text{ mol CaCl}_2}\right) = 1.00 \, M \text{ Cl}^-$

    (c) $NaOH + HCl \rightarrow NaCl + H_2O$

    $(35.0 \text{ mL})\left(\dfrac{0.333 \text{ mol NaOH}}{1000 \text{ mL}}\right) = 0.0117 \text{ mol NaOH}$

    $(22.5 \text{ mL})\left(\dfrac{0.250 \text{ mol HCl}}{1000 \text{ mL}}\right) = 0.00563 \text{ mol HCl}$

    0.00563 mol HCl reacts with 0.00563 mol NaOH. 0.0061 mol NaOH remains uareacted and 0.00563 mol NaCl is produced. The final volume is 0.0575 L and contains 0.0061 mol NaOH and

0.00563 mol NaCl. Moles of ions are: $(0.0061 \text{ mol Na}^+ + 0.00563 \text{ mol Na}^+) = 0.0117 \text{ mol Na}^+$, 0.0061 mol OH$^-$, and 0.00563 mol Cl$^-$. Concentrations of ions are:

$$\frac{0.0177 \text{ mol Na}^+}{0.0575 \text{ L}} = 0.203 \; M \; \text{Na}^+ \qquad \frac{0.0061 \text{ mol OH}^-}{0.0575 \text{ L}} = 0.11 \; M \; \text{OH}^-$$

$$\frac{0.00563 \text{ mol Cl}^-}{0.0575 \text{ L}} = 0.0979 \; M \; \text{Cl}^-$$

(d) $H_2SO_4 + 2\,NaOH \rightarrow Na_2SO_4 + 2\,H_2O$

$$(12.5 \text{ mL})\left(\frac{0.500 \text{ mol H}_2\text{SO}_4}{1000 \text{ mL}}\right) = 0.00625 \text{ mol H}_2\text{SO}_4$$

$$(23.5 \text{ mL})\left(\frac{0.175 \text{ mol NaOH}}{1000 \text{ mL}}\right) = 0.00411 \text{ mol NaOH}$$

$$(0.00411 \text{ mol NaOH})\left(\frac{1 \text{ mol H}_2\text{SO}_4}{2 \text{ mol NaOH}}\right) = 0.00206 \text{ mol H}_2\text{SO}_4 \text{ reacted}$$

0.00206 mol H_2SO_4 reacts with 0.00411 mol NaOH. 0.00419 mol H_2SO_4 remains unreacted and 0.00206 mol Na_2SO_4 is produced. The final volume is 0.0360 L and contains 0.00206 mol Na_2SO_4 and 0.00419 mol H_2SO_4. Moles of ions are 0.00412 mol Na$^+$, 0.00838 mol H$^+$, and $(0.00206 + 0.00419) = 0.00625$ mol SO_4^{2-}. Concentration of ions are:

$$\frac{0.0412 \text{ mol Na}^+}{0.0360 \text{ L}} = 0.114 \; M \; \text{Na}^+ \qquad \frac{0.00838 \text{ mol H}^+}{0.0360 \text{ L}} = 0.233 \; M \; \text{H}^+$$

$$\frac{0.00625 \text{ mol SO}_4^{2-}}{0.0360 \text{ L}} = 0.174 \; M \; \text{SO}_4^{2-}$$

20. (a) $(45.5 \text{ mL})\left(\dfrac{0.10 \text{ mol NaCl}}{1000 \text{ mL}}\right) = 0.0046 \text{ mol NaCl}$

$$(60.5 \text{ mL})\left(\frac{0.35 \text{ mol NaCl}}{1000 \text{ mL}}\right) = 0.021 \text{ mol NaCl}$$

Total mol NaCl $= 0.0046 \text{ mol} + 0.021 \text{ mol} = 0.026 \text{ mol NaCl}$

Total volume $= 0.0455 \text{ L} + 0.0605 \text{ L} = 0.1060 \text{ L}$

$$\frac{0.026 \text{ mol NaCl}}{0.1060 \text{ L}} = 0.25 \; M \; \text{NaCl}$$

$$(0.25 \; M \; \text{NaCl})\left(\frac{1 \text{ mol Na}^+}{1 \text{ mol NaCl}}\right) = 0.25 \; M \; \text{Na}^+$$

$$(0.25 \; M \; \text{NaCl})\left(\frac{1 \text{ mol Cl}^-}{1 \text{ mol NaCl}}\right) = 0.25 \; M \; \text{Cl}^-$$

(b) $(95.5 \text{ mL})\left(\dfrac{1.25 \text{ mol HCl}}{1000 \text{ mL}}\right) = 0.119 \text{ mol HCl}$

$$(125.5 \text{ mL})\left(\frac{2.50 \text{ mol HCl}}{1000 \text{ mL}}\right) = 0.314 \text{ mol HCl}$$

Total mol HCl $= 0.119 \text{ mol} + 0.314 \text{ mol} = 0.433 \text{ mol HCl}$

Total volume $= 0.0955\,L + 0.1255\,L = 0.2210\,L$

$$\frac{0.433\ \text{mol HCl}}{0.2210\,L} = 1.96\,M\ \text{HCl}$$

$$(1.96\,M\ \text{HCl})\left(\frac{1\ \text{mol H}^+}{1\ \text{mol HCl}}\right) = 1.96\,M\ \text{H}^+$$

$$(1.96\,M\ \text{HCl})\left(\frac{1\ \text{mol Cl}^-}{1\ \text{mol HCl}}\right) = 1.96\,M\ \text{Cl}^-$$

(c) $(15.5\ \text{mL})\left(\dfrac{0.10\ \text{mol Ba(NO}_3)_2}{1000\ \text{mL}}\right) = 0.0016\,M\ \text{Ba(NO}_3)_2$

$(10.5\ \text{mL})\left(\dfrac{0.20\ \text{mol AgNO}_3}{1000\ \text{mL}}\right) = 0.0021\,M\ \text{AgNO}_3$

Number of moles of each substance: $0.0016\ \text{mol Ba}^{2+}$, $0.0021\ \text{mol Ag}^+$, and $(0.0032\ \text{mol} + 0.0021\ \text{mol}) = 0.0053\ \text{mol NO}_3^-$

Total volume $= 0.0155\,L + 0.0105\,L = 0.0260\,L$

$$\frac{0.0016\ \text{mol Ba}^{2+}}{0.0260\,L} = 0.062\,M\ \text{Ba}^{2+}$$

$$\frac{0.0021\ \text{mol Ag}^+}{0.0260\,L} = 0.081\,M\ \text{Ag}^+$$

$$\frac{0.0053\ \text{mol NO}_3^-}{0.0260\,L} = 0.20\,M\ \text{NO}_3^-$$

(d) $(25.5\ \text{mL})\left(\dfrac{0.25\ \text{mol NaCl}}{1000\ \text{mL}}\right) = 0.0064\ \text{mol NaCl}$

$(15.5\ \text{mL})\left(\dfrac{0.15\ \text{mol Ca(C}_2\text{H}_3\text{O}_2)_2}{1000\ \text{mL}}\right) = 0.0023\ \text{mol Ca(C}_2\text{H}_3\text{O}_2)_2$

Number of moles of each substance: $0.0064\ \text{mol Na}^+$, $0.0064\ \text{mol Cl}^-$, $0.0023\ \text{mol Ca}^{2+}$, $0.0046\ \text{mol C}_2\text{H}_3\text{O}_2^-$.

Total volume $= 0.0255\,L + 0.0155\,L = 0.0410\,L$

$$\frac{0.0064\ \text{mol Na}^+}{0.0410\,L} = 0.16\,M\ \text{Na}^+$$

$$\frac{0.0064\ \text{mol Cl}^-}{0.0410\,L} = 0.16\,M\ \text{Cl}^-$$

$$\frac{0.0023\ \text{mol Ca}^{2+}}{0.0410\,L} = 0.056\,M\ \text{Ca}^{2+}$$

$$\frac{0.0046\ \text{mol C}_2\text{H}_3\text{O}_2^-}{0.0410\,L} = 0.11\,M\ \text{C}_2\text{H}_3\text{O}_2^-$$

21. $HNO_3(aq) + H_2O(l) \rightarrow H_3O^+(aq) + NO_3^-(aq)$

Or

$HNO_3(aq) \xrightarrow{H_2O} H^+(aq) + NO_3^-(aq)$

Because nitric acid ionizes completely it would be both a strong electrolyte and a strong acid.

22. $HCN(aq) + H_2O(l) \rightleftharpoons H_3O^+(aq) + CN^-(aq)$

Or

$HCN(aq) \overset{H_2O}{\rightleftharpoons} H^+(aq) + CN^-(aq)$

HCN is only partially ionized and so it would be a poor electrolyte and a weak acid.

23. The reaction of HCl and NaOH occurs on a 1:1 mole ratio.

$HCl + NaOH \longrightarrow NaCl + H_2O$

At the endpoint in these titration reactions, equal moles of HCl and NaOH will have reacted. Moles = (molarity) (volume). At the endpoint, mol HCl = mol NaOH.
Therefore, at the endpoint,

$$M_A V_A = M_B V_B \qquad M_A = \frac{M_B V_B}{V_A}$$

(a) $\dfrac{(37.70\,\text{mL})(0.728\,M)}{40.3\,\text{mL}} = 0.681\,M\,\text{HCl}$

(b) $\dfrac{(33.66\,\text{mL})(0.306\,M)}{19.00\,\text{mL}} = 0.542\,M\,\text{HCl}$

(c) $\dfrac{(18.00\,\text{mL})(0.555\,M)}{27.25\,\text{mL}} = 0.367\,M\,\text{HCl}$

24. The reaction of HCl and NaOH occurs on a 1:1 mole ratio.

$HCl + NaOH \rightarrow NaCl + H_2O$

At the endpoint in these titration reactions, equal moles of HCl and NaOH will have reacted. Moles = (molarity)(volume). At the endpoint, mol HCl = mol NaOH.
Therefore, at the endpoint,

$$M_A V_A = M_B V_B \qquad M_B = \frac{M_A V_A}{V_B}$$

(a) $\dfrac{(37.19\,\text{mL})(0.126\,M)}{31.91\,\text{mL}} = 0.147\,M\,\text{NaOH}$

(b) $\dfrac{(48.04\,\text{mL})(0.482\,M)}{24.02\,\text{mL}} = 0.964\,M\,\text{NaOH}$

(c) $\dfrac{(13.13\,\text{mL})(1.425\,M)}{39.39\,\text{mL}} = 0.4750\,M\,\text{NaOH}$

25. (a) $2\,PO_4^{3-}(aq) + 3\,Ca^{2+}(aq) \rightarrow Ca_3\left(PO_4^{3-}\right)_2(s)$
    (b) $2\,Al(s) + 6\,H^+(aq) \rightarrow 3\,H_2(g) + 2\,Al^{3+}(aq)$
    (c) $CO_3^{2-}(aq) + 2\,H^+(aq) \rightarrow H_2O(aq) + CO_2(g)$

26. (a) $Mg(s) + Cu^{2+}(aq) \rightarrow Cu(s) + Mg^{2+}(aq)$
    (b) $H^+(aq) + OH^-(aq) \rightarrow H_2O(l)$
    (c) $SO_3^{2-}(aq) + 2\,H^+(aq) \rightarrow H_2O(l) + SO_2(g)$

27. (a) 1 molar H_2SO_4 is more acidic. The concentration of H^+ in 1 M H_2SO_4 is greater than 1 M since there are two ionizable hydrogens per mole of H_2SO_4. In HCl the concentration of H^+ will be 1 M, since there is only one ionizable hydrogen per mole HCl.
    (b) 1 molar HCl is more acidic. HCl is a strong electrolyte, producing more H^+ than $HC_2H_3O_2$ which is a weak electrolyte.

28. (a) 2 molar HCl is more acidic. 2 M HCl will yield 2 M H^+ concentration. 1 M HCl will yield 1 M H^+ concentration.
    (b) 1 molar H_2SO_4 is more acidic. Both are strong acids. The concentration of H^+ in 1 M H_2SO_4 is greater than in 1 M HNO_3 because H_2SO_4 has two ionizable hydrogens per mole whereas HNO_3 has only one ionizable hydrogen per mole.

29. $2\,HClO_4(aq) + Ca(OH)_2(s) \rightarrow Ca(ClO_4)_2(aq) + 2\,H_2O(l)$

    g $Ca(OH)_2$ → mol $Ca(OH)_2$ → mol $HClO_4$ → mL $HClO_4$

    $$\left(50.25\text{ g }Ca(OH)_2\right)\left(\frac{mol}{74.10\text{ g}}\right)\left(\frac{2\text{ mol }HClO_4}{1\text{ mol }Ca(OH)_2}\right)\left(\frac{1000\text{ mL}}{0.525\text{ mol}}\right) = 2.58 \times 10^3\text{ mL }HClO_4$$

30. $3\,HCl(aq) + Al(OH)_3(s) \rightarrow AlCl_3(aq) + 3\,H_2O(l)$

    mL HCl → mol HCl → mol $Al(OH)_3$ → g $Al(OH)_3$

    $$\left(275\text{ mL HCl}\right)\left(\frac{0.125\text{ mol}}{1000\text{ mL}}\right)\left(\frac{1\text{ mol }Al(OH)_3}{3\text{ mol HCl}}\right)\left(\frac{78.00\text{ g}}{mol}\right) = 0.894\text{ g }Al(OH)_3$$

31. $NaOH + HCl \rightarrow NaCl + H_2O$
    First calculate the grams of NaOH in the sample.
    L HCl → mol HCl → mol NaOH → g NaOH

    $$\left(0.01825\text{ L HCl}\right)\left(\frac{0.2406\text{ mol}}{L}\right)\left(\frac{1\text{ mol NaOH}}{1\text{ mol HCl}}\right)\left(\frac{40.00\text{ g}}{mol}\right) = 0.1756\text{ g NaOH in the sample}$$

    $$\left(\frac{0.1756\text{ g NaOH}}{0.200\text{ g sample}}\right)(100) = 87.8\%\text{ NaOH}$$

32. $NaOH + HCl \rightarrow NaCl + H_2O$

    L HCl → mol HCl → mol NaOH → g NaOH

    $$\left(0.04990\text{ L HCl}\right)\left(\frac{0.466\text{ mol}}{L}\right)\left(\frac{1\text{ mol NaOH}}{1\text{ mol HCl}}\right)\left(\frac{40.00\text{ g}}{mol}\right) = 0.930\text{ g NaOH in the sample}$$

1.00 g sample $-$ 0.930 g NaOH $=$ 0.070 g NaCl in the sample

$$\left(\frac{0.070 \text{ g NaCl}}{1.00 \text{ g sample}}\right)(100) = 7.0\% \text{ NaCl in the sample}$$

33. $Zn + 2 HCl \rightarrow ZnCl_2 + H_2$

This is a limiting reactant problem. First find the moles of Zn and HCl from the given data and then identify the limiting reactant.

g Zn $\rightarrow$ mol Zn $(5.00 \text{ g Zn})\left(\dfrac{1 \text{ mol}}{65.39 \text{ g}}\right) = 0.0765 \text{ mol Zn}$

$(0.100 \text{ L HCl})\left(\dfrac{0.350 \text{ mol}}{\text{L}}\right) = 0.0350 \text{ mol HCl}$

Therefore Zn is in excess and HCl is the limiting reactant.

$(0.0350 \text{ mol HCl})\left(\dfrac{1 \text{ mol H}_2}{2 \text{ mol HCl}}\right) = 0.0175 \text{ mol H}_2 \text{ produced in the reaction}$

$T = 27°C = 300. \text{ K}$ $P = (700. \text{ torr})\left(\dfrac{1 \text{ atm}}{760 \text{ torr}}\right) = 0.921 \text{ atm}$

$PV = nRT$

$V = \dfrac{nRT}{P} = \dfrac{(0.0175 \text{ mol H}_2)(0.0821 \text{ L atm/mol K})(300. \text{ K})}{0.921 \text{ atm}} = 0.468 \text{ L H}_2$

34. $Zn + 2 HCl \rightarrow ZnCl_2 + H_2$

This is a limiting reactant problem. First find moles of Zn and HCl from the given data and then identify the limiting reactant.

g Zn $\rightarrow$ mol Zn $(5.00 \text{ g Zn})\left(\dfrac{1 \text{ mol}}{65.39 \text{ g}}\right) = 0.0765 \text{ mol Zn}$

$(0.200 \text{ L HCl})\left(\dfrac{0.350 \text{ mol}}{\text{L}}\right) = 0.0700 \text{ mol HCl}$

Zn is in excess and HCl is the limiting reactant.

$(0.0700 \text{ mol HCl})\left(\dfrac{1 \text{ mol H}_2}{2 \text{ mol HCl}}\right) = 0.0350 \text{ mol H}_2$

$T = 27°C = 300. \text{ K}$ $P = (700. \text{ torr})\left(\dfrac{1 \text{ atm}}{760 \text{ torr}}\right) = 0.921 \text{ atm}$

$PV = nRT$

$V = \dfrac{nRT}{P} = \dfrac{(0.0350 \text{ mol H}_2)(0.0821 \text{ L atm/mol K})(300. \text{ K})}{0.921 \text{ atm}} = 0.936 \text{ L H}_2$

35. $pH = -\log[H^+]$

(a) $[H^+] = 0.35 \, M$; $pH = -\log(0.35) = 0.46$
(b) $[H^+] = 1.75 \, M$; $pH = -\log(1.75) = -0.243$
(c) $[H^+] = 2.0 \times 10^{-5} \, M$; $pH = -\log(2.0 \times 10^{-5}) = 4.70$

36. $pH = -\log[H^+]$

    (a) $[H^+] = 0.0020\,M;$ $pH = -\log(0.0020) = 2.70$

    (b) $[H^+] = 7.0 \times 10^{-8}\,M;$ $pH = -\log(7.0 \times 10^{-8}) = 7.15$

    (c) $[H^+] = 3.0\,M;$ $pH = -\log(3.0) = -0.48$

37. (a)   Orange juice $= 3.7 \times 10^{-4}\,M\,H^+$

          $pH = -\log(3.7 \times 10^{-4}) = 3.43$

    (b)   Vinegar $= 2.8 \times 10^{-3}\,M\,H^+$

          $pH = -\log(2.8 \times 10^{-3}) = 2.55$

    (c)   shampoo $= 2.4 \times 10^{-6}\,M\,H^+$

          $pH = -\log(2.4 \times 10^{-6}) = 5.62$

    (d)   dishwashing detergent $= 3.6 \times 10^{-8}\,M\,H^+$

          $pH = -\log(3.6 \times 10^{-8}) = 7.44$

38. (a)   Black coffee $= 5.0 \times 10^{-5}\,M\,H^+$

          $pH = -\log(5.0 \times 10^{-5}) = 4.30$

    (b)   Limewater $= 3.4 \times 10^{-11}\,M\,H^+$

          $pH = -\log(3.4 \times 10^{-11}) = 10.47$

    (c)   fruit punch $= 2.1 \times 10^{-4}\,M\,H^+$

          $pH = -\log(2.1 \times 10^{-4}) = 3.68$

    (d)   cranberry apple drink $= 1.3 \times 10^{-3}\,M\,H^+$

          $pH = -\log(1.3 \times 10^{-3}) = 2.89$

39. (a) NH_3 is a weak base $NH_3(aq) \overset{H_2O}{\rightleftharpoons} NH_4^+(aq) + OH^-(aq)$

    (b) HCl is a strong acid $HCl(aq) \overset{H_2O}{\longrightarrow} H^+(aq) + Cl^-(aq)$

    (c) KOH is a strong base $KOH \overset{H_2O}{\longrightarrow} K^+(aq) + OH^-(aq)$

    (d) $HC_2H_3O_2$ is a weak acid $HC_2H_3O_2(aq) \overset{H_2O}{\rightleftharpoons} H^+(aq) + C_2H_3O_2^-(aq)$

40. (a) $H_2C_2O_4$ is a weak acid $H_2C_2O_4(aq) \overset{H_2O}{\rightleftharpoons} H^+(aq) + HC_2O_4^-(aq)$

    (b) $Ba(OH)_2$ is a strong base $Ba(OH)_2 \overset{H_2O}{\longrightarrow} Ba^{2+}(aq) + 2\,OH^-(aq)$

    (c) $HClO_4$ is a strong acid $HClO_4(aq) \overset{H_2O}{\longrightarrow} H^+(aq) + ClO_4^-(aq)$

    (d) HBr is a strong acid $HBr(aq) \rightarrow H^+(aq) + Br^-(aq)$

41. (a) $\underset{\text{base}}{CH_3NH_2} + H^+ \rightarrow \underset{\text{conjugate acid}}{CH_3NH_3^+}$

       Note that a neutral base forms a positively charged conjugate acid.

    (b) $\underset{\text{base}}{HS^-} + H^+ \rightarrow \underset{\text{conjugate acid}}{H_2S}$

       Note that a negatively charged base forms a neutral acid upon adding a positively charged hydrogen ion.

42. (a) $\underset{\text{acid}}{HBrO_3} - H^+ \rightarrow \underset{\text{conjugate base}}{BrO_3^-}$

    (b) $\underset{\text{acid}}{NH_4^+} - H^+ \rightarrow \underset{\text{conjugate base}}{NH_3}$

    (c) $\underset{\text{acid}}{H_2PO_4^-} - H^+ \rightarrow \underset{\text{conjugate base}}{HPO_4^{2-}}$

43. $\underset{\text{acid}}{HC_2H_3O_2} + \underset{\text{base}}{NH_3} \rightarrow \underset{\text{conjugate acid}}{NH_4^+} + \underset{\text{conjugate base}}{C_2H_3O_2^-}$

    Note that in any acid base reaction, the original acid and base react to form a new acid and base.

44. $S^{2-}(aq) + H_2O(l) \rightarrow HS^-(aq) + OH^-(aq)$

    The sulfide ion is able to act as a Bronsted-Lowry base by accepting a proton from a water molecule. Note that Bronsted-Lowry bases will often cause the formation of hydroxide ions in aqueous solution.

45. $Mg(s) + 2\,HCl(aq) \rightarrow MgCl_2(aq) + H_2(g)$

46. $H_2SO_4(aq) + CaCO_3(s) \rightarrow CaSO_4(s) + H_2O(l) + CO_2(g)$

47. $Na_2SO_4(aq) \xrightarrow{H_2O} 2\,Na^+(aq) + SO_4^{2-}(aq)$

48. (a) basic   (d) acidic
    (b) acidic   (e) acidic
    (c) neutral   (f) basic

49. (a) $CaCl_2(s) \longrightarrow Ca^{2+}(aq) + 2\,Cl^-(aq)$

    For each $CaCl_2$ ionic compound, 1 calcium ion and 2 chloride ions result.

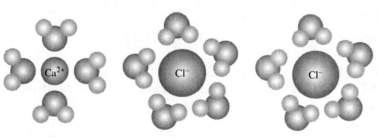

    (b) $KF(s) \longrightarrow K^+(aq) + F^-(aq)$

    For each KF ionic compound, 1 potassium ion and 1 fluoride ion result.

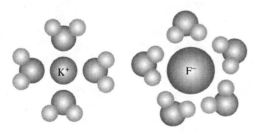

(c) $AlBr_3(s) \rightarrow Al^{3+}(aq) + 3\,Br^-(aq)$

For each $AlBr_3$ ionic compound, 1 aluminum ion and 3 bromide ions result.

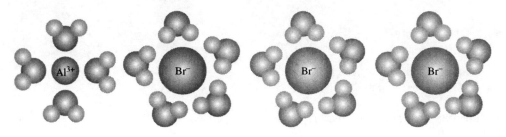

50. $AlBr_3 \rightarrow Al^{3+} + 3\,Br^-$

$$\left(\frac{0.142 \text{ mol Br}^-}{L}\right)\left(\frac{1 \text{ mol Al}^{3+}}{3 \text{ mol Br}^-}\right) = \left(\frac{0.0473 \text{ mol Al}^{3+}}{L}\right) = 0.0473 \; M \; Al^{3+}$$

51. $H_2SO_4 + 2\,NaOH \rightarrow Na_2SO_4 + 2\,H_2O$

mL NaOH $\rightarrow$ mol NaOH $\rightarrow$ mol H_2SO_4; $\left(\dfrac{\text{mol } H_2SO_4}{L}\right) = M \; H_2SO_4$

$$(35.22 \text{ mL NaOH})\left(\frac{0.313 \text{ mol}}{1000 \text{ mL}}\right)\left(\frac{1 \text{ mol } H_2SO_4}{2 \text{ mol NaOH}}\right) = 0.00551 \text{ mol } H_2SO_4$$

$$\left(\frac{0.00551 \text{ mol } H_2SO_4}{0.02522 \text{ L}}\right) = 0.218 \; M \; H_2SO_4$$

52. The acetic acid solution freezes at a lower temperature than the alcohol solution. The acetic acid ionizes slightly while the alcohol does not. The ionization of the acetic acid increases its particle concentration in solution above that of the alcohol solution, resulting in a lower freezing point for the acetic acid solution.

53. It is more economical to purchase CH_3OH at the same cost per pound as C_2H_5OH. Because CH_3OH has a lower molar mass than C_2H_5OH, the CH_3OH solution will contain more particles per pound in a given solution and therefore, have a greater effect on the freezing point of the radiator solution.
Assume 100. g of each compound.

CH_3OH: $\dfrac{100. \text{ g}}{34.04 \text{ g/mol}} = 2.84 \text{ mol}$

CH_3CH_2OH: $\dfrac{100. \text{ g}}{46.07 \text{ g/mol}} = 2.17 \text{ mol}$

54. A hydronium ion is a hydrated hydrogen ion.

$\quad\quad H^+ \quad\quad + H_2O \quad\quad\longrightarrow\quad\quad H_3O^+$
$\;$(hydrogen ion)$\quad\quad\quad\quad\quad\quad\quad\quad$(hydronium ion)

55. Freezing point depression is directly related to the concentration of particles in the solution.

1 mol$\quad\quad\quad$ 1 + mol$\quad\quad$ 2 mol$\quad\quad$ 3 mol$\quad\;$(particles in solution)
$C_{12}H_{22}O_{11} > HC_2H_3O_2 > HCl > CaCl_2$
Highest freezing point$\quad\quad\quad\quad\quad\quad\quad\;$Lowest freezing point

56. (a) $100°C$ $pH = -\log(1 \times 10^{-6}) = 6.0$
    $25°C$ $pH = -\log(1 \times 10^{-7}) = 7.0$ pH of H_2O is greater at $25°C$

    (b) $1 \times 10^{-6} > 1 \times 10^{-7}$ so, H^+ concentration is higher at $100°C$.

    (c) The water is neutral at both temperatures, because the H_2O ionizes into equal concentrations of H^+ and OH^- at any temperature.

57. As the pH changes by 1 unit, the concentration of H^+ in solution changes by a factor of 10. For example, the pH of $0.10\ M$ HCl is 1.00, while the pH of $0.0100\ M$ HCl is 2.00.

58. A $1.00\ m$ solution contains 1 mol solute plus 1000 g H_2O. We need to find the total number of moles and then calculate the mole percent of each component.

$$\left(\frac{1000 \text{ g } H_2O}{18.02 \text{ g/mol}}\right) = 55.49 \text{ mol } H_2O$$

$$55.49 \text{ mol } H_2O + 1.00 \text{ mol solute} = 56.49 \text{ total moles}$$

$$\left(\frac{1.00 \text{ mol solute}}{56.49 \text{ mol}}\right)(100) = 1.77\% \text{ solute}$$

$$\left(\frac{55.49 \text{ mol } H_2O}{56.49 \text{ mol}}\right)(100) = 98.23\% \ H_2O$$

59. $Na_2CO_3 + 2\ HCl \longrightarrow 2\ NaCl + CO_2 + H_2O$
    g $Na_2CO_3 \longrightarrow$ mol $Na_2CO_3 \longrightarrow$ mol HCl $\longrightarrow$ M HCl

$$(0.452 \text{ g } Na_2CO_3)\left(\frac{1 \text{ mol}}{106.0 \text{ g}}\right)\left(\frac{2 \text{ mol HCl}}{1 \text{ mol } Na_2CO_3}\right)\left(\frac{1}{0.0424 \text{ L}}\right) = 0.201\ M \text{ HCl}$$

60. $H_2SO_4 + 2\ KOH \rightarrow K_2SO_4 + 2\ H_2O$

    g KOH $\rightarrow$ mol KOH $\rightarrow$ mol $H_2SO_4 \rightarrow$ M H_2SO_4

$$(6.38 \text{ g KOH})\left(\frac{1 \text{ mol KOH}}{56.11 \text{ g KOH}}\right)\left(\frac{1 \text{ mol } H_2SO_4}{2 \text{ mol KOH}}\right)\left(\frac{1000 \text{ mL}}{0.4233 \text{ mol } H_2SO_4}\right) = 134 \text{ mL of } 0.4233\ M \ H_2SO_4$$

61. $KOH + HNO_3 \longrightarrow KNO_3 + H_2O$
    L $HNO_3 \longrightarrow$ mol $HNO_3 \longrightarrow$ mol KOH $\longrightarrow$ g KOH

$$(0.05000 \text{ L } HNO_3)\left(\frac{0.240 \text{ mol}}{L}\right)\left(\frac{1 \text{ mol KOH}}{1 \text{ mol } HNO_3}\right)\left(\frac{56.11 \text{ g}}{\text{mol}}\right) = 0.673 \text{ g KOH}$$

62. pH of 1.0 L solution containing 0.1 mL of $1.0\ M$ HCl

$$(0.1 \text{ mL})\left(\frac{1.0 \text{ L}}{1000 \text{ mL}}\right)\left(\frac{1 \text{ mol HCl}}{L}\right) = 1 \times 10^{-4} \text{ mol HCl added}$$

$$\frac{1 \times 10^{-4} \text{ mol HCl}}{1.0 \text{ L}} = 1 \times 10^{-4}\ M \text{ HCl}$$

$1 \times 10^{-4}\ M$ HCl produces $1 \times 10^{-4}\ M$ H^+

$pH = -\log(1 \times 10^{-4}) = 4.0$

63. Dilution problem: $V_1 M_1 = V_2 M_2$

$$M_1 = \frac{V_2 M_2}{V_1} \qquad M_1 = \frac{(10.0\text{ mL})(12\,M)}{(260.0\text{ mL})} = 0.462\,M\text{ HCl}$$

64. $NaOH + HCl \longrightarrow NaCl + H_2O$

$$(3.0\text{ g NaOH})\left(\frac{1\text{ mol}}{40.00\text{ g}}\right) = 0.075\text{ mol NaOH}$$

$$(500.\text{ mL HCl})\left(\frac{1\text{ L}}{1000\text{ mL}}\right)\left(\frac{0.10\text{ mol}}{\text{L}}\right) = 0.050\text{ mol HCl}$$

This solution is basic. The NaOH will neutralize the HCl with an excess of 0.025 mol of NaOH remaining unreacted.

65. $Ba(OH)_2(aq) + 2\,HCl(aq) \longrightarrow BaCl_2(aq) + 2\,H_2O(l)$

$$(0.38\text{ L Ba(OH)}_2)\left(\frac{0.35\text{ mol}}{\text{L}}\right) = 0.13\text{ mol Ba(OH)}_2$$

$$0.13\text{ mol Ba(OH)}_2 \longrightarrow 0.26\text{ mol OH}^-$$

$$(0.500\text{ L HCl})\left(\frac{0.65\text{ mol}}{\text{L}}\right) = 0.33\text{ mol HCl}$$

$$0.33\text{ mol HCl} \longrightarrow 0.33\text{ mol H}^+$$

0.33 mol H^+ will neutralize 0.26 mol OH^- and leave 0.07 mol $H^+(0.33 - 0.26)$ remaining in solution.
Total volume $= 500.\text{ mL} + 380\text{ mL} = 880\text{ mL}(0.88\text{ L})$

$$[H^+]\text{ in solution} = \frac{0.07\text{ mol H}^+}{0.88\text{ L}} = 0.08\,M\text{ H}^+$$

$$pH = -\log[H^+] = -\log(8 \times 10^{-2}) = 1.1$$

The solution is acidic.

66. $(0.05000\text{ L HCl})\left(\dfrac{0.2000\text{ mol}}{\text{L}}\right) = 0.01000\text{ mol HCl} = 0.01000\text{ mol H}^+\text{ in 50.00 mL HCl}$

(a) no base added: $pH = -\log(0.2000) = 0.700$

(b) 10.00 mL base added: $(0.01000\text{ L})\left(\dfrac{0.2000\text{ mol}}{\text{L}}\right) = 0.002000\text{ mol NaOH}$

$= 0.002000\text{ mol OH}^-$

$(0.01000\text{ mol H}^+) - (0.002000\text{ mol OH}^-) = 0.00800\text{ mol H}^+\text{ in 60.00 mL solution}$

$$[H^+] = \frac{0.00800\text{ mol}}{0.06000\text{ L}} \qquad pH = -\log\left(\frac{0.00800}{0.06000}\right) = 0.880$$

(c)   25.00 mL base added:

$$(0.02500 \text{ L})\left(\frac{0.2000 \text{ mol}}{\text{L}}\right) = 0.005000 \text{ mol NaOH} = \text{mol OH}^-$$

$$(0.01000 \text{ mol H}^+) - (0.005000 \text{ mol OH}^-) = 0.00500 \text{ mol H}^+ \text{ in } 75.00 \text{ solution}$$

$$[\text{H}^+] = \frac{0.00500 \text{ mol}}{0.07500 \text{ L}} \qquad \text{pH} = -\log\left(\frac{0.00500}{0.07500}\right) = 1.2$$

(d)   49.00 mL base added:

$$(0.04900 \text{ L})\left(\frac{0.2000 \text{ mol}}{\text{L}}\right) = 0.009800 \text{ mol NaOH} = \text{mol OH}^-$$

$$(0.01000 \text{ mol H}^+) - (0.009800 \text{ mol OH}^-) = 0.00020 \text{ mol H}^+ \text{ in } 99.00 \text{ mL solution}$$

$$[\text{H}^+] = \frac{0.00020 \text{ mol}}{0.09900 \text{ L}} \qquad \text{pH} = -\log\left(\frac{0.00020}{0.09900}\right) = 2.69$$

(e)   49.90 mL base added:

$$(0.04990 \text{ L})\left(\frac{0.2000 \text{ mol}}{\text{L}}\right) = 0.009980 \text{ mol NaOH} = \text{mol OH}^-$$

$$(0.01000 \text{ mol H}^+) - (0.009800 \text{ mol OH}^-) = 2 \times 10^{-5} \text{ mol H}^+ \text{ in } 99.00 \text{ mL solution}$$

$$[\text{H}^+] = \frac{2 \times 10^{-5} \text{ mol}}{0.09990 \text{ L}} \qquad \text{pH} = -\log\left(\frac{2 \times 10^{-5}}{0.09990}\right) = 3.7$$

(f)   49.99 mL base added:

$$(0.04999 \text{ L})\left(\frac{0.2000 \text{ mol}}{\text{L}}\right) = 0.009998 \text{ mol NaOH} = \text{mol OH}^-$$

$$(0.01000 \text{ mol H}^+) - (0.009998 \text{ mol OH}^-) = 2 \times 10^{-6} \text{ mol H}^+ \text{ in } 99.99 \text{ mL solution}$$

$$[\text{H}^+] = \frac{2 \times 10^{-6}}{0.09999 \text{ L}} \qquad \text{pH} = -\log\left(\frac{2 \times 10^{-6}}{9.999 \times 10^{-2}}\right) = 4.7$$

(g)   50.00 mL of 0.2000 M NaOH neutralizes 50.00 mL of 0.2000 M HCl. No excess acid or base is in the solution. Therefore, the solution is neutral with a pH = 7.0

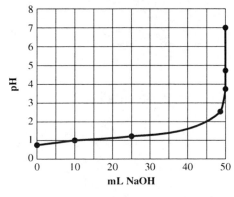

67. (a) $2 NaOH(aq) + H_2SO_4(aq) \longrightarrow Na_2SO_4(aq) + 2 H_2O(l)$

    (b) $mol H_2SO_4 \longrightarrow mol NaOH \longrightarrow mL NaOH$

    $$(0.0050 \text{ mol } H_2SO_4)\left(\frac{2 \text{ mol } NaOH}{1 \text{ mol } H_2SO_4}\right)\left(\frac{1000 \text{ mL}}{0.10 \text{ mol}}\right) = 1.0 \times 10^2 \text{ mL } NaOH$$

    (c) $(0.0050 \text{ mol } H_2SO_4)\left(\frac{1 \text{ mol } Na_2SO_4}{1 \text{ mol } H_2SO_4}\right)\left(\frac{142.1 \text{ g}}{\text{mol}}\right) = 0.71 \text{ g } Na_2SO_4$

68. $HNO_3 + KOH \longrightarrow KNO_3 + H_2O$
    $M_A V_A = M_B V_B$

    $(M_A)(25 \text{ mL}) = (0.60 \, M)(50.0 \text{ mL})$

    $M_A = 1.2 \, M$ (diluted solution)

    Dilution problem $M_1 V_1 = M_2 V_2$

    $(M_A)(10.0 \text{ mL}) = (1.2 \, M)(100.00 \text{ mL})$

    $M_A = 12 \, M \, HNO_3$ (original solution)

69. Yes, adding water changes the concentration of the acid, which changes the concentration of the $[H^+]$, and changes the pH. The pH will rise.

    No, the solution theoretically will never reach a pH of 7, but it will approach pH 7 as water is added.

70. $H_2SO_4 + 2 NaOH \rightarrow Na_2SO_4 + 2 H_2O$

    $$(0.425 \text{ L } H_2SO_4)\left(\frac{0.94 \text{ mol } H_2SO_4}{L}\right) = 0.40 \text{ mol } H_2SO_4$$

    $0.40 \text{ mol } H_2SO_4 \rightarrow 0.80 \text{ mol } H^+$

    $$(0.750 \text{ L } NaOH)\left(\frac{0.83 \text{ mol } NaOH}{L}\right) = 0.62 \text{ mol } NaOH$$

    $0.62 \text{ mol } NaOH \rightarrow 0.62 \text{ mol } OH^-$

    $0.80 \text{ mol } H^+$ will neutralize $0.62 \text{ mol } OH^-$ and leave 0.18 mol $(0.80 - 0.62)$ of H^+ remaining in solution; so the solution will be acidic.

    Total volume $= 425 \text{ mL} + 750 \text{ mL} = 1175 \text{ mL}$ (1.175 L)

    $$[H^+] \text{ in solution} = \frac{0.18 \text{ mol } H^+}{1.175 \text{ L}} = 0.15 \, M \, H^+$$

    $pH = -\log [H^+] = -\log (0.15) = 0.82$

71. (a) 1st determine kind of substance
        Copper(II) sulfate is a soluble salt so it will dissociate completely in water
        2nd write the dissociation/ionization equation

        $CuSO_4(aq) \xrightarrow{H_2O} Cu^{2+}(aq) + SO_4^{2-}(aq)$

        3rd Analyze the equation to determine number of ions formed.
        Each $CuSO_4$ will produce 1 Cu^{2+} ion and 1 SO_4^{2-} ion, so $[Cu^{2+}] = [SO_4^{2-}] = 1 \, M$

(b) 1st determine kind of substance

Nitric acid is a strong acid so it will ionize completely in water

2nd write the dissociation/ionization equation

$$HNO_3(aq) \xrightarrow{H_2O} H^+(aq) + NO_3^-(aq)$$

3rd Analyze the equation to determine number of ions formed.

Each HNO_3 will produce 1 H^+ ion and 1 NO_3^- ion, so $[H^+] = [NO_3^-] = 1\ M$

(c) 1st determine kind of substance

Sulfuric acid is a strong acid so it will ionize completely in water

2nd write the dissociation/ionization equation

$$H_2SO_4(aq) \xrightarrow{H_2O} 2\,H^+(aq) + SO_4^{2-}(aq)$$

3rd Analyze the equation to determine number of ions formed.

Each H_2SO_4 will produce 2 H^+ ions and 1 SO_4^{2-} ion, so $[H^+] = 2\ M$ and $[SO_4^{2-}] = 1\ M$

(d) 1st determine kind of substance

Calcium sulfide is an insoluble salt so it will not dissociate in water.

2nd write the dissociation/ionization equation

$$CaS(s) \xrightarrow{H_2O} \text{no reaction}$$

3rd Analyze the equation to determine number of ions formed.

CaS is insoluble so very few of the particles will dissociate and the concentration of all ions will be close to 0 M.

(e) 1st determine kind of substance

Acetic acid is a weak acid so it will ionize slightly in water

2nd write the dissociation/ionization equation

$$HC_2H_3O_2(aq) \underset{}{\overset{H_2O}{\rightleftharpoons}} H^+(aq) + C_2H_3O_2^-(aq)$$

3rd Analyze the equation to determine number of ions formed.

Each $HC_2H_3O_2$ will produce some H^+ ions and $C_2H_3O_2^-$ ions, so $[H^+]$ and $[C_2H_3O_2^-]$ will be between 0 and 1 M.

72. (a) Formula equation $HNO_3(aq) + LiOH(aq) \rightarrow H_2O(l) + LiNO_3(aq)$

Total ionic equation $H^+(aq) + NO_3^-(aq) + Li^+(aq) + OH^-(aq) \rightarrow$
$$H_2O(l) + Li^+(aq) + NO_3^-(aq)$$

Net ionic equation $H^+(aq) + OH^-(aq) \rightarrow H_2O(l)$

(b) Formula equation $2\,HBr(aq) + Ba(OH)_2(aq) \rightarrow 2\,H_2O(l) + BaBr_2(aq)$

Total ionic equation. $2\,H^+(aq) + 2Br^-(aq) + Ba^{2+}(aq) + 2\,OH^-(aq) \rightarrow$
$$2\,H_2O(l) + Ba^{2+}(aq) + 2\,Br^-(aq)$$

Net ionic equation $H^+(aq) + OH^-(aq) \rightarrow H_2O(l)$

(c) Formula equation $HF(aq) + NaOH(aq) \rightarrow H_2O(l) + NaF(aq)$
Total ionic equation. $HF(aq) + Na^+(aq) + OH^-(aq) \rightarrow H_2O(l) + Na^+(aq) + F^-(aq)$

(Note that HF is a weak acid and so it does not ionize to any appreciable extent and is written in its unionized form.)

Net ionic equation
$HF(aq) + OH^-(aq) \rightarrow H_2O(l) + F^-(aq)$

73. 1st write a balanced chemical equation for the reaction of the H^+ in the rain with sodium hydroxide

$$H^+(aq) + NaOH(aq) \rightarrow Na^+(aq) + H_2O(l)$$

2nd calculate moles of sodium hydroxide needed to react with all of the H^+ in the rain.

$$\text{mol NaOH} = (7.2 \text{ mL NaOH})\left(\frac{1 \text{ L NaOH}}{1000 \text{ mL NaOH}}\right)\left(\frac{0.125 \text{ mol NaOH}}{1 \text{ L NaOH}}\right) = 0.00090 \text{ mol NaOH}$$

3rd determine moles of H^+

$$\text{mol H}^+ = (0.00090 \text{ mol NaOH})\left(\frac{1 \text{ mol H}^+}{1 \text{ mol NaOH}}\right) = 0.00090 \text{ mol H}^+$$

4th determine molarity of H^+

$$M\,H^+ = \frac{\text{mol H}^+}{\text{L rain}} = \frac{0.00090 \text{ mol H}^+}{25 \, L \text{ rain}} = 3.6 \times 10^{-5} \, M\,H^+$$

5th determine pH of rain

$$pH = -\log[H^+] = -\log\left(3.6 \times 10^{-5} \, M\,H^+\right) = 4.44$$

74. $[H^+]$ needs to be $0.00158\,M$ to give the final solution of sodium rhidizonate a pH of 2.800.

$$V_1 M_1 = V_2 M_2$$

$$(500.0 \text{ mL})(0.00158\,M) = V_2(0.60\,M)$$

$$V_2 = \frac{(500.0 \text{ mL})(0.00158\,M)}{(0.60\,M)} = 1.3 \text{ mL } 0.60\,M \text{ HCl}$$

To make the solution put 1.3 mL of 0.60 M HCl in a graduated cylinder and fill the cylinder to the 500 mL mark with sodium rhidizonate solution.

75. $V_1 M_1 = V_2 M_2$

$$(3.00 \text{ L})(3.25\,M) = V_2(12.1\,M)$$

$$V_2 = \frac{(3.00 \text{ L})(3.25\,M)}{(12.1\,M)} = 0.806 \text{ L } 12.1\,M \text{ HCl or } 806 \text{ mL } 12.1\,M \text{ HCl}$$

76. $CaCO_3(s) + 2\,HCl(aq) \rightarrow CaCl_2(aq) + H_2O(l) + CO_2(g)$

77. First determine the molarity of the two HCl solutions. Take the antilog of the pH value to obtain the $[H^+]$.

pH = 0.300; $H^+ = 2.00\,M = 2.00\,M$ HCl
pH = 0.150; $H^+ = 1.41\,M = 1.41\,M$ HCl

Now treat the calculation as a dilution problem.

$$V_1M_1 = V_2M_2 \qquad V_2 = \frac{V_1M_1}{M_2}$$

$$\frac{(200 \text{ mL HCl})(2.00\,M)}{1.41\,M} = 284 \text{ mL solution}$$

$$284 \text{ mL} - 200 \text{ mL} = 84 \text{ mL H}_2\text{O to be added}$$

78. mol acid = mol base   (lactic acid has one acidic H)

$$\frac{1.0 \text{ g acid}}{\text{molar mass}} = (0.017 \text{ L})\left(\frac{0.65 \text{ mol NaOH}}{\text{L}}\right) = 0.011 \text{ mol NaOH}$$

mol acid = mol base

mol $HC_3H_5O_3 = 0.011$ mol

$$\frac{1.0 \text{ g}}{0.011 \text{ mol}} = 91 \text{ g/mol} = \text{molar mass}$$

molar mass (91 g/mol) = mass of empirical formula (90.08 g/mol)

Therefore the molecular formula, $HC_3H_5O_3$, is the same as the empirical formula.

CHAPTER 16

CHEMICAL EQUILIBRIUM

SOLUTIONS TO REVIEW QUESTIONS

1. At equilibrium, the rate of the forward reaction equals the rate of the reverse reaction.

2. The rate of a reaction increases when the concentration of one of the reactants increases. The increase in concentration causes the number of collisions between the reactants to increase. The rate of a reaction, being proportional to the frequency of such collisions, as a result, will increase.

3. An increase in temperature causes the rate of reaction to increase, because it increases the velocity of the molecules. Faster moving molecules increase the number and effectiveness of the collisions between molecules resulting in an increase in the rate of the reaction.

4. At 25°C both tubes would appear the same.

5. The reaction is endothermic because the increased temperature increases the concentration of product (NO_2) present at equilibrium.

6. If pure HI is placed in a vessel at 700 K, some of it will decompose. Since the reaction is reversible ($H_2 + I_2 \rightleftharpoons 2\,HI$) HI molecules will react to produce H_2 and I_2.

7. Acids stronger than acetic acid are: benzoic, cyanic, formic, hydrofluoric, and nitrous acids (all equilibrium constants are greater than the equilibrium constant for acetic acid). Acids weaker than acetic acid are: carbolic, hydrocyanic, and hypochlorous acids (all have equilibrium constants smaller than the equilibrium constant for acetic acid). All have one ionizable hydrogen atom.

8. In an endothermic process heat is absorbed or used by the system so it should be placed on the reactant side of a chemical equation. In an exothermic process heat is given off by the system so it belongs on the product side of a chemical equation.

9. A catalyst speeds up the rate of a reaction by lowering the activation energy. A catalyst is not used up in the reaction.

10. It is very important to specify the temperature because equilibrium constants change with changing temperatures. For example K_w for H_2O at 25°C = 14.00 and at 50°C = 13.26.

11. A very large equilibrium constant means the equilibrium lies far to the right.

12. A very small equilibrium constant means the equilibrium lies far to the left.

13. Free protons (H^+) do not exist in water because they are hydrated forming H_3O^+.

14. The sum of the pH and the pOH is 14. A solution whose pH is -1 would have a pOH of 15.

15. At different temperatures, the degree of ionization of water varies, being higher at higher temperatures. Consequently, the pH of water can be different at different temperatures.

16. In pure water, H^+ and OH^- are produced in equal quantities by the ionization of the water molecules, $H_2O \rightleftharpoons H^+ + OH^-$. Since pH $= -\log[H^+]$, and pOH $= -\log[OH^-]$, they will always be identical for pure water. At 25°C, they each have the value of 7, but at higher temperatures, the degree of ionization is greater, so the pH and pOH would both be less than 7, but still equal.

17. $HC_2H_3O_2 + H_2O \rightleftharpoons H_3O^+ + C_2H_3O_2^-$
    As water is added (diluting the solution from 1.0 M to 0.10 M), the equilibrium shifts to the right, giving more ions and thus yielding a higher percent ionization.

18. The statement does not contradict Le Chatelier's Principle. The previous question deals with the case of dilution. If pure acetic acid is added to a dilute solution, the reaction will shift to the right, producing more ions in accordance with Le Chatelier's Principle. But, the concentration of the un-ionized acetic acid will increase faster than the concentration of the ions, thus yielding a smaller percent ionization.

19. In water the silver acetate dissociates until the equilibrium concentration of ions is reached. In nitric acid solution, the acetate ions will react with hydrogen ions to form acetic acid molecules. The HNO_3 removes acetate ions from the silver acetate equilibrium allowing more silver acetate to dissolve. If HCl is used, a precipitate of silver chloride would be formed, since silver chloride is less soluble than silver acetate. Thus, more silver acetate would dissolve in HCl than in pure water.
    $AgC_2H_3O_2(s) \rightleftharpoons Ag^+(aq) + C_2H_3O_2{}^-(aq)$

20. When the salt, sodium acetate, is dissolved in water, the solution becomes basic. The dissolving reaction is

    $NaC_2H_3O_2(s) \xrightarrow{H_2O} Na^+(aq) + C_2H_3O_2{}^-(aq)$

    The acetate ion reacts with water. The reaction does not go to completion, but some OH^- ions are produced and at equilibrium the solution is basic.
    $C_2H_3O_2{}^-(aq) + H_2O(l) \rightleftharpoons OH^-(aq) + HC_2H_3O_2(aq)$

21. The order of solubility will correspond to the order of the values of the solubility product constants of the salts being compared. This occurs because each salt in the comparison produces the same number of ions (two in this case) for each formula unit of salt that dissolves. This type of comparison would not necessarily be valid if the salts being compared gave different numbers of ions per formula unit of salt dissolving. The order is: $AgC_2H_3O_2$, $PbSO_4$, $BaSO_4$, $AgCl$, $BaCrO_4$, $AgBr$, AgI, PbS.

22. (a) K_{sp} $Mn(OH)_2 = 2.0 \times 10^{-13}$; K_{sp} $Ag_2CrO_4 = 1.9 \times 10^{-12}$.
    Each salt gives 3 ions per formula unit of salt dissolving. Therefore, the salt with the largest K_{sp}, (in this case Ag_2CrO_4) is more soluble.

(b) K_{sp} $BaCrO_4 = 8.5 \times 10^{-11}$; Ksp $Ag_2CrO_4 = 1.9 \times 10^{-12}$. Ag_2CrO_4 has a greater molar solubility than $BaCrO_4$, even though its K_{sp} is smaller, because the Ag_2CrO_4 produces more ions per formula unit of salt dissolving than $BaCrO_4$.

$$BaCrO_4(s) \rightleftharpoons Ba^{2+} + CrO_4^{2-} \qquad K_{sp} = [Ba^{2+}][CrO_4^{2-}]$$

Let y = molar solubility of $BaCrO_4$

$$K_{sp} = [y][y] = 8.5 \times 10^{-11}$$

$$y = \sqrt{8.5 \times 10^{-11}} = 9.2 \times 10^{-6} \text{ mol } BaCrO_4/L$$

$$Ag_2CrO_4(s) \rightleftharpoons 2\,Ag^+ + CrO_4^{2-} \qquad K_{sp} = [Ag^+]^2[CrO_4^{2-}]$$

Let y = molar solubility of Ag_2CrO_4

$$K_{sp} = [2y]^2[y] = 1.9 \times 10^{-12}$$

$$y = \sqrt{\frac{1.9 \times 10^{-12}}{4}} = 7.8 \times 10^{-5} \text{ mol } Ag_2CrO_4/L$$

Ag_2CrO_4 has the greater solubility.

23. In a saturated sodium chloride solution, the equilibrium is

$$NaCl(s) \rightleftharpoons Na^+(aq) + Cl^-(aq)$$

Bubbling in HCl gas increases the concentration of Cl^-, creating a stress, which will cause the equilibrium to shift to the right, precipitating solid NaCl.

24. All four K_{eq} expressions are equilibrium constants. They describe the ratio between the concentrations of products and reactants for different types of reactions when at equilibrium. K_a is the equilibrium constant expression for the ionization of a weak acid, K_b is the equilibrium constant expression for the ionization of a weak base, K_w is the equilibrium constant expression for the ionization of water and K_{sp} is the equilibrium constant expression for a slightly soluble salt.

25. $HC_2H_3O_2 \rightleftharpoons H^+ + C_2H_3O_2^-$

	Initial Concentrations	Added	Concentration After Equilibrium Shifts
$HC_2H_3O_2$	0.10 M	-----	0.11 M
H^+	1.8×10^{-5} M	0.010 mol	1.9×10^{-5} M
$C_2H_3O_2^-$	0.10 M	-----	0.09 M
pH	4.74		4.72

The initial concentration of H^+ in the buffer solution is very low $(1.8 \times 10^{-5}\,M)$ because of the large excess of acetate ions. 0.010 mol of HCl is added to one liter of the buffer solution. This will supply 0.010 M H^+. The added H^+ creates a stress on the right side of the equation. The equilibrium shifts to the left, using up almost all the added H^+, reducing the acetate ion by approximately 0.010 M, and increasing the acetic acid by approximately 0.010 M. The concentration of H^+ will not increase significantly and the pH is maintained relatively constant.

26. A buffer solution contains a weak acid or base plus a salt of that weak acid or base, such as dilute acetic acid and sodium acetate.

$$HC_2H_3O_2(aq) \rightleftharpoons H^+(aq) + C_2H_3O_2^-(aq)$$
$$NaC_2H_3O_2(aq) \longrightarrow Na^+(aq) + C_2H_3O_2^-(aq)$$

When a small amount of a strong acid (H^+) is added to this buffer solution, the H^+ reacts with the acetate ions to form un-ionized acetic acid, thus neutralizing the added acid. When a strong base, OH^-, is added it reacts with un-ionized acetic acid to neutralize the added base. As a result, in both cases, the approximate pH of the solution is maintained.

27. $A + B \rightleftharpoons C + D$
When A and B are initially mixed, the rate of the forward reaction to produce C and D is at its maximum. As the reaction proceeds, the rate of production of C and D decreases because the concentrations of A and B decrease. As C and D are produced, some of the collisions between C and D will result in the reverse reaction, forming A and B. Finally, an equilibrium is achieved in which the forward rate exactly equals the reverse rate.

SOLUTIONS TO EXERCISES

1. Reversible systems.

    (a) $KMnO_4(s) \rightleftharpoons K^+(aq) + MnO_4^-(aq)$

    (b) $CO_2(s) \rightleftharpoons CO_2(g)$

2. Reversible systems.

    (a) $I_2(s) \rightleftharpoons I_2(g)$

    (b) $NaNO_3(s) \rightleftharpoons Na^+(aq) + NO_3^-(aq)$

3. Equilibrium system.

    $SiF_4(g) + 2\,H_2O(g) + 103.8\,kJ \rightleftharpoons SiO_2(s) + 4\,HF(g)$

    (a)   The reaction is endothermic with heat being absorbed.

    (b)   The addition of HF will shift the reaction to the left until equilibrium is reestablished. The concentrations of SiF_4, H_2O, and HF will be increased. The concentration of SiO_2 will be decreased.

    (c)   The addition of heat will shift the reaction to the right.

4. Equilibrium system.

    $4\,HCl(g) + O_2(g) \rightleftharpoons 2\,H_2O(g) + 2\,Cl_2(g) + 114.4\,kJ$

    (a)   The reaction is exothermic with heat being released.

    (b)   The addition of O_2 will shift the reaction to the right until equilibrium is reestablished. The concentrations of O_2, H_2O, and Cl_2 will be increased. The concentration of HCl will be decreased.

    (c)   The addition of heat will cause the reaction to shift to the left.

5. $N_2(g) + 3\,H_2(g) \rightleftharpoons 2\,NH_3(g) + 92.5\,kJ$

Change or stress imposed on the system at equilibrium	Direction of reaction, left or right, to reestablish equilibrium	Changes in number of moles		
		N_2	H_2	NH_3
(a) Add N_2	right	I	D	I
(b) Remove H_2	left	I	D	D
(c) Decrease volume of reaction vessel	right	D	D	I
(d) Increase temperature	left	I	I	D

I = Increase;   D = Decrease;   N = No Change;
? = insufficient information to determine

6. $N_2(g) + 3\,H_2(g) \rightleftharpoons 2\,NH_3(g) + 92.5\,kJ$

Change or stress imposed on the system at equilibrium	Direction of reaction, left or right, to reestablish equilibrium	Changes in number of moles		
		N_2	H_2	NH_3
(a) Add NH_3	left	I	I	I
(b) Increase volume of reaction vessel	left	I	I	D
(c) Add a catalyst	no change	N	N	N
(d) Add H_2 and NH_3	?	?	I	I

I = Increase;   D = Decrease;   N = No Change;
? = insufficient information to determine

7. Direction of shift in equilibrium:

Reaction	Increased Temperature	Increased Pressure (Volume Decreases)	Add Catalyst
(a)	right	right	no change
(b)	left	no change	no change
(c)	left	right	no change

8. Direction of shift in equilibrium:

Reaction	Increased Temperature	Increased Pressure (Volume Decreases)	Add Catalyst
(a)	right	left	no change
(b)	left	left	no change
(c)	left	left	no change

9. Equilibrium shifts

$CH_4(g) + 2\,O_2(g) \rightleftharpoons CO_2(g) + 2\,H_2O(g) + 802.3\,kJ$

(a)   left
(b)   none
(c)   right
(d)   none

10. Equilibrium shifts

$2\,CO_2(g) + N_2(g) + 1095.9\,kJ \rightleftharpoons C_2N_2(g) + 2\,O_2(g)$

(a)   none
(b)   left
(c)   right
(d)   right

11. (a) $K_{eq} = \dfrac{[NH_3]^2[H_2O]^4}{[NO_2]^2[H_2]^7}$ (c) $K_{eq} = \dfrac{[CO_2][CF_4]}{[COF_2]^2}$

(b) $K_{eq} = \dfrac{[H^+][HCO_3^-]}{[H_2CO_3]}$

12. (a) $K_{eq} = \dfrac{[H^+][H_2PO_4^-]}{[H_3PO_4]}$ (c) $K_{eq} = \dfrac{[N_2O_5]^2}{[NO_2]^4[O_2]}$

(b) $K_{eq} = \dfrac{[CH_4][H_2S]^2}{[CS_2][H_2]^4}$

13. (a) $K_{sp} = [Ag^+][Cl^-]$ (c) $K_{sp} = [Zn^{2+}][OH^-]^2$

(b) $K_{sp} = [Pb^{2+}][CrO_4^{2-}]$ (d) $K_{sp} = [Ca^{2+}]^3[PO_4^{3-}]^2$

14. (a) $K_{sp} = [Mg^{2+}][CO_3^{2-}]$ (c) $K_{sp} = [Tl^{3+}][OH^-]^3$

(b) $K_{sp} = [Ca^{2+}][C_2O_4^{2-}]$ (d) $K_{sp} = [Pb^{2+}]^3[AsO_4^{3-}]^2$

15. If the H^+ ion concentration is decreased:
   (a) pH is increased
   (b) pOH is decreased
   (c) $[OH^-]$ is increased
   (d) K_w remains the same. K_w is a constant at a given temperature.

16. If the H^+ ion concentration is increased:
   (a) pH is decreased (pH of 1 is more acidic than that of 4)
   (b) pOH is increased
   (c) $[OH^-]$ is decreased
   (d) K_w remains unchanged. K_w is a constant at a given temperature.

17. When excess acid (H^+) gets into the blood stream it reacts with HCO_3^- to form un-ionized H_2CO_3, thus neutralizing the acid and maintaining the approximate pH of the blood.

18. When excess base gets into the bloodstream it reacts with H^+ to form water. Then H_2CO_3 ionizes to replace H^+, thus maintaining the approximate pH of the blood.

19. (a) $H_2CO_3(aq) \rightleftharpoons H^+(aq) + HCO_3^-(aq)$

   Let x = molarity of H^+

   $[H^+] = [HCO_3^-] = x$

   $[H_2CO_3] = 1.25\,M - x = 1.25$ (since x is small)

   $K_a = \dfrac{[H^+][HCO_3^-]}{[H_2CO_3]} = \dfrac{x^2}{1.25} = 4.4 \times 10^{-7}$

   $x^2 = (1.25)(4.4 \times 10^{-7})$

   $x = \sqrt{(1.25)(4.4 \times 10^{-7})} = 7.4 \times 10^{-4}\,M = [H^+]$

   (b) $pH = -\log[H^+] = -\log(7.4 \times 10^{-4}\,M) = 3.13$

   (c) Percent ionization

   $\dfrac{[H^+]}{[H_2CO_3]}(100) = \left(\dfrac{7.4 \times 10^{-4}}{1.25}\right)(100) = 0.059\%$

20. (a) $HC_3H_5O_2(aq) \rightleftharpoons H^+(aq) + C_3H_5O_2^-(aq)$

   Let x = molarity of H^+

   $[H^+] = [C_3H_5O_2^-] = x$

   $[H_2CO_3] = 0.025\,M - x = 0.025$ (since x is small)

   $K_a = \dfrac{[H^+][C_3H_5O_2^-]}{[HC_3H_5O_2]} = \dfrac{x^2}{0.025} = 8.4 \times 10^{-4}$

   $x^2 = (0.025)(8.4 \times 10^{-4})$

   $x = \sqrt{(0.025)(8.4 \times 10^{-4})} = 4.6 \times 10^{-3}\,M = [H^+]$

   (b) $pH = -\log[H^+] = -\log(4.6 \times 10^{-3}\,M) = 2.34$
   (c) Percent ionization

   $\dfrac{[H^+]}{[HC_3H_5O_2]}(100) = \left(\dfrac{4.6 \times 10^{-3}}{0.025}\right)(100) = 18\%$

21. $HA \rightleftharpoons H^+ + A^-$
   $[H^+] = [A^-] = (0.025\,M)(0.0045) = 1.1 \times 10^{-4}\,M$
   $[HA] = 0.025\,M - 0.00011\,M = 0.025\,M$

   $K_a = K_{eq} = \dfrac{[H^+][A^-]}{[HA]} = \dfrac{(1.1 \times 10^{-4})^2}{0.025} = 4.8 \times 10^{-7}$

22. $HA \rightleftharpoons H^+ + A^-$

$[H^+] = [A^-] = (0.500\,M)(0.0068) = 3.4 \times 10^{-3}\,M$

$[HA] = 0.500\,M - 0.0034\,M = 0.497\,M$

$K_a = \dfrac{[H^+][A^-]}{[HA]} = \dfrac{(3.4 \times 10^{-3})^2}{0.497} = 2.3 \times 10^{-5}$

23. $HC_6H_5O(aq) \rightleftharpoons H^+(aq) + C_6H_5O^-(aq)$

$K_a = \dfrac{[H^+][C_6H_5O^-]}{[HC_6H_5O]} = 1.3 \times 10^{-10}$

Let x = molarity of H^+

$[HC_6H_5O]$ = initial concentration $- x$ = initial concentration

Since K_a is small, the degree of ionization is small. Therefore, the approximation, initial concentration $- x$ = initial concentration is valid.

(a) $[H^+] = [C_6H_5O^-] = x \quad [HC_6H_5O] = 1.0\,M$

$\dfrac{(x)(x)}{1.0} = 1.3 \times 10^{-10}$

$x^2 = 1.3 \times 10^{-10}$

$x = \sqrt{(1.0)(1.3 \times 10^{-10})} = 1.1 \times 10^{-5}\,M$

$\left(\dfrac{1.1 \times 10^{-5}M}{1.0\,M}\right)(100) = 0.0011\%$ ionized

$pH = -\log(1.1 \times 10^{-5}) = 4.96$

(b) $[HC_6H_5O] = 0.10\,M$

$\dfrac{(x)(x)}{0.10} = 1.3 \times 10^{-10}$

$x^2 = (0.10)(1.3 \times 10^{-10})$

$x = \sqrt{(0.10)(1.3 \times 10^{-10})} = 3.6 \times 10^{-6}\,M$

$\left(\dfrac{3.6 \times 10^{-6}M}{1.0\,M}\right)(100) = 0.00036\%$ ionized

$pH = -\log(3.6 \times 10^{-6}) = 5.44$

(c) $[HC_6H_5O] = 0.010\,M$

$\dfrac{(x)(x)}{0.010} = 1.3 \times 10^{-10}$

$x^2 = (0.010)(1.3 \times 10^{-10})$

$x = \sqrt{(0.010)(1.3 \times 10^{-10})} = 1.1 \times 10^{-6}\,M$

$\left(\dfrac{1.1 \times 10^{-6}M}{0.010\,M}\right)(100) = 0.011\%$ ionized

$pH = -\log(1.1 \times 10^{-6}) = 5.96$

24. $HC_7H_5O_2(aq) \rightleftharpoons H^+(aq) + C_7H_5O_2^-(aq)$

   (a) $[H^+] = [C_7H_5O_2^-] = x$ $[HC_7H_5O_2] = 1.0\,M$

$$\frac{(x)(x)}{1.0} = 6.3 \times 10^{-5}$$

$$x^2 = (1.0)(6.3 \times 10^{-5})$$

$$x = \sqrt{(1.0)(6.3 \times 10^{-5})} = 7.9 \times 10^{-3}\,M$$

$$\left(\frac{7.9 \times 10^{-3}\,M}{1.0\,M}\right)(100) = 0.79\%\ \text{ionized}$$

$$pH = -\log(7.9 \times 10^{-3}) = 2.10$$

   (b) $[HC_7H_5O_2] = 0.10\,M$

$$\frac{(x)(x)}{0.10} = 6.3 \times 10^{-5}$$

$$x^2 = (0.10)(6.3 \times 10^{-5})$$

$$x = \sqrt{(0.10)(6.3 \times 10^{-5})} = 2.5 \times 10^{-3}\,M$$

$$\left(\frac{2.5 \times 10^{-3}\,M}{0.10\,M}\right)(100) = 2.5\%\ \text{ionized}$$

$$pH = -\log(2.5 \times 10^{-3}) = 2.60$$

   (c) $[HC_7H_5O_2] = 0.010\,M$

$$\frac{(x)(x)}{0.010} = 6.3 \times 10^{-5}$$

$$x^2 = (0.010)(6.3 \times 10^{-5})$$

$$x = \sqrt{(0.010)(6.3 \times 10^{-5})} = 7.9 \times 10^{-4}\,M$$

$$\left(\frac{7.9 \times 10^{-4}\,M}{0.010\,M}\right)(100) = 7.9\%\ \text{ionized}$$

$$pH = -\log(7.9 \times 10^{-4}) = 3.10$$

25. $HA \rightleftharpoons H^+ + A^-$ $K_a = \dfrac{[H^+][A^-]}{[HA]}$

First, find the $[H^+]$. This is calculated from the pH expression, $pH = -\log[H^+] = 3.7$.

$[H^+] = 2 \times 10^{-4}$

$[H^+] = [A^-] = 2 \times 10^{-4}$ $[HA] = 0.37$

$$K_a = \frac{[H^+][A^-]}{[HA]} = \frac{(2 \times 10^{-4})(2 \times 10^{-4})}{0.37} = 1 \times 10^{-7}$$

26. $HA \rightleftharpoons H^+ + A^-$ $K_a = \dfrac{[H^+][A^-]}{[HA]}$ $pH = 2.89$

$-\log[H^+] = 2.89$ $[H^+] = 1.3 \times 10^{-3}$

$[H^+] = 1.3 \times 10^{-3} = [A^-]$ $[HA] = 0.23$

$K_a = \dfrac{[H^+][A^-]}{[HA]} = \dfrac{(1.3 \times 10^{-3})(1.3 \times 10^{-3})}{0.23} = 7.3 \times 10^{-6}$

27. $1.0\,M$ NaOH yields $[OH^-] = 1.0\,M$ (100% ionized)

$pOH = -\log 1.0 = 0.00$

$pH = 14 - pOH = 14.00$

$[H^+] = \dfrac{K_w}{[OH^-]} = \dfrac{1.0 \times 10^{-14}}{1.0} = 1 \times 10^{-14}$

28. $3.0\,M$ HNO$_3$ yields $[H^+] = 3.0\,M$ (100% ionized)

$pH = -\log 3.0 = -0.48$

$pOH = 14 - pH = 14 - (-0.48) = 14.48$

$[OH^-] = \dfrac{K_w}{[H^+]} = \dfrac{1 \times 10^{-14}}{3.0} = 3.3 \times 10^{-15}$

29. $pH + pOH = 14.0$ $pOH = 14.0 - pH$

(a) $0.250\,M$ HBr yields $[H^+] = 0.250\,M$ (100% ionized)
$pH = -\log[H^+] = -\log(0.250) = 0.602$ $pOH = 14.0 - 0.602 = 13.4$

(b) $0.333\,M$ KOH yields $[OH^-] = 0.333\,M$ (100% ionized)
$pOH = -\log[OH^+] = -\log(0.333) = 0.478$ $pH = 14.0 - 0.478 = 13.5$

(c) $HC_2H_3O_2 \rightleftharpoons H^+ + C_2H_3O_2^-$
     $0.895\,M$ x   x

$K_a = \dfrac{[H^+][C_2H_3C_2^-]}{[HC_2H_3O_2]} = 1.8 \times 10^{-5}$

$\dfrac{(x)(x)}{0.895} = 1.8 \times 10^{-5}$

$x^2 = (0.895)(1.8 \times 10^{-5})$ $x = \sqrt{1.6 \times 10^{-5}}$

$x = 4.0 \times 10^{-3} = [H^+]$

$pH = -\log(4.0 \times 10^{-3}) = 2.40$

$pOH = 14.0 - 2.40 = 11.6$

30. $pH + pOH = 14.0$ $pOH = 14.0 - pH$

(a) $0.0010\,M$ NaOH yields $[OH^-] = 0.0010\,M$ (100% ionized)
$pOH = -\log[OH^-] = -\log(0.0010) = 3.00$ $pH = 14.00 - 3.00 = 11.00$

(b) 0.125 M HCl yields $[H^+] = 0.125\ M(100\%$ ionized)

pH $= -\log[H^+] = -\log(0.125) = 0.903$ pOH $= 14.00 - 0.903 = 13.10$

(c) $HC_6H_5O \rightleftharpoons H^+ + C_6H_5O^-$

0.0250 M x   x

$K_a = \dfrac{[H^+][C_6H_5O^-]}{[HC_6H_5O]} = 1.3 \times 10^{-10}$

$\dfrac{(x)(x)}{0.0250} = 1.3 \times 10^{-10}$

$x^2 = (0.0250)(1.3 \times 10^{-10})$ $x = \sqrt{3.3 \times 10^{-12}}$

$x = 1.8 \times 10^{-6} = [H^+]$

pH $= -\log(1.8 \times 10^{-6}) = 5.74$

pOH $= 14.00\quad 5.74 = 8.26$

31. Calculate the $[OH^-]$ $[OH^-] = \dfrac{K_w}{[H^+]}$

(a) $[H^+] = 1.0 \times 10^{-2}$ $[OH^-] = \dfrac{1.0 \times 10^{-14}}{1.0 \times 10^{-2}} = 1.0 \times 10^{-12}M$

(b) $[H^+] = 3.2 \times 10^{-7}$ $[OH^-] = \dfrac{1.0 \times 10^{-14}}{3.2 \times 10^{-7}} = 3.1 \times 10^{-8}M$

(c) KOH is a strong base; $1.25\ M = [OH^-] = 1.25\ M$

(d) First find $[H^+]$ of 0.75 M $HC_2H_3O_2$ $K_a = 1.8 \times 10^{-5}$

$[H^+] = [C_2H_3O_2^-] = x$ $[HC_2H_3O_2] = 0.75\ M$

$\dfrac{x^2}{0.75} = 1.8 \times 10^{-5}$

$x^2 = (0.75)(1.8 \times 10^{-5})$

$x = \sqrt{(0.75)(1.8 \times 10^{-5})} = 3.7 \times 10^{-3}\ M$

$[H^+] = 3.7 \times 10^{-3}$ $[OH^-] = \dfrac{1.0 \times 10^{-14}}{3.7 \times 10^{-3}} = 2.7 \times 10^{-12}\ M$

32. Calculate the $[OH^-]$. $[OH^-] = \dfrac{K_w}{[H^+]}$

(a) $[H^+] = 4.0 \times 10^{-9}$ $[OH^-] = \dfrac{1.0 \times 10^{-14}}{4.0 \times 10^{-9}} = 2.5 \times 10^{-6}$

(b) $[H^+] = 1.2 \times 10^{-5}$ $[OH^-] = \dfrac{1.0 \times 10^{-14}}{1.2 \times 10^{-5}} = 8.3 \times 10^{-10}$

(c) First find $[H^+]$ of 1.25 M HCN $K_a = 4.0 \times 10^{-10}$

$[H^+] = [CN^-] = x$ $[HCN] = 1.25\ M$

$$\frac{x^2}{1.25} = 4.8 \times 10^{-10}$$

$$x^2 = (1.25)(4.0 \times 10^{-10})$$

$$x = \sqrt{(1.25)(4.0 \times 10^{-10})} = 2.2 \times 10^{-5} \, M$$

$$[H^+] = 2.2 \times 10^{-5} \quad [OH^-] = \frac{1.0 \times 10^{-14}}{2.2 \times 10^{-5}} = 4.5 \times 10^{-10} \, M$$

(d)   NaOH is a strong base; $0.333 \, M = [OH^-] = 0.333 \, M$

33.   Calculate the $[H^+]$ $\qquad\qquad\qquad [H^+] = \dfrac{K_w}{[OH^-]}$

(a) $[OH^-] = 1.0 \times 10^{-8} \qquad [H^+] = \dfrac{1.0 \times 10^{-14}}{1.0 \times 10^{-8}} = 1.0 \times 10^{-6}$

(b) $[OH^-] = 2.0 \times 10^{-4} \qquad [H^+] = \dfrac{1.0 \times 10^{-14}}{2.0 \times 10^{-4}} = 5.0 \times 10^{-11}$

34.   Calculate the $[H^+]$ $\qquad\qquad\qquad [H^+] = \dfrac{K_w}{[OH^-]}$

(a) $[OH^-] = 4.5 \times 10^{-2} \qquad [H^+] = \dfrac{1.0 \times 10^{-14}}{4.5 \times 10^{-2}} = 2.2 \times 10^{-13}$

(b) $[OH^-] = 5.2 \times 10^{-9} \qquad [H^+] = \dfrac{1.0 \times 10^{-14}}{5.2 \times 10^{-9}} = 1.9 \times 10^{-6}$

35.   The molar solubilities of the salts and their ions are indicated below the formulas in the equilibrium equations.

(a) $BaSO_4(s) \rightleftharpoons \qquad Ba^{2+} \quad + \quad SO_4^{2-}$
       $\qquad\qquad\qquad\qquad 3.9 \times 10^{-5} \qquad 3.9 \times 10^{-5}$

   $K_{sp} = [Ba^{2+}][SO_4^{2-}] \quad (3.9 \times 10^{-5})^2 = 1.5 \times 10^{-9}$

(b) $Ag_2CrO_4(s) \rightleftharpoons \qquad 2\,Ag^+ \quad + \quad CrO_4^{2-}$
       $\qquad\qquad\qquad\qquad 2(7.8 \times 10^{-5}) \qquad 7.8 \times 10^{-5}$

   $K_{sp} = [Ag^+]^2[CrO_4^{2-}] \quad (15.6 \times 10^{-5})^2 (7.8 \times 10^{-5}) = 1.9 \times 10^{-12}$

(c)   First change g/L $\longrightarrow$ mol/L

   $$\left(\frac{0.67 \text{ g CaSO}_4}{L}\right)\left(\frac{1 \text{ mol}}{136.1 \text{ g}}\right) = 4.9 \times 10^{-3} \, M \text{ CaSO}_4$$

   $CaSO_4(s) \rightleftharpoons \qquad Ca^{2+} \quad + \quad SO_4^{2-}$
   $\qquad\qquad\qquad\qquad 4.9 \times 10^{-3} \qquad 4.9 \times 10^{-3}$

   $K_{sp} = [Ca^{2+}]^2[SO_4^{2-}] = (4.9 \times 10^{-3})^2 = 2.4 \times 10^{-5}$

(d)   First change g/L $\longrightarrow$ mol/L

$$\left(\frac{0.0019 \text{ g AgCl}}{L}\right)\left(\frac{1 \text{ mol}}{143.4 \text{ g}}\right) = 1.3 \times 10^{-5} \, M \text{ AgCl}$$

$$
\begin{array}{ccccc}
\text{AgCl}(s) & \rightleftharpoons & \text{Ag}^+ & + & \text{Cl}^- \\
& & 1.3 \times 10^{-5} & & 1.3 \times 10^{-5}
\end{array}
$$

$$K_{sp} = [\text{Ag}^+][\text{Cl}^-] = \left(1.3 \times 10^{-5}\right)^2 = 1.7 \times 10^{-10}$$

36.   The molar solubilities of the salts and their ions are indicated below the formulas in the equilibrium equations.

(a)
$$
\begin{array}{ccccc}
\text{ZnS}(s) & \rightleftharpoons & \text{Zn}^{2+} & + & \text{S}^{2-} \\
& & 3.5 \times 10^{-12} & & 3.5 \times 10^{-12}
\end{array}
$$

$$K_{sp} = [\text{Zn}^{2+}][\text{S}^{2-}] - \left(3.5 \times 10^{-12}\right)^2 = 1.2 \times 10^{-23}$$

(b)
$$
\begin{array}{ccccc}
\text{Pb(IO}_3)_2(s) & \rightleftharpoons & \text{Pb}^{2+} & + & 2\,\text{IO}_3^- \\
& & 4.0 \times 10^{-5} & & 2\left(4.0 \times 10^{-5}\right)
\end{array}
$$

$$K_{sp} = \left[\text{Pb}^{2+}\right]\left[\text{IO}_3^-\right]^2 = \left(4.0 \times 10^{-5}\right)\left(8.0 \times 10^{-5}\right)^2 = 2.6 \times 10^{-13}$$

(c)   First change g/L $\longrightarrow$ mol/L

$$\left(\frac{6.73 \times 10^{-3} \text{ g Ag}_3\text{PO}_4}{L}\right)\left(\frac{1 \text{ mol}}{418.7 \text{ g}}\right) = 1.61 \times 10^{-5} \, M \text{ Ag}_3\text{PO}_4$$

$$
\begin{array}{ccccc}
\text{Ag}_3\text{PO}_4(s) & \rightleftharpoons & 3\,\text{Ag}^+ & + & \text{PO}_4^{3-} \\
& & 3\left(1.61 \times 10^{-5}\right) & & 1.61 \times 10^{-5}
\end{array}
$$

$$K_{sp} = [\text{Ag}^+]^3\left[\text{PO}_4^{3-}\right] = \left(4.83 \times 10^{-5}\right)^3\left(1.61 \times 10^{-5}\right) = 1.81 \times 10^{-18}$$

(d)   First change g/L $\longrightarrow$ mol/L

$$\left(\frac{2.33 \times 10^{-4} \text{ g Zn(OH)}_2}{L}\right)\left(\frac{1 \text{ mol}}{99.43 \text{ g}}\right) = 2.34 \times 10^{-6} \, M \text{ Zn(OH)}_2$$

$$
\begin{array}{ccccc}
\text{Zn(OH)}_2(s) & \rightleftharpoons & \text{Zn}^{2+} & + & 2\,\text{OH}^- \\
& & 2.34 \times 10^{-6} & & 2\left(2.34 \times 10^{-6}\right)
\end{array}
$$

$$K_{sp} = [\text{Zn}^{2+}][\text{OH}^-]^2 = \left(2.34 \times 10^{-6}\right)\left(4.68 \times 10^{-6}\right)^2 = 5.13 \times 10^{-17}$$

37.   The molar solubilities of the salts and their ions will be represented in terms of x below their formulas in the equilibrium equations.

(a)
$$
\begin{array}{ccccc}
\text{CaF}_2 & \rightleftharpoons & \text{Ca}^{2+} & + & 2\,\text{F}^- \\
& & x & & 2x
\end{array}
$$

$$K_{sp} = [\text{Ca}^{2+}][\text{F}^-]^2 = (x)(2x)^2 = 4x^3 = 3.9 \times 10^{-11}$$

$$x = \sqrt[3]{\frac{3.9 \times 10^{-11}}{4}} = 2.1 \times 10^{-4} \, M$$

(b) $\quad Fe(OH)_3 \quad \rightleftharpoons \quad Fe^{3+} \quad + \quad 3\,OH^-$
$$ x \qquad\qquad 3x$$

$$K_{sp} = [Fe^{3+}][OH^-]^3 = (x)(3x)^3 = 27x^4 = 6.1 \times 10^{-38}$$

$$x = \sqrt[4]{\frac{6.1 \times 10^{-38}}{27}} = 2.2 \times 10^{-10}\,M$$

38. The molar solubilities of the salts and their ions will be represented in terms of x below their formulas in the equilibrium equations.

(a) $\quad PbSO_4 \quad \rightleftharpoons \quad Pb^{2+} \quad + \quad SO_4^{2-}$
$$ x \qquad\quad x$$

$$K_{sp} = [Pb^{2+}][SO_4^{2-}] = (x)(x) = x^2 = 1.3 \times 10^{-8}$$

$$x = \sqrt{1.3 \times 10^{-8}} = 1.1 \times 10^{-4}\,M$$

(b) $\quad BaCrO_4 \quad \rightleftharpoons \quad Ba^{2+} \quad + \quad CrO_4^{2-}$
$$ x \qquad\quad x$$

$$K_{sp} = [Ba^{2+}][CrO_4^{2-}] = (x)(x) = x^2 = 8.5 \times 10^{-11}$$

$$x = \sqrt{8.5 \times 10^{-11}} = 9.2 \times 10^{-6}\,M$$

39. (a) $\quad \left(\dfrac{2.1 \times 10^{-4}\,mol\,CaF_2}{L}\right)(0.100\,L)\left(\dfrac{78.08\,g\,CaF_2}{mol}\right) = 1.6 \times 10^{-3}\,g\,CaF_2$

(b) $\quad \left(\dfrac{2.2 \times 10^{-10}\,mol\,Fe(OH)_3}{L}\right)(0.100\,L)\left(\dfrac{106.9\,Fe(OH)_3}{mol}\right) = 2.4 \times 10^{-9}\,g\,Fe(OH)_3$

40. (a) $\quad \left(\dfrac{1.1 \times 10^{-4}\,PbSO_4}{L}\right)(0.100\,L)\left(\dfrac{303.3\,g\,PbSO_4}{mol}\right) = 3.3 \times 10^{-3}\,g\,PbSO_4$

(b) $\quad \left(\dfrac{9.2 \times 10^{-6}\,mol\,BaCrO_4}{L}\right)(0.100\,L)\left(\dfrac{253.3\,g\,BaCrO_4}{mol}\right) = 2.3 \times 10^{-4}\,g\,BaCrO_4$

41. The molar concentrations of ions, after mixing, are calculated and these concentrations are substituted into the equilibrium expression. The value obtained is compared to the K_{sp} of the salt. If the value is greater than the K_{sp}, precipitation occurs. If the value is less than the K_{sp}, no precipitation occurs.

100. mL 0.010 M $Na_2SO_4 \longrightarrow$ 100. mL 0.010 M SO_4^{2-}

100. mL 0.001 M $Pb(NO_3)_2 \longrightarrow$ 100. mL 0.001 M Pb^{2+}

Volume after mixing = 200. mL
Concentrations after mixing: $SO_4^{2-} = 0.0050\,M \quad Pb^{2+} = 0.0005\,M$
$[Pb^{2+}][SO_4^{2-}] = (5.0 \times 10^{-3})(5 \times 10^{-4}) = 3 \times 10^{-6}$
$K_{sp} = 1.3 \times 10^{-8}$ which is less than 3×10^{-6}, therefore, precipitation occurs.

42. The molar concentrations of ions, after mixing, are calculated and these concentrations are substituted into the equilibrium expression. The value obtained is compared to the K_{sp} of the salt. If the value is greater than the K_{sp}, precipitation occurs. If the value is less than the K_{sp}, no precipitation occurs.

$50.0 \text{ mL } 1.0 \times 10^{-4} M \text{ AgNO}_3 \longrightarrow 50.0 \text{ mL } 1.0 \times 10^{-4} M \text{ Ag}^+$

$100. \text{ mL } 1.0 \times 10^{-4} M \text{ NaCl} \longrightarrow 100. \text{ mL } 1.0 \times 10^{-4} M \text{ Cl}^-$

Volume after mixing $= 150.$ mL

Concentrations after mixing:

$$\left(1.0 \times 10^{-4} M \text{ Ag}^+\right)\left(\frac{50.0 \text{ mL}}{150. \text{ mL}}\right) = 3.3 \times 10^{-5} M \text{ Ag}^+$$

$$\left(1.0 \times 10^{-4} M \text{ Cl}^-\right)\left(\frac{100. \text{ mL}}{150. \text{ mL}}\right) = 6.7 \times 10^{-5} M \text{ Cl}^-$$

$$[\text{Ag}^+][\text{Cl}^-] = \left(3.3 \times 10^{-5}\right)\left(6.7 \times 10^{-5}\right) = 2.2 \times 10^{-9}$$

2.2×10^{-9} is greater than the K_{sp} of 1.7×10^{-10} therefore, precipitation occurs.

43. The concentration of $\text{Br}^- = 0.10 M$ in 1.0 L of $0.10 M$ NaBr. Substitute this Br^- concentration in the K_{sp} expression and solve for the $[\text{Ag}^+]$ in equilibrium with $0.10 M \text{ Br}^-$.

$$K_{sp} = [\text{Ag}^+][\text{Br}^-] = 5.2 \times 10^{-13}$$

$$[\text{Ag}^+] = \frac{5.2 \times 10^{-13}}{[\text{Br}^-]} = \frac{5.2 \times 10^{-13}}{0.10} = 5.2 \times 10^{-12} M$$

$$\left(\frac{5.2 \times 10^{-12} \text{ mol Ag}^+}{\text{L}}\right)\left(\frac{1 \text{ mol AgBr}}{1 \text{ mol Ag}^+}\right)(1.0 \text{ L}) = 5.2 \times 10^{-12} \text{ mol AgBr will dissolve}$$

44. $$\left(\frac{0.10 \text{ mol MgBr}_2}{\text{L}}\right)\left(\frac{2 \text{ mol Br}^-}{1 \text{ mol MgBr}_2}\right) = \left(\frac{0.20 \text{ mol Br}^-}{\text{L}}\right) = 0.20 M \text{ Br}^- \text{ in solution.}$$

Substitute the Br^- concentration in the K_{sp} expression and solve for $[\text{Ag}^+]$ in equilibrium with $0.20 M \text{ Br}^-$.

$$[\text{Ag}^+] = \frac{5.2 \times 10^{-13}}{[\text{Br}^-]} = \frac{5.2 \times 10^{-13}}{0.20} = 2.6 \times 10^{-12} M$$

$$\left(\frac{2.6 \times 10^{-12} \text{ mol Ag}^+}{\text{L}}\right)\left(\frac{1 \text{ mol AgBr}}{1 \text{ mol Ag}^+}\right)(1.0 \text{ L}) = 2.6 \times 10^{-12} \text{ mol AgBr will dissolve}$$

45. $\text{HC}_2\text{H}_3\text{O}_2 \rightleftharpoons \text{H}^+ + \text{C}_2\text{H}_3\text{O}_2^-$

$$K_a = \frac{[\text{H}^+]\left[\text{C}_2\text{H}_3\text{O}_2^-\right]}{[\text{HC}_2\text{H}_3\text{O}_2]} = 1.8 \times 10^{-5}$$

$$[\text{H}^+] = K_a\left(\frac{[\text{HC}_2\text{H}_3\text{O}_2]}{\left[\text{C}_2\text{H}_3\text{O}_2^-\right]}\right) \qquad [\text{HC}_2\text{H}_3\text{O}_2] = 0.20 M \qquad \left[\text{C}_2\text{H}_3\text{O}_2^-\right] = 0.10 M$$

$$[H^+] = (1.8 \times 10^{-5})\left(\frac{0.20}{0.10}\right) = 3.6 \times 10^{-5}\,M$$

$$pH = -\log(3.6 \times 10^{-5}) = 4.44$$

46. $HC_2H_3O_2 \rightleftharpoons H^+ + C_2H_3O_2^-$

$$K_a = \frac{[H^+][C_2H_3O_2^-]}{[HC_2H_3O_2]} = 1.8 \times 10^{-5}$$

$$[H^+] = K_a\left(\frac{[HC_2H_3O_2]}{[C_2H_3O_2^-]}\right) \qquad [HC_2H_3O_2] = 0.20\,M \qquad [C_2H_3O_2^-] = 0.20\,M$$

$$[H^+] = (1.8 \times 10^{-5})\left(\frac{0.20}{0.20}\right) = 1.8 \times 10^{-5}\,M$$

$$pH = -\log(1.8 \times 10^{-5}) = 4.74$$

47. Initially, the solution of NaCl is neutral. $[H^+] = 1 \times 10^{-7}$

$$pH = -\log(1 \times 10^{-7}) = 7.0$$

Final $H^+ = 2.0 \times 10^{-2}\,M$

$$pH = -\log(2.0 \times 10^{-2}) = 1.70$$

Change in pH $= 7.0 - 1.70 = 5.3$ units in the unbuffered solution

48. Initially, $[H^+] = 1.8 \times 10^{-5}$ $\qquad [H^+] = \dfrac{K_a[HC_2H_3O_2]}{[C_2H_3O_2^-]}$

$$pH = -\log(1.8 \times 10^{-5}) = 4.74$$

Final $[H^+] = 1.9 \times 10^{-5}$

$$pH = -\log(1.9 \times 10^{-5}) = 4.72$$

Change in pH $= 4.74 - 4.72 = 0.02$ units in the buffered solution

49. The concentration of solid salt is not included in the K_{sp} equilibrium constant because the concentration of solid does not change. It is constant and part of the K_{sp}.

50. The energy diagram represents an exothermic reaction because the energy of the products is lower than the energy of the reactants. This means that energy was given off during the reaction.

51.

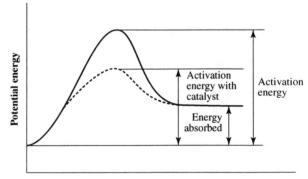

52. $H_2 + I_2 \rightleftharpoons 2\,HI$

The reaction is a 1 to 1 mole ratio of hydrogen to iodine. The data given indicates that hydrogen is the limiting reactant.

$$(2.10 \text{ mol } H_2)\left(\frac{2 \text{ mol } HI}{1 \text{ mol } H_2}\right) = 4.20 \text{ mol } HI$$

53. $H_2 + I_2 \rightleftharpoons 2\,HI$
    (a) 2.00 mol H_2 and 2.00 mol I_2 will produce 4.00 mol HI assuming 100% yield. However, at 79% yield you get
    4.00 mol HI × 0.79 = 3.16 mol HI
    (b) The addition of 0.27 mol I_2 makes the iodine present in excess and the 2.00 mol H_2 the limiting reactant. The yield increases to 85%.

$$(2.00 \text{ mol } H_2)\left(\frac{2 \text{ mol } HI}{1 \text{ mol } H_2}\right)(0.85) = 3.4 \text{ mol } HI$$

There will be 15% unreacted H_2 and I_2 plus the extra I_2 added.
$(0.15)(2.0 \text{ mol } H_2) = 0.30 \text{ mol } H_2$ present; also 0.30 mol I_2.
In addition to the 0.30 mol of unreacted I_2, will be the 0.27 mol I_2 added.
0.27 mol + 0.30 mol = 0.57 mol I_2 present.

(c) $K_{eq} = \dfrac{[HI]^2}{[H_2][I_2]}$

The formation of 3.16 mol HI required the reaction of 1.58 mol I_2 and 1.58 mol H_2. At equilibrium, the concentrations are:

3.16 mol HI; $2.00 - 1.58 = 0.42 \text{ mol } H_2 = 0.42 \text{ mol } I_2$

$$K_{eq} = \frac{(3.16)^2}{(0.42)(0.42)} = 57$$

In the calculation of the equilibrium constant, the actual number of moles of reactants and products present at equilibrium can be used in the calculation in place of molar concentrations. This occurs because the reaction is gaseous and the liters of HI produced equals the sum of the liters of H_2 and I_2 reacting. In the equilibrium expression, the volumes will cancel.

54. $H_2 + I_2 \rightleftharpoons 2\,HI$

$$(64.0 \text{ g HI})\left(\frac{1 \text{ mol}}{127.9 \text{ g}}\right) = 0.500 \text{ mol } HI \text{ present}$$

$$(0.500 \text{ mol } HI)\left(\frac{1 \text{ mol } I_2}{2 \text{ mol } HI}\right) = 0.250 \text{ mol } I_2 \text{ reacted}$$

$$(0.500 \text{ mol } HI)\left(\frac{1 \text{ mol } H_2}{2 \text{ mol } HI}\right) = 0.250 \text{ mol } H_2 \text{ reacted}$$

$$(6.00 \text{ g } H_2)\left(\frac{1 \text{ mol}}{2.016 \text{ g}}\right) = 2.98 \text{ mol } H_2 \text{ initially present}$$

$$(200. \text{ g } I_2)\left(\frac{1 \text{ mol}}{253.8 \text{ g}}\right) = 0.788 \text{ mol } I_2 \text{ initially present}$$

At equilibrium, moles present are:

0.500 mol HI; $2.98 - 0.250 = 2.73$ mol H_2

$0.788 - 0.250 = 0.538$ mol I_2

55. $PCl_3(g) + Cl_2(g) \rightleftharpoons PCl_5(g)$

$$K_{eq} = \frac{[PCl_5]}{[PCl_3][Cl_2]}$$

The concentrations are:

$$PCl_5 = \frac{0.22 \text{ mol}}{20. \text{ L}} = 0.011 \, M$$

$$PCl_3 = \frac{0.10 \text{ mol}}{20. \text{ L}} = 0.0050 \, M$$

$$Cl_2 = \frac{1.50 \text{ mol}}{20. \text{ L}} = 0.075 \, M$$

$$K_{eq} = \frac{0.011}{(0.0050)(0.075)} = 29$$

56. $100°C - 30°C = 70°C$ temperature increase. This increase is equal to seven $10°C$ increments. The reaction rate will be increased by $2^7 = 128$ times.

57. $NH_4^+ \rightleftharpoons NH_3 + H^+$ $K_{eq} = 5.6 \times 10^{-10}$

   0.30 M y y

$y = [H^+] = [NH_3]$

$$\frac{[NH_3][H^+]}{[NH_4^+]} = \frac{[y][y]}{[0.30]} = 5.6 \times 10^{-10}$$

$$y^2 = (5.6 \times 10^{-10})(0.30) = 1.7 \times 10^{-10}$$

$$y = [H^+] = 1.3 \times 10^{-5}$$

$$pH = -\log(1.3 \times 10^{-5}) = 4.89 \quad \text{(an acidic solution)}$$

58.   pH of an acetic acid-acetate buffer

$HC_2H_3O_2 \rightleftharpoons H^+ + C_2H_3O_2^-$

$$K_a = \frac{[H^+][C_2H_3O_2^-]}{[HC_2H_3O_2]} = \frac{[H^+][0.20]}{0.30} = 1.8 \times 10^{-5}$$

$$H^+ = \frac{(1.8 \times 10^{-5})(0.30)}{0.20} = 2.7 \times 10^{-5}$$

$$pH = -\log(2.7 \times 10^{-5}) = 4.57$$

59. Concentration of Ba^{2+} in solution

$$BaCl_2(aq) + Na_2CrO_4(aq) \rightleftharpoons BaCrO_4(s) + NaCl(aq)$$

$$BaCrO_4 \rightleftharpoons Ba^{2+} + CrO_4^{2-} \qquad K_{sp} = 8.5 \times 10^{-11}$$

Determine the moles of Ba^{2+} and CrO_4^{2-} in solution

$$\left(\frac{0.10 \text{ mol } Ba^{2+}}{L}\right)(0.050 \text{ L}) = 0.0050 \text{ mol } Ba^{2+}$$

$$\left(\frac{0.15 \text{ mol } CrO_4^{2-}}{L}\right)(0.050 \text{ L}) = 0.0075 \text{ mol } CrO_4^{2-}$$

Excess CrO_4^{2-} in solution $= 0.0025 \text{ mol}(2.5 \times 10^{-3})$ after 0.005 mol $BaCrO_4$ precipitate.
Concentration of CrO_4^{2-} in solution (total volume = 100 mL)

$$\frac{2.5 \times 10^{-3} \text{ mol } CrO_4^{2-}}{0.100 \text{ L}} = 2.5 \times 10^{-2} M \text{ } CrO_4^{2-}$$

Now using the K_{sp}, calculate the Ba^{2+} remaining in solution.

$$[Ba^{2+}][CrO_4^{2-}] = [Ba^{2+}][2.5 \times 10^{-2}] = 8.5 \times 10^{-11}$$

$$[Ba^{2+}] = \frac{8.5 \times 10^{-11}}{2.5 \times 10^{-2}} = 3.4 \times 10^{-9} \text{ mol/L}$$

60. **Hypochlorous acid** $\qquad HOCl \rightleftharpoons H^+ + OCl^-$

Equilibrium concentrations:

$$[H^+] - [OCl^-] = 5.9 \times 10^{-5} M$$

$$[HOCl] = 0.1 - 5.9 \times 10^{-5} = 0.10 M \text{ (neglecting } 5.9 \times 10^{-5})$$

$$K_a = \frac{[H^+][OCl^-]}{[HOCl]} = \frac{(5.9 \times 10^{-5})(5.9 \times 10^{-5})}{0.10} = 3.5 \times 10^{-8}$$

Propanoic acid $\qquad HC_3H_5O_2 \rightleftharpoons H^+ + C_3H_5O_2^-$

Equilibrium concentrations:

$$[H^+] = [C_3H_5O_2^-] = 1.4 \times 10^{-3} M$$

$$[HC_3H_5O_2] = 0.15 - 1.4 \times 10^{-3} = 0.15 M \text{ (neglecting } 1.4 \times 10^{-3})$$

$$K_a = \frac{[H^+][C_3H_5O_2^-]}{[HC_3H_5O_2]} = \frac{(1.4 \times 10^{-3})(1.4 \times 10^{-3})}{0.15} = 1.3 \times 10^{-5}$$

Hydrocyanic acid $\qquad HCN \rightleftharpoons H^+ + CN^-$

Equilibrium concentrations:

$$[H^+] = [CN^-] = 8.9 \times 10^{-6} M$$

$$[HCN] = 0.20 - 8.9 \times 10^{-6} = 0.20 M \text{ (neglecting } 8.9 \times 10^{-6})$$

$$K_a = \frac{[H^+][CN^-]}{[HCN]} = \frac{(8.9 \times 10^{-6})^2}{0.20} = 4.0 \times 10^{-10}$$

61. Let $y = M$ CaF$_2$ dissolved

$$CaF_2(s) \rightleftharpoons Ca^{2+} + 2F^-$$
$$\qquad\qquad\quad y \qquad 2y$$

y = molar solubility

(a) $K_{sp} = [Ca^{2+}][F^-]^2 = (y)(2y)^2 = 4y^3 = 3.9 \times 10^{-11}$

$$y = \sqrt[3]{\frac{3.9 \times 10^{-11}}{4}} = 2.1 \times 10^{-4}\, M\,(CaF_2 \text{ dissolved})$$

$$\left(\frac{2.1 \times 10^{-4}\text{ mol CaF}_2}{L}\right)\left(\frac{1\text{ mol Ca}^{2+}}{1\text{ mol CaF}_2}\right) = 2.1 \times 10^{-4}\, M\, Ca^{2+}$$

$$\left(\frac{2.1 \times 10^{-4}\text{ mol CaF}_2}{L}\right)\left(\frac{2\text{ mol F}^-}{1\text{ mol CaF}_2}\right) = 4.2 \times 10^{-4}\, M\, F^-$$

(b) $\left(\dfrac{2.1 \times 10^{-4}\text{ mol CaF}_2}{L}\right)(0.500\text{ L})\left(\dfrac{78.08\text{ g}}{\text{mol}}\right) = 8.2 \times 10^{-3}\text{ g CaF}_2$

62. The molar concentrations of ions, after mixing, are calculated and these concentrations are substituted into the equilibrium expression. The value obtained is compared to the K_{sp} of the salt. If the value is greater than the K_{sp}, precipitation occurs. If the value is less than the K_{sp}, no precipitation occurs.

(a) 100. mL 0.010 M Na$_2$SO$_4$ $\longrightarrow$ 100. mL 0.010 M SO$_4^{2-}$

100. mL 0.001 M Pb(NO$_3$)$_2$ $\longrightarrow$ 100. mL 0.001 M Pb^{2+}

Volume after mixing = 200. mL

Concentrations after mixing: $\quad$ SO$_4^{2-}$ = 0.0050 M $\quad$ Pb^{2+} = 0.0005 M

$$\left[Pb^{2+}\right]\left[SO_4^{2-}\right] = \left(5.0 \times 10^{-3}\right)\left(5 \times 10^{-4}\right) = 3 \times 10^{-6}$$

$K_{sp} = 1.3 \times 10^{-8}$ which is less than 3×10^{-6}, therefore, precipitation occurs.

(b) 50.0 mL $1.0 \times 10^{-4}\, M$ AgNO$_3$ $\longrightarrow$ 50.0 mL $1.0 \times 10^{-4}\, M$ Ag$^+$

100. mL $1.0 \times 10^{-4}\, M$ NaCl $\longrightarrow$ 100. mL $1.0 \times 10^{-4}\, M$ Cl$^-$

Volume after mixing = 150. mL

Concentrations after mixing:

$$\left(1.0 \times 10^{-4}\right)\left(\frac{50.0\text{ mL}}{150.\text{ mL}}\right) = 3.3 \times 10^{-5}\, M\, Ag^+$$

$$\left(1.0 \times 10^{-4}\right)\left(\frac{100.\text{ mL}}{150.\text{ mL}}\right) = 6.7 \times 10^{-5}\, M\, Cl^-$$

$$[Ag^+][Cl^-] = \left(3.3 \times 10^{-5}\right)\left(6.7 \times 10^{-5}\right) = 2.2 \times 10^{-9}$$

$K_{sp} = 1.7 \times 10^{-10}$ which is less than 2.2×10^{-9}, therefore, precipitation occurs.

(c) Convert g $Ca(NO_3)_2$ to g Ca^{2+}

$$\left(\frac{1.0 \text{ g Ca(NO}_3)_2}{0.150 \text{ L}}\right)\left(\frac{1 \text{ mol}}{164.1 \text{ g}}\right)\left(\frac{1 \text{ mol Ca}^{2+}}{1 \text{ mol Ca(NO}_3)_2}\right) = 0.041 \ M \ Ca^{2+}$$

250 mL 0.01 M NaOH $\longrightarrow$ 250 mL 0.01 M OH^-

Final volume $= 4.0 \times 10^2$ mL

Concentration after mixing:

$$(0.041 \ M \ Ca^{2+})\left(\frac{150 \text{ mL}}{4.0 \times 10^2 \text{ mL}}\right) = 0.015 \ M \ Ca^{2+}$$

$$(0.01 \ M \ OH^-)\left(\frac{250 \text{ mL}}{4.0 \times 10^2 \text{ mL}}\right) = 0.0063 \ M \ OH^-$$

$[Ca^{2+}][OH^-]^2 = (0.015)(0.0063)^2 = 6.0 \times 10^{-7}$

$K_{sp} = 1.3 \times 10^{-6}$ which is greater than 6.7×10^{-7}, therefore, no precipitation occurs.

63. With a known Ba^{2+} concentration, the SO_4^{2-} concentration can be calculated using the K_{sp} value.

$BaSO_4(s) \rightleftharpoons Ba^{2+} + SO_4^{2-}$

$K_{sp} = [Ba^{2+}][SO_4^{2}] = 1.5 \times 10^{-9} \qquad Ba^{2+} = 0.050 \ M$

(a) $[SO_4^{2-}] = \dfrac{K_{sp}}{[Ba^{2+}]} = \dfrac{1.5 \times 10^{-9}}{0.050} = 3.0 \times 10^{-8} \ M \ SO_4^{2-}$ in solution

(b) $M \ SO_4^{2-} = M \ BaSO_4$ in solution

$$\left(\frac{3.0 \times 10^{-8} \text{ mol BaSO}_4}{L}\right)(0.100 \text{ L})\left(\frac{233.4 \text{ g}}{\text{mol}}\right)$$

$= 7.0 \times 10^{-7}$ g $BaSO_4$ remain in solution

64. If $[Pb^{2+}][Cl^-]^2$ exceeds the K_{sp}, precipitation will occur.

$K_{sp} = [Pb^{2+}][Cl^-]^2 = 2.0 \times 10^{-5}$

$0.050 \ M \ Pb(NO_3)_2 \longrightarrow 0.050 \ M \ Pb^{2+}$

$0.010 \ M \ NaCl \longrightarrow 0.010 \ M \ Cl^-$

$(0.050)(0.010)^2 = 5.0 \times 10^{-6}$

$[Pb^{2+}][Cl^-]^2$ is smaller than the K_{sp} value. Therefore, no precipitate of $PbCl_2$ will form.

65. $[Ba^{2+}][SO_4^{2-}] = 1.5 \times 10^{-9} \qquad [Sr^2][SO_4^{2-}] = 3.5 \times 10^{-7}$

Both cations are present in equal concentrations (0.10 M). Therefore, as SO_4^{2-} is added, the K_{sp} of $BaSO_4$ will be exceeded before that of $SrSO_4$. $BaSO_4$ precipitates first.

66. $2 SO_2(g) + O_2(g) \rightleftharpoons 2 SO_3(g)$

$$K_{eq} = \frac{[SO_3]^2}{[SO_2]^2[O_2]} = \frac{(11.0)^2}{(4.20)^2(0.60 \times 10^{-3})} = 1.1 \times 10^4$$

67. $(0.048 \text{ g BaF}_2)\left(\dfrac{1 \text{ mol}}{175.3 \text{ g}}\right) = 2.7 \times 10^{-4} \text{ mol BaF}_2$

$\left(\dfrac{2.7 \times 10^{-4} \text{ mol}}{0.015 \text{ L}}\right) = 1.8 \times 10^{-2} \, M \text{ BaF}_2 \text{ dissolved}$

$\text{BaF}_2(s) \rightleftharpoons \quad \text{Ba}^{2+} \quad + \quad 2\,\text{F}^-$

$\qquad\qquad\qquad 1.8 \times 10^{-2} \quad 2(1.8 \times 10^{-2}) \qquad \text{(molar concentration)}$

$K_{sp} = [\text{Ba}^{2+}][\text{F}^-]^2 = (1.8 \times 10^{-2})(3.6 \times 10^{-2})^2 = 2.3 \times 10^{-5}$

68. $\text{N}_2 + 3\,\text{H}_2 \rightleftharpoons 2\,\text{NH}_3$

$K_{eq} = \dfrac{[\text{NH}_3]^2}{[\text{N}_2][\text{H}_2]^3} = 4.0 \qquad \text{Let } y = [\text{NH}_3]$

$4.0 = \dfrac{y^2}{(2.0)(2.0)^3} \qquad y^2 = 64 \qquad y = \sqrt{64}$

$y = 8.0 \, M = [\text{NH}_3]$

69. Total volume of mixture $= 40.0 \text{ mL } (0.0400 \text{ L})$

$K_{sp} = [\text{Sr}^{2+}][\text{SO}_4^{2-}] = 7.6 \times 10^{-7}$

$[\text{Sr}^{2+}] = \dfrac{(1.0 \times 10^{-3} M)(0.0250 \text{ L})}{0.0400 \text{ L}} = 6.3 \times 10^{-4} M$

$[\text{SO}_4^{2-}] = \dfrac{(2.0 \times 10^{-3} M)(0.0150 \text{ L})}{0.0400 \text{ L}} = 7.5 \times 10^{-4} M$

$[\text{Sr}^{2+}][\text{SO}_4^{2-}] = (6.3 \times 10^{-4})(7.5 \times 10^{-4}) = 4.7 \times 10^{-7}$

$4.7 \times 10^{-7} < 7.6 \times 10^{-7}$ no precipitation should occur.

70. First change g $\text{Hg}_2\text{I}_2 \longrightarrow$ mol Hg_2I_2

$\left(\dfrac{3.04 \times 10^{-7} \text{g Hg}_2\text{I}_2}{\text{L}}\right)\left(\dfrac{1 \text{ mol}}{655.0 \text{ g}}\right) = 4.64 \times 10^{-10} M \text{ Hg}_2\text{I}_2 \qquad \text{(molar solubility)}$

$\text{Hg}_2\text{I}_2 \rightleftharpoons \text{Hg}_2^{2+} \quad + \quad 2\,\text{I}^-$

$\qquad 4.64 \times 10^{-10} M \quad 2(4.64 \times 10^{-10} M)$

$K_{sp} = [\text{Hg}_2^{2+}][\text{I}^-]^2 = (4.64 \times 10^{-10})(9.28 \times 10^{-10})^2 = 4.00 \times 10^{-28}$

71. $3\,\text{O}_2(g) + \text{heat} \rightleftharpoons 2\,\text{O}_3(g)$

Three ways to increase ozone
(a) increase heat
(b) increase amount of O_2
(c) increase pressure
(d) remove O_3 as it is made

72. $H_2O(l) \rightleftharpoons H_2O(g)$

    Conditions on the second day
    (a)   the temperature could have been cooler
    (b)   the humidity in the air could have been higher
    (c)   the air pressure could have been greater

73. $CO(g) + H_2O(g) \rightleftharpoons CO_2(g) + H_2(g)$

    (c)   is the correct answer

    $$K_{eq} = \frac{[CO_2][H_2]}{[CO][H_2O]} = 1$$

    With equal concentrations of products and reactants, the K_{eq} value will equal 1.

74. (a) $K_{eq} = \dfrac{[O_3]^2}{[O_2]^3}$ (c) $K_{eq} = [CO_2(g)]$

    (b) $K_{eq} = \dfrac{[H_2O(l)]}{[H_2O(g)]}$ (d) $K_{eq} = \dfrac{[H^+]^6}{[Bi^{3+}]^2[H_2S]^3}$

75.
2A	+	B	$\rightleftharpoons$	C	
1.0 M		1.0 M		0	Initial conditions
1.0 − 2(0.30)		1.0 − 0.30		0.30	Equilibrium concentrations
0.4 M		0.7 M		0.30 M	

$$K_{eq} = \frac{[C]}{[A]^2[B]} = \frac{0.30}{(0.4)^2(0.7)} = 3$$

76. Since the second reaction is the reverse of the first, the K_{eq} value of the second reaction will be the reciprocal of the K_{eq} value of the first reaction.

    $$K_{eq} = \frac{[I_2][Cl_2]}{[ICl]^2} = 2.2 \times 10^{-3} \quad \text{(first reaction)}$$

    $$K_{eq} = \frac{[ICl]^2}{[I_2][Cl_2]} \qquad K_{eq} = \frac{1}{2.2 \times 10^{-3}} = 450$$

77. $HNO_2(aq) \rightleftharpoons H^+(aq) + NO_2^-(aq)$

    OH^- reacts with H^+ and equilibrium shifts to the right.

    (a)   After an initial increase, $[OH^-]$ will be neutralized and equilibrium shifts to the right.
    (b) $[H^+]$ will be reduced (reacts with OH^-). Equilibrium shifts to the right.
    (c) $[NO_2^-]$ increases as equilibrium shifts to the right.
    (d) $[HNO_2]$ decreases and equilibrium shifts to the right.

78. $CaSO_4(s) \rightleftharpoons Ca^{2+}(aq) + SO_4^{2-}(aq)$

$K_{sp} = [Ca^{2+}][SO_4^{2-}] = 2.0 \times 10^{-4}$

Let x = moles $CaSO_4$ that dissolve per L = $[Ca^{2+}]$ = $[SO_4^{2-}]$

$(x)(x) = 2.0 \times 10^{-4}$ $x = \sqrt{2.0 \times 10^{-4}}$

$x = 0.014\ M\ CaSO_4$

$M \longrightarrow$ moles $\longrightarrow$ grams

$\left(\dfrac{0.014\ \text{mol}\ CaSO_4}{L}\right)(0.600\ L)\left(\dfrac{136.2\ g}{\text{mol}}\right) = 1.1\ g\ CaSO_4$

79. $PbF_2(s) \rightleftharpoons Pb^{2+} + 2\,F^-$

change g $PbF_2 \longrightarrow$ mol PbF_2

$\left(\dfrac{0.098\ g\ PbF_2}{0.400\ L}\right)\left(\dfrac{1\ \text{mol}}{245.2\ g}\right) = 1.0 \times 10^{-3}\ \text{mol/L} = 1.0 \times 10^{-3}\ M\ PbF_2$

$K_{sp} = (Pb^{2+})(F^-)^2$

$[Pb^2] = 1.0 \times 10^{-3}$; $[F^-] = 2(1.0 \times 10^{-3}) = 2.0 \times 10^{-3}$

$K_{sp} = (1.0 \times 10^{-3})(2.0 \times 10^{-3})^2 = 4.0 \times 10^{-9}$

80. Equilibrium shifts

$Co(H_2O)_6{}^{2+} + 4\,Cl^- \rightleftharpoons [CoCl_4]^{2-} + 6\,H_2O$

(a) Equilibrium would shift to the right and humidity indicator would turn blue
(b) Equilibrium would shift to the left and humidity indicator would remain pink.
(c) Equilibrium would shift to the left and humidity indicator would remain pink.
(d) Equilibrium would shift to the right and humidity indicator would turn blue.

81. Equilibrium shifts

$HCOOH(g) \rightleftharpoons CO(g) + H_2O(g)$

(a) Reaction will shift to the right
(b) Reaction will shift to the left
(c) Reaction will shift to the right
(d) Reaction will shift to the left

82. $C_7H_6O_3 + C_4H_6O_3 \rightleftharpoons C_9H_8O_4 + C_2H_4O_2$

Molar cost of salicylic acid

$$\left(\frac{\$58.90}{500 \text{ g } C_7H_6O_3}\right)\left(\frac{138.1 \text{ g } C_7H_6O_3}{1 \text{ mol } C_7H_6O_3}\right) = \frac{\$16.27}{\text{mol } C_7H_6O_3}$$

Molar cost of acetic anhydride

$$\left(\frac{\$53.50}{1000 \text{ g } C_4H_6O_3}\right)\left(\frac{102.1 \text{ g } C_4H_6O_3}{1 \text{ mol } C_4H_6O_3}\right) = \frac{\$5.46}{\text{mol } C_4H_6O_3}$$

It is more economical to use the acetic anhydride to drive the reaction because it has a lower cost per mole.

83. (a) $HCO_3^-(aq) + H^+(aq) \rightleftharpoons H_2CO_3(aq)$
     (b) $HCO_3^-(aq) + OH^-(aq) \rightleftharpoons H_2O(l) + CO_3^{2-}(aq)$

84. Treat this is an equilibrium where W = whole nuts, S = shell halves, and K = kernels

W	$\rightleftharpoons$	2 S	+	K	
144		0		0	amount before cracking
144 − x		2x		x	x = number of kernels after cracking

$144 - x + 2x + x = 194$ total pieces

$144 + 2x = 194; \quad 2x = 50$

$x - 25$ kernels; 50 shell halves; 119 whole nuts left

$$K_{eq} = \frac{(2x)^2(x)}{144 - x} = \frac{(50)^2(25)}{119} = 5.3 \times 10^2$$

85.
$SO_2(g)$	+	$NO_2(g)$	$\rightleftharpoons$	$SO_3(g)$	+	$NO(g)$	
0.50 M		0.50 M		0		0	Initial conditions
0.50 − x		0.50 − x		x		x	Equilibrium concentrations

$$K_{eq} = \frac{[SO_3][NO]}{[SO_2][NO_2]} = \frac{x^2}{(0.50 - x)^2} = 81$$

Take the square root of both sides

$$\frac{x}{0.50 - x} = 9.0 \qquad x = 0.45 \, M$$

$[SO_3] = [NO] = 0.45 \, M$

$[SO_2] = [NO_2] = 0.05 \, M$

CHAPTER 17

OXIDATION-REDUCTION

SOLUTIONS TO REVIEW QUESTIONS

1. (a) Iodine is oxidized. Its oxidation number increases from 0 to +5.
   (b) Chlorine is reduced. Its oxidation number decreases from 0 to −1.

2. The oxidation number for an atom in an ionic compound is the same as the charge of the ion that resulted when that atom lost or gained electrons to form an ionic bond. In a covalently bonded compound electrons are shared between the two atoms making up the bond. Those shared electrons are assigned to the atom in the bond with a higher electronegativity giving it a negative oxidation number.

3. Oxidation and reduction are complementary processes because one does not occur without the other. The loss of e^- in oxidation is accompanied by a gain of e^- in reduction.

4. Redox reactions are usually very difficult to balance by inspection so other more efficient methods are utilized to balance them.

5. $Mn^{2+} \rightarrow Mn^{+6}$ 4 electrons are needed on the right-hand side of the equation.

6. $O^0 \rightarrow O^{2-}$ 2 electrons are needed on the left-hand side of the equation.

7. Break the reaction into half reactions; one for the oxidation reaction and one for the reduction reaction.

8. $2 IO_3^- + 12 H^+ + 10 e^- \rightarrow I_2 + 6 H_2O$

   To balance this equation in base you would add $12 OH^-$ ions to both sides of the reaction. On the left side, these OH^-s would combine with the H^+ to form water. Cancel 6 H_2O from each side of the equation.

   $2 IO_3^- + 6 H_2O + 10 e^- \rightarrow I_2 + 12 OH^-$

9. Oxidation of a metal occurs when the metal loses electrons. The easier it is for a metal to lose electrons, the more active the metal is.

10. The higher metal on the activity series list is more reactive.
    (a) Ca   (b) Fe   (c) Zn

11. If the free element is higher on the activity series than the ion with which it is paired, the reaction occurs. The higher metal on the activity series is more reactive.
    (a) $2 Al(s) + 3 ZnCl_2(aq) \rightarrow 3 Zn(s) + 2 AlCl_3(aq)$
    (b) $Sn(s) + 2 HCl(aq) \rightarrow H_2(g) + SnCl_2(aq)$
    (c) $Ag(s) + H_2SO_4(aq) \rightarrow$ no reaction H_2 is more reactive than Ag
    (d) $Fe(s) + 2 AgNO_3(aq) \rightarrow 2 Ag(s) + Fe(NO_3)_2(aq)$
    (e) $2 Cr(s) + 3 Ni^{2+}(aq) \rightarrow 3 Ni(s) + 2 Cr^{3+}(aq)$
    (f) $Mg(s) + Ca^{2+}(aq) \rightarrow$ no reaction   Ca is more reactive than Mg
    (g) $Cu(s) + H^+(aq) \rightarrow$ no reaction H_2 is more reactive than Cu
    (h) $Ag(s) + Al^{3+}(aq) \rightarrow$ no reaction   Al is more reactive than Ag

12. If a copper wire is placed into a solution of lead (II) nitrate, no reaction will occur. Lead is more active than copper and undergoes oxidation more readily than copper. Lead is already oxidized, therefore will stay oxidized in the presence of copper.

13. (a) $2\,Al + Fe_2O_3 \longrightarrow Al_2O_3 + 2\,Fe + Heat$
    (b) Al is above Fe in the activity series, which indicates Al is more active than Fe.
    (c) No. Iron is less active than aluminum and will not displace aluminum from its compounds.
    (d) Yes. Aluminum is above chromium in the activity series and will displace Cr^{3+} from its compounds.

14. (a) $2\,Al(s) + 6\,HCl(aq) \longrightarrow 2\,AlCl_3(aq) + 3\,H_2(g)$
       $2\,Al(s) + 3\,H_2SO_4(aq) \longrightarrow Al_2(SO_4)_3(aq) + 3\,H_2(g)$
    (b) $2\,Cr(s) + 6\,HCl(aq) \longrightarrow 2\,CrCl_3(aq) + 3\,H_2(g)$
       $2\,Cr(s) + 3\,H_2SO_4(aq) \longrightarrow Cr_2(SO_4)_3(aq) + 3\,H_2(g)$
    (c) $Au(s) + HCl(aq) \longrightarrow$ no reaction
       $Au(s) + H_2SO_4(aq) \longrightarrow$ no reaction
    (d) $Fe(s) + 2\,HCl(aq) \longrightarrow FeCl_2(aq) + H_2(g)$
       $Fe(s) + H_2SO_4(aq) \longrightarrow FeSO_4(aq) + H_2(g)$
    (e) $Cu(s) + HCl(aq) \longrightarrow$ no reaction
       $Cu(s) + H_2SO_4(aq) \longrightarrow$ no reaction
    (f) $Mg(s) + 2\,HCl(aq) \longrightarrow MgCl_2(aq) + H_2(g)$
       $Mg(s) + H_2SO_4(aq) \longrightarrow MgSO_4(aq) + H_2(g)$
    (g) $Hg(l) + HCl(aq) \longrightarrow$ no reaction
       $Hg(l) + H_2SO_4(aq) \longrightarrow$ no reaction
    (h) $Zn(s) + 2HCl(aq) \longrightarrow ZnCl_2(aq) + H_2(g)$
       $Zn(s) + H_2SO_4(aq) \longrightarrow ZnSO_4(aq) + H_2(g)$

15. In an electrolytic cell the anode is the positively charged electrode and attracts negatively charged ions (anions). The cathode is the negatively charged electrode and attracts positively charge ions (cations). In a voltaic cell the anode is the negatively changed electrode where oxidation occurs. The cathode is the positively charged electrode where reduction occurs.

16. (a) Oxidation occurs at the anode. The reaction is

    $$2\,Cl^-(aq) \longrightarrow Cl_2(g) + 2\,e^-$$

    (b) Reduction occurs at the cathode. The reaction is

    $$Ni^{2+}(aq) + 2\,e^- \longrightarrow Ni(s)$$

    (c) The net chemical reaction is

    $$Ni^{2+}(aq) + 2\,Cl^-(aq) \xrightarrow[\text{energy}]{\text{electrical}} Ni(s) + Cl_2(g)$$

17. In Figure 17.4, electrical energy is causing chemical reactions to occur. In Figure 17.5, chemical reactions are used to produce electrical energy.

18. (a) It would not be possible to monitor the voltage produced, but the reactions in the cell would still occur.
    (b) If the salt bridge were removed, the reaction would stop. Ions must be mobile to maintain an electrical neutrality of ions in solution. The two solutions would be isolated with no complete electrical circuit.

19. $Ca^{2+} + 2e^- \longrightarrow Ca$ cathode reaction, reduction
     $2\,Br^- \longrightarrow Br_2 + 2\,e^-$ anode reaction, oxidation

20. During electroplating of metals, the metal is plated by reducing the positive ions of the metal in the solution. The plating will occur at the cathode, the source of the electrons. With an alternating current, the polarity of the electrode would be constantly changing, so at one instant the metal would be plating and the next instant the metal would be dissolving.

21. Since lead dioxide and lead(II) sulfate are insoluble, it is unnecessary to have salt bridges in the cells of a lead storage battery.

22. The electrolyte in a lead storage battery is dilute sulfuric acid. In the discharge cycle, SO_4^{2-}, is removed from solution as it reacts with PbO_2 and H^+ to form $PbSO_4(s)$ and H_2O. Therefore, the electrolyte solution contains less H_2SO_4 and becomes less dense.

23. If Hg^{2+} ions are reduced to metallic mercury, this would occur at the cathode, because reduction takes place at the cathode.

24. In both electrolytic and voltaic cells, oxidation and reduction reactions occur. In an electrolytic cell an electric current is forced through the cell causing a chemical change to occur. In voltaic cells, spontaneous chemical changes occur, generating an electric current.

25. In some voltaic cells, the reactants at the electrodes are in solution. For the cell to function, these reactants must be kept separated. A salt bridge permits movement of ions in the cell. This keeps the solution neutral with respect to the charged particles (ions) in the solution.

SOLUTIONS TO EXERCISES

1. Oxidation numbers of each element in the compound
   (a) $CuCO_3$ $\quad$ $Cu = +2$ $\quad$ $C = +4$ $\quad$ $O = -2$
   (b) CH_4 $\quad\quad$ $C = -4$ $\quad$ $H = +1$
   (c) IF $\quad\quad\quad$ $I = +1$ $\quad\;\;$ $F = -1$
   (d) CH_2Cl_2 $\quad\;$ $C = 0$ $\quad\;\;$ $H = +1$ $\quad$ $Cl = -1$
   (e) SO_2 $\quad\quad$ $S = +4$ $\quad$ $O = -2$
   (f) Rb_2CrO_4 $\quad$ $Rb = +1$ $\quad$ $Cr = +6$ $\quad$ $O = -2$

2. Oxidation numbers of each element in the compound.
   (a) CHF_3 $\quad\quad$ $C = +2$ $\quad$ $H = +1$ $\quad$ $F = -1$
   (b) P_2O_5 $\quad\quad$ $P = +5$ $\quad$ $O = -2$
   (c) SF_6 $\quad\quad\;$ $S = +6$ $\quad$ $F = -1$
   (d) $SnSO_4$ $\quad\;$ $Sn = +2$ $\quad$ $S = +6$ $\quad$ $O = -2$
   (e) CH_3OH $\quad$ $C = -2$ $\quad$ $H = +1$ $\quad$ $O = -2$
   (f) H_3PO_4 $\quad$ $H = +1$ $\quad$ $P = +5$ $\quad$ $O = -2$

3. The oxidation number of the underlined element is indicated by the number following the formula.
   (a) $\underline{P}O_3{}^{3-} \;+3$ $\quad\quad$ (c) $NaH\underline{C}O_3 + 4$
   (b) $Ca\underline{S}O_4 + 6$ $\quad\quad$ (d) $\underline{Br}O_4{}^- \;+7$

4. The oxidation number of the underlined element is indicated by the number following the formula.
   (a) $\underline{C}O_3{}^{2-} \;+4$ $\quad\quad$ (c) $NaH_2\underline{P}O_4 + 5$
   (b) $H_2\underline{S}O_4 + 6$ $\quad\quad$ (d) $\underline{Cr}_2O_7{}^{2-} \;+6$

5. (a) Na_2CrO_4 $\quad\quad\quad\quad\quad\quad\quad$ $Cr + 6$
   (b) $K_3[Fe(CN)_6]$ $\quad\quad\quad\quad\quad$ $Fe + 3$
   (c) $CoCl_2$ $\quad\quad\quad\quad\quad\quad\quad$ $Co + 2; Cl - 1$
   (d) $NiCl_2 \cdot 6H_2O$ $\quad\quad\quad\quad\;$ $Ni + 2$

6. (a) $CuSO_4 \cdot 5H_2O$ $\quad\quad\quad\quad$ $Cu + 2; S + 6$
   (b) $KMnO_4$ $\quad\quad\quad\quad\quad\quad$ $Mn + 7$
   (c) MnO_2 $\quad\quad\quad\quad\quad\quad\;$ $Mn + 4$
   (d) $K_2Cr_2O_7$ $\quad\quad\quad\quad\quad\;$ $Cr + 6$

7. Balanced half-reaction

	Element Changing	Type of Reaction
(a) $Na \rightarrow Na^+ + 1e^-$	Na	oxidation
(b) $C_2O_4{}^{2-} \rightarrow 2CO_2 + 2e^-$	C	oxidation
(c) $2I^- \rightarrow I_2 + 2e^-$	I	oxidation
(d) $Cr_2O_7{}^{2-} + 14H^+ + 6e^- \rightarrow 2Cr^{3+} + 7H_2O$	Cr	reduction

8. Balanced half-reaction

	Element Changing	Type of Reaction
(a) $Cu^{2+} + 1e^- \rightarrow Cu^+$	Cu	reduction
(b) $F_2 + 2e^- \rightarrow 2F^-$	F	reduction
(c) $2IO_4{}^- + 16H^+ + 14e^- \rightarrow I_2 + 8H_2O$	I	reduction
(d) $Mn \rightarrow Mn^{2+} + 2e^-$	Mn	oxidation

9. (a) Cu is oxidized, Ag is reduced;
      Cu is the reducing agent, $AgNO_3$ is the oxidizing agent
  (b) Zn is oxidized, H is reduced
      Zn is the reducing agent, HCl is the oxidizing agent

10. (a) C is oxidized, O is reduced
      CH_4 is the reducing agent, O_2 is the oxidizing agent
  (b) Mg is oxidized, Fe is reduced
      Mg is the reducing agent, $FeCl_3$ is the oxidizing agent

11. (a) correctly balanced
  (b) correctly balanced
  (c) incorrectly balanced

$$Mg(s) + 2\,HCl(aq) \longrightarrow Mg^{2+}(aq) + 2\,Cl^-(aq) + H_2(g)$$

  (d) incorrectly balanced

$$3\,CH_3OH(aq) + Cr_2O_7^{2-}(aq) + 8\,H^+(aq) \longrightarrow 2\,Cr^{3+}(aq) + 3\,CH_2O(aq) + 7\,H_2O(l)$$

12. (a) incorrectly balanced

$$3\,MnO_2(s) + 4\,Al(s) \longrightarrow 3\,Mn(s) + 2\,Al_2O_3(s)$$

  (b) correctly balanced
  (c) correctly balanced
  (d) incorrectly balanced

$$8\,H_2O(l) + 2\,MnO_4^-(aq) + 7\,S^{2-}(aq) \longrightarrow 2\,MnS(s) + 16\,OH^-(aq) + 5\,S(s)$$

13. Balancing oxidation-reduction equations using the change-in-oxidation number method:
  (a) $Cu + O_2 \rightarrow CuO$

    ox $Cu^0 \rightarrow Cu^{2+} + 2\,e^-$ Multiply by 2;
    red $\underline{O_2{}^0 + 4\,e^- \rightarrow 2\,O^{2-}}$ Add the equations; the $4\,e^-$ cancel

    $2\,Cu^0 + O_2{}^0 \rightarrow 2\,Cu^{2+} + 2\,O^{2-}$

    Transfer the coefficients to the original equation appropriately.

    $2\,Cu + O_2 \rightarrow 2\,CuO$

  (b) $KClO_3 \rightarrow KCl + O_2$

    ox $2\,O^{2-} \rightarrow O_2{}^0 + 4\,e^-$ Multiply by 3;
    red $\underline{Cl^{5+} + 6\,e^- \rightarrow Cl^-}$ Multiply by 2, add, the $12\,e^-$ cancel

    $2\,Cl^{5+} + 6\,O^{2-} \rightarrow 2\,Cl^- + 3\,O_2{}^0$

    Transfer the coefficients to the original equation appropriately.

    $2\,KClO_3 \rightarrow 2\,KCl + 3\,O_2$

  (c) $Ca + H_2O \rightarrow Ca(OH)_2 + H_2$

    ox $Ca \rightarrow Ca^{2+} + 2\,e^-$
    red $\underline{2\,H^+ + 2\,e^- \rightarrow H_2}$ Add equations together; the $2\,e^-$ cancel.

    $Ca + 2\,H^+ \rightarrow Ca^{2+} + H_2$

Balance the equation by inspection.

$$Ca + 2\,H_2O \rightarrow Ca(OH)_2 + H_2$$

(d) $PbS + H_2O_2 \rightarrow PbSO_4 + H_2O$

    ox $S^{2-} \rightarrow S^{6+} + 8\,e^-$

    red $\underline{2\,O^- + 2\,e^- \rightarrow 2\,O^{2-}}$ Multiply by 4, add, the $8\,e^-$ cancel.

                $S^{2-} + 8\,O^- \rightarrow S^{6+} + 8\,O^{2-}$

Transfer the coefficients to the original equation and complete the balancing by inspection.

$$PbS + 4\,H_2O_2 \rightarrow PbSO_4 + 4\,H_2O$$

(e) $CH_4 + NO_2 \rightarrow N_2 + CO_2 + H_2O$

    ox $C^{4-} \rightarrow C^{4+} + 8\,e^-$

    red $\underline{2\,N^{4+} + 8\,e^- \rightarrow N_2^{\,0}}$ Add equations together; the $8\,e^-$ cancel.

                $C^{4-} + 2\,N^{4+} \rightarrow C^{4+} + N_2^{\,0}$

Transfer the coefficients to the original equation and complete the balancing by inspection.

$$CH_4 + 2\,NO_2 \rightarrow N_2 + CO_2 + 2\,H_2O$$

14. Balancing oxidation-reduction equations using the change-in-oxidation number method:

(a) $Cu + AgNO_3 \rightarrow Ag + Cu(NO_3)_2$

    ox $Cu^0 \rightarrow Cu^{2+} + 2\,e^-$

    red $\underline{Ag^+ + 1\,e^- \rightarrow Ag^0}$ Multiply by 2, add, the $2\,e^-$ cancel

                $Cu^0 + 2\,Ag^+ \rightarrow Cu^{2+} + 2\,Ag^0$

Transfer the coefficients to the original equation.

$$Cu + 2\,AgNO_3 \rightarrow 2\,Ag + Cu(NO_3)_2$$

(b) $MnO_2 + HCl \rightarrow MnCl_2 + Cl_2 + H_2O$

    ox $2\,Cl^- \rightarrow Cl_2^{\,0} + 2\,e^-$ Add equations together; the $2\,e^-$ cancel

    red $\underline{Mn^{4+} + 2\,e^- \rightarrow Mn^{2+}}$

                $2\,Cl^- + Mn^{4+} \rightarrow Cl_2^{\,0} + Mn^{2+}$

Transfer the coefficients to the original equation and complete the balancing by inspection.

$$MnO_2 + 4\,HCl \rightarrow MnCl_2 + Cl_2 + 2\,H_2O$$

(c) $HCl + O_2 \rightarrow Cl_2 + H_2O$

    red $O_2^{\,0} + 4\,e^- \rightarrow 2\,O^{2-}$

    ox $\underline{2\,Cl^- \rightarrow Cl_2^{\,0} + 2\,e^-}$ Multiply by 2, add, the $4\,e^-$ cancel

                $4\,Cl^- + 2\,O^0 \rightarrow 2\,Cl_2^{\,0} + 2\,O^{2-}$

Transfer the coefficients to the original equation and complete the balancing by inspection.

$$4\,HCl + O_2 \rightarrow 2\,Cl_2 + 2\,H_2O$$

(d) $Ag + H_2S + O_2 \rightarrow Ag_2S + H_2O$

red $O_2^0 + 4\,e^- \rightarrow 2\,O^{2-}$

ox $\underline{Ag \rightarrow Ag^+ + 1\,e^-}$ Multiply by 4, add, the $4\,e^-$ cancel

$4\,Ag + O_2^0 \rightarrow 4\,Ag^+ + 2\,O^{2-}$

Transfer the coefficients to the original equation and complete the balancing by inspection.

$$4\,Ag + 2\,H_2S + O_2 \rightarrow 2\,Ag_2S + 2\,H_2O$$

(e) $KMnO_4 + CaC_2O_4 + H_2SO_4 \rightarrow K_2SO_4 + MnSO_4 + CaSO_4 + CO_2 + H_2O$

red $Mn^{7+} + 5\,e^- \rightarrow Mn^{2+}$ Multiply by 2;

ox $\underline{2\,C^{3+} \rightarrow 2\,C^{4+} + 2\,e^-}$ Multiply by 5, add, the $10\,e^-$ cancel

$10\,C^{3+} + 2\,Mn^{7+} \rightarrow 10\,C^{4+} + 2\,Mn^{2+}$

Transfer the coefficients to the original equation and complete the balancing by inspection.

$$2\,KMnO_4 + 5\,CaC_2O_4 + 8\,H_2SO_4 \rightarrow K_2SO_4 + 2\,MnSO_4 + 5\,CaSO_4 + 10\,CO_2 + 8\,H_2O$$

15.   Balancing ionic redox equations
(a) $Zn + NO_3^- \longrightarrow Zn^{2+} + NH_4^+$ (acidic solution)

Step 1   Write half-reaction equations. Balance except H and O.

$Zn \longrightarrow Zn^{2+}$

$NO_3^- \longrightarrow NH_4^+$

Step 2   Balance H and O using H_2O and H^+

$Zn \longrightarrow Zn^{2+}$

$10\,H^+ + NO_3^- \longrightarrow NH_4^+ + 3\,H_2O$

Step 3   Balance electrically with electrons

$Zn \longrightarrow Zn^{2+} + 2\,e^-$

$10\,H^+ + NO_3^- + 8\,e^- \longrightarrow NH_4^+ + 3\,H_2O$

Step 4   Equalize the loss and gain of electrons

$4(Zn \longrightarrow Zn^{2+} + 2\,e^-)$

$10\,H^+ + NO_3^- + 8\,e^- \longrightarrow NH_4^+ + 3\,H_2O$

Step 5   Add the half-reactions; electrons cancel

$10\,H^+ + 4\,Zn + NO_3^- \longrightarrow 4\,Zn^{2+} + NH_4^+ + 3\,H_2O$

(b) $NO_3^- + S \longrightarrow NO_2 + SO_4^{2-}$ (acidic solution)

Step 1   Write half-reaction equations. Balance except H and O.

$$S \longrightarrow SO_4^{2-}$$

$$NO_3^- \longrightarrow NO_2$$

Step 2   Balance H and O using H_2O and H^+

$$4\,H_2O + S \longrightarrow SO_4^{2-} + 8\,H^+$$

$$2\,H^+ + NO_3^- \longrightarrow NO_2 + H_2O$$

Step 3   Balance electrically with electrons

$$4\,H_2O + S \longrightarrow SO_4^{2-} + 8\,H^+ + 6\,e^-$$

$$2\,H^+ + NO_3^- + e^- \longrightarrow NO_2 + H_2O$$

Step 4 and 5   Equalize the loss and gain of electrons; add the half-reactions

$$4\,H_2O + S \longrightarrow SO_4^{2-} + 8\,H^+ + 6\,e^-$$

$$\underline{6\,(2\,H^+ + NO_3^- + e^- \longrightarrow NO_2 + H_2O)}$$

$$4\,H^+ + S + 6\,NO_3^- \longrightarrow 6\,NO_2 + SO_4^{2-} + 2\,H_2O$$

$4\,H_2O$, $8\,H^+$ and $6\,e^-$ canceled from each side

(c) $PH_3 + I_2 \longrightarrow H_3PO_2 + I^-$ (acidic solution)

Step 1   Write half-reaction equations. Balance except H and O.

$$PH_3 \longrightarrow H_3PO_2$$

$$I_2 \longrightarrow 2\,I^-$$

Step 2   Balance H and O using H_2O and H^+

$$2\,H_2O + PH_3 \longrightarrow H_3PO_2 + 4\,H^+$$

$$I_2 \longrightarrow 2\,I^-$$

Step 3   Balance electrically with electrons

$$2\,H_2O + PH_3 \longrightarrow H_3PO_2 + 4\,H^+ + 4\,e^-$$

$$I_2 + 2\,e^- \longrightarrow 2\,I^-$$

Step 4 and 5   Equalize the loss and gain of electrons; add the half-reactions

$$2\,H_2O + PH_3 \longrightarrow H_3PO_2 + 4\,H^+ + 4\,e^-$$

$$\underline{2\,(I_2 + 2\,e^- \longrightarrow 2\,I^-)}$$

$$PH_3 + 2\,H_2O + 2\,I_2 \longrightarrow H_3PO_2 + 4\,I^- + 4\,H^+$$

(d) $Cu + NO_3^- \longrightarrow Cu^{2+} + NO$ (acidic solution)

Step 1   Write half-reaction equations. Balance except H and O.

$$Cu \longrightarrow Cu^{2+}$$

$$NO_3^- \longrightarrow NO$$

Step 2   Balance H and O using H_2O and H^+

$$Cu \longrightarrow Cu^{2+}$$

$$4\,H^+ + NO_3^- \longrightarrow NO + 2\,H_2O$$

Step 3   Balance electrically with electrons

$$Cu \longrightarrow Cu^{2+} + 2\,e^-$$

$$4\,H^+ + NO_3^- + 3\,e^- \longrightarrow NO + 2\,H_2O$$

Step 4 and 5   Equalize the loss and gain of electrons; add the half-reactions

$$3\,(Cu \longrightarrow Cu^{2+} + 2\,e^-)$$
$$2\,(4\,H^+ + NO_3^- + 3\,e^- \longrightarrow NO + 2\,H_2O)$$
$$\overline{3\,Cu + 8\,H^+ + 2\,NO_3^- \longrightarrow 3\,Cu^{2+} + 2\,NO + 4\,H_2O}$$

(e) $ClO_3^- + Cl^- \longrightarrow Cl_2$ (acidic solution)

Step 1   Write half-reaction equations. Balance except H and O.

$$Cl^- \longrightarrow Cl^0$$
$$ClO_3^- \longrightarrow Cl^0$$

Step 2   Balance H and O using H_2O and H^+

$$Cl^- \longrightarrow Cl^0$$
$$6\,H^+ + ClO_3^- \longrightarrow Cl^0 + 3\,H_2O$$

Step 3   Balance electrically with electrons

$$Cl^- \longrightarrow Cl^0 + e^-$$
$$6\,H^+ + ClO_3^- + 5\,e^- \longrightarrow Cl^0 + 3\,H_2O$$

Step 4 and 5   Equalize the loss and gain of electrons; add the half-reactions

$$5(Cl^- \longrightarrow Cl^0 + e^-)$$
$$6\,H^+ + ClO_3^- + 5\,e^- \longrightarrow Cl^0 + 3\,H_2O$$
$$\overline{6\,H^+ + ClO_3^- + 5\,Cl^- \longrightarrow 3\,Cl_2 + 3\,H_2O}$$

16. (a) $ClO_3^- + I^- \longrightarrow I_2 + Cl^-$ (acidic solution)

Step 1   Write half-reaction equations. Balance except H and O.

$$2\,I^- \longrightarrow I_2$$
$$ClO_3^- \longrightarrow Cl^-$$

Step 2   Balance H and O using H_2O and H^+

$$2\,I^- \longrightarrow I_2$$
$$6\,H^+ + ClO_3^- \longrightarrow Cl^- + 3\,H_2O$$

Step 3   Balance electrically with electrons

$$2\,I^- \longrightarrow I_2 + 2\,e^-$$
$$6\,H^+ + ClO_3^- + 6\,e^- \longrightarrow Cl^- + 3\,H_2O$$

Step 4 and 5   Equalize the loss and gain of electrons; add the half-reactions

$$3\,(2\,I^- \longrightarrow I_2 + 2\,e^-)$$
$$6\,H^+ + ClO_3^- + 6\,e^- \longrightarrow Cl^- + 3\,H_2O$$
$$\overline{6\,H^+ + ClO_3^- + 6\,I^- \longrightarrow 3\,I_2 + Cl^- + 3\,H_2O}$$

(b) $Cr_2O_7^{2-} + Fe^{2+} \longrightarrow Cr^{3+} + Fe^{3+}$ (acidic solution)

Step 1   Write half-reaction equations. Balance except H and O.

$$Fe^{2+} \longrightarrow Fe^{3+}$$
$$Cr_2O_7^{2-} \longrightarrow 2\,Cr^{3+}$$

Step 2   Balance H and O using H_2O and H^+

$$Fe^{2+} \longrightarrow Fe^{3+}$$

$$14\,H^+ + Cr_2O_7{}^{2-} \longrightarrow 2\,Cr^{3+} + 7\,H_2O$$

Step 3   Balance electrically with electrons

$$Fe^{2+} \longrightarrow Fe^{3+} + e^-$$

$$14\,H^+ + Cr_2O_7{}^{2-} + 6\,e^- \longrightarrow 2\,Cr^{3+} + 7\,H_2O$$

Step 4 and 5   Equalize the loss and gain of electrons; add the half-reactions

$$6(Fe^{2+} \longrightarrow Fe^{3+} + e^-)$$

$$\underline{14\,H^+ + Cr_2O_7{}^{2-} + 6\,e^- \longrightarrow 2\,Cr^{3+} + 7\,H_2O}$$

$$14\,H^+ + Cr_2O_7{}^{2-} + 6\,Fe^{2+} \longrightarrow 2\,Cr^{3+} + 6\,Fe^{3+} + 7\,H_2O$$

(c) $MnO_4{}^- + SO_2 \longrightarrow Mn^{2+} + SO_4{}^{2-}$ (acidic solution)

Step 1   Write half-reaction equations. Balance except H and O.

$$SO_2 \longrightarrow SO_4{}^{2-}$$

$$MnO_4{}^- \longrightarrow Mn^{2+}$$

Step 2   Balance H and O using H_2O and H^+

$$2\,H_2O + SO_2 \longrightarrow SO_4{}^{2-} + 4\,H^+$$

$$8\,H^+ + MnO_4{}^- \longrightarrow Mn^{2+} + 4\,H_2O$$

Step 3   Balance electrically with electrons

$$2\,H_2O + SO_2 \longrightarrow SO_4{}^{2-} + 4\,H^+ + 2\,e^-$$

$$8\,H^+ + MnO_4{}^- + 5\,e^- \longrightarrow Mn^{2+} + 4\,H_2O$$

Step 4 and 5   Equalize the loss and gain of electrons; add the half-reactions

$$5(2\,H_2O + SO_2 \longrightarrow SO_4{}^{2-} + 4\,H^+ + 2\,e^-)$$

$$\underline{2(8\,H^+ + MnO_4{}^- + 5\,e^- \longrightarrow Mn^{2+} + 4\,H_2O)}$$

$$2\,H_2O + 2\,MnO_4{}^- + 5\,SO_2 \longrightarrow 4\,H^+ + 2\,Mn^{2+} + 5\,SO_4{}^{2-}$$

$8\,H_2O$, $16\,H^+$, and $10\,e^-$ canceled from each side

(d) $H_3AsO_3 + MnO_4{}^- \longrightarrow H_3AsO_4 + Mn^{2+}$ (acidic solution)

Step 1   Write half-reaction equations. Balance except H and O.

$$H_3AsO_3 \longrightarrow H_3AsO_4$$

$$MnO_4{}^- \longrightarrow Mn^{2+}$$

Step 2   Balance H and O using H_2O and H^+

$$H_2O + H_3AsO_3 \longrightarrow 2\,H^+ + H_3AsO_4$$

$$8\,H^+ + MnO_4{}^- \longrightarrow Mn^{2+} + 4\,H_2O$$

Step 3   Balance electrically with electrons

$$H_2O + H_3AsO_3 \longrightarrow 2\,H^+ + H_3AsO_4 + 2\,e^-$$

$$8\,H^+ + MnO_4{}^- + 5\,e^- \longrightarrow Mn^{2+} + 4\,H_2O$$

Step 4 and 5   Equalize the loss and gain of electrons; add the half-reactions

$$5(H_2O + H_3AsO_3 \longrightarrow 2H^+ + H_3AsO_4 + 2e^-)$$

$$\underline{2(8H^+ + MnO_4^- + 5e^- \longrightarrow Mn^{2+} + 4H_2O)}$$

$$6H^+ + 5H_3AsO_3 + 2MnO_4^- \longrightarrow 5H_3AsO_4 + 2Mn^{2+} + 3H_2O$$

$5H_2O$, $10H^+$, and $10e^-$ canceled from each side

(e) $Cr_2O_7^{2-} + H_3AsO_3 \longrightarrow Cr^{3+} + H_3AsO_4$ (acidic solution)

Step 1   Write half-reaction equations. Balance except H and O.

$$H_3AsO_3 \longrightarrow H_3AsO_4$$

$$Cr_2O_7^{2-} \longrightarrow 2Cr^{3+}$$

Step 2   Balance H and O using H_2O and H^+

$$H_2O + H_3AsO_3 \longrightarrow 2H^+ + H_3AsO_4$$

$$14H^+ + Cr_2O_7^{2-} \longrightarrow 2Cr^{3+} + 7H_2O$$

Step 3   Balance electrically with electrons

$$H_2O + H_3AsO_3 \longrightarrow 2H^+ + H_3AsO_4 + 2e^-$$

$$14H^+ + Cr_2O_7^{2-} + 6e^- \longrightarrow 2Cr^{3+} + 7H_2O$$

Step 4 and 5   Equalize the loss and gain of electrons; add the half-reactions

$$3(H_2O + H_3AsO_3 \longrightarrow 2H^+ + H_3AsO_4 + 2e^-)$$

$$\underline{14H^+ + Cr_2O_7^{2-} + 6e^- \longrightarrow 2Cr^{3+} + 7H_2O}$$

$$8H^+ + Cr_2O_7^{2-} + 3H_3AsO_3 \longrightarrow 2Cr^{3+} + 3H_3AsO_4 + 4H_2O$$

$3H_2O$, $6H^+$, and $6e^-$ canceled from each side

17.   (a) $Cl_2 + IO_3^- \longrightarrow Cl^- + IO_4^-$ (basic solution)

Step 1   Write half-reaction equations. Balance except H and O.

$$IO_3^- \longrightarrow IO_4^-$$
$$Cl_2 \longrightarrow 2Cl^-$$

Step 2   Balance H and O using H_2O and H^+

$$H_2O + IO_3^- \longrightarrow IO_4^- + 2H^+$$
$$Cl_2 \longrightarrow 2Cl^-$$

Step 3   Add OH^- ions to both sides (same number as H^+ ions)

$$2OH^- + H_2O + IO_3^- \longrightarrow IO_4^- + 2H^+ + 2OH^-$$
$$Cl_2 \longrightarrow 2Cl^-$$

Step 4   Combine H^+ and OH^- to form H_2O; cancel H_2O where possible

$$2OH^- + H_2O + IO_3^- \longrightarrow IO_4^- + 2H_2O$$
$$Cl_2 \longrightarrow 2Cl^-$$

$$2OH^- + IO_3^- \longrightarrow IO_4^- + H_2O \qquad \text{(1 } H_2O \text{ cancelled)}$$
$$Cl_2 \longrightarrow 2Cl^-$$

Step 5   Balance electrically with electrons

$$2\,OH^- + IO_3^- \longrightarrow IO_4^- + H_2O + 2\,e^-$$

$$Cl_2 + 2\,e^- \longrightarrow 2\,Cl^-$$

Step 6   Electron loss and gain is balanced

Step 7   Add half-reactions

$$2\,OH^- + IO_3^- + Cl_2 \longrightarrow IO_4^- + 2\,Cl^- + H_2O$$

(b) $MnO_4^- + ClO_2^- \longrightarrow MnO_2 + ClO_4^-$ (basic solution)

Step 1   Write half-reaction equations. Balance except H and O.

$$ClO_2^- \longrightarrow ClO_4^-$$

$$MnO_4^- \longrightarrow MnO_2$$

Step 2   Balance H and O using H_2O and H^+

$$2\,H_2O + ClO_2^- \longrightarrow ClO_4^- + 4\,H^+$$

$$MnO_4^- + 4\,H^+ \longrightarrow MnO_2 + 2\,H_2O$$

Step 3   Add OH^- ions to both sides (same number as H^+ ions)

$$4\,OH^- + 2\,H_2O + ClO_2^- \longrightarrow ClO_4^- + 4\,H^+ + 4\,OH^-$$

$$4\,OH^- + MnO_4^- + 4\,H^+ \longrightarrow MnO_2 + 2\,H_2O + 4\,OH^-$$

Step 4   Combine H^+ and OH^- to form H_2O; cancel H_2O where possible

$$4\,OH^- + 2\,H_2O + ClO_2^- \longrightarrow ClO_4^- + 4\,H_2O$$

$$4\,H_2O + MnO_4^- \longrightarrow MnO_2 + 2\,H_2O + 4\,OH^-$$

$$4\,OH^- + ClO_2^- \longrightarrow ClO_4^- + 2\,H_2O \quad (2\,H_2O \text{ cancelled})$$

$$2\,H_2O + MnO_4^- \longrightarrow MnO_2 + 4\,OH^- \quad (2\,H_2O \text{ cancelled})$$

Step 5   Balance electrically with electrons

$$4\,OH^- + ClO_2^- \longrightarrow ClO_4^- + 2\,H_2O + 4\,e^-$$

$$2\,H_2O + MnO_4^- + 3\,e^- \longrightarrow MnO_2 + 4\,OH^-$$

Step 6 and 7   Equalize gain and loss of electrons; add half-reactions

$$3\,(4\,OH^- + ClO_2^- \longrightarrow ClO_4^- + 2\,H_2O + 4\,e^-)$$

$$\underline{4\,(2\,H_2O + MnO_4^- + 3\,e^- \longrightarrow MnO_2 + 4\,OH^-)}$$

$$2\,H_2O + 4\,MnO_4^- + 3\,ClO_2^- \longrightarrow 4\,MnO_2 + 3\,ClO_4^- + 4\,OH^-$$

6 H_2O, 12 OH^-, and 12 e^- canceled from each side

(c) $Se \longrightarrow SeO_3^{2-} + Se^{2-}$ (basic solution)

Step 1   Write half-reaction equations. Balance except H and O.

$$Se \longrightarrow SeO_3^{2-}$$

$$Se \longrightarrow Se^{2-}$$

Step 2   Balance H and O using H_2O and H^+

$$3\,H_2O + Se \longrightarrow SeO_3^{2-} + 6\,H^+$$

$$Se \longrightarrow Se^{2-}$$

Step 3   Add OH^- ions to both sides (same number as H^+ ions)

$$6\,OH^- + 3\,H_2O + Se \longrightarrow SeO_3^{2-} + 6\,H^+ + 6\,OH^-$$

$$Se \longrightarrow Se^{2-}$$

Step 4   Combine H^+ and OH^- to form H_2O; cancel H_2O where possible

$$6\,OH^- + 3\,H_2O + Se \longrightarrow SeO_3^{2-} + 6\,H_2O$$

$$Se \longrightarrow Se^{2-}$$

$$6\,OH^- + Se \longrightarrow SeO_3^{2-} + 3\,H_2O \qquad (3\,H_2O \text{ cancelled})$$

Step 5   Balance electrically with electrons

$$6\,OH^- + Se \longrightarrow SeO_3^{2-} + 3\,H_2O + 4\,e^-$$

$$Se + 2\,e^- \longrightarrow Se^{2-}$$

Step 6 and 7   Equalize gain and loss of electrons; add half-reactions

$$6\,OH^- + Se \longrightarrow SeO_3^{2-} + 3\,H_2O + 4\,e^-$$

$$\underline{2(Se + 2\,e^- \longrightarrow Se^{2-})}$$

$$6\,OH^- + 3\,Se \longrightarrow SeO_3^{2-} + 2\,Se^{2-} + 3\,H_2O$$

(d) $Fe_3O_4 + MnO_4^- \longrightarrow Fe_2O_3 + MnO_2$ (basic solution)

Step 1   Write half-reaction equations. Balance except H and O.

$$2\,Fe_3O_4 \longrightarrow 3\,Fe_2O_3$$

$$MnO_4^- \longrightarrow MnO_2$$

Step 2   Balance H and O using H_2O and H^+

$$H_2O + 2\,Fe_3O_4 \longrightarrow 3\,Fe_2O_3 + 2\,H^+$$
$$4\,H^+ + MnO_4^- \longrightarrow MnO_2 + 2\,H_2O$$

Step 3   Add OH^- ions to both sides (same number as H^+ ions)

$$2\,OH^- + H_2O + 2\,Fe_3O_4 \longrightarrow 3\,Fe_2O_3 + 2\,H^+ + 2\,OH^-$$
$$4\,OH^- + 4\,H^+ + MnO_4^- \longrightarrow MnO_2 + 2\,H_2O + 4\,OH^-$$

Step 4   Combine H^+ and OH^- to form H_2O; cancel H_2O where possible

$$2\,OH^- + H_2O + 2\,Fe_3O_4 \longrightarrow 3\,Fe_2O_3 + 2\,H_2O$$

$$4\,H_2O + MnO_4^- \longrightarrow MnO_2 + 2\,H_2O + 4\,OH^-$$

$$2\,OH^- + 2\,Fe_3O_4 \longrightarrow 3\,Fe_2O_3 + H_2O \qquad (1\,H_2O \text{ cancelled})$$

$$2\,H_2O + MnO_4^- \longrightarrow MnO_2 + 4\,OH^- \qquad (2\,H_2O \text{ cancelled})$$

Step 5   Balance electrically with electrons

$$2\,OH^- + 2\,Fe_3O_4 \longrightarrow 3\,Fe_2O_3 + H_2O + 2\,e^-$$
$$2\,H_2O + MnO_4^- + 3\,e^- \longrightarrow MnO_2 + 4\,OH^-$$

Step 6 and 7   Equalize gain and loss of electrons; add half-reactions

$$3(2\,OH^- + 2\,Fe_3O_4 \longrightarrow 3\,Fe_2O_3 + H_2O + 2\,e^-)$$

$$\underline{2(2\,H_2O + MnO_4^- + 3\,e^- \longrightarrow MnO_2 + 4\,OH^-)}$$

$$H_2O + 6\,Fe_3O_4 + 2\,MnO_4^- \longrightarrow 9\,Fe_2O_3 + 2\,MnO_2 + 2\,OH^-)$$

$3\,H_2O$, $6\,OH^-$, and $6\,e^-$ canceled from each side

(e) $BrO^- + Cr(OH)_4^- \longrightarrow Br^- + CrO_4^{2-}$ (basic solution)

Step 1   Write half-reaction equations. Balance except H and O.

$$Cr(OH)_4^- \longrightarrow CrO_4^{2-}$$

$$BrO^- \longrightarrow Br^-$$

Step 2   Balance H and O using H_2O and H^+

$$Cr(OH)_4^- \longrightarrow CrO_4^{2-} + 4H^+$$

$$2H^+ + BrO^- \longrightarrow Br^- + H_2O$$

Step 3   Add OH^- ions to both sides (same number as H^+ ions)

$$4OH^- + Cr(OH)_4^- \longrightarrow CrO_4^{2-} + 4H^+ + 4OH^-$$

$$2OH^- + 2H^+ + BrO^- \longrightarrow Br^- + H_2O + 2OH^-$$

Step 4   Combine H^+ and OH^- to form H_2O; cancel H_2O where possible

$$4OH^- + Cr(OH)_4^- \longrightarrow CrO_4^{2-} + 4H_2O$$

$$2H_2O + BrO^- \longrightarrow Br^- + H_2O + 2OH^-$$

$$H_2O + BrO^- \longrightarrow Br^- + 2OH^- \quad \text{(1 } H_2O \text{ cancelled)}$$

Step 5   Balance electrically with electrons

$$4OH^- + Cr(OH)_4^- \longrightarrow CrO_4^{2-} + 4H_2O + 3e^-$$

$$H_2O + BrO^- + 2e^- \longrightarrow Br^- + 2OH^-$$

Step 6 and 7   Equalize gain and loss of electrons; add half-reactions

$$2(4OH^- + Cr(OH)_4^- \longrightarrow CrO_4^{2-} + 4H_2O + 3e^-)$$

$$3(H_2O + BrO^- + 2e^- \longrightarrow Br^- + 2OH^-)$$

$$\overline{2OH^- + 3BrO^- + 2Cr(OH)_4^- \longrightarrow 3Br^- + 2CrO_4^{2-} + 5H_2O}$$

$3 H_2O$, $6 OH^-$ and $6 e^-$ canceled from each side

18. (a) $MnO_4^- + SO_3^{2-} \longrightarrow MnO_2 + SO_4^{2-}$ (basic solution)

Step 1   Write half-reaction equations. Balance except H and O.

$$SO_3^{2-} \longrightarrow SO_4^{2-}$$

$$MnO_4^- \longrightarrow MnO_2$$

Step 2   Balance H and O using H_2O and H^+

$$H_2O + SO_3^{2-} \longrightarrow SO_4^{2-} + 2H^+$$

$$MnO_4^- + 4H^+ \longrightarrow MnO_2 + 2H_2O$$

Step 3   Add OH^- ions to both sides (same number as H^+ ions)

$$2OH^- + H_2O + SO_3^{2-} \longrightarrow SO_4^{2-} + 2H^+ + 2OH^-$$

$$4OH^- + MnO_4^- + 4H^+ \longrightarrow MnO_2 + 2H_2O + 4OH^-$$

Step 4 $\qquad$ Combine H^+ and OH^- to form H_2O; cancel H_2O where possible

$$2\,OH^- + H_2O + SO_3^{2-} \longrightarrow SO_4^{2-} + 2\,H_2O$$

$$MnO_4^- + 4\,H_2O \longrightarrow MnO_2 + 2\,H_2O + 4\,OH^-$$

$$2\,OH^- + SO_3^{2-} \longrightarrow SO_4^{2-} + H_2O \quad (1\ H_2O\ cancelled)$$

$$MnO_4^- + 2\,H_2O \longrightarrow MnO_2 + 4\,OH^- \quad (2\ H_2O\ cancelled)$$

Step 5 $\qquad$ Balance electrically with electrons

$$2\,OH^- + SO_3^{2-} \longrightarrow SO_4^{2-} + H_2O + 2\,e^-$$

$$3\,e^- + MnO_4^- + 2\,H_2O \longrightarrow MnO_2 + 4\,OH^-$$

Step 6 and 7 $\quad$ Equalize gain and loss of electrons; add half-reactions

$$3\,(2\,OH^- + SO_3^{2-} \longrightarrow SO_4^{2-} + H_2O + 2\,e^-)$$

$$\underline{2\,(MnO_4^- + 2\,H_2O + 3\,e^- \longrightarrow MnO_2 + 4\,OH^-)}$$

$$H_2O + 2\,MnO_4^- + 3\,SO_3^{2-} \longrightarrow 2\,MnO_2 + 3\,SO_4^{2-} + 2\,OH^-$$

$3\,H_2O$, $4\,OH^-$, and $6\,e^-$ canceled from each side

(b) $\quad ClO_2 + SbO_2^- \longrightarrow ClO_2^- + Sb(OH)_6^- \quad$ (basic solution)

Step 1 $\qquad$ Write half-reaction equations. Balance except H and O.

$$SbO_2^- \longrightarrow Sb(OH)_6^-$$

$$ClO_2 \longrightarrow ClO_2^-$$

Step 2 $\qquad$ Balance H and O using H_2O and H^+

$$4\,H_2O + SbO_2^- \longrightarrow Sb(OH)_6^- + 2\,H^+$$

$$ClO_2 \longrightarrow ClO_2^-$$

Step 3 $\qquad$ Add OH^- ions to both sides (same number as H^+ ions)

$$2\,OH^- + 4\,H_2O + SbO_2^- \longrightarrow Sb(OH)_6^- + 2\,H^+ + 2\,OH^-$$

$$ClO_2 \longrightarrow ClO_2^-$$

Step 4 $\qquad$ Combine H^+ and OH^- to form H_2O; cancel H_2O where possible

$$2\,OH^- + 4\,H_2O + SbO_2^- \longrightarrow Sb(OH)_6^- + 2\,H_2O$$

$$ClO_2 \longrightarrow ClO_2^-$$

$$2\,OH^- + 2\,H_2O + SbO_2^- \longrightarrow Sb(OH)_6^- \quad (2\ H_2O\ cancelled)$$

Step 5 $\qquad$ Balance electrically with electrons

$$2\,OH^- + 2\,H_2O + SbO_2^- \longrightarrow Sb(OH)_6^- + 2\,e^-$$

$$ClO_2 + e^- \longrightarrow ClO_2^-$$

Step 6 and 7 $\quad$ Equalize gain and loss of electrons; add half-reactions

$$2\,H_2O + 2\,OH^- + SbO_2^- \longrightarrow Sb(OH)_6^- + 2\,e^-$$

$$\underline{2\,(ClO_2 + e^- \longrightarrow ClO_2^-)}$$

$$2\,H_2O + 2\,ClO_2 + 2\,OH^- + SbO_2^- \longrightarrow 2\,ClO_2^- + Sb(OH)_6^-$$

(c) $Al + NO_3^- \longrightarrow NH_3 + Al(OH)_4^-$ (basic solution)

Step 1	Write half-reaction equations. Balance except H and O.
	$Al \longrightarrow Al(OH)_4^-$
	$NO_3^- \longrightarrow NH_3$
Step 2	Balance H and O using H_2O and H^+
	$4\,H_2O + Al \longrightarrow Al(OH)_4^- + 4\,H^+$
	$9\,H^+ + NO_3^- \longrightarrow NH_3 + 3\,H_2O$
Step 3	Add OH^- ions to both sides (same number as H^+ ions)
	$4\,OH^- + 4\,H_2O + Al \longrightarrow Al(OH)_4^- + 4\,H^+ + 4\,OH^-$
	$9\,OH^- + 9\,H^+ + NO_3^- \longrightarrow NH_3 + 3\,H_2O + 9\,OH^-$
Step 4	Combine H^+ and OH^- to form H_2O; cancel H_2O where possible
	$4\,OH^- + 4\,H_2O + Al \longrightarrow Al(OH)_4^- + 4\,H_2O$
	$9\,H_2O + NO_3^- \longrightarrow NH_3 + 3\,H_2O + 9\,OH^-$
	$4\,OH^- + Al \longrightarrow Al(OH)_4^-$ (4 H_2O cancelled)
	$6\,H_2O + NO_3^- \longrightarrow NH_3 + 9\,OH^-$ (3 H_2O cancelled)
Step 5	Balance electrically with electrons
	$4\,OH^- + Al \longrightarrow Al(OH)_4^- + 3\,e^-$
	$6\,H_2O + NO_3^- + 8\,e^- \longrightarrow NH_3 + 9\,OH^-$
Step 6 and 7	Equalize gain and loss of electrons; add half-reactions

$$8\left(4\,OH^- + Al \longrightarrow Al(OH)_4^- + 3\,e^-\right)$$
$$\underline{3\left(6\,H_2O + NO_3^- + 8\,e^- \longrightarrow NH_3 + 9\,OH^-\right)}$$
$$8\,Al + 3\,NO_3^- + 18\,H_2O + 5\,OH^- \longrightarrow 3\,NH_3 + 8\,Al(OH)_4^-$$

27 OH^- and 24 e^- canceled from each side

(d) $P_4 \longrightarrow HPO_3^{2-} + PH_3$ (basic solution)

Step 1	Write half-reaction equations. Balance except H and O.
	$P_4 \longrightarrow 4\,HPO_3^{2-}$
	$P_4 \longrightarrow 4\,PH_3$
Step 2	Balance H and O using H_2O and H^+
	$12\,H_2O + P_4 \longrightarrow 4\,HPO_3^{2-} + 20\,H^+$
	$12\,H^+ + P_4 \longrightarrow 4\,PH_3$
Step 3	Add OH^- ions to both sides (same number as H^+ ions)
	$20\,OH^- + 12\,H_2O + P_4 \longrightarrow 4\,HPO_3^{2-} + 20\,H^+ + 20\,OH^-$
	$12\,OH^- + 12\,H^+ + P_4 \longrightarrow 4\,PH_3 + 12\,OH^-$

Step 4   Combine H^+ and OH^- to form H_2O; cancel H_2O where possible

$$20\,OH^- + 12\,H_2O + P_4 \longrightarrow 4\,HPO_3^{2-} + 20\,H_2O$$
$$12\,H_2O + P_4 \longrightarrow 4\,PH_3 + 12\,OH^-$$
$$20\,OH^- + P_4 \longrightarrow 4\,HPO_3^{2-} + 8\,H_2O \;(12\,H_2O\text{ cancelled})$$

Step 5   Balance electrically with electrons

$$20\,OH^- + P_4 \longrightarrow 4\,HPO_3^{2-} + 8\,H_2O + 12\,e^-$$
$$12\,H_2O + P_4 + 12\,e^- \longrightarrow 4\,PH_3 + 12\,OH^-$$

Step 6 and 7   Loss and gain of electrons are equal; add half-reactions

$$8\,OH^- + 4\,H_2O + 2\,P_4 \longrightarrow 4\,HPO_3^{2-} + 4\,PH_3$$

Divide equation by 2

$$4\,OH^- + 2\,H_2O + P_4 \longrightarrow 2\,HPO_3^{2-} + 2\,PH_3$$

(e) $Al + OH^- \longrightarrow Al(OH)_4^- + H_2$ (basic solution)

Step 1   Write half-reaction equations. Balance except H and O.

$$Al \longrightarrow Al(OH)_4^-$$
$$OH^- \longrightarrow H_2$$

Step 2   Balance H and O using H_2O and H^+

$$4\,H_2O + Al \longrightarrow Al(OH)_4^- + 4\,H^+$$
$$3\,H^+ + OH^- \longrightarrow H_2 + H_2O$$

Step 3   Add OH^- ions to both sides (same number as H^+ ions)

$$4\,OH^- + 4\,H_2O + Al \longrightarrow Al(OH)_4^- + 4\,H^+ + 4\,OH^-$$
$$3\,OH^- + 3\,H^+ + OH^- \longrightarrow H_2 + H_2O + 3\,OH^-$$

Step 4   Combine H^+ and OH^- to form H_2O; cancel H_2O where possible

$$4\,OH^- + 4\,H_2O + Al \longrightarrow Al(OH)_4^- + 4\,H_2O$$
$$3\,H_2O + OH^- \longrightarrow H_2 + H_2O + 3\,OH^-$$
$$4\,OH^- + Al \longrightarrow Al(OH)_4^- \;(4\,H_2O\text{ cancelled})$$
$$2\,H_2O + OH^- \longrightarrow H_2 + 3\,OH^- \;(1\,H_2O\text{ cancelled})$$

Step 5   Balance electrically with electrons

$$4\,OH^- + Al \longrightarrow Al(OH)_4^- + 3\,e^-$$
$$2\,H_2O + OH^- + 2\,e^- \longrightarrow H_2 + 3\,OH^-$$

Step 6 and 7   Equalize gain and loss of electrons; and half-reactions

$$2\,(4\,OH^- + Al \longrightarrow Al(OH)_4^- + 3\,e^-)$$
$$3\,(2\,H_2O + OH^- + 2\,e^- \longrightarrow H_2 + 3\,OH^-)$$
$$\overline{2\,Al + 6\,H_2O + 2\,OH^- \longrightarrow 2\,Al(OH)_4^- + 3\,H_2}$$

$9\,OH^-$ and $6\,e^-$ canceled on each side

19. (a) $IO_3^- + I^- \longrightarrow I_2$ (acidic solution)

Step 1	Write half-reaction equations. Balance except H and O.

$$2\,IO_3^- \longrightarrow I_2$$
$$2\,I^- \longrightarrow I_2$$

Step 2	Balance H and O using H_2O and H^+

$$12\,H^+ + 2\,IO_3^- \longrightarrow I_2 + 6\,H_2O$$
$$2\,I^- \longrightarrow I_2$$

Step 3	Balance electrically with electrons

$$12\,H^+ + 2\,IO_3^- + 10\,e^- \longrightarrow I_2 + 6\,H_2O$$
$$2\,I^- \longrightarrow I_2 + 2\,e^-$$

Step 4 and 5	Equalize the loss and gain of electrons; add the half reactions.

$$12\,H^+ + 2\,IO_3^- + 10\,e^- \longrightarrow I_2 + 6\,H_2O$$
$$\underline{5\,(2\,I^- \longrightarrow I_2 + 2\,e^-)}$$
$$12\,H^+ + 2\,IO_3^- + 10\,I^- \longrightarrow 6\,I_2 + 6\,H_2O$$

Each side has 12 H, 6 O, 12 I, and no charge

(b) $Mn^{2+} + S_2O_8^{2-} \longrightarrow MnO_4^- + SO_4^{2-}$ (acid solution)

Step 1	Write half-reaction equations. Balance except H and O

$$Mn^{2+} \longrightarrow MnO_4^-$$
$$S_2O_8^{2-} \longrightarrow 2\,SO_4^{2-}$$

Step 2	Balance H and O using H_2O and H^+

$$4\,H_2O + Mn^{2+} \longrightarrow MnO_4^- + 8\,H^+$$
$$S_2O_8^{2-} \longrightarrow 2\,SO_4^{2-}$$

Step 3	Balance electrically with electrons

$$4\,H_2O + Mn^{2+} \longrightarrow MnO_4^- + 8\,H^+ + 5\,e^-$$
$$2\,e^- + S_2O_8^{2-} \longrightarrow 2\,SO_4^{2-}$$

Step 4 and 5	Equalize the loss and gain of electrons; add the half-reactions

$$2\,(4\,H_2O + Mn^{2+} \longrightarrow MnO_4^- + 8\,H^+ + 5\,e^-)$$
$$\underline{5\,(2\,e^- \longrightarrow S_2O_8^{2-} \longrightarrow 2\,SO_4^{2-})}$$
$$2\,Mn^{2+} + 5\,S_2O_8^{2-} + 8\,H_2O \longrightarrow 2\,MnO_4^- + 10\,SO_4^{2-} + 16\,H^+$$

Each side has 2 Mn, 10 S, 16 H, and 48 O and a -6 charge.

(c) $Co(NO_2)_6^{3-} + MnO_4^- \longrightarrow Co^{2+} + Mn^{2+} + NO_3^-$ (acidic solution)

Step 1	Write half-reaction equations. Balance except H and O.

$$Co(NO_2)_6^{3-} \longrightarrow Co^{2+} + 6\,NO_3^-$$
$$MnO_4^- \longrightarrow Mn^{2+}$$

Step 2 Balance H and O using H_2O and H^+

$$6\,H_2O + Co(NO_2)_6^{3-} \longrightarrow Co^{2+} + 6\,NO_3^- + 12\,H^+$$
$$8\,H^+ + MnO_4^- \longrightarrow Mn^{2+} + 4\,H_2O$$

Step 3 Balance electrically with e^-

$$6\,H_2O + Co(NO_2)_6^{3-} \longrightarrow Co^{2+} + 6\,NO_3^- + 12\,H^+ + 11\,e^-$$
$$5\,e^- + 8\,H^+ + MnO_4^- \longrightarrow Mn^{2+} + 4\,H_2O$$

Step 4 Equalize the loss and gain of electrons.

$$5\,(6\,H_2O + Co(NO_2)_6^{3-} \longrightarrow Co^{2+} + 6\,NO_3^- + 12\,H^+ + 11\,e^-)$$
$$11\,(5\,e^- + 8\,H^+ + MnO_4^- \longrightarrow Mn^{2+} + 4\,H_2O)$$

Step 5 Add the half-reactions

$$5\,Co(NO_2)_6^{3-} + 11\,MnO_4^- + 28\,H^+ \longrightarrow 5\,Co^{2+} + 30\,NO_3^- + 11\,Mn^{2+} + 14\,H_2O$$

Each side has 5 Co, 30 N, 11 Mn, 28 H, 104 O and a $+2$ charge.

20. (a) $Mo_2O_3 + MnO_4^- \longrightarrow MoO_3 + Mn^{2+}$ (acid solution)

Step 1 Write half-reactions equations. Balance except H and O

$$Mo_2O_3 \longrightarrow 2\,MoO_3$$
$$MnO_4^- \longrightarrow Mn^{2+}$$

Step 2 Balance H and O using H_2O and H^+

$$3\,H_2O + Mo_2O_3 \longrightarrow 2\,MoO_3 + 6\,H^+$$
$$8\,H^+ + MnO_4^- \longrightarrow Mn^{2+} + 4\,H_2O$$

Step 3 Balance electrically with electrons

$$3\,H_2O + Mo_2O_3 \longrightarrow 2\,MoO_3 + 6\,H^+ + 6\,e^-$$
$$5\,e^- + 8\,H^+ + MnO_4^- \longrightarrow Mn^{2+} + 4\,H_2O$$

Steps 4 and 5 Equalize the loss and gain of electrons; add the half-reactions.

$$5\,(3\,H_2O + Mo_2O_3 \longrightarrow 2\,MoO_3 + 6\,H^+ + 6\,e^-)$$
$$\underline{6\,(5\,e^- + 8\,H^+ + MnO_4^- \longrightarrow Mn^{2+} + 4\,H_2O)}$$
$$5\,Mo_2O_3 + 6\,MnO_4^- + 18\,H^+ \longrightarrow 10\,MoO_3 + 6\,Mn^{2+} + 9\,H_2O$$

Each side has 18 H, 39 O, 10 Mo, 6 Mn, and a $+12$ charge.

(b) $BrO^- + Cr(OH)_4^- \longrightarrow Br^- + CrO_4^{2-}$ (basic solution)

Step 1 Write half-reaction equation. Balance except H and O

$$BrO^- \longrightarrow Br^-$$
$$Cr(OH)_4^- \longrightarrow CrO_4^{2-}$$

Step 2 Balance H and O using H_2O and H^+

$$2\,H^+ + BrO^- \longrightarrow Br^- + H_2O$$
$$Cr(OH)_4^- \longrightarrow CrO_4^{2-} + 4\,H^+$$

Step 3   Add OH^- ions to both sides (same number as H^+)

$$2\,OH^- + 2\,H^+ + BrO^- \longrightarrow Br^- + H_2O + 2\,OH^-$$

$$4\,OH^- + Cr(OH)_4^- \longrightarrow CrO_4^{2-} + 4\,H^+ + 4\,OH^-$$

Step 4   Combine H^+ and OH^- to form H_2O; cancel H_2O where possible

$$2\,H_2O + BrO^- \longrightarrow Br^- + H_2O + 2\,OH^-$$

$$4\,OH^- + Cr(OH)_4^- \longrightarrow CrO_4^{2-} + 4\,H_2O$$

$$H_2O + BrO^- \longrightarrow Br^- + 2\,OH^- \quad (\text{1 } H_2O \text{ cancelled})$$

Step 5   Balance electrically with electrons

$$2\,e^- + H_2O + BrO^- \longrightarrow Br^- + 2\,OH^-$$

$$4\,OH^- + Cr(OH)_4^- \longrightarrow CrO_4^{2-} + 4\,H_2O + 3\,e^-$$

Steps 6 and 7   Equalize loss and gain of electrons; add the half-reactions

$$3\,(2\,e^- + H_2O + BrO \longrightarrow Br^- + 2\,OH^-)$$

$$\underline{2\,(4\,OH^- + Cr(OH)_4^- \longrightarrow CrO_4^{2-} + 4\,H_2O + 3\,e^-)}$$

$$3\,BrO^- + 2\,Cr(OH)_4^- + 2\,OH^- \longrightarrow 3\,Br^- + 2\,CrO_4^{2-} + 5\,H_2O$$

Each side has 3 Br, 13 O, 2 Cr, 10 H, and a -7 charge.

(c) $S_2O_3^{2-} + MnO_4^- \longrightarrow SO_4^{2-} + MnO_2$ (basic solution)

Step 1   Write half-reaction equations. Balance except H and O.

$$S_2O_3^{2-} \longrightarrow 2\,SO_4^{2-}$$

$$MnO_4^- \longrightarrow MnO_2$$

Step 2   Balance H and O using H_2O and H^+

$$5\,H_2O + S_2O_3^{2-} \longrightarrow 2\,SO_4^{2-} + 10\,H^+$$

$$4\,H^+ + MnO_4^- \longrightarrow MnO_2 + 2\,H_2O$$

Step 3   Add OH^- ions to both sides (same number as H^+)

$$10\,OH^- + 5\,H_2O + S_2O_3^{2-} \longrightarrow 2\,SO_4^{2-} + 10\,H^+ + 10\,OH^-$$

$$4\,OH^- + 4\,H^+ + MnO_4^- \longrightarrow MnO_2 + 2\,H_2O + 4\,OH^-$$

Step 4   Combine H^+ and OH^- to form H_2O; cancel H_2O where possible

$$10\,OH^- + 5\,H_2O + S_2O_3^{2-} \longrightarrow 2\,SO_4^{2-} + 10\,H_2O$$

$$4\,H_2O + MnO_4^- \longrightarrow MnO_2 + 2\,H_2O + 4\,OH^-$$

$$10\,OH^- + S_2O_3^{2-} \longrightarrow 2\,SO_4^{2-} + 5\,H_2O \quad (\text{5 } H_2O \text{ cancelled})$$

$$2\,H_2O + MnO_4^- \longrightarrow MnO_2 + 4\,OH^- \quad\quad (\text{2 } H_2O \text{ cancelled})$$

Step 5   Balance electrically with electrons

$$10\,OH^- + S_2O_3^{2-} \longrightarrow 2\,SO_4^{2-} + 5\,H_2O + 8\,e^-$$

$$3\,e^- + 2\,H_2O + MnO_4^- \longrightarrow MnO_2 + 4\,OH^-$$

Step 6 and 7   Equalize loss and gain of electrons; add half-reactions.

$$3\,(10\,OH^- + S_2O_3^{2-} \longrightarrow 2\,SO_4^{2-} + 5\,H_2O + 8\,e^-)$$

$$\underline{8\,(3\,e^- + 2\,H_2O + MnO_4^- \longrightarrow MnO_2 + 4\,OH^-)}$$

$$3\,S_2O_3^{2-} + 8\,MnO_4^- + H_2O \longrightarrow 6\,SO_4^{2-} + 8\,MnO_2 + 2\,OH^-$$

Each side has 6 S, 8 Mn, 2 H, 42 O and a -14 charge.

21. $2 C_6H_8O_6 + O_2 \rightarrow 2 C_6H_6O_6 + 2 H_2O$

22.

$$\overset{+1\ +5\ -2}{4\,KNO_3(s)} + \overset{0}{7\,C(s)} + \overset{0}{S(s)} \rightarrow \overset{+4\ -2}{3\,CO_2(g)} + \overset{+2\ -2}{3\,CO(g)} + \overset{0}{2\,N_2(g)} + \overset{+1\ +4\ -2}{K_2CO_3(s)} + \overset{+1\ -2}{K_2S(s)}$$

Nitrogen and sulfur are reduced. Carbon is oxidized.

23.

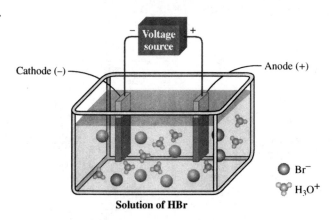

Cathode (–)

– Voltage source +

Anode (+)

Br⁻

H₃O⁺

Solution of HBr

24. (a) $Pb + SO_4{}^{2-} \longrightarrow PbSO_4 + 2\,e^-$

$PbO_2 + SO_4{}^{2-} + 4\,H^+ + 2\,e^- \longrightarrow PbSO_4 + 2\,H_2O$

(b)   The first reaction is oxidation (Pb^0 is oxidized to Pb^{2+}).
The second reaction is reduction (Pb^{4+} is reduced to Pb^{2+}).

(c)   The first reaction (oxidation) occurs at the anode of the battery.

25. (a)   The oxidizing agent is $KMnO_4$.

(b)   The reducing agent is HCl.

(c)   5 moles of electrons $5\,e^- + Mn^{7+} \longrightarrow Mn^{2+}$

$$\left(\frac{5\ mol\ e^-}{mol\ KMnO_4}\right)\left(\frac{6.022 \times 10^{23}\,e^-}{mol\ e^-}\right) = 3.011 \times 10^{24}\ \frac{electrons}{mol\ KMnO_4}$$

26. Zinc is a more reactive metal than copper so when corrosion occurs the zinc preferentially reacts. Zinc is above hydrogen in the Activity series of metals; copper is below hydrogen.

27. $3\,Cu + 8\,HNO_3 \longrightarrow 3\,Cu(NO_3)_2 + 2\,NO + 4\,H_2O$

$$(75.5\ g\ Cu)\left(\frac{1\ mol\ Cu}{63.55\ g}\right)\left(\frac{2\ mol\ NO}{3\ mol\ Cu}\right)\left(\frac{22.4\ L}{mol}\right) = 17.7\ L\ NO$$

$$(55.0\ g\ HNO_3)\left(\frac{1\ mol\ HNO_3}{63.02\ g}\right)\left(\frac{2\ mol\ NO}{8\ mol\ HNO_3}\right)\left(\frac{22.4\ L}{mol}\right) = 4.89\ L\ NO$$

4.89 L of NO gas at STP will be produced; HNO_3 is limiting.
Cu is oxidized; N is reduced.

28. $K_2S_2O_8 + H_2C_2O_4 \longrightarrow K_2SO_4 + H_2SO_4 + 2\,CO_2$

$$(25.5 \text{ g } K_2S_2O_8)\left(\frac{1 \text{ mol } K_2S_2O_8}{270.3 \text{ g}}\right)\left(\frac{2 \text{ mol } CO_2}{1 \text{ mol } HNO_3}\right)\left(\frac{22.4 \text{ L}}{\text{mol}}\right) = 4.23 \text{ L } CO_2$$

$$(35.5 \text{ g } H_2C_2O_4)\left(\frac{1 \text{ mol } H_2C_2O_4}{90.04 \text{ g}}\right)\left(\frac{2 \text{ mol } CO_2}{1 \text{ mol } H_2C_2O_4}\right)\left(\frac{22.4 \text{ L}}{\text{mol}}\right) = 17.7 \text{ L } CO_2$$

4.23 L CO_2 will be produced; $K_2S_2O_8$ is limiting.
C in $H_2C_2O_4$ is oxidized and S in $K_2S_2O_8$ is reduced.

29. $C_6H_8O_6(aq) + KI_3(aq) \rightarrow C_6H_6O_6(aq) + KI(aq) + 2\,HI(aq)$

$$0.03261 \text{ L}\left(\frac{0.03741 \text{ mol } KI_3}{1 \text{ L}}\right)\left(\frac{1 \text{ mol } C_6H_8O_6}{1 \text{ mol } KI_3}\right)\left(\frac{176.1 \text{ g}}{1 \text{ mol}}\right) = 0.2148 \text{ g in 100 g pepper}$$

$$\left(\frac{0.2148 \text{ g } C_6H_8O_6}{100 \text{ g pepper}}\right)\left(\frac{1000 \text{ mg}}{1 \text{ g}}\right) = \frac{2.148 \text{ mg } C_6H_8O_6}{\text{g pepper}}$$

30. $CH_3CH_2OH + Cr_2O_7{}^{2-} \rightarrow CH_3COOH + Cr^{3+}$

$3(CH_3CH_2OH + H_2O \rightarrow CH_3COOH + 4\,H^+ + 4\,e^-)$
$2(Cr_2O_7{}^{2-} + 14\,H^+ + 6\,e^- \rightarrow 2\,Cr^{3+} + 7\,H_2O)$
$$\overline{3\,CH_3CH_2OH + 3\,H_2O + 2\,Cr_2O_7{}^{2-} + 28\,H^+ \rightarrow 3\,CH_3CO_2H + 12\,H^+ + 4\,Cr^{3+} + 14\,H_2O}$$

Which simplifies to

$3\,CH_3CH_2OH + 2\,Cr_2O_7{}^{2-} + 16\,H^+ \rightarrow 3\,CH_3COOH + 4\,Cr^{3+} + 11\,H_2O$

31. $5\,H_2O_2 + 2\,KMnO_4 + 3\,H_2SO_4 \longrightarrow 5\,O_2 + 2\,MnSO_4 + K_2SO_4 + 8\,H_2O$

mL $H_2O_2 \longrightarrow$ g $H_2O_2 \longrightarrow$ mol $H_2O_2 \longrightarrow$ mol $KMnO_4 \longrightarrow$ g $KMnO_4$

$$(100. \text{ mL } H_2O_2 \text{ solution})\left(\frac{1.031 \text{ g}}{\text{mL}}\right)\left(\frac{9.0 \text{ g } H_2O_2}{100. \text{ g } H_2O_2 \text{ solution}}\right)\left(\frac{1 \text{ mol}}{34.02 \text{ g}}\right)$$

$$\left(\frac{2 \text{ mol } KMnO_4}{5 \text{ mol } H_2O_2}\right)\left(\frac{158.0 \text{ g}}{\text{mol}}\right) = 17 \text{ g } KMnO_4$$

32. $3\,Zn + 2\,Fe^{3+} \longrightarrow 3\,Zn^{2+} + 2\,Fe$

$$(0.0250 \text{ L } FeCl_3)\left(\frac{1.2 \text{ mol } FeCl_3}{\text{L}}\right)\left(\frac{3 \text{ mol } Zn}{2 \text{ mol } FeCl_3}\right)\left(\frac{65.39 \text{ g}}{\text{mol}}\right) = 2.94 \text{ g } Zn$$

33. $Cr_2O_7{}^{2-} + 6\,Fe^{2+} + 14\,H^+ \longrightarrow 2\,Cr^{3+} + 6\,Fe^{3+} + 7\,H_2O$

mL $FeSO_4 \longrightarrow$ mol $FeSO_4 \longrightarrow$ mol $Cr_2O_7{}^{2-} \longrightarrow$ mL $Cr_2O_7{}^{2-}$

$$(60.0 \text{ mL } FeSO_4)\left(\frac{0.200 \text{ mol}}{1000 \text{ mL}}\right)\left(\frac{1 \text{ mol } Cr_2O_7{}^{2-}}{6 \text{ mol } FeSO_4}\right)\left(\frac{1000 \text{ mL}}{0.200 \text{ mol}}\right)$$

$$= 10.0 \text{ mL of 0.200 M } K_2Cr_2O_7$$

34. $2\,Al + 2\,OH^- + 6\,H_2O \longrightarrow 2\,Al(OH)_4^- + 3\,H_2$

    $g\,Al \longrightarrow mol\,Al \longrightarrow mol\,H_2$

    $(100.0\,g\,Al)\left(\dfrac{1\,mol\,Al}{26.98\,g}\right)\left(\dfrac{3\,mol\,H_2}{2\,mol\,Al}\right) = 5.560\,mol\,H_2$

35. (a) $Cu^+ \longrightarrow Cu^{2+}$ is an oxidation, but when electrons are gained reduction should occur.

    $Cu^+ + e^- \longrightarrow Cu^0$ or $Cu^+ \longrightarrow Cu^{2+} + e^-$

    (b) When Pb^{2+} is reduced, it requires two individual electrons. $Pb^{2+} + 2\,e^- \longrightarrow Pb^0$.
    An electron has only a single negative charge (e^-).

36. The electrons lost by the species undergoing oxidation must be gained (or attracted) by another species which then undergoes reduction.

37. $A(s) + B^{2+}(aq) \longrightarrow NR$ B^{2+} cannot take e^- from A
    $A(s) + C^+(aq) \longrightarrow NR$ C^+ cannot take e^- from A
    $D(s) + 2\,C^+(aq) \longrightarrow 2C(s) + D^{2+}(aq)$ C^+ takes e^- from D
    $B(s) + D^{2+}(aq) \longrightarrow D(s) + B^{2+}(aq)$ D^{2+} takes $2\,e^-$ from B
    Therefore, B^{2+} is least able to attract e^-, then D^{2+}, then C^+, then A^+

38. Sn^{4+} can only be an oxidizing agent.

    Sn^0 can only be a reducing agent.

    Sn^{2+} can be both oxidizing and reducing.

    $Sn^{4+} + 2\,e^- \longrightarrow Sn^{2+}$
    $Sn^{4+} + 4\,e^- \longrightarrow Sn^0$
    $Sn^0 \longrightarrow Sn^{2+} + 2\,e^-$
    $Sn^0 \longrightarrow Sn^{4+} + 4\,e^-$
    $Sn^{2+} + 2\,e^- \longrightarrow Sn^0$ (oxidizing agent)
    $Sn^{2+} \longrightarrow Sn^{4+} + 2\,e^-$ (reducing agent)

39. $Mn(OH)_2$ +2 $KMnO_4$ is the best oxidizing agent of the group, since its greater
    MnF_3 +3   oxidation number (+7) makes it very attractive to electrons.
    MnO_2 +4
    K_2MnO_4 +6
    $KMnO_4$ +7

40. Equations (a) and (b) represent oxidation
    (a) $Mg \longrightarrow Mg^{2+} + 2\,e^-$
    (b) $SO_2 \longrightarrow SO_3;\ (S^{4+} \longrightarrow S^{6+} + 2\,e^-)$

41. (a) $MnO_2 + 2\,Br^- + 4\,H^+ \longrightarrow Mn^{2+} + Br_2 + 2\,H_2O$
    (b) $mL\,Mn^{2+} \longrightarrow mol\,Mn^{2+} \longrightarrow mol\,MnO_2 \longrightarrow g\,MnO_2$

    $(100.0\,mL\,Mn^{2+})\left(\dfrac{0.05\,mol}{1000\,mL}\right)\left(\dfrac{1\,mol\,MnO_2}{1\,mol\,Mn^{2+}}\right)\left(\dfrac{86.94\,g}{mol}\right) = 0.4\,g\,MnO_2$

    (c) $(100.0\,mL\,Mn^{2+})\left(\dfrac{0.05\,mol}{1000\,mL}\right)\left(\dfrac{1\,mol\,Br_2}{1\,mol\,Mn^{2+}}\right) = 0.005\,mol\,Br_2$

$$PV = nRT \qquad V = \frac{nRT}{P}$$

$$V = \left(\frac{0.005 \text{ mol}}{1.4 \text{ atm}}\right)\left(\frac{0.0821 \text{ L atm}}{\text{mol K}}\right)(323 \text{ K}) = 0.09 \text{ L Br}_2 \text{ vapor}$$

42. (a) $F_2 + 2\,Cl^- \longrightarrow 2\,F^- + Cl_2$
    (b) $Br_2 + Cl^- \longrightarrow NR$
    (c) $I_2 + Cl^- \longrightarrow NR$
    (d) $Br_2 + 2\,I^- \longrightarrow 2\,Br^- + I_2$

43. $Mn(s) + 2\,HCl(aq) \longrightarrow Mn^{2+}(aq) + H_2(g) + 2\,Cl^-(aq)$

44. $4\,Zn + NO_3^- + 10\,H^+ \longrightarrow 4\,Zn^{2+} + NH_4^+ + 3\,H_2O$ See Exercise 15(a).

45.

	Equation 1	2	3	4	5
a	C oxidized	S oxidized	N oxidized	S oxidized	O_2^{2-} oxidized
b	O_2 reduced	N reduced	Cu reduced	O_2^{2-} reduced	O_2^{2-} reduced
c	O_2, O.A.	HNO_3, O.A.	CuO, O.A.	H_2O_2, O.A.	H_2O_2, O.A.
d	C_3H_8, R.A.	H_2S, R.A.	NH_3, R.A.	Na_2SO_3, R.A.	H_2O_2, R.A.
e	$C^{2\frac{2}{3}+} \longrightarrow C^{4+}$	$S^{2-} \longrightarrow S^0$	$N^{3-} \longrightarrow N_2^0$	$S^{4+} \longrightarrow S^{6+}$	$O_2^{2-} \longrightarrow O_2^0$
f	$O^0 \longrightarrow O^{2-}$	$N^{5+} \longrightarrow N^{2+}$	$Cu^{2+} \longrightarrow Cu^0$	$O_2^{2-} \longrightarrow O^{2-}$	$O_2^{2-} \longrightarrow O^{2-}$

O.A. = Oxidizing agent
R.A. = Reducing agent

46. $Pb + 2\,Ag^+ \longrightarrow 2\,Ag + Pb^{2+}$
    (a) Pb is the anode
    (b) Ag is the cathode
    (c) Oxidation occurs at Pb (anode)
    (d) Reduction occurs at Ag (cathode)
    (e) Electrons flow from the lead electrode through the wire to the silver electrode.
    (f) Positive ions flow through the salt bridge towards the negatively charged cathode of silver; negative ions flow toward the positively charged anode of lead.

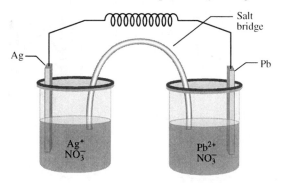

47. $8\,KI + 5\,H_2SO_4 \longrightarrow 4\,I_2 + H_2S + 4\,K_2SO_4 + 4\,H_2O$
    start with grams of I_2 and work towards g of KI

    $g\,I_2 \longrightarrow mol\,I_2 \longrightarrow mol\,KI \longrightarrow g\,KI$

    $(2.79\,g\,I_2)\left(\dfrac{1\,mol}{253.8\,g}\right)\left(\dfrac{8\,mol\,KI}{4\,mol\,I_2}\right)\left(\dfrac{166.0\,g}{mol}\right) = 3.65\,g\,KI$ in sample

    $\left(\dfrac{3.65\,g\,KI}{4.00\,g\,sample}\right)(100) = 91.3\%\,KI$

48. $3\,Ag + 4\,HNO_3 \longrightarrow 3\,AgNO_3 + NO + 2\,H_2O$

    $mol\,Ag \longrightarrow mol\,NO$

    $(0.500\,mol\,Ag)\left(\dfrac{1\,mol\,NO}{3\,mol\,Ag}\right) = 0.167\,mol\,NO$

    $PV = nRT \qquad V = \dfrac{nRT}{P}$

    $P = (744\,torr)\left(\dfrac{1\,atm}{760.\,torr}\right) = 0.979\,atm$

    $T = 301\,K$

    $V = \dfrac{(0.167\,mol\,NO)(0.0821\,L\,atm/mol\,K)(301\,K)}{(0.979\,atm)} = 4.22\,L\,NO$

CHAPTER 18

NUCLEAR CHEMISTRY

SOLUTIONS TO REVIEW QUESTIONS

1. Contributions to the early history of radioactivity include:
   (a) Henri Becquerel: He discovered radioactivity.
   (b) Marie and Pierre Curie: They discovered the elements polonium and radium.
   (c) Wilhelm Roentgen: He discovered X rays and developed the technique of producing them. While this was not a radioactive phenomenon, it triggered Becquerel's discovery of radioactivity.
   (d) Ernest Rutherford: He discovered alpha and beta particles, established the link between radioactivity and transmutation, and produced the first successful man-made transmutation.
   (e) Otto Hahn and Fritz Strassmann: They were first to produce nuclear fission.

2. Chemical reactions are caused by atoms or ions coming together, so are greatly influenced by temperature and concentration, which affect the number of collisions. Radioactivity is a spontaneous reaction of an individual nucleus, and is independent of such influences.

3. The term isotope is used with reference to atoms of the same element that contains different masses. For example, $^{12}_{6}C$ and $^{14}_{6}C$. The term nuclide is used in nuclear chemistry to infer any isotope of any atom.

4. $(5 \times 10^9 \text{ years}) \left(\dfrac{1 \text{ half-life}}{7.6 \times 10^7 \text{ year}} \right) = 70 \text{ half-lives}$

   Even if plutonium-244 had been present in large quantities five billion years ago, no measureable amount would survive after 70 half-lives.

5. In living species, the ratio of carbon-14 to carbon-12 is constant due to the constant C-14/C-12 ratio in the atmosphere and food sources. When a species dies, life processes stop. The C-14/C-12 ratio decreases with time because C-14 is radioactive and decays according to its half-life, while the amount of C-12 in the species remains constant. Thus, the age of an archaeological artifact containing carbon can be calculated by comparing the C-14/C-12 ratio in the artifact with the C-14/C-12 ratio in the living species.

6. The half-life of carbon-14 is 5730 years.

   $(4 \times 10^6 \text{ years}) \left(\dfrac{1 \text{ half-life}}{5730 \text{ years}} \right) = 7 \times 10^2 \text{ half-lives}$

   700 half-lives would pass in 4 million years. Not enough C-14 would remain to allow detection with any degree of reliability. C-14 dating would not prove useful in this case.

7. (a) Gamma radiation requires the most shielding.
   (b) Alpha radiation requires the least shielding.

8. Alpha particles are deflected less than beta particles while passing through a magnetic field, because they are much heavier (more than 7,000 times heavier) than beta particles.

9.

	charge	mass	nature of particles	penetrating power
Alpha	$+2$	4 amu	He nucleus	low
Beta	-1	$\dfrac{1}{1837\ \text{amu}}$	electron	moderate
Gamma	0	0	electromagnetic radiation	high

10. Decay of bismuth-211

$$^{211}_{83}\text{Bi} \longrightarrow {}^{4}_{2}\text{He} + {}^{207}_{81}\text{Tl} \qquad {}^{207}_{81}\text{Tl} \longrightarrow {}^{0}_{-1}\text{e} + {}^{207}_{82}\text{Pb}$$

11. Pairs of nuclides that would be found in the fission reaction of U-235. Any two nuclides, whose atomic numbers add up to 92 and mass numbers (in the range of 70-160) add up to 230-234. Examples include:

$$^{90}_{38}\text{Sr} \quad \text{and} \quad {}^{141}_{54}\text{Xe} \qquad\qquad {}^{139}_{56}\text{Ba} \quad \text{and} \quad {}^{94}_{36}\text{Kr} \qquad\qquad {}^{101}_{42}\text{Mo} \quad \text{and} \quad {}^{131}_{50}\text{Sn}$$

12. Natural radioactivity is the spontaneous disintegration of those radioactive nuclides found in nature. Artificial radioactivity is the spontaneous disintegration of radioactive isotopes produced synthetically, otherwise it is the same as natural radioactivity.

13. A radioactive disintegration series starts with a particular radionuclide and progresses stepwise by alpha and beta emissions to other radionuclides, ending at a stable nuclide. For example:

$$^{238}_{92}\text{U} \xrightarrow{\ 14\ \text{steps}\ } {}^{206}_{82}\text{Pb (stable)}$$

14. Transmutation is the conversion of one element into another by natural or artificial means. The nucleus of an atom is bombarded by various particles (alpha, beta, protons, etc.). The fast moving particles are captured by the nucleus, forming an unstable nucleus, which decays to another kind of atom. For example:

$$^{9}_{4}\text{Be} + {}^{4}_{2}\text{He} \longrightarrow {}^{12}_{6}\text{C} + {}^{1}_{0}\text{n}$$

15. $^{232}_{90}\text{Th} \xrightarrow{-\alpha} {}^{228}_{88}\text{Ra} \xrightarrow{-\beta} {}^{228}_{89}\text{Ac} \xrightarrow{-\beta} {}^{228}_{90}\text{Th} \xrightarrow{-\alpha} {}^{224}_{88}\text{Ra} \xrightarrow{-\alpha} {}^{220}_{86}\text{Rn} \xrightarrow{-\alpha}$

$^{216}_{84}\text{Po} \xrightarrow{-\alpha} {}^{212}_{82}\text{Pb} \xrightarrow{-\beta} {}^{212}_{83}\text{Bi} \xrightarrow{-\beta} {}^{212}_{84}\text{Po} \xrightarrow{-\alpha} {}^{208}_{82}\text{Pb}$

16. $^{237}_{93}\text{Np}$ loses seven alpha particles and four beta particles.
Determination of the final product: $^{209}_{83}\text{Bi}$

nuclear charge $= 93 - 7(2) + 4(1) = 83$
mass $= 237 - 7(4) = 209$

17. Radioactivity could be used to locate a leak in an underground pipe by using a water soluble tracer element. Dissolve the tracer in water and pass the water through the pipe. Test the ground along the path of the pipe with a Geiger counter until radioactivity from the leak is detected. Then dig.

18. A scintillation counter is a radiation detector which contains molecules that emit light when they are struck by ionizing radiation. The number of light flashes are recorded as a numerical output of the radiation level.

19. The curie describes the amount of radioactivity produced by an element. One curie is equal to 3.7×10^{10} disintegrations/sec

20. REM stands for roentgen equivalent to man and measures the effective exposure to ionizing radiation.

21. Two Germans, Otto Hahn and Fritz Strassmann, were the first scientists to report nuclear fission. The fission resulted from bombarding uranium nuclei with neutrons.

22. Natural uranium is 99+% U-238. Commercial nuclear reactors use U-235 enriched uranium as a fuel. Slow neutrons will cause the fission of U-235, but not U-238. Fast neutrons are capable of a nuclear reaction with U-238 to produce fissionable Pu-239. A breeder reactor converts nonfissionable U-238 to fissionable Pu-239, and in the process, manufactures more fuel than it consumes.

23. The fission reaction in a nuclear reactor and in an atomic bomb are essentially the same. The difference is that the fissioning is "wild" or uncontrolled in the bomb. In a nuclear reactor, the fissioning rate is controlled by means of moderators, such as graphite, to slow the neutrons and control rods of cadmium or boron to absorb some of the neutrons.

24. A certain amount of fissionable material (a critical mass) must be present before a self-sustaining chain reaction can occur. Without a critical mass, too many neutrons from fissions will escape, and the reaction cannot reach a chain reaction status, unless at least one neutron is captured for every fission that occurs.

25. The disadvantages of nuclear power include the danger of contamination from radioactive material and the radioactive waste products that accumulate, some having half-lives of thousands of years.

26. The hazards associated with an atomic bomb explosion include shock waves, heat and radiation from alpha particles, beta particles, gamma rays, and ultraviolet rays. Gamma rays and X-rays cause burns, sterilization, and gene mutation. If the bomb explodes close to the ground radioactive material is carried by dust particles and is spread over wide areas.

27. Heavy elements undergo fission and lighter elements undergo fusion.

28. In a nuclear power plant, a controlled nuclear fission reaction provides heat energy that is used to produce steam. The steam turns a turbine that generates electricity.

29. The mass defect is the difference between the mass of an atom and the sum of the masses of the number of protons, neutrons, and electrons in that atom. The energy equivalent of this mass defect is known as the nuclear binding energy.

30. When radioactive rays pass through normal matter, they cause that matter to become ionized (usually by knocking out electrons). Therefore, the radioactive rays are classified as ionizing radiation.

31. Some biological hazards associated with radioactivity are:
    (a) High levels of radiation can cause nausea, vomiting, diarrhea, and death. The radiation produces ionization in the cells, particularly in the nucleus of the cells.
    (b) Long-term exposure to low levels of radiation can weaken the body and cause malignant tumors.
    (c) Radiation can damage DNA molecules in the body causing mutations, which by reproduction, can be passed on to succeeding generations.

32. Strontium-90 has two characteristics that create concern. Its half-life is 28 years, so it remains active for a long period of time (disintegrating by emitting β radiation). The other characteristic is that Sr-90 is chemically similar to calcium, so when it is present in milk Sr-90 is deposited in bone tissue along with calcium. Red blood cells are produced in the bone marrow. If the marrow is subjected to beta radiation from strontium-90, the red blood cells will be destroyed, increasing the incidence of leukemia and bone cancer.

33. A radioactive "tracer" is a radioactive material, whose presence is traced by a Geiger counter or some other detecting device. Tracers are often injected into the human body, animals, and plants to determine chemical pathways, rates of circulation, etc. For example, use of a tracer could determine the length of time for material to travel from the root system to the leaves in a tree.

SOLUTIONS TO EXERCISES

1.

		Protons	Neutrons	Nucleons
(a)	$^{207}_{82}Pb$	82	125	207
(b)	$^{70}_{31}Ga$	31	39	70

2.

		Protons	Neutrons	Nucleons
(a)	$^{128}_{52}Te$	52	76	128
(b)	$^{32}_{16}S$	16	16	32

3. When a nucleus loses an alpha particle, its atomic number decreases by two, and its mass number decreases by four.

4. When a nucleus loses a beta particle, its atomic number increases by one, and its mass number remains unchanged.

5. Equations for alpha decay:
   (a) $^{210}_{83}Bi \longrightarrow {}^{4}_{2}He + {}^{206}_{81}Ra$
   (b) $^{238}_{92}U \longrightarrow {}^{4}_{2}He + {}^{234}_{90}Th$

6. Equations for alpha decay:
   (a) $^{238}_{90}Th \longrightarrow {}^{4}_{2}He + {}^{234}_{88}Ra$
   (b) $^{239}_{94}Pu \longrightarrow {}^{4}_{2}He + {}^{235}_{92}U$

7. Equations for beta decay:
   (a) $^{13}_{7}N \longrightarrow {}^{0}_{-1}e + {}^{13}_{8}O$
   (b) $^{234}_{90}Th \longrightarrow {}^{0}_{-1}e + {}^{234}_{91}Pa$

8. Equations for beta decay:
   (a) $^{28}_{13}Al \longrightarrow {}^{0}_{-1}e + {}^{28}_{14}Si$
   (b) $^{239}_{93}Np \longrightarrow {}^{0}_{-1}e + {}^{239}_{94}Pu$

9. (a) alpha-emission   (b) beta-emission then gamma-emission   (c) beta-emission

10. (a) gamma-emission   (b) alpha-emission then beta-emission
    (c) alpha-emission then gamma-emission

11. $^{26}_{13}Al \longrightarrow {}^{0}_{+1}e + {}^{26}_{12}Mg$

12. $^{32}_{15}P \longrightarrow {}^{0}_{-1}e + {}^{32}_{16}S$

13. (a) $^{66}_{29}\text{Cu} \longrightarrow ^{66}_{30}\text{Zn} + ^{0}_{-1}\text{e}$

(b) $^{0}_{-1}\text{e} + ^{7}_{4}\text{Be} \longrightarrow ^{7}_{3}\text{Li}$

(c) $^{27}_{13}\text{Al} + ^{4}_{2}\text{He} \longrightarrow ^{30}_{14}\text{Si} + ^{1}_{1}\text{H}$

(d) $^{85}_{37}\text{Rb} + ^{1}_{0}\text{n} \longrightarrow ^{82}_{35}\text{Br} + ^{4}_{2}\text{He}$

14. (a) $^{27}_{13}\text{Al} + ^{4}_{2}\text{He} \longrightarrow ^{30}_{15}\text{P} + ^{1}_{0}\text{n}$

(b) $^{27}_{14}\text{Si} \longrightarrow ^{0}_{+1}\text{e} + ^{27}_{13}\text{Al}$

(c) $^{12}_{6}\text{C} + ^{2}_{1}\text{H} \longrightarrow ^{13}_{7}\text{N} + ^{1}_{0}\text{n}$

(d) $^{82}_{35}\text{Br} \longrightarrow ^{82}_{36}\text{Kr} + ^{0}_{-1}\text{e}$

15. $(112 \text{ years})\left(\dfrac{1 \text{ half-life}}{28 \text{ years}}\right) = 4 \text{ half-lives}$

In 4 half-lives $1/16$th or $(^1/_2)^4$ of the starting amount would remain.

$\dfrac{1.00 \text{ mg Sr-90}}{16} = 0.0625 \text{ mg Sr-90 remains after 112 years.}$

16. $\dfrac{240 \text{ Cts/min}}{2} = 120 \text{ Cts/min};\quad \dfrac{120 \text{ Cts/min}}{2} = 60 \text{ Cts/min};\quad \dfrac{60 \text{ Cts/min}}{2} = 30 \text{ Cts/min};$

3 half-lives are required to reduce the count from 240 to 30 counts/min.

$1980 + (3 \times 28) = 2064$. After 1 half-life, 1/2 of the sample remains, after 2 half-lives 1/4 of the sample remains, and after 3 half-lives 1/8 of the sample remains.

17. Loss of mass: $233 - 225 = 8$, equivalent to 2 alpha particles.

Loss in atomic number: $91 - 89 = 2$, equivalent to 1 alpha particle. With a loss of 2 alpha particles the loss of atomic number should be 4. Therefore, there must also be a loss of 2 beta particles to increase the atomic number by 2. So one possible series is:

$^{233}_{91}\text{Pa} \xrightarrow{-\beta} ^{233}_{92}\text{U} \xrightarrow{-\alpha} ^{229}_{90}\text{Th} \xrightarrow{-\alpha} ^{225}_{88}\text{Ra} \xrightarrow{-\beta} ^{225}_{89}\text{Ac}$

18. Loss of mass: $228 - 212 = 16$, which is equivalent to 4 alpha particles.

Loss in atomic number $90 - 82 = 8$, which is equivalent to 4 alpha particles. This looks like a total loss of 4 alpha particles. Therefore the series is:

$^{228}_{90}\text{Th} \xrightarrow{-\alpha} ^{224}_{88}\text{Ra} \xrightarrow{-\alpha} ^{220}_{86}\text{Rn} \xrightarrow{-\alpha} ^{216}_{84}\text{Po} \xrightarrow{-\alpha} ^{212}_{82}\text{Pb}$

19. (a) $^{235}_{92}U + ^{1}_{0}n \longrightarrow ^{94}_{38}Sr + ^{139}_{54}Xe + 3\,^{1}_{0}n + energy$

Mass loss = mass of reactants − mass of products

Mass of reactants = 235.0439 amu + 1.0087 amu = 236.0526 amu

Mass of products = 93.9154 amu + 138.9179 amu + 3(1.0087 amu) = 235,8594 amu

Mass lost = 236.0526 amu − 235.8594 amu = 0.1932 amu

$(0.1932\ amu)\left(\dfrac{1.000\ g}{6.022 \times 10^{23}\ amu}\right)\left(\dfrac{9.0 \times 10^{13}\ J}{1.00\ g}\right) = 2.9 \times 10^{-11}\ J/atom\ U\text{-}235$

(b) $\left(\dfrac{2.9 \times 10^{-11}\ J}{atom}\right)\left(\dfrac{6.022 \times 10^{23}\ atoms}{mol}\right) = 1.7 \times 10^{13}\ J/mol$

(c) $\left(\dfrac{0.1932\ amu}{236.0526\ amu}\right)(100) = 0.08185\%$ mass loss

20. (a) $^{1}_{1}H + ^{2}_{1}H \longrightarrow ^{3}_{2}He + energy$

Mass loss = mass of reactants − mass of products

Mass of reactants = 1.00794 g/mol + 2.01410 g/mol = 3.02204 g/mol

Mass of products = 3.01603 g/mol

Mass lost = 3.02204 g/mol − 3.01603 g/mol = 0.00601 g/mol

$\left(\dfrac{0.00601\ g}{mol}\right)\left(\dfrac{9.0 \times 10^{13}\ J}{g}\right) = 5.4 \times 10^{11} J/mol$

(b) $\left(\dfrac{0.00601\ g}{3.02204\ g}\right)(100) = 0.199\%$ mass loss

21. (a) Chromium-51: 24 protons; 27 neutrons; 24 electrons
    (b) Holmium-166: 67 protons; 99 neutrons; 67 electrons
    (c) Palladium-103: 46 protons; 57 neutrons; 46 electrons
    (d) Strontium-89: 38 protons; 51 neutrons; 38 electrons

22. $^{232}_{90}Th \longrightarrow ^{208}_{82}Pb$

$^{232}_{90}Th \xrightarrow{-\alpha} ^{228}_{88}Ra \xrightarrow{-\beta} ^{228}_{89}Th \xrightarrow{-\beta} ^{228}_{90}Ac \xrightarrow{-\alpha} ^{224}_{88}Ra \xrightarrow{-\alpha} ^{220}_{86}Rn$

$\xrightarrow{-\alpha} ^{216}_{84}Po \xrightarrow{-\beta} ^{216}_{85}At \xrightarrow{-\alpha} ^{212}_{83}Bi \xrightarrow{-\beta} ^{212}_{84}Po \xrightarrow{-\alpha} ^{208}_{82}Pb$

23. $\dfrac{1}{2} \times 25.0 = 12.5$ g left after one half of the sample disintegrates.

$(1\ half\text{-}life)(1.25 \times 10^9\ years)\left(\dfrac{12\ mo}{1\ yr}\right) = 1.50 \times 10^{10}$ months.

24. $^{249}_{98}\text{Cf} + {}^{15}_{7}\text{N} \longrightarrow + 4^1_0\text{n} + {}^{260}_{105}\text{Db}$

25. $^{226}_{88}\text{Ra}$ contains 138 neutrons and 88 electrons

    mass of neutron $= 1.0087$ amu $\qquad$ mass of electron $= 0.00055$ amu

    $\dfrac{(138)(1.0087 \text{ amu})}{226 \text{ amu}} (100) = 61.59\%$ neutrons by mass

    $\dfrac{(88)(0.00055 \text{ amu})}{226 \text{ amu}} (100) = 0.021\%$ electrons by mass

26. $(0.0100 \text{ g RaCl}_2)\left(\dfrac{226.0 \text{ g Ra}}{296.9 \text{ g RaCl}_2}\right)\left(\dfrac{\$90,000}{1 \text{ g Ra}}\right) = \$685$

27. 100% to 25% requires 2 half-lives. The half-life of C-14 is 5730 years. The specimen will be the age of two half-lives:

    $(2)(5730 \text{ years}) = 11,460$ years old.

28. $16.0 \text{ g} \longrightarrow 8.0 \text{ g} \longrightarrow 4.0 \text{ g} \longrightarrow 2.0 \text{ g} \longrightarrow 1.0 \text{ g} \longrightarrow 0.50 \text{ g}$

    16.0 g to 0.50 g requires five half-lives.

    $\dfrac{90 \text{ minutes}}{5 \text{ half-lives}} = 18$ minutes/half-life

29. (a) ^7_3Li is made up of 3 protons, 4 neutrons, and 3 electrons.

    Calculated mass

3 protons	$3(1.0073 \text{ g})$	=	3.0219 g
4 neutrons	$4(1.0087 \text{ g})$	=	4.0348 g
3 electrons	$3(0.00055 \text{ g})$	=	0.0017 g
calculated mass			7.0584 g

    Mass defect $=$ calculated mass $-$ actual mass

    Mass defect $= 7.0584 \text{ g} - 7.0160 \text{ g} = 0.0424$ g/mol

    (b) Binding energy

    $\left(\dfrac{0.0424 \text{ g}}{\text{mol}}\right)\left(\dfrac{9.0 \times 10^{13}\text{J}}{\text{g}}\right) = 3.8 \times 10^{12}\text{J/mol}$

30. $^{235}_{92}\text{U} \longrightarrow {}^{207}_{82}\text{Pb}$

    Mass loss: $235 - 207 = 28$

    Net proton loss (atomic number): $92 \text{ p} - 82 \text{ p} = 10 \text{ p}$

    The mass loss is equivalent to 7 alpha particles (28/4). A loss of 7 alpha particles gives a loss of 14 protons. A decrease in the atomic number to 78 (14 protons) is due to the loss of 7 alpha particles ($92 - 14 = 78$). Therefore, a loss of 4 beta particles is required to increase the atomic number from 78 to 82.

    The total loss $=$ 7 alpha particles and 4 beta particles.

31. (a) Geiger counter: Radiation passes through a thin glass window into a chamber filled with argon gas and containing two electrodes. Some of the argon ionizes, sending a momentary electrical impulse between the electrodes to the detector. This signal is amplified electronically and read out on a counter or as a series of clicks.
    (b) Scintillation counter: Radiation strikes a scintillator, which is composed of molecules that emit light in the presence of ionizing radiation. A light sensitive detector counts the flashes and converts them into a digital readout.
    (c) Film badge: Radiation penetrates a film holder. The silver grains in the film darken when exposed to radiation. The film is developed at regular intervals.

32. (3 days)(24 hours/day) = 72 hours

    72 hr + 6 hr = 78 hr

    $$\frac{78 \text{ hr}}{13 \frac{\text{hr}}{t_{\frac{1}{2}}}} = 6 \text{ half-lives} \qquad \left(\frac{1}{2}\right)^6 (10 \text{ mg}) = 0.16 \text{ mg remaining}$$

33. First change micrograms to nanograms.

    15.0 μg (1000 ng/μg) = 15,000 ng

    15,000 ng → 7,500 ng → 3,750 ng → 1,875 ng → 937 ng → 468 ng → 234 ng → 117 ng → 59 ng → 29 ng → 15 ng → 7 ng → 4 ng →< 2 ng →< 1 ng

    Each arrow represents 1 half-life. There are 14 half-lives for a total of 112 days, (14 half-lives)(8 days/half-life) = 112 days, for the iodine−131 to decay to less than a nanogram.

34. If you were to start with a 100 μg sample more than 99% would decay after 7 half-lives. This would take 5 ½ hours for the bismuth-213, 28 days for the rhenium-186, and 35 years for the cobalt-60. The best choice is the rhenium-186 because it will be detectable long enough for the termites to move the sample, but not so long that it becomes a hazard.

35. $^{241}_{95}\text{Am} \rightarrow {}^{4}_{2}\alpha + {}^{237}_{93}\text{Np}$

36. Fission is the splitting of a heavy nuclide into two or more intermediate-sized fragments with the conversion of some mass into energy. Fission occurs in nuclear reactors, or atomic bombs.

    Example: $^{235}_{92}\text{U} + {}^{1}_{0}\text{n} \longrightarrow {}^{143}_{54}\text{Xe} + {}^{90}_{38}\text{Sr} + 3\,{}^{1}_{0}\text{n}$

    Fusion is the process of combining two relatively small nuclei to form a single larger nucleus. Fusion occurs on the sun, or in a hydrogen bomb.

    Example: $^{3}_{1}\text{H} + {}^{2}_{1}\text{H} \longrightarrow {}^{4}_{2}\text{He} + {}^{1}_{0}\text{n} + \text{energy}$

37.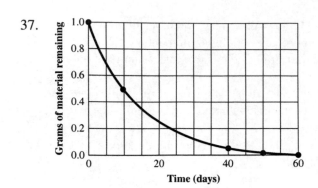

The graph produces a curve for radioactive decay which never actually crosses the x-axis (where mass $= 0$), it simply approaches that point.

38. (a) $^{235}_{92}\text{U} + ^1_0\text{n} \longrightarrow ^{143}_{54}\text{Xe} + 3\,^1_0\text{n} + ^{90}_{38}\text{Sr}$

    (b) $^{235}_{92}\text{U} + ^1_0\text{n} \longrightarrow ^{102}_{39}\text{Y} + 3\,^1_0\text{n} + ^{131}_{53}\text{I}$

    (c) $^{14}_{7}\text{N} + ^1_0\text{n} \longrightarrow ^1_1\text{H} + ^{14}_{6}\text{C}$

39. (a) $\text{H}_2\text{O}(l) \longrightarrow \text{H}_2\text{O}(g)$

    Energy$_2$: Weakest bond changes requires the least energy.

    (b) $2\,\text{H}_2(g) + \text{O}_2(g) \longrightarrow 2\,\text{H}_2\text{O}(g)$

    Energy$_1$: medium-sized value involved in interatomic bonds.

    (c) $^2_1\text{H} + ^2_1\text{H} \longrightarrow ^3_1\text{H} + ^1_1\text{H}$

    Energy$_3$: Nuclear process; greatest amount of energy involved.

40. $^{236}_{92}\text{U} \longrightarrow ^{90}_{38}\text{Sr} + 3\,^1_0\text{n} + ^{143}_{54}\text{Xe}$

41. (a) beta emission: $\qquad ^{29}_{12}\text{Mg} \longrightarrow \,^{0}_{-1}\text{e} + ^{29}_{13}\text{Al}$

    (b) alpha emission: $\qquad ^{150}_{60}\text{Nd} \longrightarrow ^4_2\text{He} + ^{146}_{58}\text{Ce}$

    (c) positron emission: $\qquad ^{72}_{33}\text{As} \longrightarrow \,^{0}_{+1}\text{e} + ^{72}_{32}\text{Ge}$

42. $(270 \text{ years})\left(\dfrac{1 \text{ half-life}}{30 \text{ years}}\right) = 9 \text{ half-lives}$ \qquad Work down from 270 years.

$t_{\frac{1}{2}}$, years	270	240	210	180	150	120	90	60	30	0
Amount, g	15.0	30.0	60.0	120.	240.	480.	960.	1920	3840	7680

There would have been 7680 g originally or 15.0 g $(2^9) = 7680$ g

43. 1 Curie $= 3.7 \times 10^{10}$ disintegrations/sec

   1 becquerel $= 1$ disintegration/sec

   Therefore there are 3.7×10^{10} becquerels/1 Curie

   $$\left(\frac{3.7 \times 10^{10} \text{ becquerel}}{1 \text{ Curie}}\right)(1.24 \text{ Curies}) = 4.6 \times 10^{10} \text{ becquerels}$$

44. 1.00 g Co-60

   (a) one half-life: $\dfrac{1.00 \text{ g}}{2} = 0.500 \text{ g left}$

   (b) two half-lives: $\dfrac{0.500 \text{ g}}{2} = 0.250 \text{ g left}$

   (c) four half-lives: $2^4 = 16$; $\dfrac{1}{16}$ left $\quad \dfrac{1.00 \text{ g}}{16} = 0.0625 \text{ g left}$

   (d) ten half-lives: $2^{10} = 1024$; $\dfrac{1}{1024}$ left $\quad \dfrac{1.00 \text{ g}}{1024} = 9.77 \times 10^{-4} \text{ g left}$

45. (a) $^{11}_{5}\text{B} \longrightarrow ^{4}_{2}\text{He} + ^{7}_{3}\text{Li}$

   (b) $^{88}_{38}\text{Sr} \longrightarrow _{-1}^{0}\text{e} + ^{88}_{39}\text{Y}$

   (c) $^{107}_{47}\text{Ag} + ^{1}_{0}\text{n} \longrightarrow ^{108}_{47}\text{Ag}$

   (d) $^{41}_{19}\text{K} \longrightarrow ^{1}_{1}\text{H} + ^{40}_{18}\text{Ar}$

   (e) $^{116}_{51}\text{Sb} + _{-1}^{0}\text{e} \longrightarrow ^{116}_{50}\text{Sn}$

46. C-14 content is 1/16 of that in living plants. This means that four half-lives have passed. ^{14}C half-life is 5730 years.

   $$\left(\frac{5730 \text{ years}}{\text{half-life}}\right)(4 \text{ half-lives}) = 22,920 \text{ years} \ (2.29 \times 10^4 \text{ years})$$

47. Ionizing radiation can change the genetic material (DNA). These changes can then be passed on to future generations.

48. Long-term exposure to low-level ionizing radiation can cause tumors, cancer, and damage to blood producing cells.

49. Scientists can feed a plant with radiophosphorus labeled phosphates and by recording the rate of increase of radiation emitted from the plant can gauge the rate of uptake of these phosphates by the plant.

50. (a) $^{87}_{37}\text{Rb} \longrightarrow _{-1}^{0}\text{e} + ^{87}_{38}\text{Sr}$

   (b) $^{87}_{38}\text{Sr} \longrightarrow _{+1}^{0}\text{e} + ^{87}_{37}\text{Rb}$

51.

$t_{\frac{1}{2}}$, hours	0	12.5	25.0	37.5	50.0	62.5	75.0	87.5	100.
Amount, mg	15.4	7.7	3.85	1.93	0.965	0.483	0.242	0.121	0.0605

Fraction of K-42 remaining: $\dfrac{0.0605\,\text{mg}}{15.4\,\text{mg}} = 0.00393\,(\text{or }0.393\%)$

No. After an additional eight half-lives there would be less than one microgram (0.000001 g) remaining.

$$(200\,\text{hrs})\left(\frac{1\,\text{half-life}}{12.5\,\text{hrs}}\right) = 16\,\text{half-lives}$$

$$\text{Amount remaining} = \left(\frac{1}{2}\right)^{16}(15.4\,\text{mg})\left(\frac{10^3\,\mu\text{g}}{\text{mg}}\right) = 0.235\,\mu\text{g after 16 half-lives}$$

CHAPTER 19

ORGANIC CHEMISTRY: SATURATED HYDROCARBONS

SOLUTIONS TO REVIEW QUESTIONS

1. Two of the major reasons for the large number of organic compounds is the ability of carbon to form short or very long chains of atoms covalently bonded together and isomerism.

2. The carbon atom has only two unshared electrons, making two covalent bonds logical, but in CH_4, carbon forms four equivalent bonds. Promoting one 2s electron to the empty 2p orbital would make four bonds possible, but without hybridization, we could not explain the fact that all four bonds in CH_4 are identical, and the bond angles are equal (109.5°).

3. The first ten normal alkanes:

methane	CH_4	hexane	C_6H_{14}
ethane	C_2H_6	heptane	C_7H_{16}
propane	C_3H_8	octane	C_8H_{18}
butane	C_4H_{10}	nonane	C_9H_{20}
pentane	C_5H_{12}	decane	$C_{10}H_{22}$

4. Aromatic hydrocarbons contain benzene rings. "Aliphatic" refers to all other hydrocarbons including alkanes, alkenes, alkynes, and cycloalkanes.

5. Advantages to alkyl halide anesthetics include nonflammability, a pleasant odor, and a much reduced hepatotoxicity

6. Cyclopropane is more reactive than cyclohexane because cyclopropane's carbon–carbon bond angles are substantially smaller than the normal tetrahedral angle.

7. (a) A substitution reaction allows an exchange of atoms or groups of atoms between reactants while in an elimination reaction a single reactant is split into two products.

    (b) Two reactants combine together in an addition reaction while one reactant is split into two products in an elimination reaction.

8. Alkanes are hydrocarbons containing only the elements hydrogen and carbon. Alkyl halides also contain halogens and, thus, cannot be classified as alkanes.

9. E85 is a gasoline that contains 85% ethanol and 15% petroleum. This mixture reduces the use of petroleum, a nonrenewable resource. E85 is the cleanest burning gasoline now available.

10. Renewable gasoline differs in several ways from the gasoline we buy at the pump today: (1) Renewable gasoline is produced by bacteria instead of being pumped from the ground; (2) Renewable gasoline differs in composition depending on which bacteria are used.

SOLUTIONS TO EXERCISES

1. Lewis structures:

   (a) CCl_4

   $$:\overset{..}{\underset{..}{Cl}}:$$
   $$:\overset{..}{\underset{..}{Cl}}:\overset{..}{\underset{..}{C}}:\overset{..}{\underset{..}{Cl}}:$$
   $$:\overset{..}{\underset{..}{Cl}}:$$

   (b) C_2Cl_6

   $$:\overset{..}{\underset{..}{Cl}}::\overset{..}{\underset{..}{Cl}}:$$
   $$:\overset{..}{\underset{..}{Cl}}:\overset{..}{\underset{..}{C}}:\overset{..}{C}:\overset{..}{\underset{..}{Cl}}:$$
   $$:\overset{..}{\underset{..}{Cl}}::\overset{..}{\underset{..}{Cl}}:$$

   (c) $CH_3CH_2CH_3$

   $$\begin{array}{ccc} H & H & H \\ H:\overset{..}{\underset{..}{C}}:\overset{..}{\underset{..}{C}}:\overset{..}{\underset{..}{C}}:H \\ H & H & H \end{array}$$

2. Lewis structures:

   (a) CH_4

   $$\begin{array}{c} H \\ H:\overset{..}{\underset{..}{C}}:H \\ H \end{array}$$

   (b) C_3H_8

   $$\begin{array}{ccc} H & H & H \\ H:\overset{..}{\underset{..}{C}}:\overset{..}{\underset{..}{C}}:\overset{..}{\underset{..}{C}}:H \\ H & H & H \end{array}$$

   (c) C_5H_{12}

   $$\begin{array}{ccccc} H & H & H & H & H \\ H:\overset{..}{\underset{..}{C}}:\overset{..}{\underset{..}{C}}:\overset{..}{\underset{..}{C}}:\overset{..}{\underset{..}{C}}:\overset{..}{\underset{..}{C}}:H \\ H & H & H & H & H \end{array}$$

3. The formulas (a), (c), (f), and (g) represent isomers.

4. The same compound is represented by formulas (a), (c), and (f).

5. The number of methyl groups in each formula in Exercise #3 is as follows:

   (a) 2   (b) 2   (c) 3   (d) 4   (e) 1   (f) 3   (g) 4   (h) 3

6. The number of methyl groups in each formula in Exercise #4 is as follows:

   (a) 3   (b) 2   (c) 3   (d) 4   (e) 4   (f) 3

7. Isomers of heptane

   $CH_3CH_2CH_2CH_2CH_2CH_2CH_3$

   $$CH_3CH_2CH_2CH_2\underset{\underset{CH_3}{|}}{C}HCH_3$$

   $$CH_3CH_2CH_2\underset{\underset{CH_3}{|}}{C}HCH_2CH_3$$

   $$CH_3CH_2CH_2\underset{\overset{|}{CH_3}}{\overset{\overset{CH_3}{|}}{C}}CH_3$$

   $$CH_3CH_2\underset{\overset{|}{CH_3}}{\overset{\overset{CH_3}{|}}{C}}CH_2CH_3$$

   $$CH_3CH_2\underset{\underset{CH_3}{|}}{C}H\overset{\overset{CH_3}{|}}{C}HCH_3$$

   $$CH_3\underset{\underset{CH_3}{|}}{C}HCH_2\underset{\underset{CH_3}{|}}{C}HCH_3$$

   $$CH_3\underset{\underset{CH_3}{|}}{C}H-\underset{\underset{CH_3}{|}}{\overset{\overset{CH_3}{|}}{C}}CH_3$$

   $$CH_3CH_2\underset{\underset{CH_2CH_3}{|}}{C}HCH_2CH_3$$

8. Isomers of hexane

$CH_3CH_2CH_2CH_2CH_2CH_3$ $CH_3CH_2CH_2CHCH_3$
$\quad\qquad\qquad\qquad\qquad\qquad\qquad\qquad\qquad\quad |$
$\quad\qquad\qquad\qquad\qquad\qquad\qquad\qquad\qquad CH_3$

$\qquad\qquad\qquad\qquad\qquad\quad CH_3 \qquad\qquad\quad CH_3$
$\qquad\qquad\qquad\qquad\qquad\quad | \qquad\qquad\qquad\quad |$
$CH_3CH_2CHCH_2CH_3 \qquad CH_3CH_2CCH_3 \qquad CH_3CHCHCH_3$
$\qquad\quad |\qquad\qquad\qquad\qquad\qquad\quad |\qquad\qquad\qquad\qquad\quad |$
$\qquad\quad CH_3 \qquad\qquad\qquad\qquad CH_3 \qquad\qquad\qquad CH_3$

9. (a) CH_2Cl_2, one CH_2Cl_2

    (b) C_3H_7Br, two $CH_3CH_2CH_2Br$ $CH_3CHBrCH_3$

    (c) $C_3H_6Cl_2$, four $CH_3CH_2CHCl_2$ $CH_3CHClCH_2Cl$
        $CH_2ClCH_2CH_2Cl$ $CH_3CCl_2CH_3$

    (d) $C_4H_8Cl_2$, nine $CH_3CH_2CH_2CHCl_2$ $CH_3CH_2CHClCH_2Cl$
        $CH_3CHClCH_2CH_2Cl$ $CH_2ClCH_2CH_2CH_2Cl$ $CH_3CH_2CCl_2CH_3$
        $CH_3CHClCHClCH_3$ CH_3CHCH_2Cl CH_3CClCH_2Cl $CH_3CHCH_2Cl_2$
                              CH_2Cl CH_3 CH_3

10. (a) CH_3Br, one CH_3Br

    (b) C_2H_5Cl, one CH_3CH_2Cl

    (c) C_4H_9I, four $CH_3CH_2CH_2CH_2I$ $CH_3CH_2CHICH_3$
        CH_3CHCH_2I CH_3CICH_3
           $|$ $|$
          CH_3 CH_3

    (d) C_3H_6BrCl, five $CH_3CH_2CHBrCl$ $CH_3CHClCH_2Br$
        $CH_3CHBrCH_2Cl$ $CH_2ClCH_2CH_2Br$ $CH_3CBrClCH_3$

11. (a) 5   (b) 6   (c) 5

12. (a) 7   (b) 7   (c) 6

13. IUPAC names
    (a) 1-chloropropane
    (b) 2-chloropropane
    (c) 2-chloro-2-methylpropane
    (d) 2-methylbutane
    (e) 2,3-dimethylhexane

14. IUPAC names
    (a) chloroethane
    (b) 1-chloro-2-methylpropane
    (c) 2-chlorobutane
    (d) methylcyclopropane
    (e) 2,4-dimethylpentane

15. The tertiary carbons are circled in the following structures from Exercise #11:

(a) 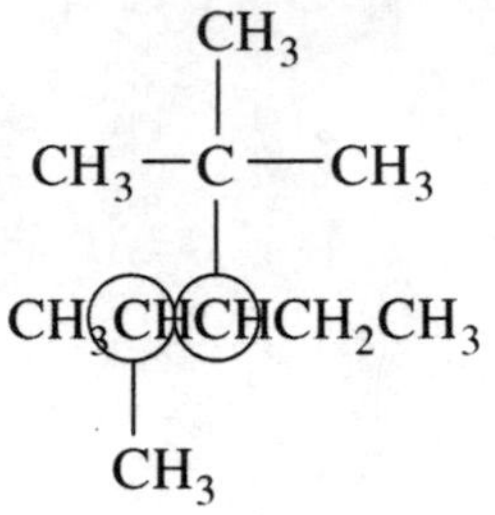
$$CH_3$$
$$|$$
$$CH_3-C-CH_3$$
$$|$$
CH₃CHCHCH₂CH₃
$$|$$
$$CH_3$$

(b)
$$CH_2CH_3$$
$$|$$
CH₃CHCHCH₃
$$|$$
$$CH_2CH_3$$

(c)
$$CH_3 \quad CH_3 \quad CH_3$$
$$| \quad\quad | \quad\quad |$$
$$CH_3-C-C-C-CH_3$$
$$| \quad\quad | \quad\quad |$$
$$CH_3 \quad CH_3 \quad CH_3$$

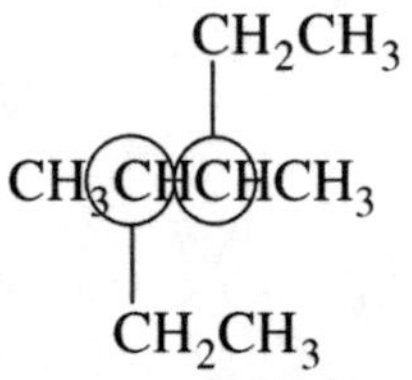

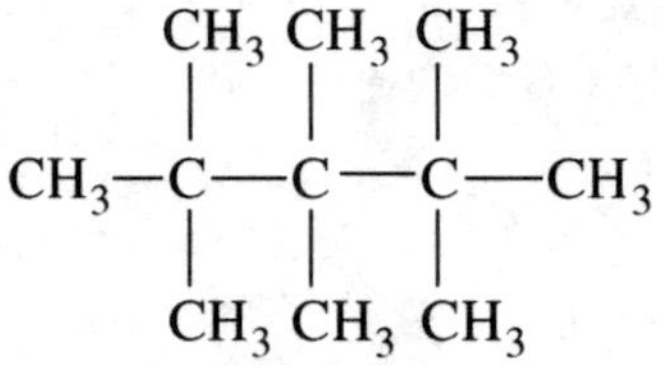

16. The tertiary carbons are circled in the following structures from Exercise #12:

(a) 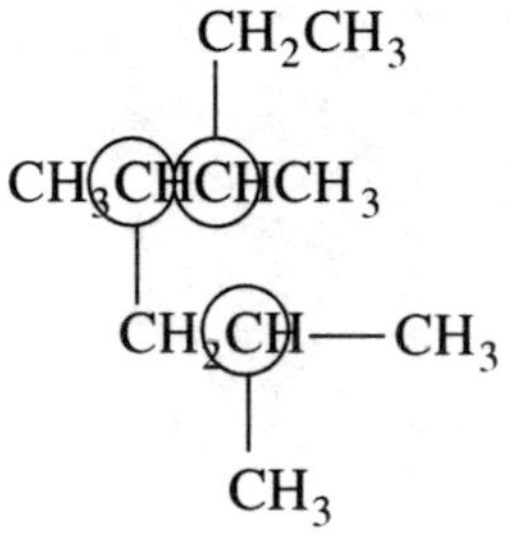
$$CH_2CH_3$$
$$|$$
CH₃CHCHCH₃
$$|$$
CH₂CH—CH₃
$$|$$
$$CH_3$$

(b) 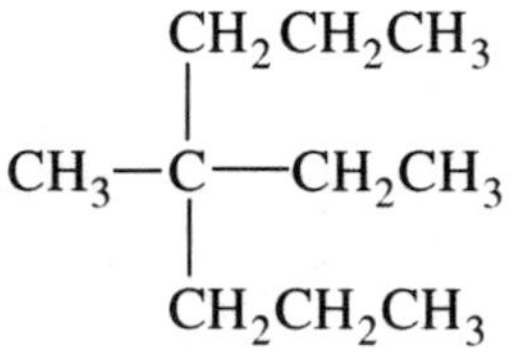
$$CH_2CH_2CH_3$$
$$|$$
$$CH_3-C-CH_2CH_3$$
$$|$$
$$CH_2CH_2CH_3$$

(c)
CH₃CHCH₂CH₃
$$|$$
$$CH_3-C-CH_3$$
$$|$$
$$CH_2CH_3$$

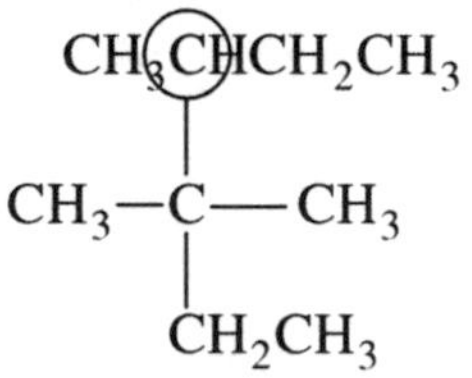

17. Structural formulas:

(a) 2,4-dimethylpentane

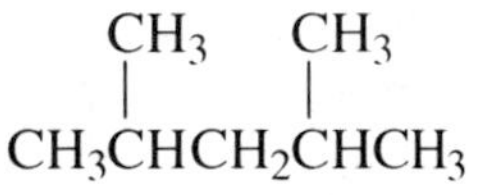

$$\begin{array}{ccc} & CH_3 & CH_3 \\ & | & | \\ CH_3CHCH_2CHCH_3 \end{array}$$

(b) 2, 2-dimethylpentane

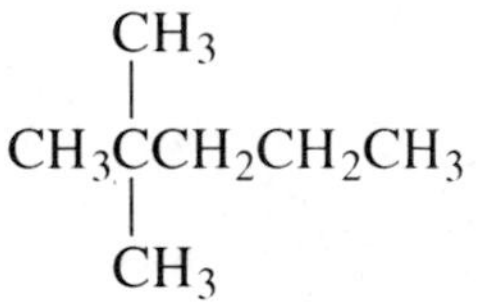

$$\begin{array}{c} CH_3 \\ | \\ CH_3CCH_2CH_2CH_3 \\ | \\ CH_3 \end{array}$$

(c) 3-isopropyloctane

$$\begin{array}{c} CH_3CH_2CHCH_2CH_2CH_2CH_2CH_3 \\ | \\ CH(CH_3)_2 \end{array}$$

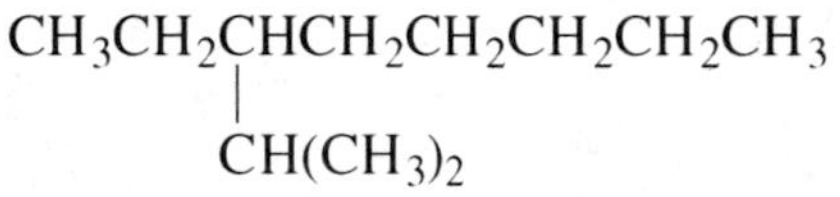

(d) 5,6-diethyl-2,7-dimethyl-5-propylnonane

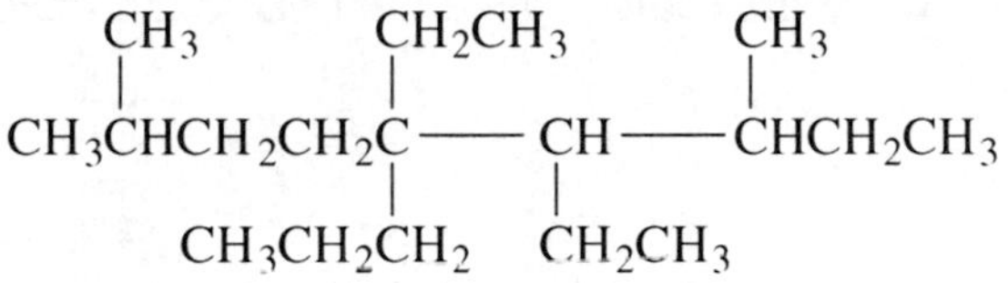

$$\begin{array}{ccccc} CH_3 & & CH_2CH_3 & & CH_3 \\ | & & | & & | \\ CH_3CHCH_2CH_2C & \!\!-\!\!\!-\!\! & CH & \!\!-\!\!\!-\!\! & CHCH_2CH_3 \\ & | & & | & \\ & CH_3CH_2CH_2 & & CH_2CH_3 & \end{array}$$

(e) 2-ethyl-1,3-dimethylcyclohexane

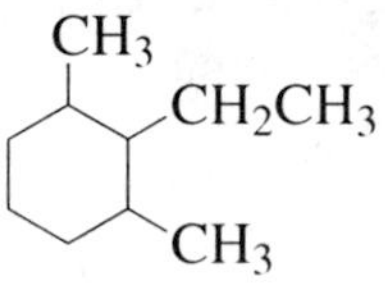

18. (a) 4-ethyl-2-methylhexane

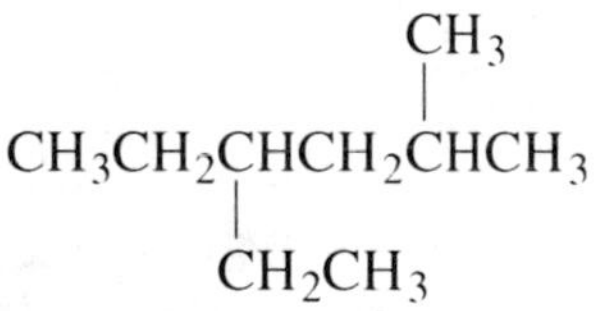

$$\begin{array}{c} CH_3 \\ | \\ CH_3CH_2CHCH_2CHCH_3 \\ | \\ CH_2CH_3 \end{array}$$

(b) 4-t-butylheptane

$$\begin{array}{c} CH_3CH_2CH_2CHCH_2CH_2CH_3 \\ | \\ C(CH_3)_3 \end{array}$$

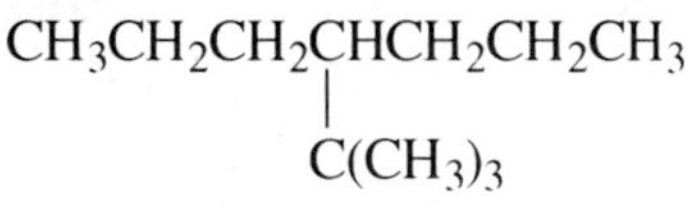

(c) 4-ethyl-7-isopropyl-2,4,8-trimethyldecane

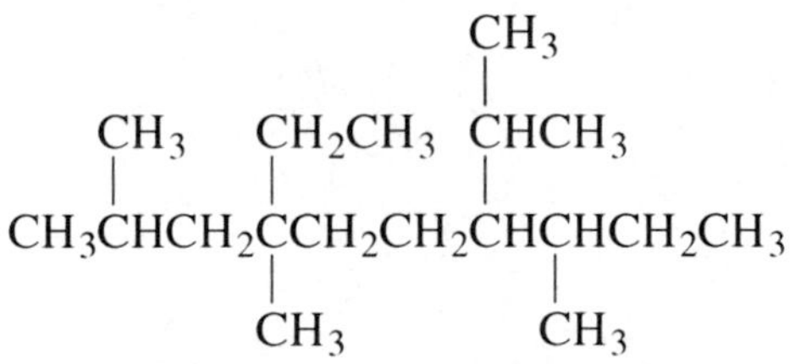

$$\begin{array}{ccc} & & CH_3 \\ & & | \\ CH_3 & CH_2CH_3 & CHCH_3 \\ | & | & | \\ CH_3CHCH_2CCH_2CH_2CHCHCH_2CH_3 \\ & | & | \\ & CH_3 & CH_3 \end{array}$$

(d) 3-ethyl-2,2-dimethyloctane

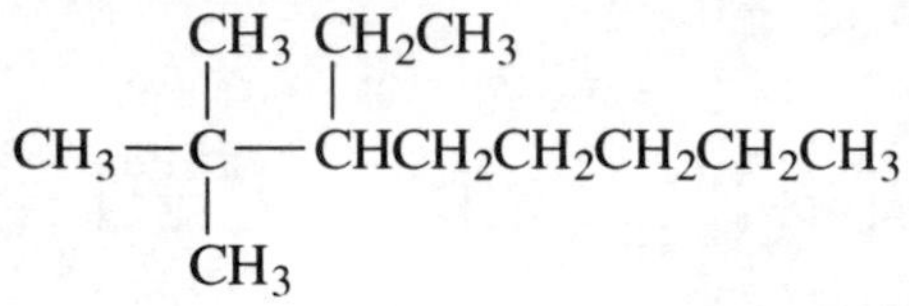

$$CH_3-\overset{\overset{\displaystyle CH_3}{|}}{\underset{\underset{\displaystyle CH_3}{|}}{C}}-\overset{\overset{\displaystyle CH_2CH_3}{|}}{CH}CH_2CH_2CH_2CH_2CH_3$$

(e) 1,3-diethylcyclohexane

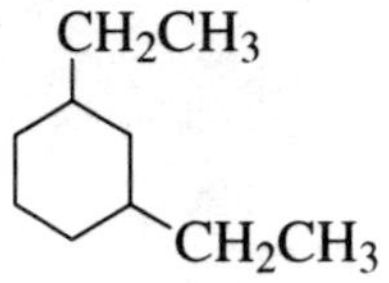

CH₂CH₃ and CH₂CH₃ substituents on cyclohexane ring

19. (a) 3-methylbutane $CH_3CH_2\overset{\overset{\displaystyle CH_3}{|}}{CH}CH_3$

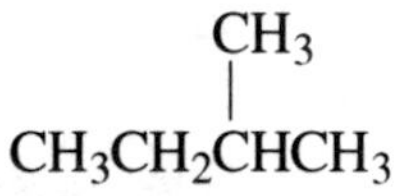

Numbering was done from the wrong end of the molecule. The correct name is 2-methylbutane.

(b) 2-ethylbutane $CH_3\overset{\underset{\underset{\displaystyle CH_2CH_3}{|}}{}}{CH}CH_2CH_3$

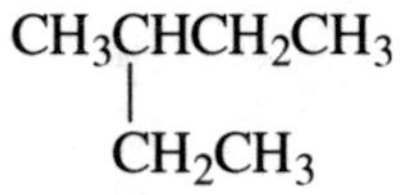

The name is not based on the longest carbon chain (5 carbons). The correct name is 3-methylpentane.

(c) 2-dimethylpentane. Each methyl group needs to be numbered. Depending on the structure, the correct name is 2,2-dimethylpentane; 2,3-dimethylpentane; or 2,4-dimethylpentane.

(d) 1,4-dimethylcyclopentane

CH₃ and CH₃ substituents on cyclopentane ring

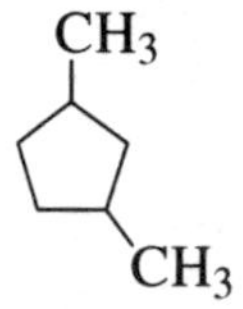

The ring was numbered in the wrong direction. The correct name is 1,3-dimethylcyclopentane.

20. (a) 3-methyl-5-ethyloctane $CH_3CH_2\overset{\overset{}{\underset{\underset{\displaystyle CH_3}{|}}{CH}}}{}CH_2\overset{\overset{}{\underset{\underset{\displaystyle CH_2CH_3}{|}}{CH}}}{}CH_2CH_2CH_3$

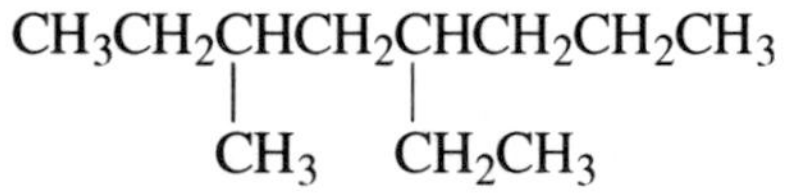

Ethyl should be named before methyl (alphabetical order). The numbering is correct. The correct name is 5-ethyl-3-methyloctane.

(b) 3,5,5-triethylhexane $CH_3CH_2\overset{\underset{\underset{\underset{\underset{\displaystyle CH_3}{|}}{CH_2}}{|}}{CH}}{}CH_2\overset{\overset{\displaystyle CH_2CH_3}{|}}{\underset{\underset{\displaystyle CH_2CH_3}{|}}{C}}CH_3$

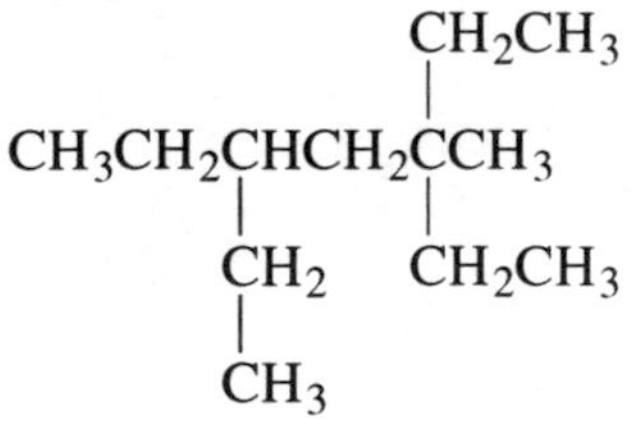

The name is not based on the longest carbon chain (7 carbons). The correct name is 3,5-diethyl-3-methylheptane.

(c)   4,4-dimethyl-3-ethylheptane

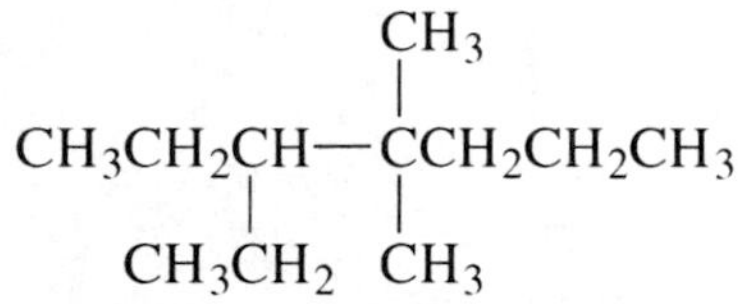

$$CH_3CH_2CH—\overset{\overset{\displaystyle CH_3}{|}}{\underset{\underset{\displaystyle CH_3}{|}}{C}}CH_2CH_2CH_3$$

with CH_3CH_2 below the left carbon.

Ethyl should be named before dimethyl (alphabetical order). The correct name is 3-ethyl-4,4-dimethylheptane.

(d)   1,6-dimethylcyclohexane

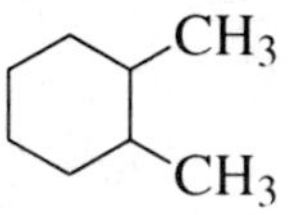

The ring was numbered in the wrong direction. The correct name is 1,2-dimethylcyclohexane.

21.   (a)   butane, $CH_3CH_2CH_2CH_3$

(b)   2-methylpropane, $CH_3\overset{\overset{\displaystyle CH_3}{|}}{C}HCH_3$

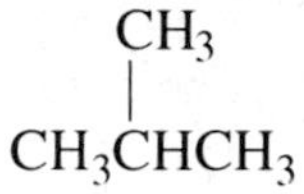

22.   (a)   2-methylbutane, $CH_3\overset{\overset{\displaystyle CH_3}{|}}{C}HCH_2CH_3$

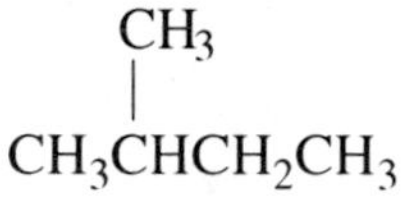

(b)   2,2-dimethylbutane, $CH_3\overset{\overset{\displaystyle CH_3}{|}}{\underset{\underset{\displaystyle CH_3}{|}}{C}}CH_2CH_3$

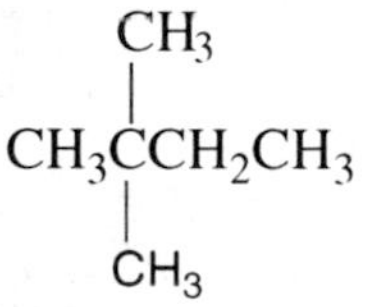

23.   An isomer is shown in (b).

24.   Isomers are shown in (a) and (b).

25.   (a) 2-chlorobutane   (b) 2-chloro-2-methylpropane

26.   (a) 2-bromo-2,3-dimethylbutane   (b) 2-bromo-2-methylbutane

27.   (a) $CH_3CH(CH_3)CH_2CH_2CH(CH_3)CH_3$, 2,5-dimethylhexane

(b) 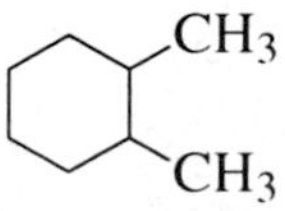   1,2-dimethylcyclohexane

28.   (a) $CH_3C(CH_3)_2CH_3$, 2,2-dimethylpropane
(b) $CH_3CH_2CH_2CH_2CH_3$, pentane

29. (a)   Five isomers:

H
|
$CH_3—C—CH_2—CHCl—CH_3$
|
CH_3

2-chloro-4-methylpentane

H
|
$CH_2Cl—C—CH_2—CH_2—CH_3$
|
CH_3

1-chloro-2-methylpentane

H
|
$CH_3—C—CH_2—CH_2—CH_2Cl$
|
CH_3

1-chloro-4-methylpentane

$CH_3—CClCH_2—CH_2—CH_3$
|
CH_3

2-chloro-2-methylpentane

H
|
$CH_3—C—CHCl—CH_2—CH_3$
|
CH_3

3-chloro-2-methylpentane

(b)   Five isomers:

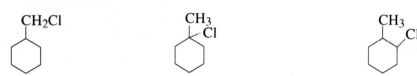

chloromethylcyclohexane   1-chloro-1-methylcyclohexane   2-chloro-1-methylcyclohexane

3-chloro-1-methylcyclohexane   4-chloro-1-methylcyclohexane

30. (a)   Three isomers:

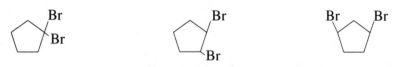

1,1-dibromocyclopentane   1,2-dibromocyclopentane   1,3-dibromocyclopentane

(b)   Seven isomers:

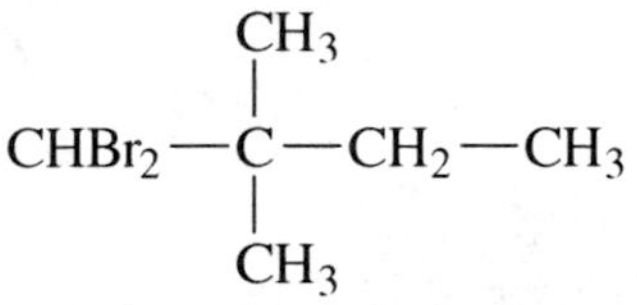

$$CHBr_2-\underset{\underset{CH_3}{|}}{\overset{\overset{CH_3}{|}}{C}}-CH_2-CH_3$$

1,1-dibromo-2,2-dimethylbutane

$$H_3C-\underset{\underset{CH_3}{|}}{\overset{\overset{CH_3}{|}}{C}}-CBr_2-CH_3$$

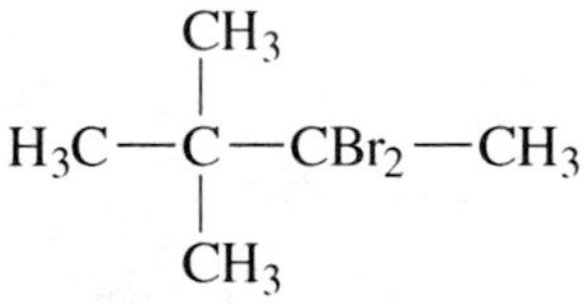

2,2-dibromo-3,3-dimethylbutane

$$H_3C-\underset{\underset{CH_3}{|}}{\overset{\overset{CH_3}{|}}{C}}-CH_2-CHBr_2$$

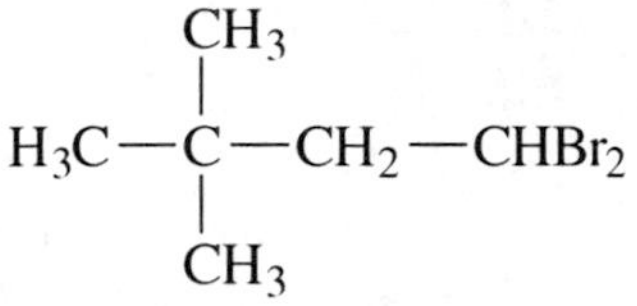

1,1-dibromo-3,3-dimethylbutane

$$CH_2Br-\underset{\underset{CH_3}{|}}{\overset{\overset{CH_2Br}{|}}{C}}-CH_2-CH_3$$

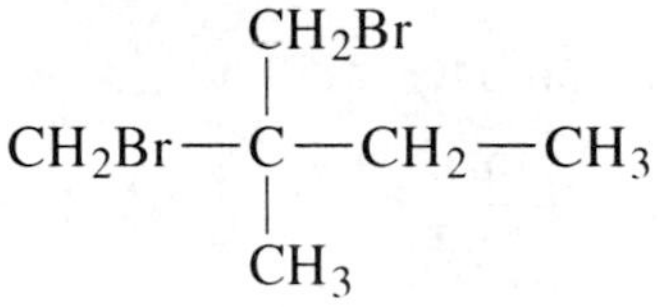

1-bromo-2-bromomethyl-2-methylbutane

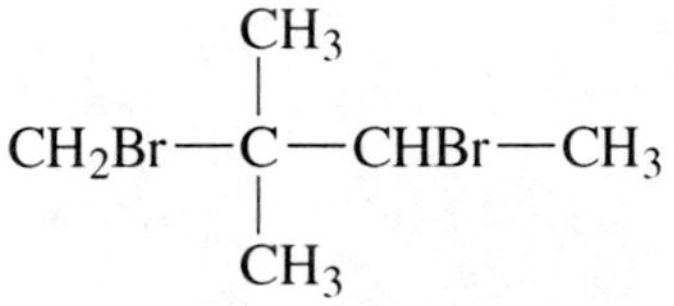

$$CH_2Br-\underset{\underset{CH_3}{|}}{\overset{\overset{CH_3}{|}}{C}}-CHBr-CH_3$$

1,3-dibromo-2,2-dimethylbutane

$$CH_2Br-\underset{\underset{CH_3}{|}}{\overset{\overset{CH_3}{|}}{C}}-CH_2-CH_2Br$$

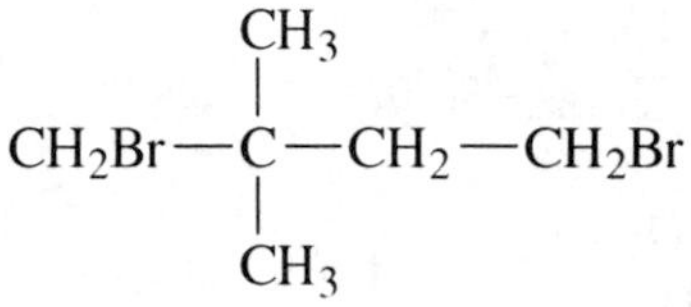

1,4-dibromo-2,2-dimethylbutane

$$H_3C-\underset{\underset{CH_3}{|}}{\overset{\overset{CH_3}{|}}{C}}-CHBr-CH_2Br$$

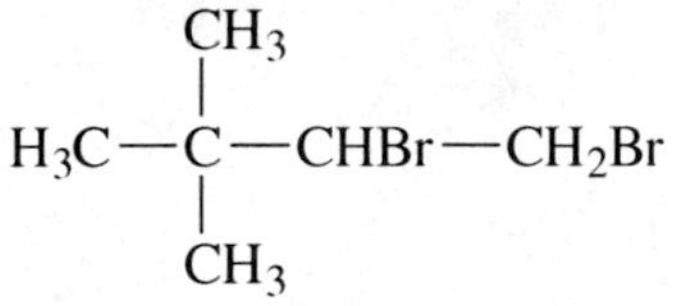

1,2-dibromo-3,3-dimethylbutane

31. Names:
    (a) 1-chloro-2-ethylcyclohexane
    (b) 1-chloro-3-ethyl-1-methylcyclohexane
    (c) 1,4-diisopropylcyclohexane

32. Structural formulas:
    (a)

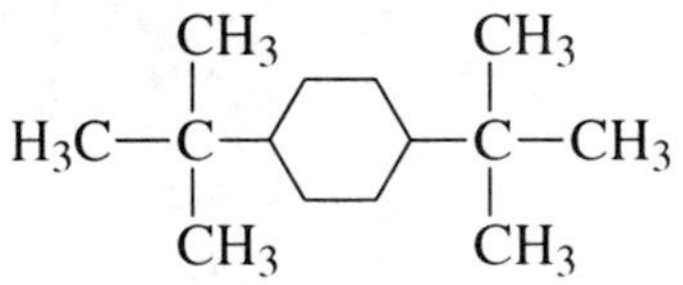

    (b)

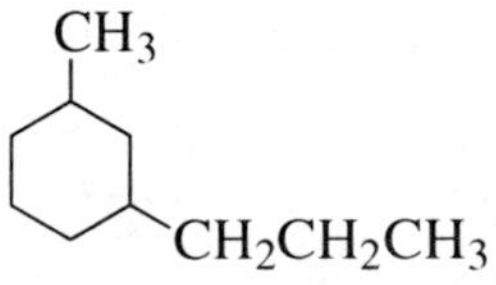

    (c)

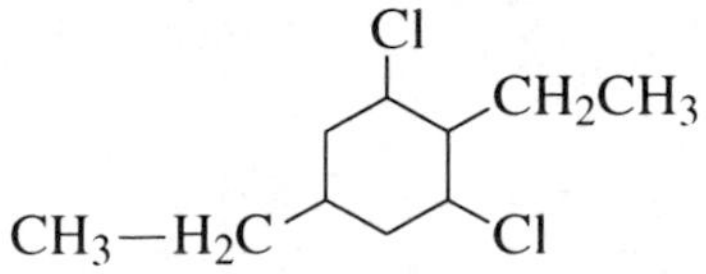

33. (a) chair
    (b) chair

34. (a) chair
    (b) boat

35. The formula for dodecane is $C_{12}H_{26}$.

36. $FCH_2CH_2F + Cl_2 \longrightarrow FCH_2CHClF + HCl$

37. Data: $\dfrac{1 \text{ gal}}{60 \text{ mi}}$; 60 mi traveled; $\dfrac{19 \text{ mol } C_8H_{18}}{\text{gal}}$; $T = 293K$

    $2 C_8H_{18} + 25 O_2 \longrightarrow 16 CO_2 + 18 H_2O$

    $\dfrac{1 \text{ gal}}{60 \text{ mi}} \times 60 \text{ mi} = 1 \text{ gal gasoline used}$

    $19 \text{ mol } C_8H_{18} \times \dfrac{16 \text{ mol } CO_2}{2 \text{ mol } C_8H_{18}} = 1.5 \times 10^2 \text{ mol } CO_2$

    $PV = nRT \qquad V = \dfrac{nRT}{P}$

    $V = \dfrac{(1.5 \times 10^2 \text{ mol } CO_2)(0.0821 \text{ L} - \text{atm})(293 \text{ K})}{1 \text{ atm} \qquad \text{mol-K}} = 3.6 \times 10^3 \text{ L } CO_2$

38. (a) elimination
    (b) substitution
    (c) addition

39. It is not possible to distinguish hexane from 3-methylheptane based on solubility in water because both compounds are nonpolar and, thus, insoluble in water.

40. (a) The compounds are isomers.
    (b) The compounds are not the same and are not isomers.
    (c) The compounds are not the same and are not isomers.
    (d) The compounds are not the same and are not isomers.

41. (a) Initiation: $Cl_2 \xrightarrow{\text{uv}} 2Cl\cdot$

    Propagation: $Cl\cdot + CH_3Cl \longrightarrow CH_2Cl\cdot + HCl$

    $CH_2Cl\cdot + Cl_2 \longrightarrow CH_2Cl_2 + Cl\cdot$

    Termination: $Cl\cdot + Cl\cdot \longrightarrow Cl_2$

    $CH_2Cl\cdot + CH_2Cl\cdot \longrightarrow CH_2ClCH_2Cl$

    $CH_2Cl\cdot + Cl\cdot \longrightarrow CH_2Cl_2$

(b)   Initiation: $Cl_2 \xrightarrow{uv} 2\,Cl\cdot$

       Propagation: $Cl\cdot + CH_3\underset{\underset{CH_3}{|}}{C}HCH_3 \longrightarrow (CH_3)_3C\cdot + HCl$

                       $(CH_3)_3C\cdot + Cl_2 \longrightarrow (CH_3)_3CCl + Cl\cdot$

       Termination: $Cl\cdot + Cl\cdot \longrightarrow Cl_2$

                       $(CH_3)_3C\cdot + (CH_3)_3C\cdot \longrightarrow (CH_3)_3CC(CH_3)_3$

                       $(CH_3)_3C\cdot + Cl\cdot \longrightarrow (CH_3)_3CCl$

(c)   Initiation: $Cl_2 \xrightarrow{uv} 2\,Cl\cdot$

       Propagation: $Cl\cdot + \bigcirc\!\!\!\!\!\text{(cyclopentane)} \longrightarrow \bigcirc\!\!\!\!\!\cdot + HCl$

                   $\bigcirc\!\!\!\!\!\cdot + Cl_2 \longrightarrow \bigcirc\!\!\!\!\!^{Cl} + Cl\cdot$

       Termination: $Cl\cdot + Cl\cdot \longrightarrow Cl_2$

                   $\bigcirc\!\!\!\!\!\cdot + \bigcirc\!\!\!\!\!\cdot \longrightarrow \bigcirc\!\!\!\!\!-\!\!\!\!\!\bigcirc$

                   $\bigcirc\!\!\!\!\!\cdot + Cl\cdot \longrightarrow \bigcirc\!\!\!\!\!^{Cl}$

(d)   Initiation: $Cl_2 \xrightarrow{uv} 2\,Cl\cdot$

       Propagation: $Cl\cdot + \bigcirc\!\!\!\!\!^{CH_3} \longrightarrow \bigcirc\!\!\!\!\!\overset{CH_3}{\cdot} + HCl$

                   $\bigcirc\!\!\!\!\!\overset{CH_3}{\cdot} + Cl_2 \longrightarrow \bigcirc\!\!\!\!\!\overset{CH_3}{\underset{Cl}{}} + Cl\cdot$

       Termination: $Cl\cdot + Cl\cdot \longrightarrow Cl_2$

                   $\bigcirc\!\!\!\!\!\overset{CH_3}{\cdot} + \bigcirc\!\!\!\!\!\overset{CH_3}{\cdot} \longrightarrow \bigcirc\!\!\!\!\!\overset{CH_3}{\underset{CH_3}{}}\!\!\!\!\!\bigcirc$

                   $\bigcirc\!\!\!\!\!\overset{CH_3}{\cdot} + Cl\cdot \longrightarrow \bigcirc\!\!\!\!\!\overset{CH_3}{\underset{Cl}{}}$

42.   Using a high mole ratio of methane to chlorine will allow a chlorine free radical to react with a methane molecule rather than a chloromethane molecule and minimize the formation of di-, tri-, and tetrachloromethane.

43.   (a) $CH_3CH_2CH_2CH_2CH_2CH_2CH_2CH_2CH_2CH_2CH_3$ undecane

        $CH_3CH_2CH_2CH_2CH_2CH_2CH_2CH_2CH_2CH_2CH_2CH_2CH_3$ tridecane

    (b)   Both compounds are alkanes. They are composed of only carbon and hydrogen and have only single bonds between the carbon atoms. Both formulas agree with the general formula for alkanes, C_nH_{2n+2}.

44. Cyclopentane and cycloheptane are in the same homologous series (cycloalkanes) and share the same general formula, C_nH_{2n}. The formula for cycloheptane is C_7H_{14}.

45. (a)

R-32
$$F\underset{\underset{\displaystyle H}{|}}{\overset{\overset{\displaystyle F}{|}}{C}}H$$

R-125
$$H\underset{\underset{\displaystyle F}{|}}{\overset{\overset{\displaystyle F}{|}}{C}}\underset{\underset{\displaystyle F}{|}}{\overset{\overset{\displaystyle F}{|}}{C}}F$$

R-134a
$$H\underset{\underset{\displaystyle F}{|}}{\overset{\overset{\displaystyle H}{|}}{C}}\underset{\underset{\displaystyle F}{|}}{\overset{\overset{\displaystyle F}{|}}{C}}F$$

(b) R-32, four sigma bonds; R-125, seven sigma bonds; R-134a, seven sigma bonds.

46. (a) 2-methylpentane; (b) 2,4-dimethylheptane; (c) 4-ethyl-3-methyloctane.

47. (a)

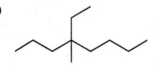

(b)

48. A mixture. A hydrocarbon of the formula, C_4H_{10}, can have two possible structures:

(I) $CH_3CH_2CH_2CH_3$ and (II) $CH_3CH(CH_3)CH_3$.

Structure I can form only two monobromo compounds:

$CH_2BrCH_2CH_2CH_3$ and $CH_3CHBrCH_2CH_3$

Structure II can form only two monobromo compounds:

$CH_2BrCH(CH_3)CH_3$. and $CH_3CBr(CH_3)CH_3$.

A mixture of the two structures gives four monobromo compounds.

CHAPTER 20

UNSATURATED HYDROCARBONS

SOLUTIONS TO REVIEW QUESTIONS

1. The sigma bond in the double bond of ethene is formed by the overlap of two sp^2 electron orbitals and is symmetrical about a line drawn between the nuclei of the two carbon atoms. The pi bond is formed by the sidewise overlap of two p orbitals which are perpendicular to the carbon-carbon sigma bond. The pi bond consists of two electron clouds, one above and one below the plane of the carbon-carbon sigma bond.

2. The restricted rotation around the carbon-carbon double bond in 1,2-dichloroethene means two different structural isomers exist, one with two chlorines on the same side of the double bond (*cis*) and when the chlorines are on opposite sides of the double bond (*trans*). For 1,2-dichloroethyne, the triple bond also has restricted rotation. However, as all atoms lie in a line, there is only one way to arrange the atoms for 1,2-dichloroethyne.

3. 1-pentene is a liquid at room temperature (25°C) because its boiling point is above room temperature (30°C). 2-methylpropene is a gas at room temperature because its boiling point (-7°C) is much lower than 25°C.

4. 1-butene has a much lower melting point (-185°C) than 2-methylpropene (-14°C) although their boiling points are almost the same (1-butene, -6°C; 2-methylpropene, -7°C). 1-butene has a much different molecular shape than 2-methylpropene. This shape difference has a great effect on melting points because the molecules fit closely together in a solid. In a liquid the molecules are rapidly moving and molecular shape differences are much less important. Thus, the boiling points for these two molecules are almost the same.

5. *Trans* fats contain double bonds in the *trans* isomer form. In contrast, most naturally occurring unsaturated fats have double bonds in the *cis* form. Our bodies can't metabolize the *trans* fats because they are the wrong shape. The *trans* fats accumulate over time and increase the risk of cardiovascular disease.

6. During the 10-year period from 1935 to 1945, the major source of aromatic hydrocarbons shifted from coal tar to petroleum due to the rapid growth of several industries which used aromatic hydrocarbons as raw material. These industries include drugs, dyes, detergents, explosives, insecticides, plastics, and synthetic rubber. Since the raw material needs far exceeded the aromatics available from coal tar, another source had to be found, and processes were developed to make aromatic compounds from alkanes in petroleum. World War II, which occurred during this period, put high demands on many of these industries, particularly explosives.

7. Benzene does not undergo the typical reactions of an alkene. Benzene does not decolorize bromine rapidly and it does not destroy the purple color of permanganate ions. The reactions of benzene are more like those of an alkane. Reaction of benzene with chlorine requires a catalyst. Benzene does not readily add Cl_2, but rather a hydrogen atom is replaced by a chlorine atom.

$$C_6H_6 + Cl_2 \xrightarrow{Fe} C_6H_5Cl + HCl$$

8. The 11-*cis* isomer of retinal combines with a protein (opsin) to form the visual pigment rhodopsin. When light is absorbed, the 11-*cis* double bond is converted to a *trans*-double bond. This process initiates the mechanism of visual excitation which our brains perceive as light or light forms.

9. Polycyclic aromatic hydrocarbons are potent carcinogens.

10. Cyclohexane carbons form bond angles of about 109° *in a tetrahedron*; this causes the ring to be either a "boat" or a "chair" shape. Benzene carbons form bond angles of 120° *in a plane*. Therefore, the benzene molecule must be planar.

11. According to the mechanism, the addition of an unsymmetrical module such as HX adds to a carbon-carbon double bond. In the first step, the H adds to the carbon of the carbon-carbon double bond that has the most hydrogen atoms on it, according to Markovnikoff's rule. The second step completes the addition by adding the more negative element, X of the HX.

12. "Cracking" means breaking into pieces, which, according to the pyrolysis products, is what occurs in the reaction.

13. H^+ adds to C-1 forming the more stable cation,

$$CH_3\overset{+}{C}HCH_2CH_3$$

A Cl^- then adds to C-2 to form the product,

$$CH_3CHClCH_2CH_3$$

14. 1-chloropropene meets the requirement of two different groups on each carbon atom of the carbon-carbon-double bond to have *cis-trans* isomers. 2-chloropropene does not meet this requirement.

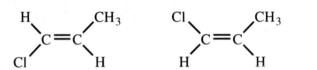

trans-1-chloropropene   *cis*-1-chloropropene   2-chloropropene

15. They are the same compound. Both structures have the same name, 2,3-dimethylcyclohexene.

SOLUTIONS TO EXERCISES

1. (a) ethane   (b) ethene   (c) ethyne

$$\begin{array}{cc} \text{H} & \text{H} \\ \text{H:}\overset{..}{\text{C}}\text{:}\overset{..}{\text{C}}\text{:H} \\ \text{H} & \text{H} \end{array}$$

$$\begin{array}{cc} \text{H} & \text{H} \\ \overset{..}{\text{C}}\text{::}\overset{..}{\text{C}} \\ \text{H} & \text{H} \end{array}$$

$$\text{H:C:::C:H}$$

2. (a) propane   (b) propene   (c) propyne

$$\begin{array}{ccc} \text{H} & \text{H} & \text{H} \\ \text{H:}\overset{..}{\text{C}}\text{:}\overset{..}{\text{C}}\text{:}\overset{..}{\text{C}}\text{:H} \\ \text{H} & \text{H} & \text{H} \end{array}$$

$$\begin{array}{ccc} & \text{H} & \\ \text{H:}\overset{..}{\text{C}}\text{:}\overset{..}{\text{C}}\text{::}\overset{..}{\text{C}}\text{:H} \\ \text{H} & \text{H} & \text{H} \end{array}$$

$$\begin{array}{c} \text{H} \\ \text{H:}\overset{..}{\text{C}}\text{:C:::C:H} \\ \text{H} \end{array}$$

3. $\overset{sp^3}{\text{CH}_3}-\overset{sp^2}{\text{CH}}=\overset{sp^2}{\text{CH}}-\overset{sp^3}{\text{CH}_3}$

4. $\overset{sp}{\text{HC}}\equiv\overset{sp}{\text{C}}-\overset{sp^3}{\text{CH}_2}\overset{sp^3}{\text{CH}_2}\overset{sp^3}{\text{CH}_3}$

5. Isomeric iodobutenes, C_4H_7I

$$\begin{array}{c} \text{H} \qquad\qquad \text{H} \\ \diagdown\quad\diagup \\ \text{C}=\text{C} \\ \diagup\quad\diagdown \\ \text{CH}_3\text{CH}_2 \qquad \text{I} \end{array}$$
cis-1-iodo-1-butene

$$\begin{array}{c} \text{H} \qquad\qquad \text{I} \\ \diagdown\quad\diagup \\ \text{C}=\text{C} \\ \diagup\quad\diagdown \\ \text{CH}_3\text{CH}_2 \qquad \text{H} \end{array}$$
trans-1-iodo-1-butene

$\text{CH}_3\text{CH}_2\text{CI}=\text{CH}_2$
2-iodo-1-butene

$\text{CH}_3\text{CHICH}=\text{CH}_2$
3-iodo-1-butene

$\text{CH}_2\text{ICH}_2\text{CH}=\text{CH}_2$
4-iodo-1-butene

$$\begin{array}{c} \text{H} \qquad\qquad \text{I} \\ \diagdown\quad\diagup \\ \text{C}=\text{C} \\ \diagup\quad\diagdown \\ \text{CH}_3 \qquad\qquad \text{CH}_3 \end{array}$$
cis-2-iodo-2-butene

$$\begin{array}{c} \text{H} \qquad\qquad \text{CH}_3 \\ \diagdown\quad\diagup \\ \text{C}=\text{C} \\ \diagup\quad\diagdown \\ \text{CH}_3 \qquad\qquad \text{I} \end{array}$$
trans-2-iodo-2-butene

$$\begin{array}{c} \text{H} \qquad\qquad \text{H} \\ \diagdown\quad\diagup \\ \text{C}=\text{C} \\ \diagup\quad\diagdown \\ \text{CH}_3 \qquad\qquad \text{CH}_2\text{I} \end{array}$$
cis-1-iodo-2-butene

$$\begin{array}{c} \text{H} \qquad\qquad \text{CH}_2\text{I} \\ \diagdown\quad\diagup \\ \text{C}=\text{C} \\ \diagup\quad\diagdown \\ \text{CH}_3 \qquad\qquad \text{H} \end{array}$$
trans-1-iodo-2-butene

$$\begin{array}{c} \text{CH}_3 \\ | \\ \text{CH}_3\text{C}=\text{CHI} \end{array}$$
1-iodo-2-methylpropene

$$\begin{array}{c} \text{CH}_3 \\ | \\ \text{CH}_2\text{IC}=\text{CH}_2 \end{array}$$
3-iodo-2-methylpropene

6. (a) C_3H_5Cl

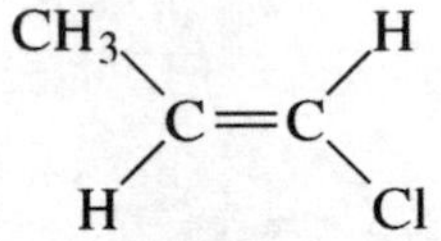

trans-1-chloropropene

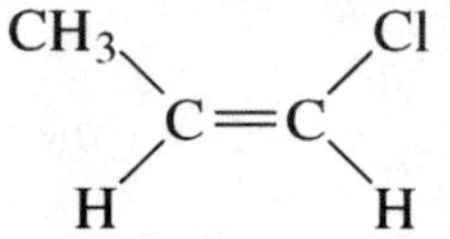

cis-1-chloropropene

$CH_3CCl=CH_2$
2-chloropropene

$CH_2ClCH=CH_2$
3-chloropropene

(b) chlorocyclopropane

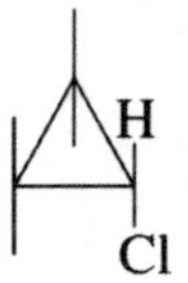

7. (a) $CH_3CHCH=CHCHCH_3$

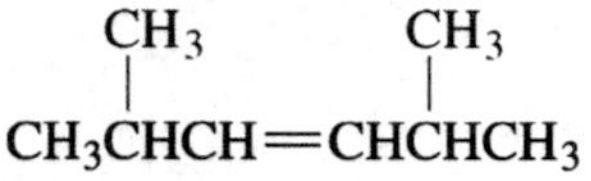

2,5-dimethyl-3-hexene

(b) cis-4-methyl-2-pentene

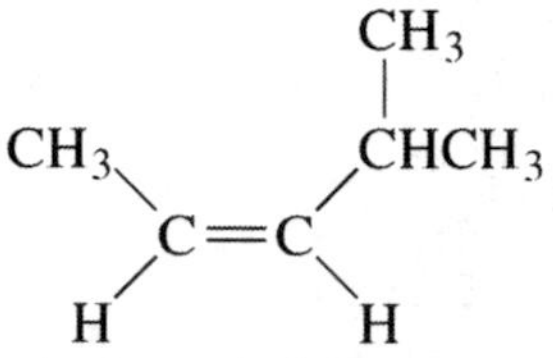

(c) $CH\equiv CCH=CHCH_3$
3-penten-1-yne

(d) trans-3-hexene

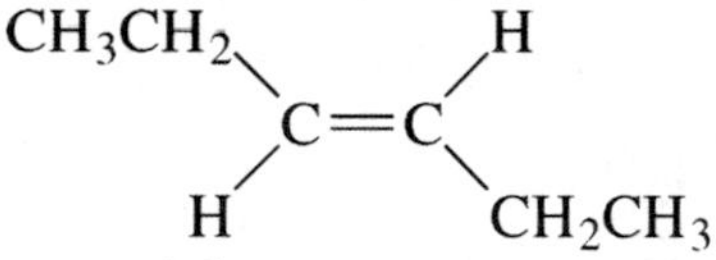

(e) $CH\equiv CCHCH_2CH_3$
     CH_3

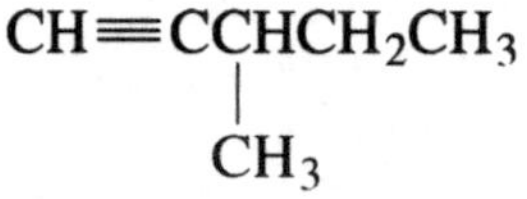

3-methyl-1-pentyne

(f) $CH_3CHCHCH_2CH_2CH_3$
     CH_3

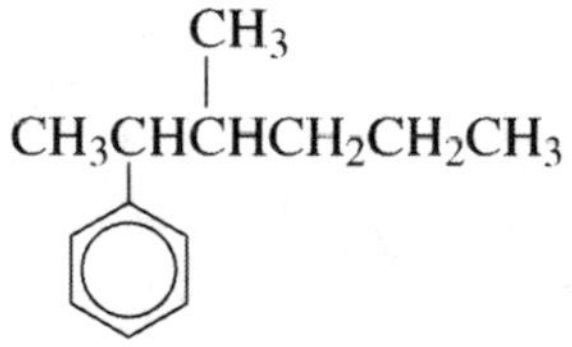

3-methyl-2-phenylhexane

8. (a) $CH_2=CHCCH_2CH_3$
     CH_2CH_3
     CH_3
3-ethyl-3-methyl-1-pentene

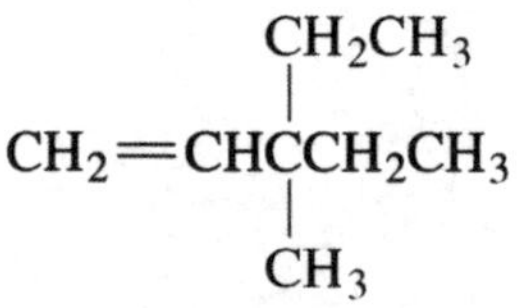

(b) cis, 1,2-diphenylethene

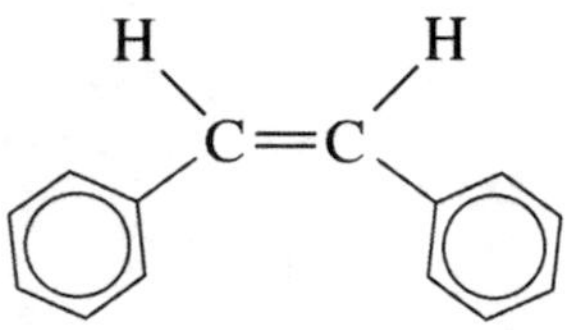

(c) $CH\equiv CCHCH_3$
3-phenyl-1-butyne

(d) cyclopentene

(e)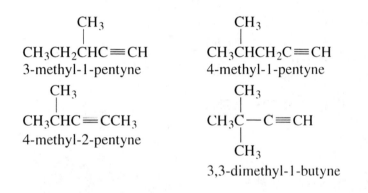

1-methylcyclohexene

(f) CH(CH₃)₂

3-isopropylcyclopentene

9. (a) 23 sigma bonds, 1 pi bond; (b) 17 sigma bonds, 1 pi bond; (c) 10 sigma bonds, 3 pi bonds; (d) 17 sigma bonds, 1 pi bond; (e) 15 sigma bonds, 2 pi bonds; (f) 33 sigma bonds, 3 pi bonds.

10. (a) 23 sigma bonds, 1 pi bond; (b) 27 sigma bonds, 7 pi bonds; (c) 20 sigma bonds, 5 pi bonds; (d) 13 sigma bonds, 1 pi bond; (e) 19 sigma bonds, 1 pi bond; (f) 21 sigma bonds, 1 pi bond.

11. (a) *trans*-6-chloro-3-heptene; (b) 4,4-dimethyl-2-pentyne; (c) 2-ethyl-1-pentene.

12. (a) 3-phenyl-1-butyne; (b) 2-methyl-2-hexene; (c) *cis*-3,4-dimethyl-3-hexene.

13. All the hexynes, C₆H₁₀

$CH_3CH_2CH_2CH_2C{\equiv}CH$ $CH_3CH_2CH_2C{\equiv}CCH_3$ $CH_3CH_2C{\equiv}CCH_2CH_3$
1-hexyne   2-hexyne   3-hexyne

$\quad\quad\quad\quad CH_3$
$\quad\quad\quad\quad |$
$CH_3CH_2CHC{\equiv}CH$
3-methyl-1-pentyne

$\quad\quad\quad\quad CH_3$
$\quad\quad\quad\quad |$
$CH_3CHCH_2C{\equiv}CH$
4-methyl-1-pentyne

$\quad\quad CH_3$
$\quad\quad |$
$CH_3CHC{\equiv}CCH_3$
4-methyl-2-pentyne

$\quad\quad CH_3$
$\quad\quad |$
$CH_3C{-}C{\equiv}CH$
$\quad\quad |$
$\quad\quad CH_3$
3,3-dimethyl-1-butyne

14. All the pentynes, C₅H₈

$CH_3CH_2CH_2C{\equiv}CH$ $CH_3CH_2C{\equiv}CCH_3$
1-pentyne   2-pentyne

$\quad\quad\quad CH_3$
$\quad\quad\quad |$
$CH_3CHC{\equiv}CH$
3-methyl-1-butyne

15. (a) 1-butyne (The smallest numbered carbon that is involved in the triple bond is used to number the triple bond.); (b) 2-hexyne (The parent chain is the longest continuous carbon chain that contains the triple bond.); (c) propyne (The triple bond has only one possible location in propyne and no number is needed.)

16. (a) 4-methyl-2-pentyne (The parent chain is numbered from the end closest to the triple bond.); (b) 4-methyl-2-hexyne (The parent chain is the longest continuous carbon chain that contains the triple bond.); (c) 3-bromo-1-butyne (The parent chain is numbered from the end closest to the triple bond.).

17. *Cis-trans* isomers exist for (c) only. Alkynes do not have *cis-trans* isomerism. Alkenes have *cis-trans* isomerism only when each double-bonded carbon is attached to two different groups.

18. *Cis-trans* isomers exist for (b) only. Alkynes do not have *cis-trans* isomerism. Alkenes have *cis-trans* isomerism only when each double-bonded carbon is attached to two different groups.

19. (a) *cis*
    (b) neither
    (c) *trans*

20. (a) *trans*
    (b) *trans*
    (c) neither

21. (a) $CH_3CH_2CH_2CH{=}CH_2 + Br_2 \longrightarrow CH_3CH_2CH_2CHBrCH_2Br$

    (b) $CH_3CH_2\underset{\underset{CH_3}{|}}{C}{=}CHCH_3 + HI \longrightarrow CH_3CH_2\underset{\underset{CH_3}{|}}{\overset{\overset{I}{|}}{C}}CH_2CH_3$

    (c) $CH_3CH_2CH{=}CH_2 + H_2O \xrightarrow{H^+} CH_3CH_2\underset{\underset{OH}{|}}{C}HCH_3$

    (d) 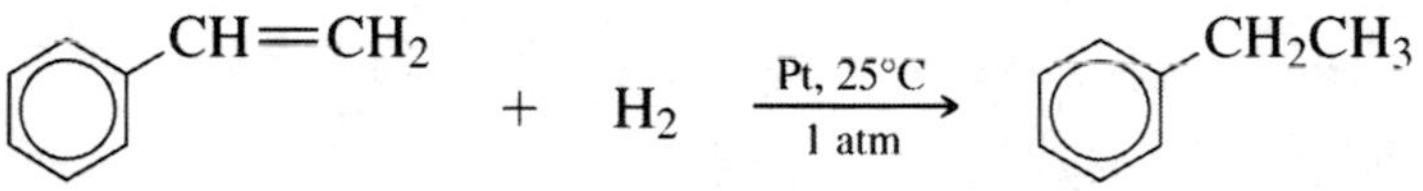 $\underset{\bigcirc}{}CH{=}CH_2 + H_2 \xrightarrow[\text{1 atm}]{\text{Pt, 25°C}} \underset{\bigcirc}{}CH_2CH_3$

    (e) $CH_3CH{=}CHCH_3 + KMnO_4 \xrightarrow[\text{cold}]{H_2O} CH_3\underset{\underset{HO}{|}}{C}H\underset{\underset{OH}{|}}{C}HCH_3$

22. (a) $CH_3CH_2CH_2CH{=}CH_2 + H_2O \xrightarrow{H^+} CH_3CH_2CH_2\underset{\underset{OH}{|}}{C}HCH_3$

    (b) $CH_3CH_2CH{=}CHCH_3 + HBr \longrightarrow CH_3CH_2CHBrCH_2CH_3 + CH_3CH_2CH_2CHBrCH_3$

    (c) $(CH_2{=}CHCl + Br_2 \longrightarrow CH_2BrCHClBr$

    (d) 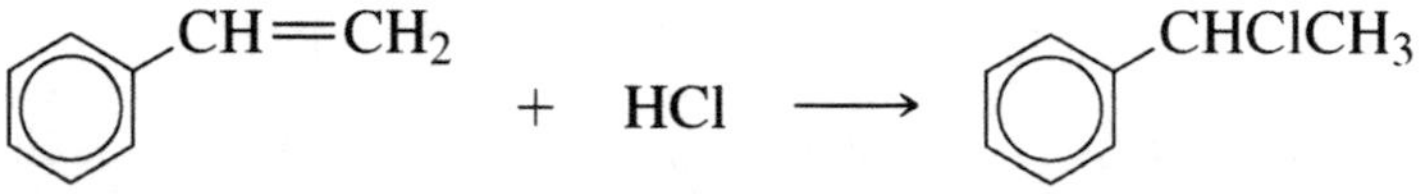 $\underset{\bigcirc}{}CH{=}CH_2 + HCl \longrightarrow \underset{\bigcirc}{}CHClCH_3$

    (e) $CH_2{=}CHCH_2CH_3 + KMnO_4 \xrightarrow[\text{cold}]{H_2O} CH_2\underset{\underset{OH}{|}}{C}H\underset{\underset{OH}{|}}{C}H_2CH_3$

23. (a) $CH_3C{\equiv}CCH_3 + Br_2 \text{ (1 mole)} \longrightarrow CH_3CBr{=}CBrCH_3$
    (b) Two-step reaction:

    $CH{\equiv}CH + HCl \longrightarrow CH_2{=}CHCl \xrightarrow{HCl} CH_3CHCl_2$
    (c) $CH_3CH_2CH_2C{\equiv}CH + H_2 \text{ (1 mole)} \xrightarrow[25°C]{Pt} CH_3CH_2CH_2CH{=}CH_2$

24. (a) $CH_3C\equiv CH + H_2$ (1 mol) $\xrightarrow[\text{1 atm}]{\text{Pt,25°C}}$ $CH_3CH=CH_2$

    (b) $CH_3C\equiv CCH_3 + Br_2$ (2 mol) $\longrightarrow$ $CH_3CBr_2CBr_2CH_3$

    (c) Two-step reaction:

$$CH_3C\equiv CH + HCl \longrightarrow CH_3CCl=CH_2 \xrightarrow{HCl} CH_3CCl_2CH_3$$

25. When cyclohexene, reacts with:

    (a) Br_2, the product is 1,2-dibromocyclohexane

    (b) HI, the product is iodocyclohexane

    (c) H_2O, H^+, the product is cyclohexanol

    (d) $KMnO_4(aq)$, the product is cyclohexene glycol or 1,2-dihydroxycyclohexane

26. When cyclopentene, reacts with:

    (a) Cl_2, the product is 1,2-dichlorocyclopentane

    (b) HBr, the product is bromocyclopentane

    (c) H_2, Pt, the product is cyclopentane

    (d) H_2O, H^+, the product is cyclopentanol

27.

(a) $CH\equiv C-\overset{\overset{\displaystyle CH_3}{|}}{C}H-CH_3 + HBr \longrightarrow CH_2=\overset{\overset{\displaystyle Br}{|}}{C}-\overset{\overset{\displaystyle CH_3}{|}}{C}H-CH_3$

(b) $CH\equiv C-\overset{\overset{\displaystyle CH_3}{|}}{C}H-CH_3 + 2\,HBr \longrightarrow CH_3-\overset{\overset{\displaystyle Br}{|}}{\underset{\underset{\displaystyle Br}{|}}{C}}-\overset{\overset{\displaystyle CH_3}{|}}{C}H-CH_3$

(c) $CH\equiv C-\overset{\overset{\displaystyle CH_3}{|}}{C}H-CH_3 + 2\,Br_2 \longrightarrow \overset{\overset{\displaystyle Br}{|}}{\underset{\underset{\displaystyle Br}{|}}{C}}H-\overset{\overset{\displaystyle Br}{|}}{\underset{\underset{\displaystyle Br}{|}}{C}}-\overset{\overset{\displaystyle CH_3}{|}}{C}H-CH_3$

28. (a) 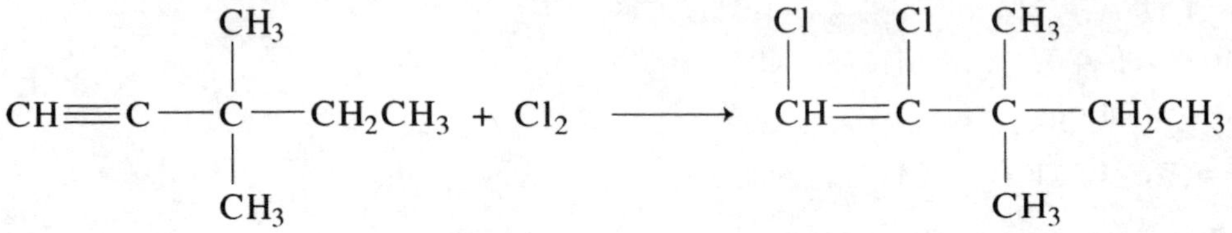 $CH\equiv C-\underset{\underset{CH_3}{|}}{\overset{\overset{CH_3}{|}}{C}}-CH_2CH_3 + Cl_2 \longrightarrow CH=\underset{\underset{CH_3}{|}}{\overset{\overset{Cl}{|}}{C}}-\overset{\overset{CH_3}{|}}{\underset{}{C}}-CH_2CH_3$ (with Cl on CH=)

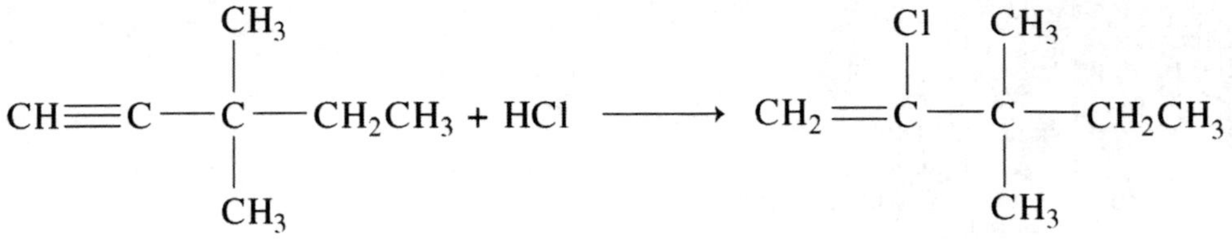

(b) $CH\equiv C-\underset{\underset{CH_3}{|}}{\overset{\overset{CH_3}{|}}{C}}-CH_2CH_3 + HCl \longrightarrow CH_2=\overset{\overset{Cl}{|}}{C}-\underset{\underset{CH_3}{|}}{\overset{\overset{CH_3}{|}}{C}}-CH_2CH_3$

(c) $CH\equiv C-\underset{\underset{CH_3}{|}}{\overset{\overset{CH_3}{|}}{C}}-CH_2CH_3 + 2\,HCl \longrightarrow CH_3-\underset{\underset{Cl}{|}}{\overset{\overset{Cl}{|}}{C}}-\underset{\underset{CH_3}{|}}{\overset{\overset{CH_3}{|}}{C}}-CH_2CH_3$

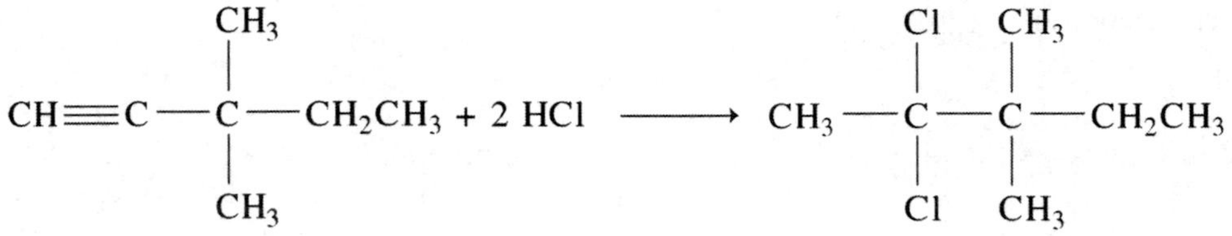

29. (a) *meta*
   (b) *meta*
   (c) *ortho*

30. (a) *para*
   (b) *meta*
   (c) *para*

31. (a) 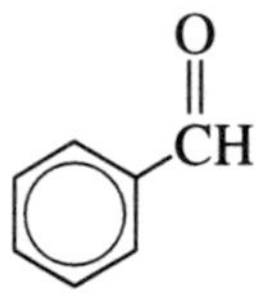 benzaldehyde
   (b) 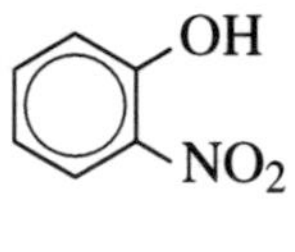 2-nitrophenol
   (c) 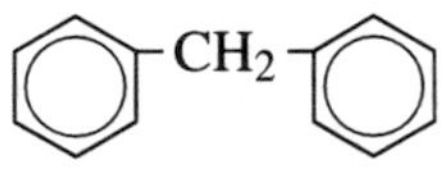 diphenylmethane
   (d) isopropylbenzene
   (e) 2,4-dibromoaniline
   (f) 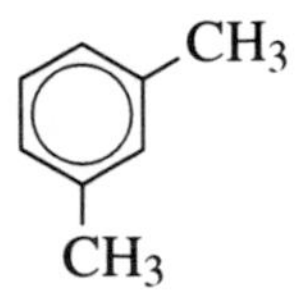 1,3-dimethylbenzene

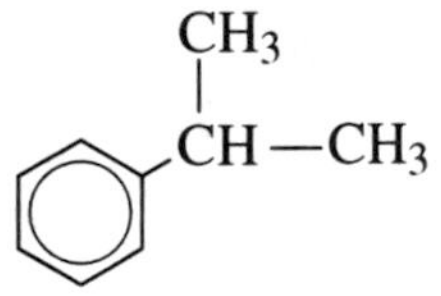

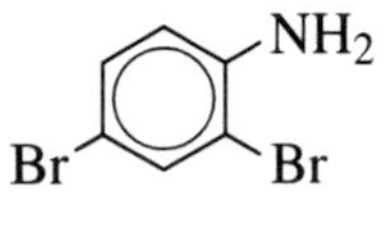

32. (a) 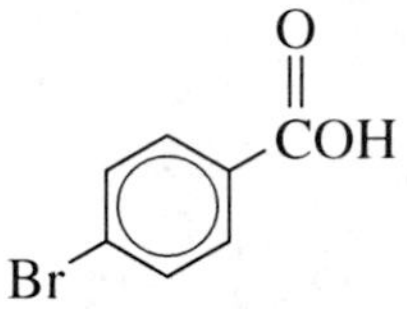   (d)

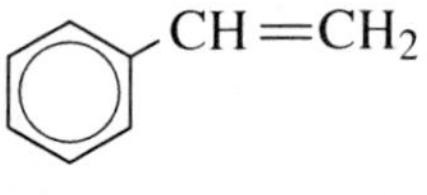

(b) 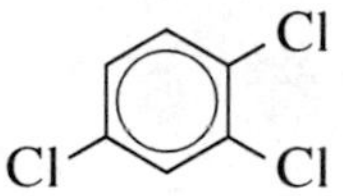   (e)

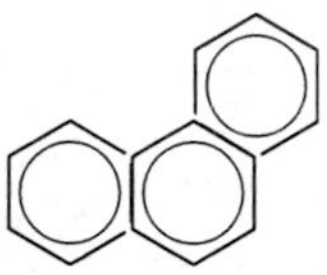

(c) 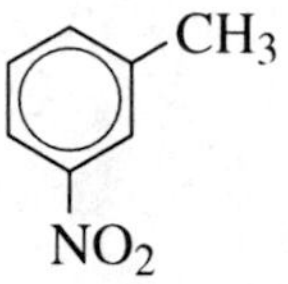   (f)

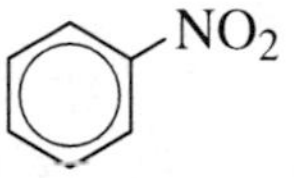

33. (a) bromodichlorobenzenes

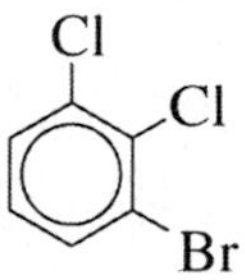

3-bromo-1,2-dichlorobenzene

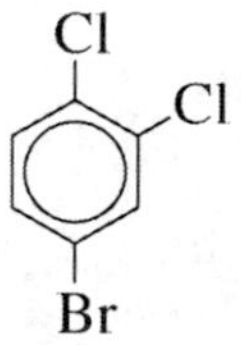

4-bromo-1,2-dichlorobenzene

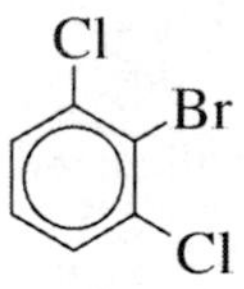

2-bromo-1,3-dichlorobenzene

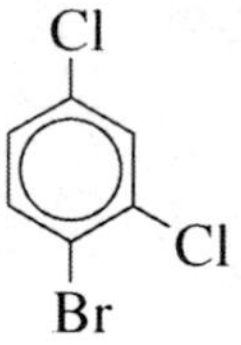

4-bromo-1,3-dichlorobenzene

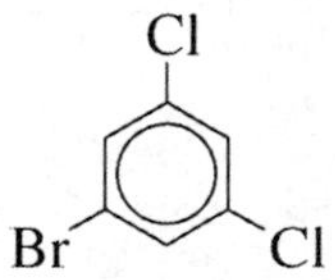

5-bromo-1,3-dichlorobenzene

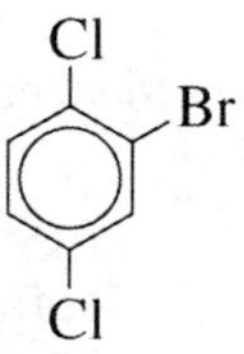

2-bromo-1,4-dichlorobenzene

(b)   The toluene derivatives of formula C₉H₁₂:

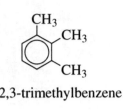

1,2,3-trimethylbenzene

1,2,4-trimethylbenzene

1,3,5-trimethylbenzene

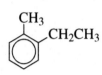

o-ethyltoluene

m-ethyltoluene

p-ethyltoluene

34.   (a)   trichlorobenzenes

1,2,3-trichlorobenzene

1,2,4-trichlorobenzene

1,3,5-trichlorobenzene

(b)   The benzene derivatives of formula C₈H₁₀:

o-xylene or 1,2-dimethylbenzene

m-xylene or 1,3-dimethylbenzene

p-xylene or 1,4-dimethylbenzene

ethylbenzene

35.   (a) *p*-chloroethylbenzene   (d) *p*-bromophenol
       (b) propylbenzene   (e) triphenylmethane
       (c) *m*-nitroaniline

36.   (a) styrene   (d) isopropylbenzene
       (b) *m*-nitrotoluene   (e) 2,4,6-tribromophenol
       (c) 2,4-dibromobenzoic acid

37. (a) 2-chloro-1,4-diiodobenzene

    (b) *o*-dibromobenzene

    (c) *m*-chloroaniline

38. (a) 2,3-dibromophenol

    (b) 1,3,5-trichlorobenzene

    (c) *m*-iodotoluene

39. (a)

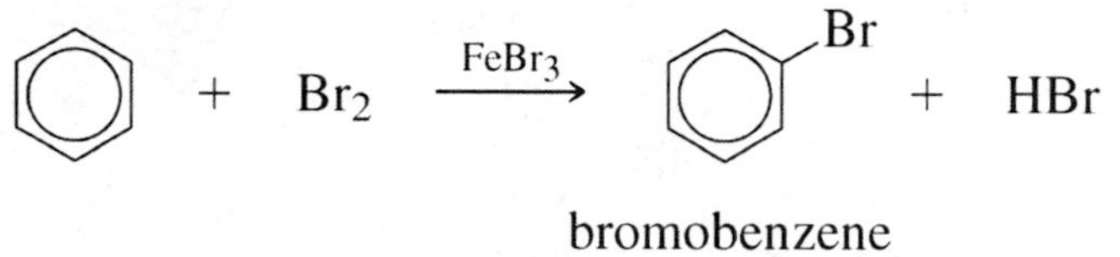

bromobenzene

    (b)

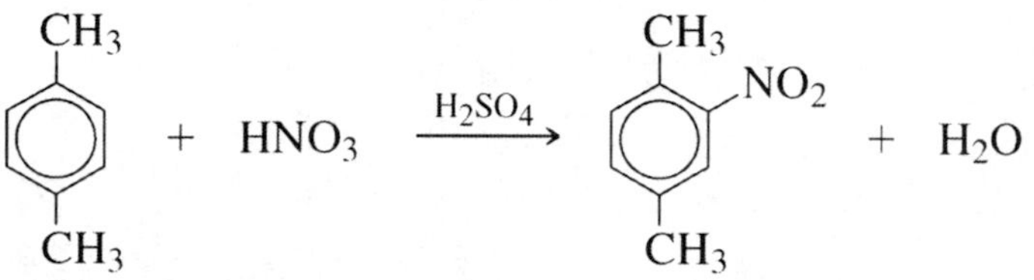

1,4-dimethyl-2-nitrobenzene

40. (a)

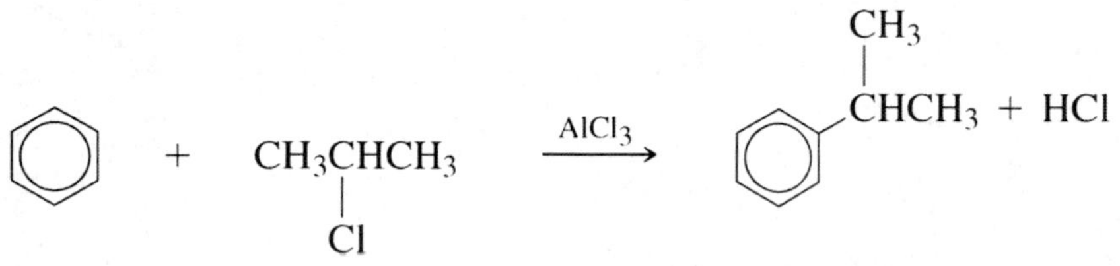

isopropylbenzene

    (b)

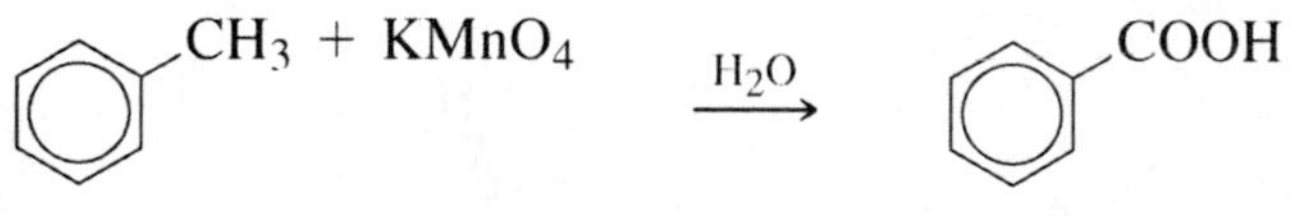

benzoic acid

41. When

$$CH_2{=}\underset{\underset{CH_3}{|}}{C}CH_2CH_2CH_3$$

reacts with HBr, two products are possible:

$$CH_3\underset{\underset{CH_3}{|}}{C}BrCH_2CH_2CH_3 \quad \text{and} \quad CH_2Br\underset{\underset{CH_3}{|}}{C}HCH_2CH_2CH_3$$

The first will strongly predominate. This is the product according to Markovnikov's rule and forms because the tertiary carbocation intermediate formed is more stable than a primary carbocation.

42. Two tests can be used. (1) Baeyer test—hexene will decolorize $KMnO_4$ solution; cyclohexane will not. (2) In the absence of sunlight, hexene will react with and decolorize bromine; cyclohexane will not.

43.

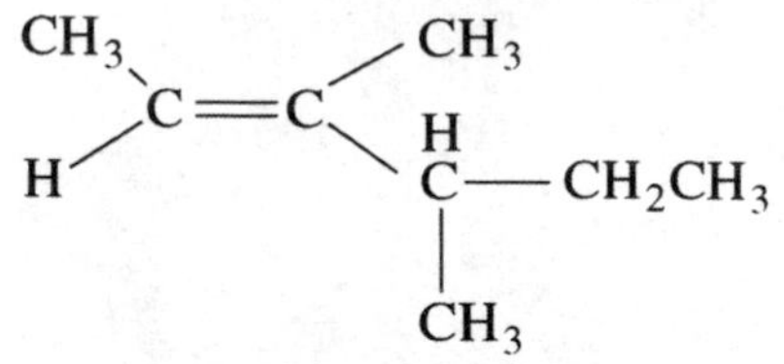

trans-3,4-dimethyl-2-hexene

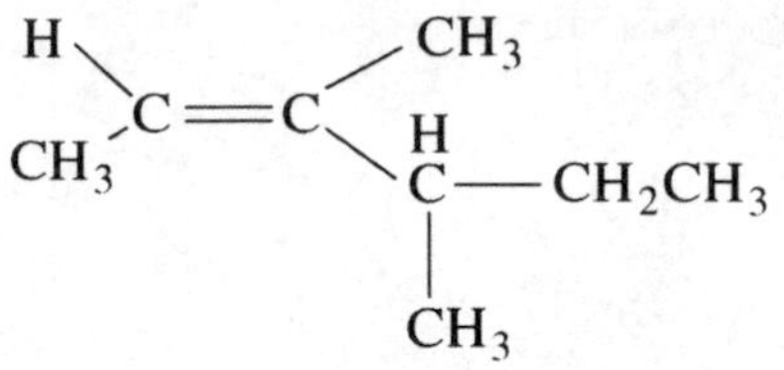

cis-3,4-dimethyl-2-hexene

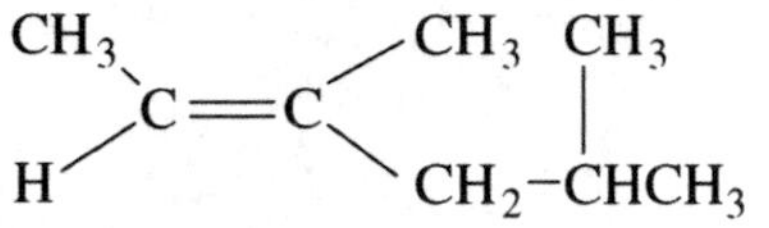

trans-3,5-dimethyl-2-hexene

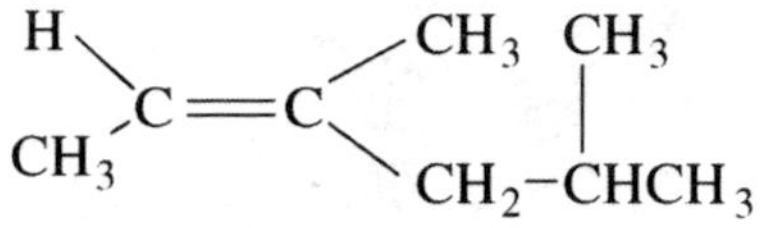

cis-3,5-dimethyl-2-hexene

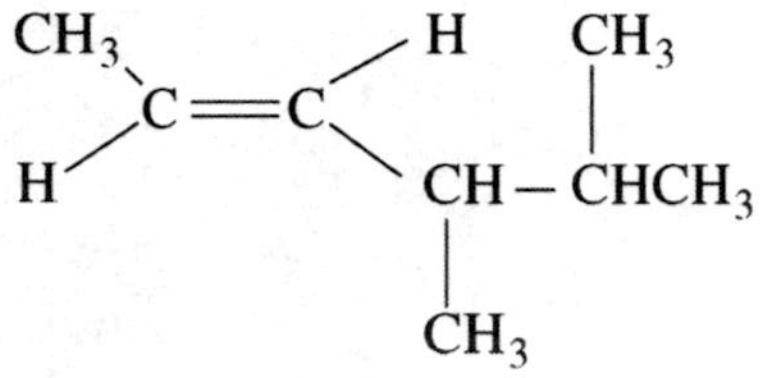

trans-4,5-dimethyl-2-hexene

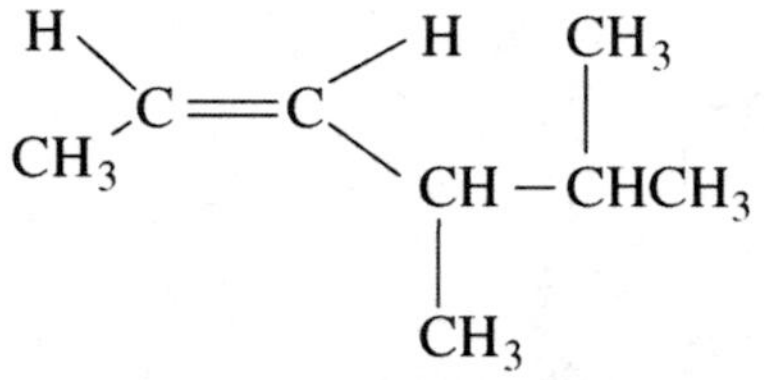

cis-4,5-dimethyl-2-hexene

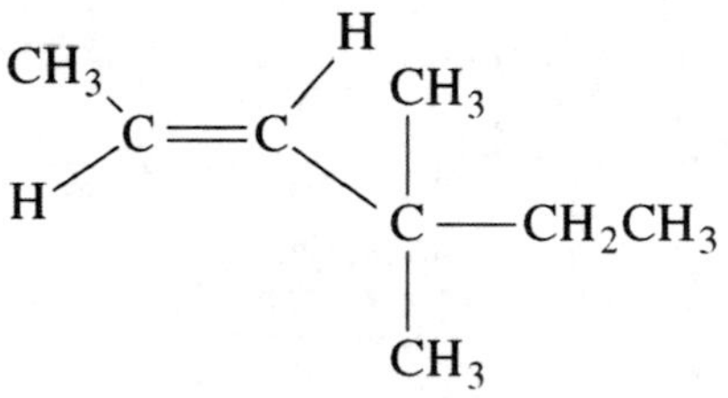

trans-4,4-dimethyl-2-hexene

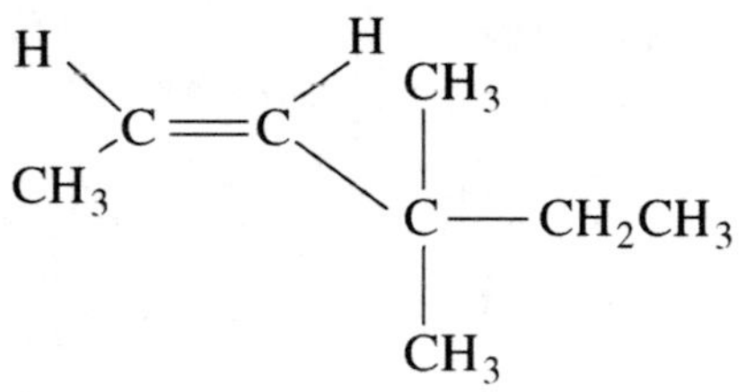

cis-4,4-dimethyl-2-hexene

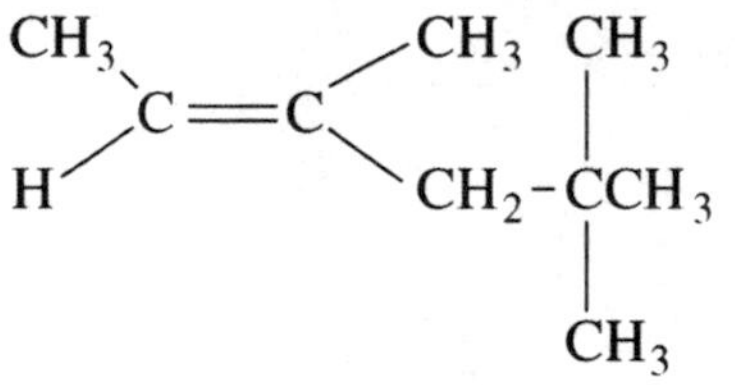

trans-3,5,5-trimethyl-2-hexene

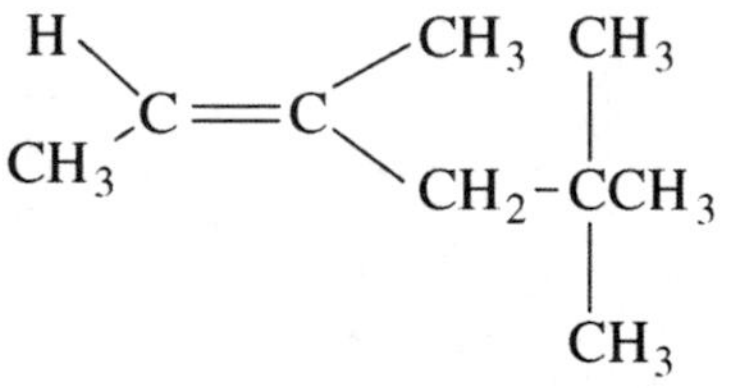

cis-3,5,5-trimethyl-2-hexene

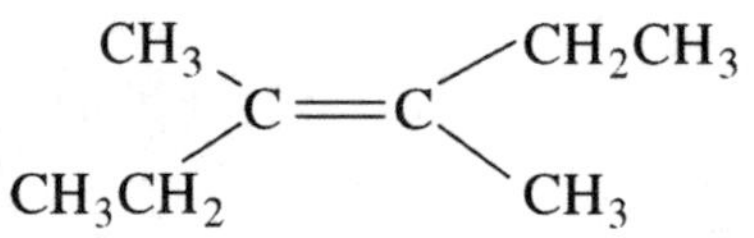

trans-3,4-dimethyl-3-hexene

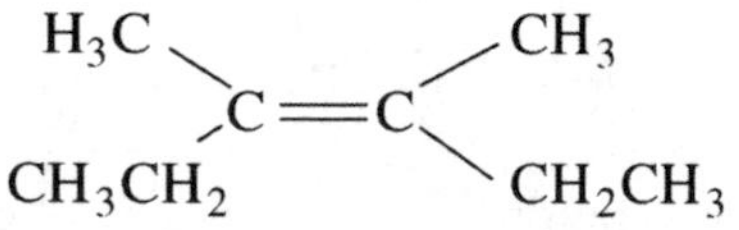

cis-3,4-dimethyl-3-hexene

43. (Cont.)

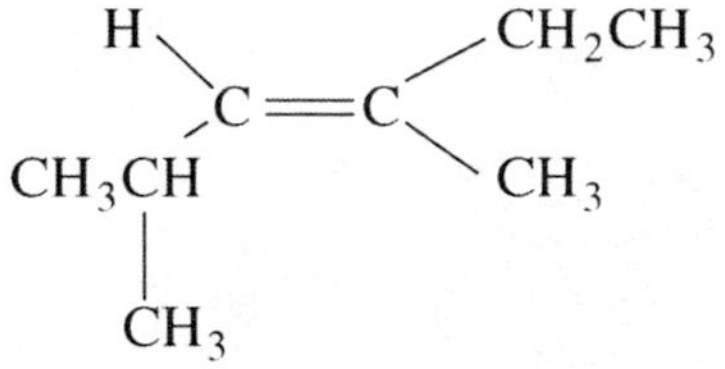

trans-2,4-dimethyl-3-hexene

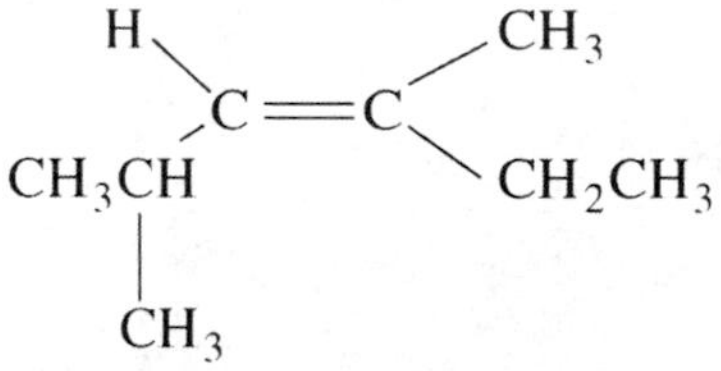

cis-2,4-dimethyl-3-hexene

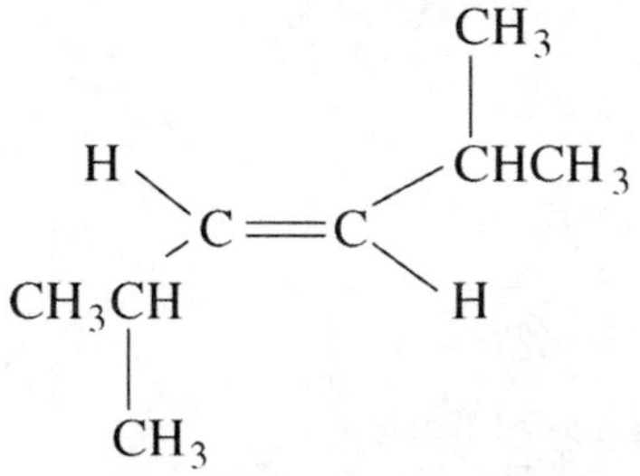

trans-2,5-dimethyl-3-hexene

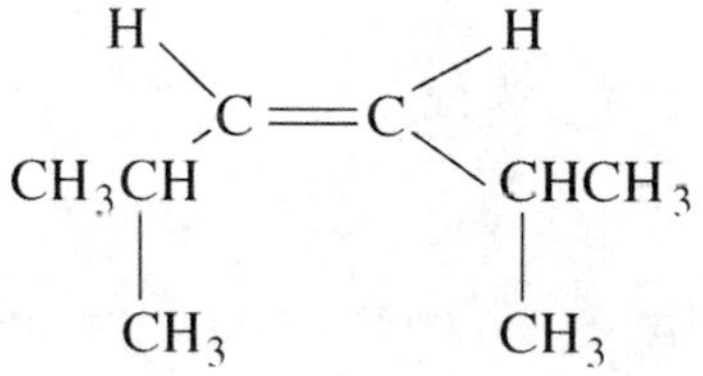

cis-2,5-dimethyl-3-hexene

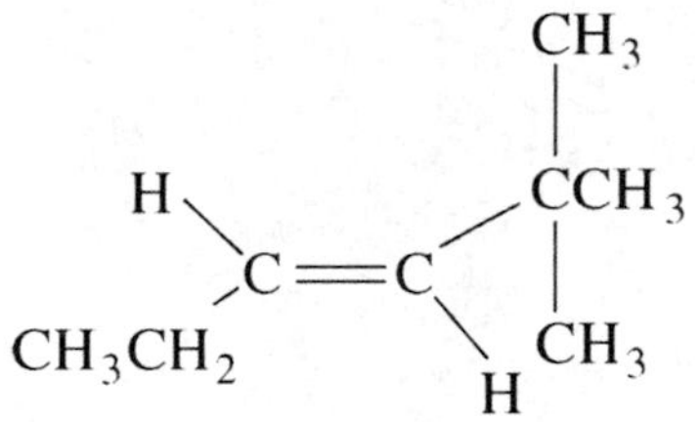

trans-2,2-dimethyl-3-hexene

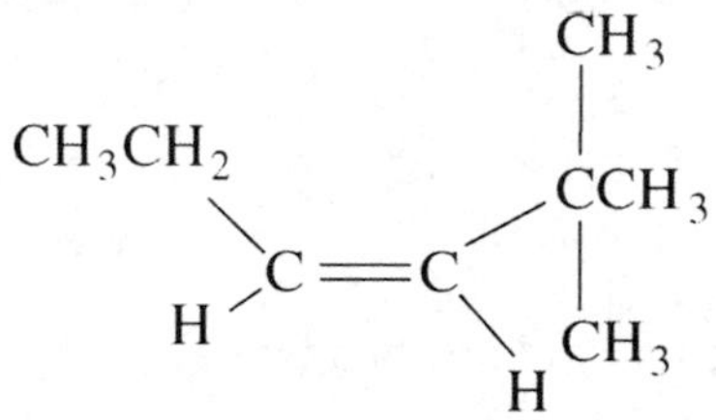

cis-2,2-dimethyl-3-hexene

44. (a) $CH_3 \overset{+}{C} H_2$ primary carbocation (a positively charged carbon bonded to one other carbon)

(b) $CH_3 \overset{+}{C} HCH_3$ secondary carbocation (a positively charged carbon bonded to two other carbons)

(c) $CH_3 - \overset{+}{\underset{|}{C}} - CH_3$ tertiary carbocation (a positively charged carbon bonded to three other carbons)
            CH_3

45. (a)

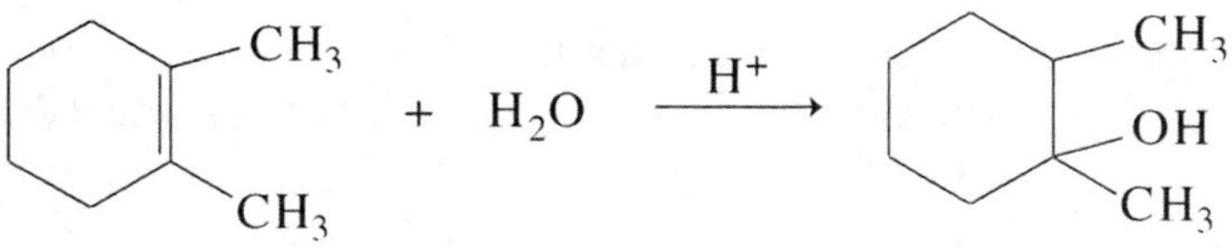

(b)

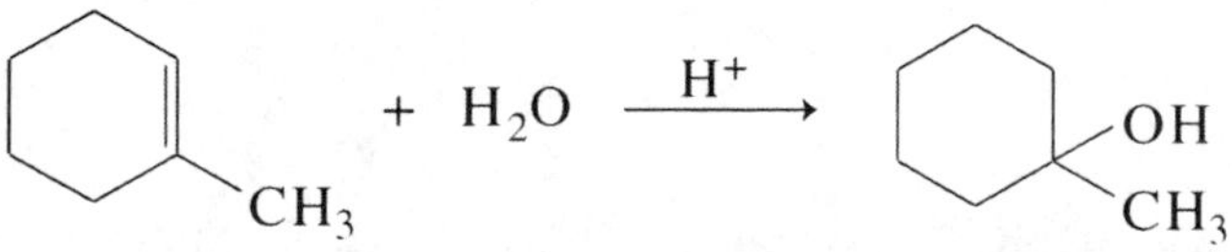

(c) 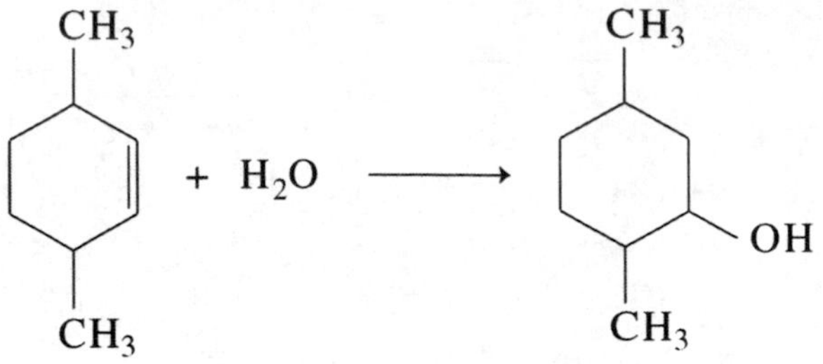 + H₂O ⟶

(d) 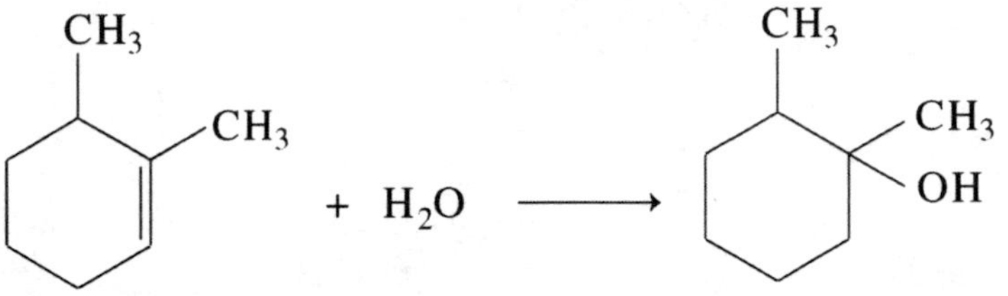 + H₂O ⟶

46. The reaction mechanism by which benzene is brominated in the presence of $FeBr_3$:

(a) $FeBr_3 + Br_2 \longrightarrow FeBr_4^- + Br^+$

Formation of a bromonium ion (Br^+), an electrophile.

(b)

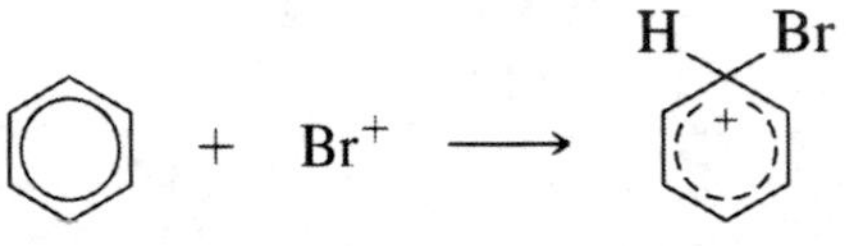

The bromonium ion adds to benzene forming a carbocation intermediate.

(c)

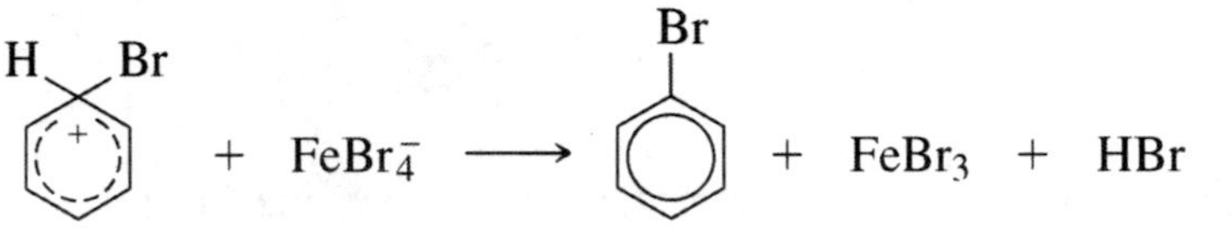

A hydrogen ion is lost from the carbocation forming the product bromobenzene.

47. (a) $CH_3CHClCH_2CH_3 \xrightarrow{-HCl} CH_2{=}CHCH_2CH_3 + CH_3CH{=}CHCH_3$

(b) $CH_2ClCH_2CH_2CH_2CH_3 \xrightarrow{-HCl} CH_2{=}CHCH_2CH_2CH_3$

(c)

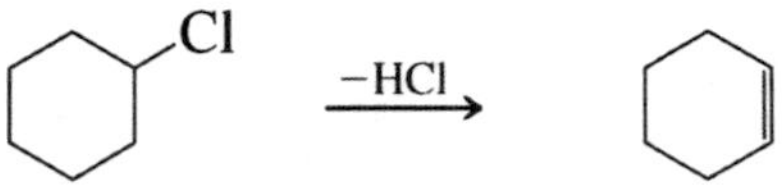

48. Yes, there will be a color change (loss of Br_2 color). The fact that there is no HBr formed indicates that the reaction is not substitution but addition. Therefore, C_4H_8 must contain a carbon-carbon double bond. Three structures are possible.

$CH_3CH_2CH{=}CH_2$ or $CH_3CH{=}CHCH_3$ or $CH_3\overset{\displaystyle CH_3}{\underset{|}{C}}{=}CH_2$

49. Baeyer test: Add $KMnO_4$ solution to each sample. The $KMnO_4$ will lose its purple color with 1-heptene. There will be no reaction (no color change) with heptane.

50. I would not expect graphene to react easily with HCl in an addition reaction. Graphene is composed of sheets of aromatic rings that do not easily undergo addition reactions.

51.

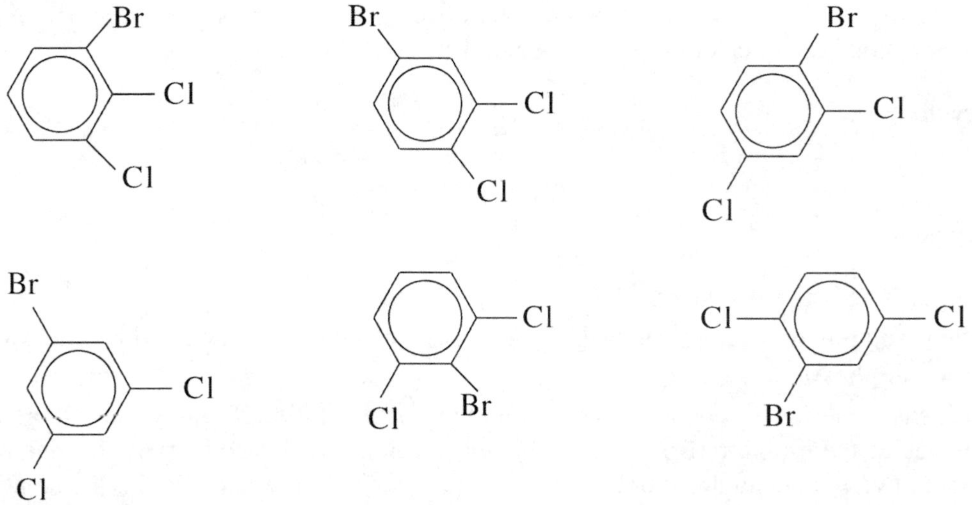

52. The double bonds in fatty acids are oxidized. This reaction adds oxygen and the double bond converts to a single bond. This reaction is thought to start the process that eventually results in "hardening of the arteries." This is an oxidation reaction.

53. (a)

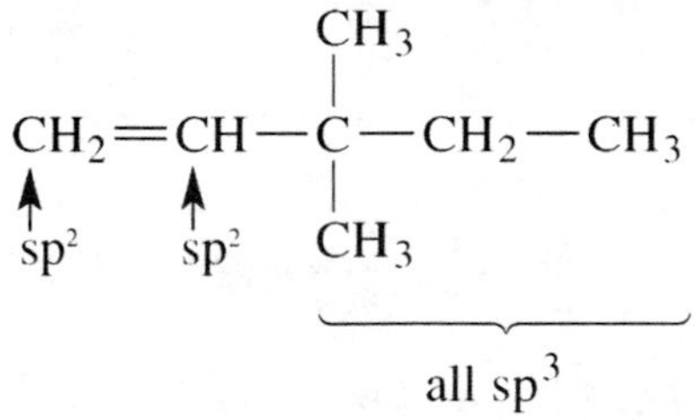

all sp^3

(b) $HC \equiv C - CH_3$
sp   sp   sp^3

(c)

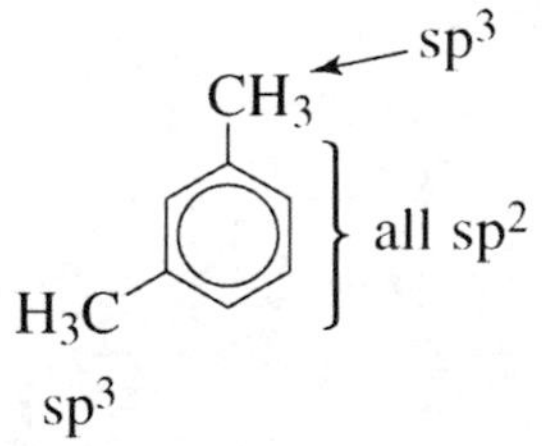

all sp^2

54. (a) Four isomers: $CH_2=CHCH_2CH_3$ $CH_3CH = CHCH_3$
C$_4$H$_8$ 1-butene   2-butene (cis and trans)
$CH_2 = C(CH_3)_2$
2-methyl-1-propene

(b) One isomer
C$_5$H$_8$

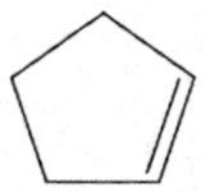

cyclopentene

55. (b) is the correct structure, (a) is an incorrect structure because the carbons where the two rings are fused have five bonds, not four bonds to each carbon.

56. Carbon-2 has two methyl groups on it. The configuration of two of the same groups on a carbon of a carbon-carbon double bond does not show cis-trans isomerism.

57. (a) alkyne or cycloalkene
    (b) alkene
    (c) alkane
    (d) alkyne or cycloalkene

58. Chemically distinguishing between benzene, 1-hexene, and 1-hexyne

    Step 1. Add $KMnO_4$ solution to a sample of each liquid. Benzene is the only one in which the $KMnO_4$ does not lose its purple color.

    Step 2. To 0.5 mL samples of 1-hexene and 1-hexyne add bromine solution dropwise until there is no more color change of the bromine (from reddish-brown to colorless). 1-hexyne (with a triple bond) will decolor about twice as many drops of bromine as 1-hexene. Thus the three liquids are identified.

59. (a) $CH_3C{\equiv}CH \xrightarrow{\text{HCl}} CH_3CCl{=}CH_2 \xrightarrow{\text{HBr}} CH_3CClBrCH_3$

    (b) $CH_3C{\equiv}CH \xrightarrow{\text{Br}_2} CH_3CBr{=}CHBr \xrightarrow{\text{HCl}} CH_3CClBrCH_2Br$

    (c) $CH_3C{\equiv}CH \xrightarrow{\text{2Cl}_2} CH_3CCl_2CHCl_2$

CHAPTER 21

POLYMERS: MACROMOLECULES

SOLUTIONS TO REVIEW QUESTIONS

1. An addition polymer is one that is produced by the successive addition of repeating monomer molecules. A condensation polymer is one that is formed from monomer molecules in a reaction which splits out water or some other simple molecule. Condensation polymerization usually involves two different monomers.

2. Those polymers which soften on reheating are thermoplastic polymers; those which set to an infusible solid and do not soften on reheating are thermosetting polymers.

3. Low-density polyethylene is used in packing material, molded articles, plastic films, garbage bags, flexible bottles, containers, and toys. High-density polyethylene is used in plastic grocery bags; bottles for milk, juice, water and laundry products; and medical products.

4. Unique problems of plastic recycling that distinguish it from aluminum or glass recycling are as follows:
    (a) Recycled plastic is not as good as new plastic. While plastic can typically be recycled once, glass and aluminum may be recycled many times;
    (b) Sorting of plastics during recycling is much more difficult than sorting glass or aluminum;
    (c) In contrast to glass and aluminum recycling, it is much cheaper to create new plastic rather than recycle.

5. The word "plastic" means capable of being shaped. Many (but *not* all) organic polymers have this capability.

6. A propagation step always reproduces a reactive compound that can continue the chain reaction. A termination step destroys the reactive compounds that are responsible for the polymerization.

7. The big problem of recycling is with the separating the different types of plastics.

8. Copolymers contain more than one monomer. Polyethylene is made from only one monomer, ethylene.

SOLUTIONS TO EXERCISES

1. Each propylene unit is composed of $CH_2=CHCH_3$ and has a molar mass of 42 g/mole of propylene. Thus,

    (20,000 g/mole of polymer)/(42.0 g/mole of propylene)
    $= 476$ moles of propylene/mole of polymer

    There are about 476 propylene units per polymer.

2. The monomer that makes up teflon is tetrafluoroethylene, $CF_2=CF_2$, 100.0 g/mole of monomer unit. There are 3500 monomer units/teflon polymer so there are 3500 moles of monomer units/mole of teflon.

    (3500 moles of monomer units/mole of teflon)(100.0 g/mole of monomer unit)
    $= 350,000$ g/mole of teflon

    The molar mass of the teflon polymer is 350,000 g/mole.

3. (a) $\left(\!\!-CH_2-CH-\right)_n$
     $\quad\quad\quad\quad |$
     $\quad\quad\quad OCCH_3$
     $\quad\quad\quad\quad \|$
     $\quad\quad\quad\quad O$

   (c) $\left(\!\!-CF_2-CF_2-\right)_n$

   (b) $\left(\!\!-CH_2-C-\right)_n$ with CH_3 above and CH_3 below

   (d) $\left(\!\!-CH_2-C-\right)_n$ with CH_3 above and $C-OCH_3$, $\|$ O below

4. (a) $\left(\!\!-CH_2-CCl_2-\right)_n$

   (c) $\left(\!\!-CH_2-CH-\right)_n$ with phenyl ring below

   (b) $\left(\!\!-CH_2-CH-\right)_n$ with CH_3 below

   (d) $\left(\!\!-CH_2-CH-\right)_n$ with CN below

5. Free radical polymerization starts with an initiation step that forms the following free radical:

    RO·

    This free radical then attacks one vinyl chloride to start polymerization:

$$ROCH_2 - \overset{\overset{\displaystyle Cl}{|}}{CH}\cdot$$

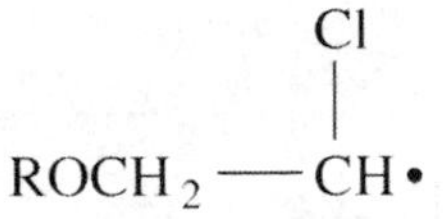

    A second vinyl chloride then reacts to give the following free radical:

$$ROCH_2 - \overset{\overset{\displaystyle Cl}{|}}{CH} - CH_2 - \overset{\overset{\displaystyle Cl}{|}}{CH}\cdot$$

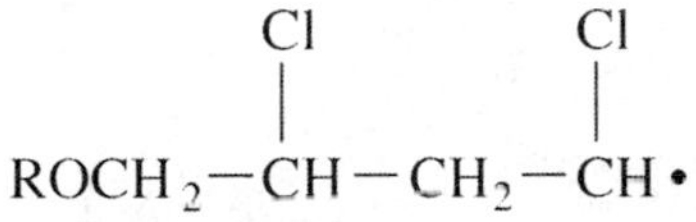

6. Free radical polymerization starts with an initiation step that forms the following free radical:

    RO·

    This free radical then attacks one propylene to start polymerization:

$$ROCH_2 - \overset{\overset{\displaystyle CH_3}{|}}{CH}\cdot$$

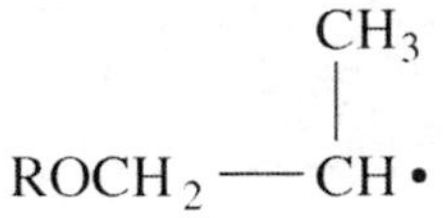

    A second propylene then reacts to give the following free radical:

$$ROCH_2 - \overset{\overset{\displaystyle CH_3}{|}}{CH} - CH_2 - \overset{\overset{\displaystyle CH_3}{|}}{CH}\cdot$$

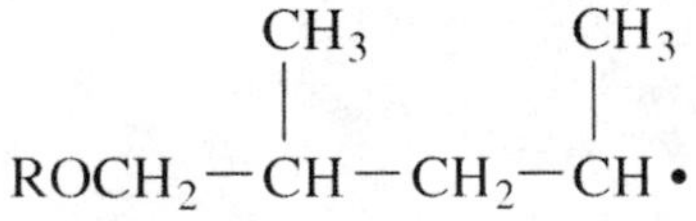

7. Formulas of the polymers that can be formed from:

    (a) propylene $\left(CH_2 - \underset{\underset{\displaystyle CH_3}{|}}{CH}\right)_n$

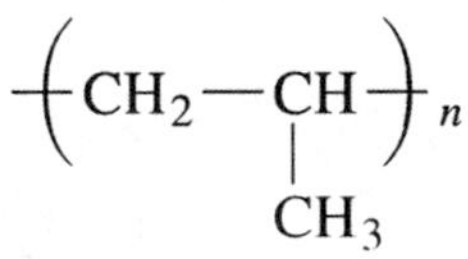

    (b) 2-methylpropene $\left(CH_2 - \underset{\underset{\displaystyle CH_3}{|}}{\overset{\overset{\displaystyle CH_3}{|}}{C}}\right)_n$

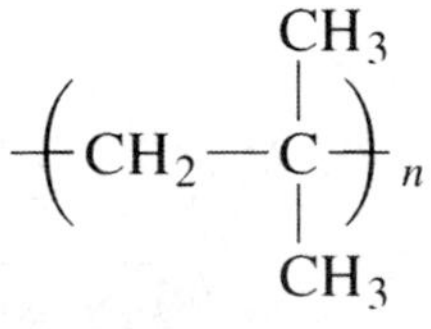

    (c) 2-butene $\left(\underset{\underset{\displaystyle CH_3}{|}}{CH} - \underset{\underset{\displaystyle CH_3}{|}}{CH}\right)_n$

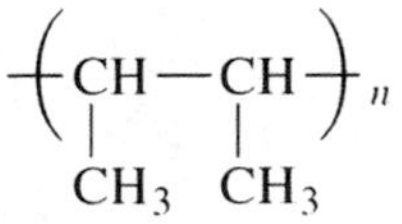

8. Formulas of the polymers that can be formed from:

   (a) ethylene $-(CH_2CH_2)_n$

   (b) chloroethene $-(CH_2CHCl)_n$

   (c) 1-butene

$$-(CH_2-\underset{\underset{CH_2CH_3}{|}}{CH})_n$$

9. Two possible ways in which vinyl chloride can polymerize to form polyvinyl chloride:

$$-(CH_2\underset{\underset{Cl}{|}}{CH}-CH_2\underset{\underset{Cl}{|}}{CH}-CH_2\underset{\underset{Cl}{|}}{CH}-CH_2\underset{\underset{Cl}{|}}{CH})_n$$

$$-(CH_2\underset{\underset{Cl}{|}}{CH}-\underset{\underset{Cl}{|}}{CH}CH_2-CH_2\underset{\underset{Cl}{|}}{CH}-\underset{\underset{Cl}{|}}{CH}CH_2)_n \quad \text{(other possibilities also)}$$

10. Two possible ways in which acrylonitrile can polymerize to form Orlon.

$$-(CH_2\underset{\underset{CN}{|}}{CH}-CH_2\underset{\underset{CN}{|}}{CH}-CH_2\underset{\underset{CN}{|}}{CH}-CH_2\underset{\underset{CN}{|}}{CH})_n$$

$$-(CH_2\underset{\underset{CN}{|}}{CH}-\underset{\underset{CN}{|}}{CH}CH_2-CH_2\underset{\underset{CN}{|}}{CH}-\underset{\underset{CN}{|}}{CH}CH_2)_n \quad \text{(other possibilities also)}$$

11. The polymer is low-density polyethylene. It is made from ethylene, $CH_2=CH_2$.

12. The polymer is polypropylene. It is made from propylene,

$$CH_2=\underset{\underset{CH_3}{|}}{CH}$$

13.

$$-(CH_2-CH_2-CH_2-\underset{\overset{CH_3}{|}}{CH})_n$$

14.

$$-(CH_2-\underset{\underset{\bigcirc}{|}}{CH}-CH_2-\underset{\underset{CH_3}{\overset{CH_3}{|}}}{C})_n$$

15. Natural rubber (all *cis*)

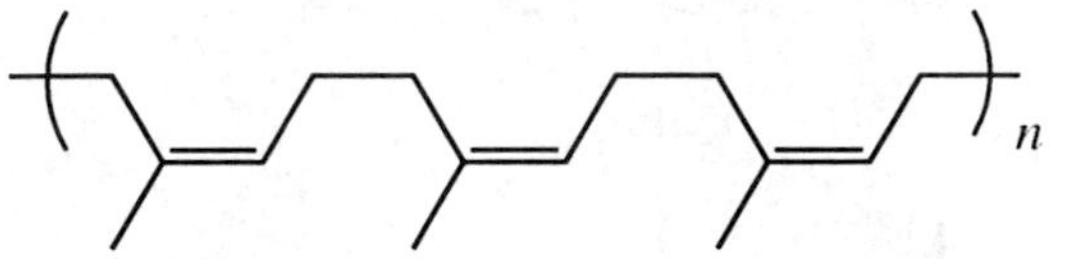

16. Gutta percha (all *trans*)

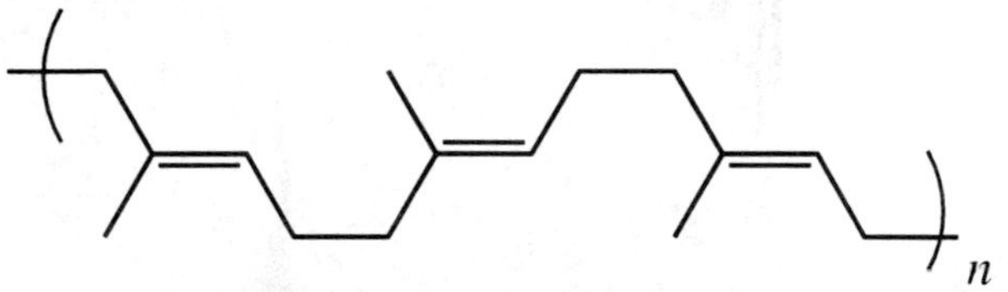

17. Yes, smog in the atmosphere contains ozone and ozone attacks natural rubber at the site of the double bond causing "age hardening" and cracking.

18. Yes, styrene-butadiene rubber contains carbon-carbon double bonds and is attacked by the ozone in smog causing "age hardening" and cracking.

19. Chemical structures for the monomers of:

    (a) natural rubber

    $$CH_2{=}CCH{=}CH_2$$
    $$\underset{\displaystyle CH_3}{|}$$

    (b) synthetic rubber (SBR)

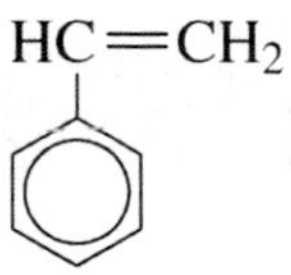

    $$HC{=}CH_2 \quad \text{and} \quad CH_2{=}CHCH{=}CH_2$$

20. Chemical structures for the monomers of:

    (a) synthetic natural rubber

    $$CH_2{=}CCH{=}CH_2$$
    $$\underset{\displaystyle CH_3}{|}$$

    (b) neoprene rubber

    $$CH_2{=}CCl{-}CH = CH_2$$

21. Syndiotactic polypropylene

22. (a) Isotactic polypropylene

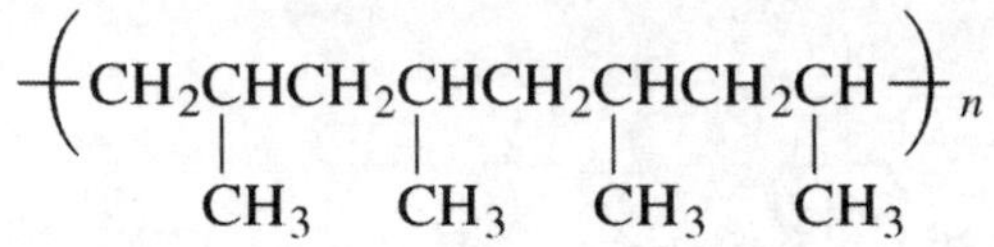

$$\left(CH_2CHCH_2CHCH_2CHCH_2CH\right)_n$$
$$\quad\quad | \quad\quad | \quad\quad | \quad\quad |$$
$$\quad\quad CH_3 \quad CH_3 \quad CH_3 \quad CH_3$$

(b) Another form of polypropylene

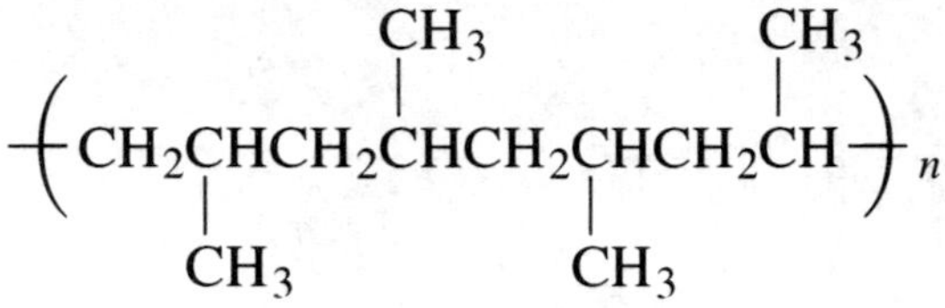

$$\quad\quad\quad\quad\quad\quad CH_3 \quad\quad\quad CH_3$$
$$\quad\quad\quad\quad\quad\quad | \quad\quad\quad\quad |$$
$$\left(CH_2CHCH_2CHCH_2CHCH_2CH\right)_n$$
$$\quad\quad | \quad\quad\quad\quad\quad | $$
$$\quad\quad CH_3 \quad\quad\quad\quad CH_3$$

(other structures are possible)

23. (a) none

(b) one,
$$\quad\quad CH_3$$
$$\quad\quad\ |$$
$$-CH_2C{=}CHCH_2-$$

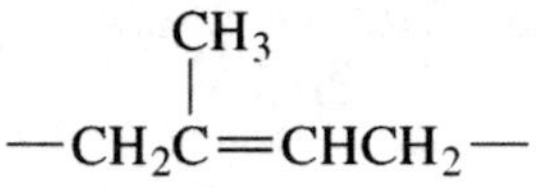

24. (a) one, $-CH_2CH{=}CHCH_2-$
    (b) none

25. (a) *trans*
    (b) *cis*
    (c) *trans*
    (d) *trans*

26. (a) *trans*
    (b) *cis*
    (c) *cis*

27. Polystyrene has the highest mass percent carbon.

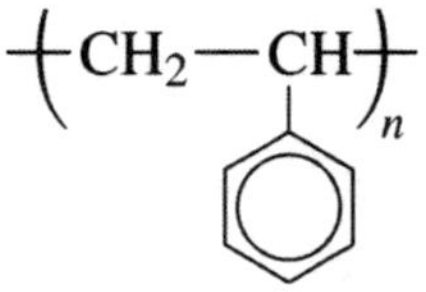

$$\left(CH_2-CH\right)_n$$

C_8H_8 molar mass $= 104.1$ g/mol

$\dfrac{96.08\ \text{g C}}{104.1\ \text{g}} \times 100 = 92.30\%$ C

28.
$$\quad\quad\quad\quad\quad\quad O$$
$$\quad\quad\quad\quad\quad\quad ||$$
Polymer of $CH_2{=}CHC-OCH_2CH_3$

$$\left(CH_2CH-----CH_2CH-----CH_2CH\right)_n$$
$$\quad\ |\quad\quad\quad\quad\quad\ |\quad\quad\quad\quad\ |$$
$$\quad C{=}O\quad\quad\quad C{=}O\quad\quad\quad C{=}O$$
$$\quad\ |\quad\quad\quad\quad\quad\ |\quad\quad\quad\quad\ |$$
$$\ OCH_2CH_3\quad OCH_2CH_3\quad OCH_2CH_3$$

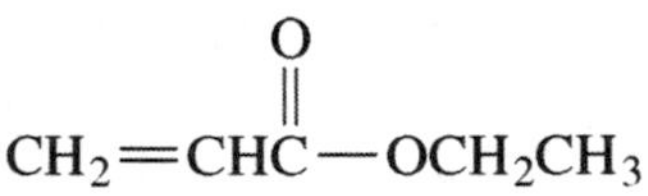

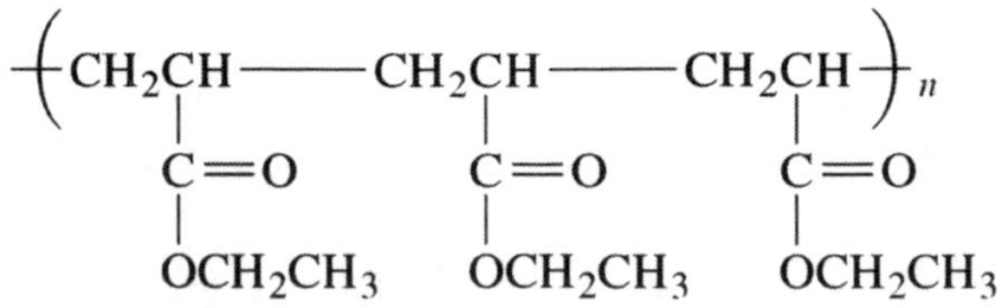

29. Cyanoacrylate ester polymer

$$\left(CH_2-\underset{\underset{O=C-OCH_3}{|}}{\overset{\overset{CN}{|}}{CH}}\right)_n$$

30. (a) From 2,3-dimethyl-1,3-butadiene, $CH_2{=}\underset{\underset{H_3C}{|}}{C}-\underset{\underset{CH_3}{|}}{C}{=}CH_2$

this polymer can be made $\left(CH_2C{=}CCH_2\right)_n$ or $\left(CH_2C{=}CCH_2\right)_n$

(b) If produced by the free-radical mechanism, a random mixture of *cis* and *trans* connections are made. It is possible, using catalysts, for the reaction to proceed by an ionic mechanism which will give a stereochemically controlled polymer.

31. It is easier to recycle a thermoplastic polymer because they are linear and can easily be reformed on heating. On the other hand, the cross linkages in thermosetting polymers are not easily broken or reformed.

32. $CH_2{=}CH_2$ $CF_2{=}CF_2$
    ethylene tetrafluoroethylene

$$+CH_2-CH_2-CF_2-CF_2-CH_2-CH_2-CF_2-CF_2+_n$$
(many other possible structures)

33. A representative structure for the polymer of methylmethacrylate follows:

$$\left(CH_2-\underset{\underset{\underset{O}{\|}}{\underset{C-OCH_3}{|}}}{\overset{\overset{CH_3}{|}}{C}}\right)_n$$

This plastic goes by the trade names Lucite and Plexiglas. It is a hard, transparent plastic that is used, for example, in contact lenses, in medical enclosures such as those used for premature babies, in furniture, etc.

34. During vulcanization, sulfur atoms cross-link rubber polymers together. In an analogous fashion, artificial polymers cross-link to form a thermosetting plastic. Both rubber and thermosetting plastic gain strength from this cross-linking.

35. Disagree. HDPE is thermoplastic and can be reused by melting and reforming without breaking down the polymer to monomers.

CHAPTER 22

ALCOHOLS, ETHERS, PHENOLS, AND THIOLS

SOLUTIONS TO REVIEW QUESTIONS

1. The question allows great freedom of choice. These shown here are very simple examples of each type.

    (a) an alkyl halide CH_3CH_2Cl

    (b) a phenol

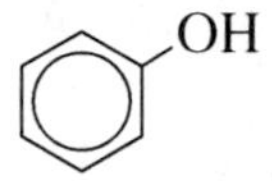

    (c) an ether $CH_3CH_2OCH_2CH_3$

    (d) an aldehyde

    $$CH_3\overset{\overset{\displaystyle H}{|}}{C}=O$$

    (e) a ketone

    $$CH_3\overset{\overset{\displaystyle O}{||}}{C}CH_3$$

    (f) a carboxylic acid CH_3COOH

    (g) an ester

    $$CH_3\overset{\overset{\displaystyle O}{||}}{C}-OCH_2CH_3$$

    (h) a thiol CH_3CH_2SH

2. Alkenes are almost never made from alcohols because the alcohols are almost always the higher value material. This is because recovering alkenes from hydrocarbon sources in an oil refinery (primarily catalytic cracking) is a relatively cheap process.

3. Oxidation of primary alcohols yields aldehydes. Further oxidation yields carboxylic acids. Examples:

$$CH_3CH_2OH \xrightarrow{[O]} CH_3\overset{\overset{\displaystyle O}{||}}{C}-H + H_2O$$

$$CH_3\overset{\overset{\displaystyle O}{||}}{C}-H \xrightarrow{[O]} CH_3\overset{\overset{\displaystyle O}{||}}{C}-OH$$

4. 1,2-Ethanediol is superior to methanol as an antifreeze because of its low volatility. Methanol is much more volatile than water. If the radiator leaks gas under pressure (normally steam), it would primarily leak methanol vapor so you would soon have no antifreeze. Ethylene glycol has a lower volatility than water, so it does not present this problem.

5. Oxidation of alcohols affects the hydroxyl carbon in two ways: (1) the carbon adds a new bond to oxygen; (2) the carbon loses a bond to hydrogen. When the hydroxyl carbon is not directly bonded to hydrogen (as in tertiary alcohols), oxidation is difficult.

6. Dietary polyphenols have a variety of health benefits. They protect cells from oxidation by acting as antioxidants. Perhaps more importantly polyphenols can serve as cellular signaling compounds. They can decrease the risk of coronary heart disease by impacting arterial cells. By interacting with brain cells, polyphenols can decrease cognitive loss. Current research suggests that polyphenols have a beneficial impact on many cell types.

7. The liver oxidizes ingested methanol, first to methanal (formaldehyde) and then to methanoic acid (formic acid). Methanoic acid is much more toxic than methanol. This acid causes metabolic acidosis and inhibition of central energy metabolism that can lead to death.

8. The following classes of organic compounds can be easily formed from alcohols: aldehydes, ketones, and carboxylic acids by oxidation; alkenes and ethers by dehydration; esters by esterification.

9.

10. Some common phenols include (a) hydroquinone, a photographic reducer and developer, (b) vanillin, a flavoring, (c) eugenol, used to make artificial vanillin, (d) thymol, used as an antiseptic in mouthwashes, (e) butylated hydroxytoluene (BHT), an antioxidant, and, (f) the cresols, used as disinfectants.

11. Low molar mass ethers present two hazards. They are very volatile and their highly flammable vapors form explosive mixtures with air. They also slowly react with oxygen in the air to form unstable explosive peroxides.

12. Ethanol (molar mass = 46.07) is a liquid at room temperature because it has a significant amount of hydrogen bonding between molecules in the liquid state, and thus has a much higher boiling point than would be predicted from molar mass alone. Dimethyl ether (molar mass = 46.07) is not capable of hydrogen bonding to itself, so has low attraction between molecules, making it a gas at room temperature.

13. Alcohols form strong intermolecular hydrogen bonds while thiols do not. Ethanol hydrogen bonding makes for a higher boiling point than that for ethanethiol which does not hydrogen bond.

14. At the boiling point of diethyl ether (35°C) the vapor pressure of diethyl ether must equal the atmospheric pressure.

15. Benzyl alcohol is a primary alcohol. It can be oxidized to form an aldehyde or a carboxylic acid.

SOLUTIONS TO EXERCISES

1. (a)
OH
|
CH₃CHCH₃

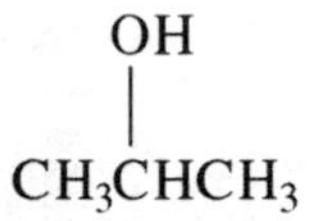

(b)
CH₂—OH
|
CH—OH
|
CH₂—OH

(c)
CH₃   CH₃
| |
CH₃CH₂CHCH₂CHCH₂OH

(d)
CH₃   CH₃
| |
CH₃CHCH₂C CH₃
          |
          OH

(e) ⬡—OH

(f) CH₃CH₂OH

2. (a) ⬠—OH

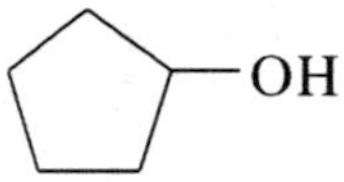

(b)
OH
|
CH₃CCH₂CH₃
|
CH₃

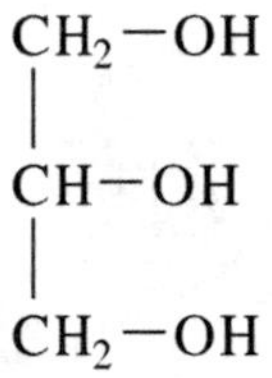

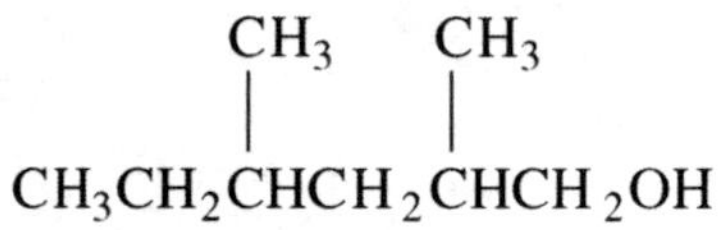

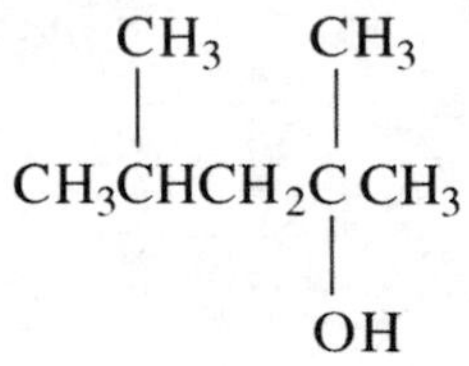

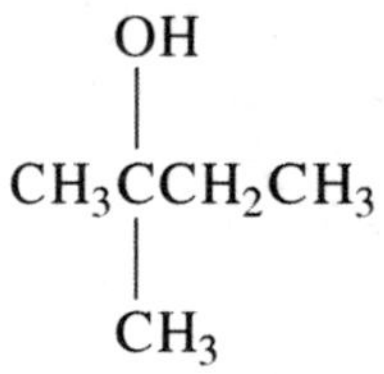

(c)
CH₃ OH
| |
CH₃C—CHCH₂CH₂CH₃
|
CH₃

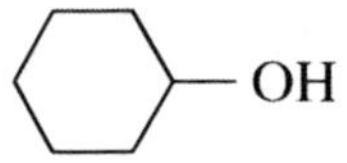

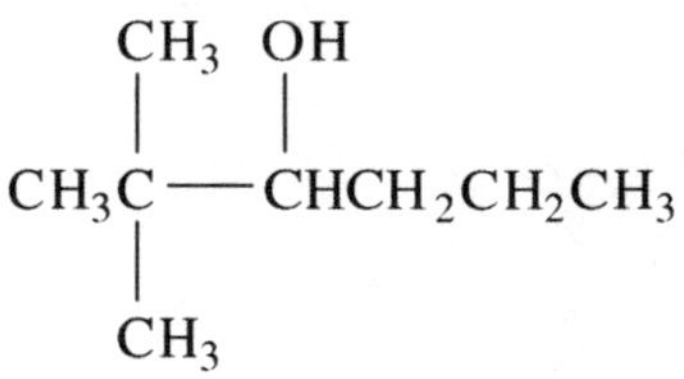

(d)
OH
|
CH₃CHCH₂CH₂OH

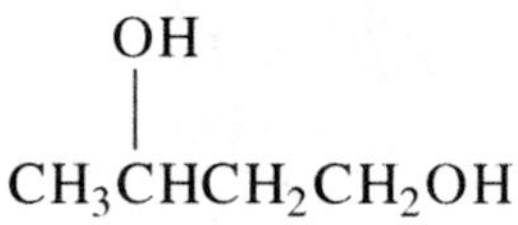

(e)
CH₂—OH
|
CH—OH
|
CH₂—OH

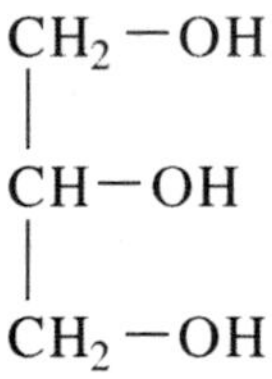

(f)
OH   CH₃
| |
CH₃C—CCH₂CH₃
| |
CH₃ CH₃

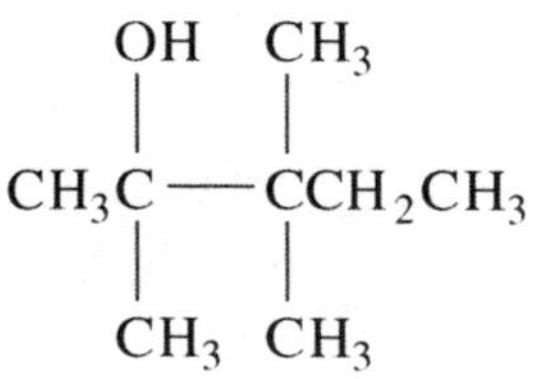

3. There are only five isomers:

$$\begin{array}{c} CH_3 \\ | \\ CH_3CCH_2CH_2OH \\ | \\ CH_3 \end{array}$$ $$\begin{array}{c} CH_3 \\ | \\ CH_3CHCHCH_2OH \\ | \\ CH_3 \end{array}$$ $$\begin{array}{c} H_3C \quad OH \\ | \quad | \\ CH_3CHCCH_3 \\ | \\ CH_3 \end{array}$$

$$\begin{array}{c} HO \quad CH_3 \\ | \quad | \\ CH_3CHCCH_3 \\ | \\ CH_3 \end{array}$$ $$\begin{array}{c} CH_3 \\ | \\ CH_3CH_2CCH_2OH \\ | \\ CH_3 \end{array}$$

4. There are only eight isomers:

$$\begin{array}{c} CH_3 \\ | \\ HOCH_2CHCH_2CH_2CH_3 \end{array}$$ $$\begin{array}{c} CH_3 \\ | \\ CH_3CCH_2CH_2CH_3 \\ | \\ OH \end{array}$$ $$\begin{array}{c} CH_3 \\ | \\ CH_3CHCHCH_2CH_3 \\ | \\ OH \end{array}$$

$$\begin{array}{c} CH_3 \quad OH \\ | \quad | \\ CH_3CHCH_2CHCH_3 \end{array}$$ $$\begin{array}{c} CH_3 \\ | \\ CH_3CHCH_2CH_2CH_2OH \end{array}$$ $$\begin{array}{c} CH_3 \\ | \\ HOCH_2CH_2CHCH_2CH_3 \end{array}$$

$$\begin{array}{c} CH_3 \\ | \\ CH_3CHCHCH_2CH_3 \\ | \\ OH \end{array}$$ $$\begin{array}{c} CH_3 \\ | \\ CH_3CH_2CCH_2CH_3 \\ | \\ OH \end{array}$$

5. (a) primary alcohols (where the hydroxyl carbon is bonded to <u>one</u> other carbon): b, c, f;
   (b) secondary alcohols (where the hydroxyl carbon is bonded to <u>two</u> other carbons): a, b, e;
   (c) tertiary alcohols (where the hydroxyl carbon is bonded to <u>three</u> other carbons): d; diol (where the compound contains <u>two</u> hydroxyl groups): none; triols (where the compound contains <u>three</u> hydroxyl groups): b.

6. (a) primary alcohols (where the hydroxyl carbon is bonded to <u>one</u> other carbon): d, e;
   (b) secondary alcohols (where the hydroxyl carbon is bonded to <u>two</u> other carbons): a, c, d, e;
   (c) tertiary alcohols (where the hydroxyl carbon is bonded to <u>three</u> other carbons): b, f; diol (where the compound contains <u>two</u> hydroxyl groups): d; triols (where the compound contains <u>three</u> hydroxyl groups): e.

7. The names of the compounds are:
   (a) 1-butanol (butyl alcohol)
   (b) 2-propanol (isopropyl alcohol)
   (c) 2-methyl-3-phenyl-1-propanol
   (d) oxirane (ethylene oxide)
   (e) 2-methyl-2-butanol
   (f) 2-methylcyclohexanol
   (g) 2,3-dimethyl-1,4-butanediol

8. The names of the compounds are
   (a) ethanol (ethyl alcohol)
   (b) 2-phenylethanol
   (c) 3-methyl-3-pentanol
   (d) 1-methylcyclopentanol
   (e) 3-pentanol
   (f) 1,2-propanediol
   (g) 4-ethyl-2-hexanol

9. Chief product of dehydration

   (a)
   $$\underset{\text{CH}_3\text{C}=\text{CHCH}_3}{\overset{\text{CH}_3}{|}}$$
   2-methyl-2-butene

   (b) $\text{CH}_3\text{CH}=\text{CHCH}_2\text{CH}_3$ 2-pentene

   (c)    cyclohexene

10. Chief product of dehydration

    (a) 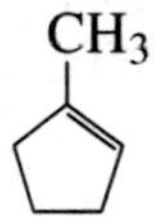   1-methylcyclopentene

    (b) $\text{CH}_3\text{CH}=\text{CHCH}_3$ 2–butene

    (c) $\underset{\text{CH}_3\text{C}=\text{CHCH}_2\text{CH}_3}{\overset{\text{CH}_3}{|}}$ 2-methyl-2-pentene

11. (a) 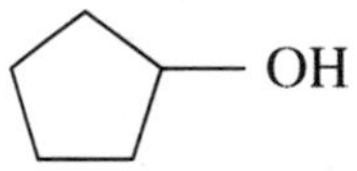   — OH

    cyclopentanol

    (b) $\text{CH}_3-\text{CH}_2-\underset{\overset{|}{\text{CH}_3}}{\text{CH}}-\text{CH}_2\text{OH}$

    2-methyl-1-butanol

    (c) 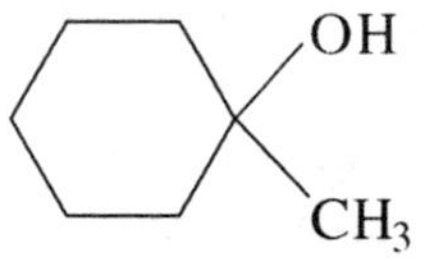   OH   CH₃

    1-methylcyclohexanol

12. (a)

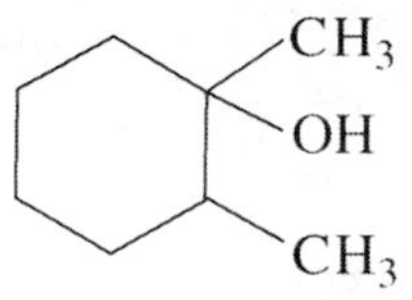

CH_3
OH
CH_3

1,2-dimethylcyclohexanol

(c)

OH
$CH_3CH_2CHCH_2CH_3$

3-pentanol

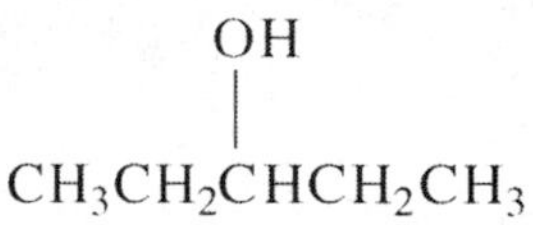

(b)

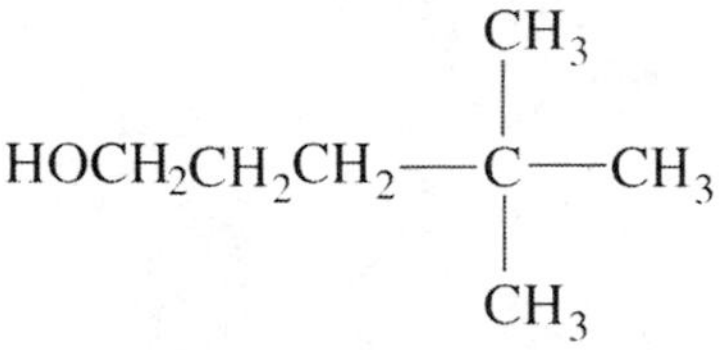

CH_3
$HOCH_2CH_2CH_2-C-CH_3$
CH_3

4,4-dimethyl-1-pentanol

13. Primary alcohols oxidize to carboxylic acids; secondary alcohols oxidize to ketones; tertiary alcohols don't easily oxidize.

(a)

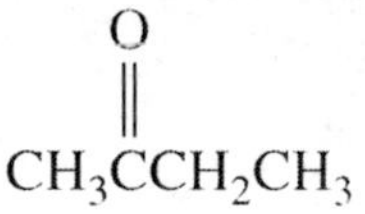

O
$\parallel$
$CH_3CCH_2CH_3$ butanone

(b) $HOOCCH_2CHCH_2CH_3$ 3-ethylpentanoic acid
CH_2CH_3

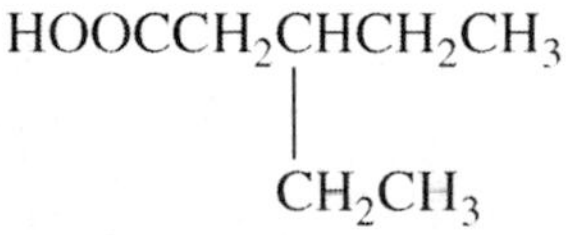

(c)

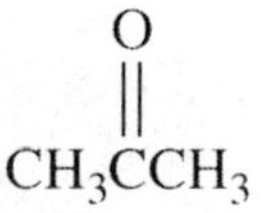

O
$\parallel$
CH_3CCH_3 acetone, propanone

(d) NR

14. Primary alcohols oxidize to carboxylic acids; secondary alcohols oxidize to ketones; tertiary alcohols don't easily oxidize.
(a) HCOOH, methanoic acid, formic acid
(b) NR
(c) CH_3COOH, ethanoic acid, acetic acid

(d)

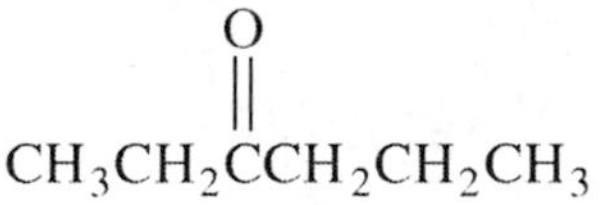

O
$\parallel$
$CH_3CH_2CCH_2CH_2CH_3$ 3-hexanone

15. Upon ester hydrolysis, the alcohol forms from the alkyl group and oxygen bonded to the C=O.
(a) $CH_3CH_2CH_2OH$ 1-butanol
(b) CH_3CH_2OH ethanol

(c) 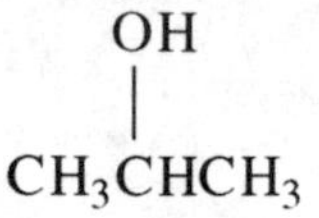   OH
        |
    CH₃CHCH₃   2-propanol

16. Upon ester hydrolysis, the alcohol forms from the alkyl group and oxygen bonded to the C=O.

(a) 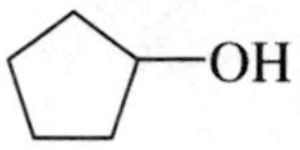 ⬠—OH   cyclopentanol

                    CH₃
                     |
(b) 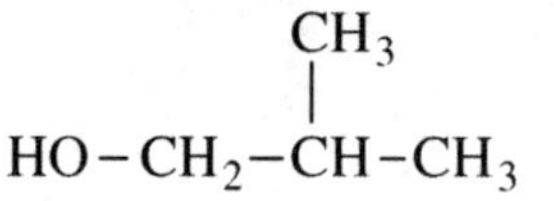 HO–CH₂–CH–CH₃   2-methyl-1-propanol

(c) CH₃OH   methanol

17. (a) CH₃CHBrCH₃   2-bromopropane
    (b) 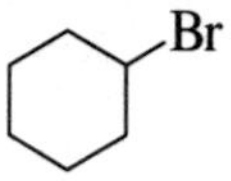 ⬡—Br   bromocyclohexane (cyclohexyl bromide)

    (c) 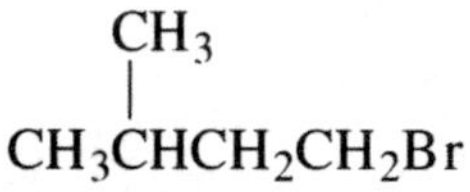   CH₃
              |
        CH₃CHCH₂CH₂Br   1-bromo-3-methylbutane

18. (a) 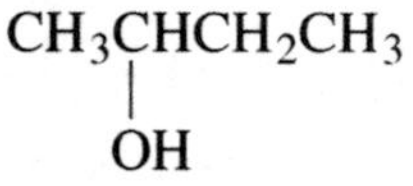 CH₃CHCH₂CH₃   2-butanol
             |
            OH

    (b) 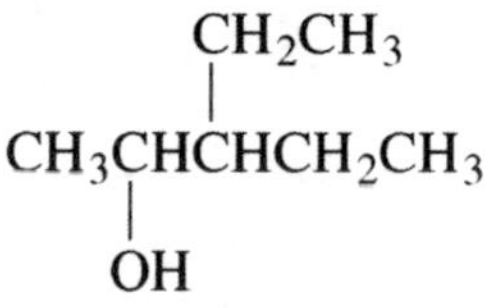   CH₂CH₃
                    |
        CH₃CHCHCH₂CH₃   3-ethyl-2-pentanol
             |
            OH

    (c) 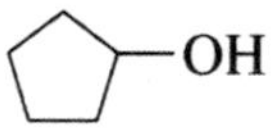 ⬠—OH   cyclopentanol

19. (a) $2\,CH_3CH_2OH + H_2SO_4 \xrightarrow{140°C} CH_3CH_2OCH_2CH_3 + H_2O$
                                    diethyl ether

    (b) $CH_3CH_2CH_2OH + H_2SO_4 \xrightarrow{180°C} CH_3CH=CH_2 + H_2O$
                                    propene

                                    O
                                    ‖
    (c) 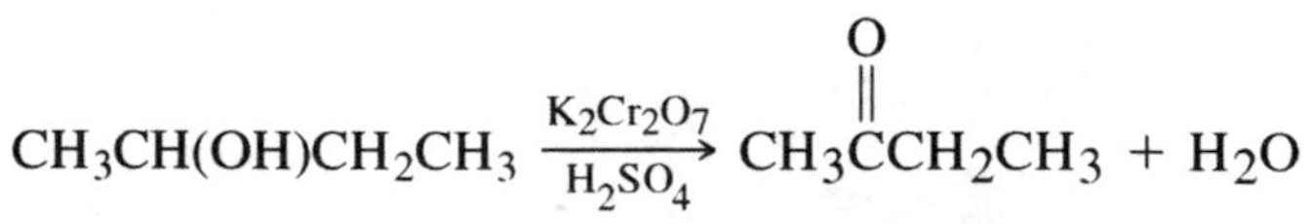 $CH_3CH(OH)CH_2CH_3 \xrightarrow[H_2SO_4]{K_2Cr_2O_7}$ CH₃CCH₂CH₃ + H₂O
                                    2-butanone

              O
              ‖
    (d) 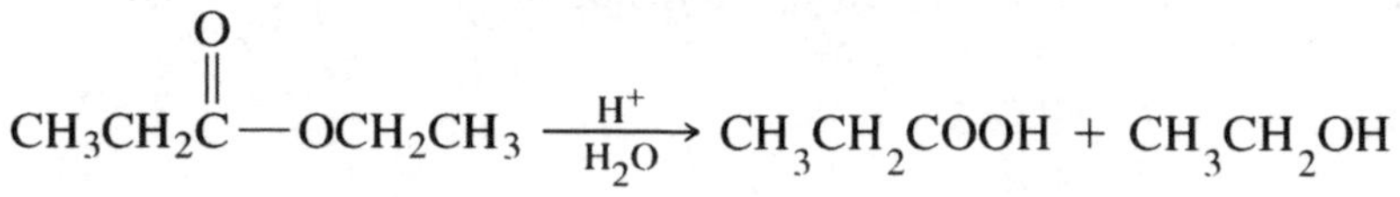 CH₃CH₂C—OCH₂CH₃ $\xrightarrow[H_2O]{H^+}$ $CH_3CH_2COOH + CH_3CH_2OH$
                            propanoic acid   ethanol

20. (a) $2\,CH_3CH_2OH + 2\,Na \longrightarrow 2\,CH_3CH_2O^-Na^+ + H_2$

(b) $CH_3CH_2CH_2CH_2OH \xrightarrow[H_2SO_4]{K_2Cr_2O_7} CH_3CH_2CH_2\overset{\overset{\displaystyle O}{\|}}{C}OH$

(c) 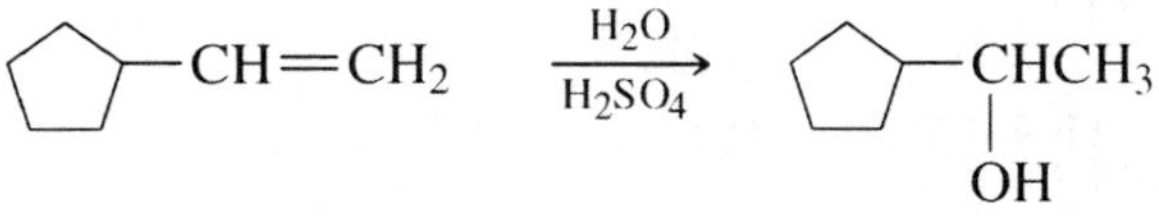
$\text{(cyclopentyl)}-CH{=}CH_2 \xrightarrow[H_2SO_4]{H_2O} \text{(cyclopentyl)}-\underset{\underset{\displaystyle OH}{|}}{C}HCH_3$

(d) $CH_3CH_2\overset{\overset{\displaystyle O}{\|}}{C}-OCH_3 + NaOH \longrightarrow CH_3CH_2\overset{\overset{\displaystyle O}{\|}}{C}-O^-Na^+ + CH_3OH$

21. (a) 2-methyl-l-propanol

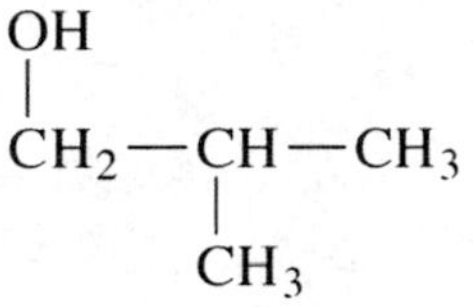

$$\underset{\underset{\displaystyle CH_3}{|}}{CH_2}-\overset{\overset{\displaystyle OH}{|}}{C}H-CH_3$$

(b) 3-methyl-2-butanol

$$CH_3-\overset{\overset{\displaystyle OH}{|}}{C}H-\underset{\underset{\displaystyle CH_3}{|}}{C}H\,-CH_3$$

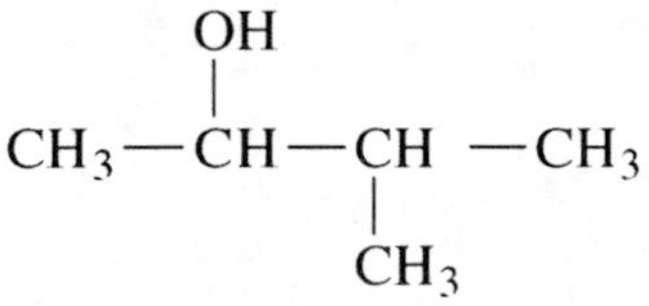

(c) cyclopentanol

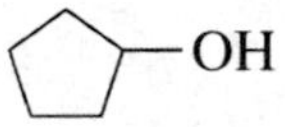

22. (a) methanol, CH_3-OH

(b) 2,2-dimethyl-l-propanol $\quad CH_3-\overset{\overset{\displaystyle CH_3}{|}}{\underset{\underset{\displaystyle CH_3}{|}}{C}}-CH_2-OH$

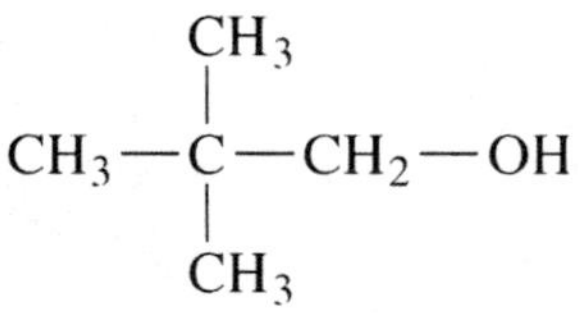

(c) 3,3-dimethyl-1-pentanol $\quad CH_3-CH_2-\overset{\overset{\displaystyle CH_3}{|}}{\underset{\underset{\displaystyle CH_3}{|}}{C}}-CH_2CH_2-OH$

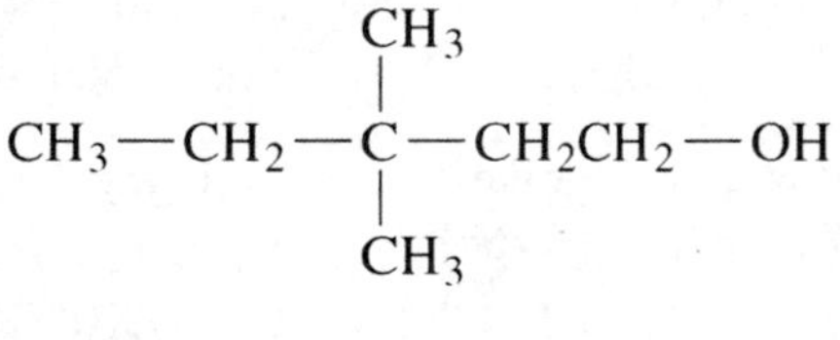

23. (a) methanol
    (b) 2-methyl-1-propanol
    (c) cyclohexanol

24. (a) cyclopentanol
    (b) 2-propanol
    (c) 2-methyl-2-propanol

25. (a) *o*-methylphenol   (b) *m*-dihydroxybenzene

(c) 4-hydroxy-3-methoxybenzaldehyde

26. (a) *p*-nitrophenol   (b) 2,6-dimethylphenol

(c) *o*-dihydroxybenzene

27. (a) phenol   (c) 2-ethyl-5-nitrophenol
    (b) *m*-methylphenol   (d) 4-bromo-2-chlorophenol

28. (a) *p*-dihydroxybenzene (hydroquinone)   (c) 2,4-dinitrophenol
    (b) 2,4-dimethylphenol   (d) *m*-hexylphenol

29. Order of increasing solubility in water [c (lowest), a, b, d (highest)]
    (c) $CH_3CH_2CH_2CH_2CH_3$ (1)   (b) $CH_3CH(OH)CH_2CH_2CH_3$ (3)
    (a) $CH_3CH_2OCH_2CH_2CH_3$ (2)   (d) $CH_3CH(OH)CH(OH)CH_2CH_3$ (4)

30. Order of decreasing solubility in water: [a (highest), d, c, b (lowest)]
    (a) $CH_3CH(OH)CH(OH)CH_2OH$ (1)   (c) $CH_3CH_2CH_2CH_2OH$ (3)
    (d) $CH_3CH(OH)CH_2CH_2OH$ (2)   (b) $CH_3CH_2OCH_2CH_3$ (4)

31. The three compounds have about the same molar mass but differ with respect to hydrogen bonding and polarity. Compound (b) (1-propanethiol) does not hydrogen bond, is nonpolar, and so, has the lowest boiling point. Compound (c) (diethyl ether) is polar but does not hydrogen bond to itself. It has a higher boiling point than compound (b). Compound a (1,2-propanediol) hydrogen bonds to itself and has the highest boiling point.

32. The three compounds have about the same molar mass but differ with respect to hydrogen bonding and polarity. Compound (c) (pentane) does not hydrogen bond, is nonpolar, and so, has the lowest boiling point. Compound (b) (diethyl ethcr) is polar but does not hydrogen bond to itself. It has a higher boiling point than Compound (c). Compound (a) (2-butanol) hydrogen bonds to itself and has the highest boiling point.

33. $CH_3-O-CH_2CH_3$ $CH_3CH_2CH_2-OH$

    ethyl methyl ether   1 – propanol

    Each of these molecules has about the same molar mass. In this case, differences in boiling point arise because of differences in intermolecular attractive forces. 1-propanol can hydrogen bond to itself while ethyl methyl ether can not. Thus, ethyl methyl ether will have the lower boiling point.
    boiling point for ethyl methyl ether $= 8°C$
    boiling point for 1-propanol $= 97.4°$ C

34.
$$CH_3$$
$$|$$
$$CH_3CH_2-O-CHCH_3 \qquad CH_3CH_2CH_2CH_2CH_2-OH$$

    ethyl isopropyl ether   1-pentanol

    Each of these molecules has about the same molar mass. In this case, differences in boiling point arise because of differences in intermolecular attractive forces. 1-pentanol can hydrogen bond to itself while ethyl isopropyl ether cannot. Thus, ethyl isopropyl ether will have the lower boiling point.
    boiling point for ethyl isopropyl ether $= 54°C$
    boiling point for 1-pentanol $= 138°C$

35. (a)   common, isopropyl methyl ether; IUPAC, 2-methoxypropane
    (b)   common, ethyl phenyl ether; IUPAC, ethoxybenzene
    (c)   common, butyl ethyl ether; IUPAC, 1-ethoxybutane

36. (a)   common, ethyl methyl ether; IUPAC, methoxyethane
    (b)   common, ethyl isobutyl ether; IUPAC, 1-ethoxy-2-methylpropane
    (c)   common, ethyl phenyl ether; IUPAC, ethoxybenzene

37. There are three isomeric saturated ethers with the formula, $C_4H_{10}O$.

$$CH_3$$
$$|$$
$$CH_3-O-CH_2CH_2CH_3 \qquad CH_3CH_2-O-CH_2CH_3 \qquad CH_3-O-CHCH_3$$

   methyl propyl ether   diethyl ether   methyl isopropyl ether
   (1-methoxypropane)   (ethoxyethane)   (2-methoxypropane)

38. Six isomeric ethers: $C_5H_{12}O$

$CH_3OCH_2CH_2CH_2CH_3$
n-butyl methyl ether
(1-methoxybutane)

$\overset{\overset{\displaystyle CH_3}{|}}{CH_3OCHCH_2CH_3}$
sec-butyl methyl ether
(2-methoxybutane)

$\overset{\overset{\displaystyle CH_3}{|}}{CH_3OCH_2CHCH_3}$
isobutyl methyl ether
(1-methoxy-2-methylpropane)

$\overset{\overset{\displaystyle CH_3}{|}}{\underset{\underset{\displaystyle CH_3}{|}}{CH_3OCCH_3}}$
t-butyl methyl ether
(2-methoxy-2-methylpropane)

$CH_3CH_2OCH_2CH_2CH_3$
ethyl *n*-propyl ether
(1-ethoxypropane)

$\overset{\overset{\displaystyle CH_3}{|}}{CH_2CH_2OCHCH_3}$
ethyl isopropyl ether
(2-ethoxypropane)

39. Possible combinations of reactants to make the following ethers by the Williamson synthesis:

(a) $CH_3CH_2OCH_3$ $CH_3ONa + CH_3CH_2Cl$ or
                            $CH_3CH_2ONa + CH_3Cl$

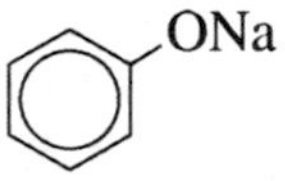

(b) 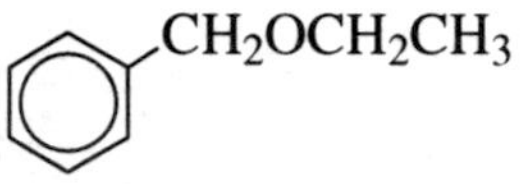 (benzene)–$CH_2OCH_2CH_3$ (benzene)–CH_2ONa + CH_3CH_2Cl

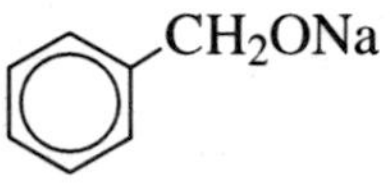

       or   (benzene)–CH_2Cl + CH_3CH_2ONa

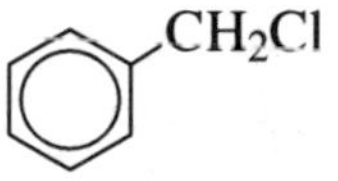

(c) 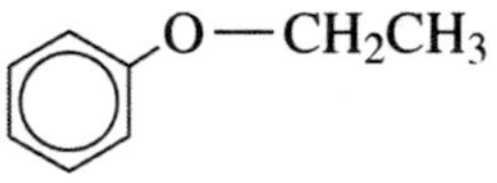 (benzene)–$O-CH_2CH_3$ (benzene)–ONa + CH_3CH_2Cl

40. Possible combinations of reactants to make the following ethers by the Williamson synthesis:

(a) $CH_3CH_2CH_2OCH_2CH_2CH_3$ $CH_3CH_2CH_2ONa + CH_3CH_2CH_2Cl$

(b) $\overset{\overset{\displaystyle CH_3}{|}}{\underset{\underset{\displaystyle CH_3}{|}}{HCOCH_2CH_2CH_3}}$ $\overset{\overset{\displaystyle CH_3}{|}}{\underset{\underset{\displaystyle CH_3}{|}}{HCONa}} + CH_3CH_2CH_2Cl$

(c)

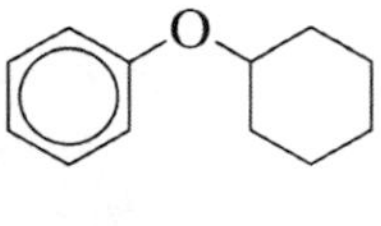

cannot be made by the Williamson synthesis. Cannot use secondary RX.

41. IUPAC names
    (a) 2-methyl-2-butanethiol
    (b) cyclohexanethiol
    (c) 2-methyl-2-propanethiol
    (d) 2-methyl-3-pentanethiol

42. IUPAC names
    (a) 2-methylcyclopentanethiol
    (b) 2,3-dimethyl-2-pentanethiol
    (c) 2-propanethiol
    (d) 3-ethyl-2,2,4-trimethyl-3-pentanethiol

43.

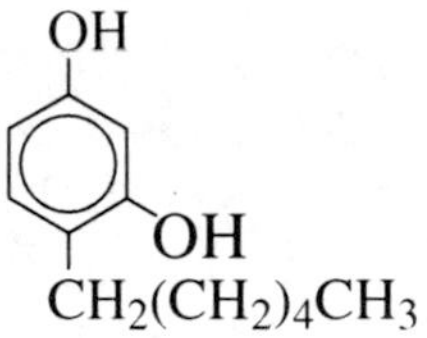

4-hexylresorcinol or 4-hexyl-1,3-dihydroxybenzene

44.

cis-1,2-cyclopentanediol    *trans*-1,2-cyclopentanediol

45. The phenol compound (3-methylphenol) is more acidic than the alcohol compound (benzyl alcohol) but is less acidic than carbonic acid.

46. The name "catecholamine" is built of two pieces, "catechol" and "amine." A catechol is a phenol derivative that contains an additional hydroxyl group (o-hydroxyphenol).

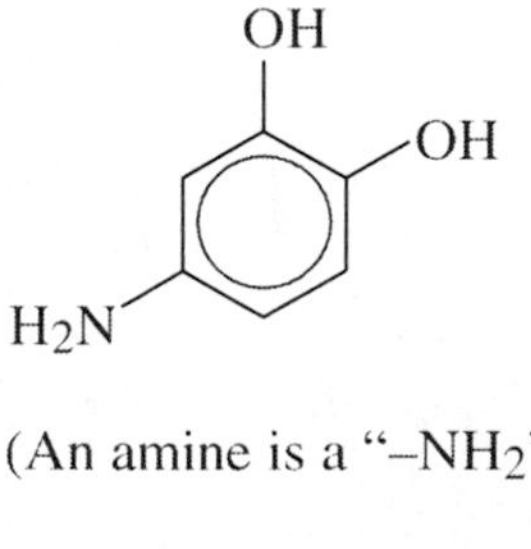

(An amine is a "–NH₂" functional group.)

47. Oxidation products from ethylene glycol include

$$\text{HOCH}_2\overset{\overset{\displaystyle O}{\|}}{-}\text{CH} \qquad \text{HC}\overset{\overset{\displaystyle O}{\|}}{-}\overset{\overset{\displaystyle O}{\|}}{}\text{CH} \qquad \text{HOCH}_2\text{COOH} \qquad \text{HOOCCOOH}$$

$$\text{HC}\overset{\overset{\displaystyle O}{\|}}{-}\text{COOH}$$

48. In each compound, the most oxidized carbon will have the most positive oxidation number and, commonly, the most bonds to oxygen. In contrast, the most reduced carbon will have the most negative oxidation number and, commonly, the most bonds to hydrogen. The most reduced carbon has been circled while an asterisk has been placed by the most oxidized carbon.

oxidation number = –1, two bonds to hydrogen

oxidation number = +1, two bonds to oxygen

oxidation number = –3, three bonds to hydrogen

oxidation number = +3, three bonds to oxygen

oxidation number = –3, three bonds to hydrogen

oxidation number = +3, three bonds to oxygen

ribose
(a sugar)

stearic acid
(a fat)

pyruvic acid
(a metabolite)

49. (a)

$$\text{CH}_3\overset{\overset{\displaystyle OH}{|}}{\text{C}}\text{HCH}_3 \xrightarrow[\text{H}^+]{\text{Cr}_2\text{O}_7^{2-}} \text{CH}_3\overset{\overset{\displaystyle O}{\|}}{\text{C}}\text{CH}_3$$

(b) $\text{CH}_3\text{CH}_2\text{CH}_2\text{CH}=\text{CH}_2 \xrightarrow[\text{H}_2\text{O}]{\text{H}^+} \text{CH}_3\text{CH}_2\text{CH}_2\overset{\overset{\displaystyle }{}}{\text{C}}\text{HCH}_3$ with OH below

(c) $2\,\text{CH}_3\text{CH}_2\text{OH} + 2\,\text{Na(metal)} \longrightarrow 2\,\text{CH}_3\text{CH}_2\text{ONa} + \text{H}_2$

(d) $\text{CH}_3\text{CH}_2\text{CH}=\text{CH}_2 \xrightarrow[\text{H}_2\text{O}]{\text{H}^+} \text{CH}_3\text{CH}_2\text{CHCH}_3 \xrightarrow[\text{H}^+]{\text{Cr}_2\text{O}_7^{2-}} \text{CH}_3\text{CH}_2\text{CCH}_3$ (OH below first product, O below second)

(e) $\text{CH}_3\text{CH}_2\text{CH}_2\text{CH}_2\text{OH} \xrightarrow[\Delta]{\text{H}_2\text{SO}_4} \text{CH}_3\text{CH}=\text{CHCH}_3 + \text{CH}_3\text{CH}_2\text{CH}=\text{CH}_2$

$$\xrightarrow{\text{HCl}} \text{CH}_3\text{CH}_2\text{CHCH}_3$$ (Cl below)

(f) $CH_3CH_2CH_2Cl \xrightarrow{NaOH} CH_3CH_2CH_2OH \xrightarrow[H^+]{Cr_2O_7^{2-}} CH_3CH_2\overset{\displaystyle O}{\overset{\|}{C}}-H$

50.

$$CH_3CH_3 + Cl_2 \xrightarrow{light} CH_3CH_2Cl + HCl$$

51. A simple chemical test to distinguish between:
   (a) ethanol and dimethyl ether. Ethanol will react readily with potassium dichromate and sulfuric acid to make acetaldehyde. Visibly, the orange color of the dichromate changes to green. Ethanol reacts with metallic sodium to produce hydrogen gas. Dimethyl ether does not react with either of these reagents.
   (b) 1-pentanol and 1-pentene. 1-pentene will rapidly decolorize bromine as it adds to the double bond. 1-pentanol does not react.
   (c) *p*-methylphenol has acidic properties so it will react with sodium hydroxide. Methoxybenzene does not have acidic properties and will not react with sodium hydroxide.

52. Isomers of $C_8H_{10}O$

53. Only compound (a) will react with NaOH.

54. Order of increasing boiling points.

1-pentanol	<	1-octanol	<	1,2-pentanediol
138°C		194°C		210°C

All three compounds are alcohols. 1-pentanol has the lowest molar mass and hence the lowest boiling point. 1-octanol has a higher molar mass and therefore a higher boiling point than 1-pentanol. 1,2-pentanediol has two —OH groups and therefore forms more hydrogen bonds than the other two alcohols which causes its higher boiling point.

55. (a) The primary carbocation that is formed first is unstable.

Thus, the primary carbocation shifts to a much more stable, tertiary carbocation

It is this intermediate that goes on to form the major product, 2-methyl-2-butene.

(b) Hydration will not form 2-methyl-1-butanol because the double bond lies between the middle two carbons in 2-methyl-2-butene. By Markovnikov's rule, the —H will add to the double bonded carbon that already has a hydrogen. Thus, 2-methyl-2-butanol will be the major product.

56.

| hydroxyethanal | ethanedial | oxoethanoic acid | hydroxyethanoic acid | ethanedioic acid |

ALDEHYDES AND KETONES

SOLUTIONS TO REVIEW QUESTIONS

1.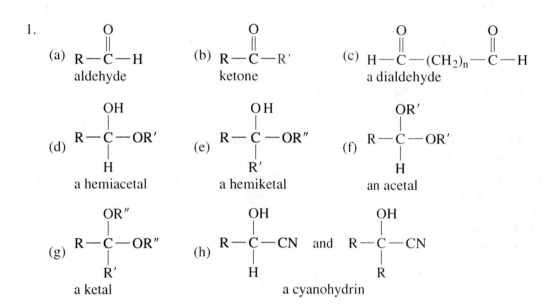

2. Propanal   propanone

$$CH_3CH_2\overset{\displaystyle H}{C}=O$$ $$CH_3-\overset{\displaystyle O}{\overset{\|}{C}}-CH_3$$

Each has the molecular formula of C_3H_6O, so aldehydes and ketones appear to be isomeric with each other.

Butanal   butanone

$$CH_3CH_2CH_2\overset{\displaystyle H}{C}=O$$ $$CH_3-\overset{\displaystyle O}{\overset{\|}{C}}-CH_2CH_3$$

Each has the molecular formula C_4H_8O, so the generalization seems to check out. The general formula for aldehydes and ketones is $C_nH_{2n}O$.

3. The strength of collagen depends on aldol condensations. After collagen is formed, aldehydes add along its length. Collagen fibers adjacent to each other undergo an aldol condensation. The cross linking bonds between collagen strands form a strong network, giving collagen its strength.

4. 1-Butanol has a boiling point of 118°C while the boiling point for butanal is 76°C and for butanone, 80°C. Of these three compounds, only 1-butanol can hydrogen bond to itself. This additional bonding holds the molecules together more tightly and accounts for the much higher boiling point.

5. The versatile aldol condensation reaction allows *Streptomyces* to synthesize a wide variety of different antibiotics.

6. Acetaldehyde carries a reactive aldehyde functional group that can bond to many different biochemicals: by reacting with amino acids, acetaldehyde slows protein synthesis; by reacting with antioxidants, acetaldehyde increases oxidative damage to the liver; by reacting with specific proteins, acetaldehyde hampers the liver's ability to export needed chemicals to the blood stream. Acetaldehyde can cause liver cirrhosis.

7. A ketone is a carbonyl functional group that is bonded to two alkyl or aryl groups. Cortisone is a ketone-containing hormone. The sugar, fructose, contains a ketone. Ketones are also found in Vitamin K_1.

8. (a) When an aldehyde reacts with Benedict solution, the blue color of copper ion disappears and a red-brown precipitate (Cu_2O) forms.

    (b) When an aldehyde reacts with Tollens solution, the silver ion in solution forms a thin silver metal mirror on the inside of the glass container used in the test.

9. The phenol-formaldehyde polymer is rigid because covalent bonds not only form the polymers but also cross-link the polymers to each other.

10. MEK is an abbreviation for methyl ethyl ketone. Its main use is as a solvent, especially for lacquers and paints.

11. A ketone group, $\diagdown C = O$, cannot be located at the end of a carbon-carbon chain. Consequently its only possible location in both propanone and butanone is on C-2 of these ketones. Therefore its location need not be numbered.

12. There must be an H atom on the alpha carbon adjacent to a carbonyl group of one of the reacting compounds in the aldol condensation. This hydrogen transfers to the carbonyl of the other reactant and breaks the carbonyl pi bond leaving intermediates in which a carbon atom of each molecule has three bonds. The two intermediates then bond together to give the final product.

SOLUTIONS TO EXERCISES

1. Names of aldehydes.
   (a) $H_2C{=}O$ methanal, formaldehyde

   (b)
   $$CH_3CHCH_2\overset{\displaystyle O}{\overset{\|}{C}}{-}H$$
   with CH_3 on the CH   3-methylbutanal

   (c) $H{-}\overset{O}{\overset{\|}{C}}CH_2CH_2\overset{O}{\overset{\|}{C}}{-}H$ butanedial

   (d) $C_6H_5{-}CH{=}CH\overset{O}{\overset{\|}{C}}{-}H$ 3-phenylpropenal

   (e)
   $$\underset{H}{\overset{CH_3}{C}}{=}\underset{H}{\overset{\overset{O}{\overset{\|}{C}}{-}H}{C}}$$
   cis-2-butenal

2. Names of aldehydes
   (a) $CH_3\overset{O}{\overset{\|}{C}}{-}H$ ethanal, acetaldehyde

   (b) $CH_3CH_2CH_2\overset{O}{\overset{\|}{C}}{-}H$ butanal

   (c) $C_6H_5\overset{O}{\overset{\|}{C}}{-}H$ benzaldehyde

   (d) benzaldehyde ring with Cl and $\overset{O}{\overset{\|}{C}}{-}H$ and CH_3CHCH_3 substituents   2-chloro-5-isopropylbenzaldehyde

(e) 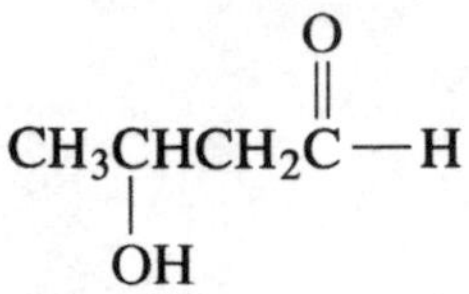
$$CH_3CHCH_2\overset{\displaystyle O}{\overset{\|}{C}}-H \qquad \text{3-hydroxybutanal}$$
$$\underset{OH}{|}$$

3. Names of ketones
   (a) propanone, acetone, dimethyl ketone
   (b) 1-phenyl-1-propanone, ethyl phenyl ketone
   (c) cyclopentanone
   (d) 4-hydroxy-4-methyl-2-pentanone

4. Names of ketones
   (a) butanone, methyl ethyl ketone (MEK)
   (b) 3,3-dimethylbutanone, *t*-butyl methyl ketone
   (c) 2,5-hexanedione
   (d) 1-phenyl-2-propanone, benzyl methyl ketone

5. Structural formulas

   (a) 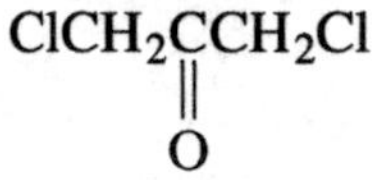 $ClCH_2\underset{\underset{O}{\|}}{C}CH_2Cl$ \qquad 1,3-dichloropropanone

   (b)  $CH_2\!=\!CHCH_2\overset{\overset{O}{\|}}{C}-H$ \qquad 3-butenal

   (c) 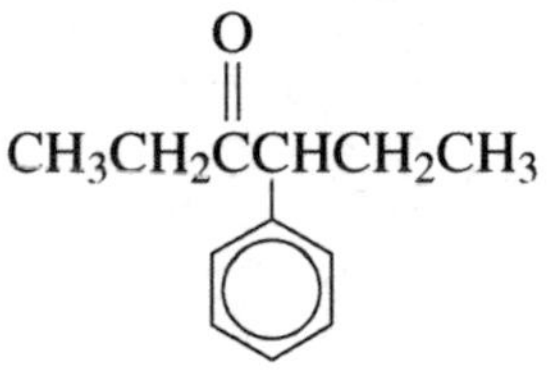 $CH_3CH_2\overset{\overset{O}{\|}}{C}CHCH_2CH_3$ \qquad 4-phenyl-3-hexanone

   (d) 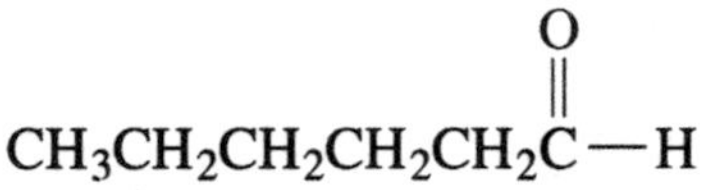 $CH_3CH_2CH_2CH_2CH_2\overset{\overset{O}{\|}}{C}-H$ \qquad hexanal

   (e) 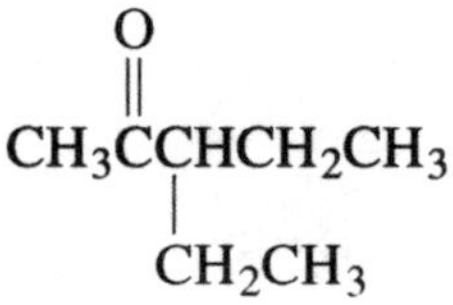 $CH_3\overset{\overset{O}{\|}}{C}CHCH_2CH_3$ \qquad 3-ethyl-2-pentanone
   $$\underset{CH_2CH_3}{|}$$

6. Structural formulas

   (a)  $HOCH_2CH_2\overset{\overset{O}{\|}}{C}-H$ \qquad 3-hydroxypropanal

(b) $\underset{\underset{CH_3}{|}}{CH_3CH_2\overset{\overset{O}{\|}}{C}CHCH_2CH_3}$ 4-methyl-3-hexanone

(c)   cyclohexanone

(d) $\underset{\underset{Cl}{|}\ \ \underset{Cl}{|}\ \ \underset{Cl}{|}}{CH_3CHCH_2CHCH_2\overset{\overset{O}{\|}}{C}—H}$ 2,4,6-trichloroheptanal

(e) $CH_3CH\!=\!CHCH_2\overset{\overset{O}{\|}}{C}—H$ 3-pentenal

7. (a) Acetaldehyde is a common name. The IUPAC name is ethanal.

(b) 2-Methyl-3-butanone is numbered incorrectly. The correct IUPAC name is 3-methyl-2-butanone.

(c) 1-Hexanal is incorrect because the aldehyde needs no number. The correct IUPAC name is hexanal.

8. (a) 2-Methyl-3-propanal is numbered incorrectly and the aldehyde needs no number. The correct IUPAC name is 2-methylpropanal.

(b) Formaldehyde is a common name. The IUPAC name is methanal.

(c) 4-Hexanone is numbered incorrectly. The correct IUPAC name is 3-hexanone.

9. Boiling points depend on molecular size (larger molecules have higher boiling points) and intermolecular bonding (molecules that hydrogen bond have higher boiling points than molecules that don't hydrogen bond; polar molecules have higher boiling points than non-polar molecules).

(a) Although 3-pentanone and 2-pentanol are about the same size, 2-pentanol can hydrogen bond to itself. 2-Pentanol has the higher boiling point.

(b) Ethanal is both larger and more polar than ethane. Ethanal has the higher boiling point.

(c) 1,3-Propanediol is larger and hydrogen bonds to itself while propanone is smaller and can't hydrogen bond to itself. 1,3-Propanediol has the higher boiling point.

(d) Pentanal is both larger and more polar than propane. Pentanal has the higher boiling point.

10. Boiling points depend on molecular size (larger molecules have a higher boiling points) and intermolecular bonding (molecules that hydrogen bond have higher boiling points than molecules that don't hydrogen bond; polar molecules have higher boiling points than non-polar molecules).

(a) Butanone is both larger and more polar than butane. Butanone has the higher boiling point.

(b) Although 2-butanol and butanone are about the same size, 2-butanol can hydrogen bond to itself while butanone can't. 2-Butanol has the higher boiling point.

(c) 2-Pentanone is both larger and more polar than propane. 2-Pentanone has the higher boiling point.

(d) 1,2-Propanediol is larger and can hydrogen bond to itself while propanal is smaller and can't hydrogen bond to itself. 1,2-Propanediol has the higher boiling point.

11. Aqueous solubility depends on the size of the alkyl chain (the smaller alkyl chain has more aqueous solubility) and bonding between the solute and water (hydrogen bonding solutes have higher aqueous solubilities; more polar solutes have higher aqueous solubilities).
    (a) Both 2-heptanone and propanone are ketones but propanone has a much smaller alkyl chain. Propanone has a higher aqueous solubility.
    (b) Pentane and pentanal have the same sized alkyl chains but pentanal can hydrogen bond to water. Pentanal has the higher water solubility.
    (c) 3-Pentanone and 2,4-pentanedione have the same sized alkyl chains but 2,4-pentanedione has two ketone functional groups while 3-pentanone has only one. 2,4-Pentanedione can hydrogen bond more strongly to water and has the higher aqueous solubility.

12. Aqueous solubility depends on the size of the alkyl chain (the smaller alkyl chain has more aqueous solubility) and bonding between the solute and water (hydrogen bonding solutes have higher aqueous solubilities; more polar solutes have higher aqueous solubilities).
    (a) 3-Hydroxypentanal has a smaller alkyl chain and two functional groups that can hydrogen bond to water while 3-hexanone has a larger alkyl chain and only one functional group that can hydrogen bond to water. 3-Hydroxypentanal has the higher aqueous solubility.
    (b) Cyclohexanone and cyclohexane have the same sized alkyl chains but cyclohexanone can hydrogen bond to water. Cyclohexanone has the higher aqueous solubility.
    (c) Propanone and 2-pentanone are both ketones but propanone has the smaller alkyl chain. Propanone has the higher aqueous solubility.

13. Equations for the oxidation of:
    (a) 3-pentanol

$$\underset{\underset{\text{CH}_3\text{CH}_2\text{CHCH}_2\text{CH}_3}{|}}{\text{OH}} \xrightarrow[\text{H}_2\text{SO}_4]{\text{K}_2\text{Cr}_2\text{O}_7} \quad \underset{\text{CH}_3\text{CH}_2\overset{\text{O}}{\overset{||}{\text{C}}}\text{CH}_2\text{CH}_3}{}$$

This is the same product as for the oxidation by Cu^{2+} or Ag^+.

    (b) 3-methyl-l-hexanol

$$\underset{\underset{\text{CH}_3}{|}}{\text{CH}_3\text{CH}_2\text{CH}_2\text{CHCH}_2\text{CH}_2\text{OH}} + \text{O}_2 \xrightarrow[\Delta]{\text{Cu}} \underset{\underset{\text{CH}_3}{|}}{\text{CH}_3\text{CH}_2\text{CH}_2\text{CHCH}_2\overset{\text{O}}{\overset{||}{\text{C}}}-\text{H}}$$

$$\underset{\underset{\text{CH}_3}{|}}{\text{CH}_3\text{CH}_2\text{CH}_2\text{CHCH}_2\text{CH}_2\text{OH}} \xrightarrow[\text{H}_2\text{SO}_4]{\text{K}_2\text{Cr}_2\text{O}_7} \underset{\underset{\text{CH}_3}{|}}{\text{CH}_3\text{CH}_2\text{CH}_2\text{CHCH}_2\overset{\text{O}}{\overset{||}{\text{C}}}-\text{H}}$$

$$\xrightarrow[\text{H}_2\text{SO}_4]{\text{K}_2\text{Cr}_2\text{O}_7} \underset{\underset{\text{CH}_3}{|}}{\text{CH}_3\text{CH}_2\text{CH}_2\text{CHCH}_2\text{COOH}}$$

14. Equations for the oxidation of:
    (a) 1-propanol

$$CH_3CH_2CH_2OH \xrightarrow[H_2SO_4]{K_2Cr_2O_7} \quad CH_2CH_2\overset{\overset{\displaystyle O}{\parallel}}{C}-H \quad or \quad CH_3CH_2COOH$$

$$CH_3CH_2CH_2OH + O_2 \xrightarrow[\Delta]{Cu} \quad CH_2CH_2\overset{\overset{\displaystyle O}{\parallel}}{C}-H$$

(b) $CH_3\overset{\overset{\displaystyle OH}{|}}{CH}-\overset{\overset{\displaystyle |}{\underset{\underset{\displaystyle CH_3}{|}}{C}}}{C}CH_3 + O_2 \xrightarrow[\Delta]{Ag}$ No reaction (3° alcohol) with either oxidizing agent

$\quad\quad\quad\;\;\underset{CH_3}{}$

15. (a) $CH_3-CH_2-\overset{\overset{\displaystyle O}{\parallel}}{C}-OH$

    (b) $^-O\overset{\overset{\displaystyle O}{\parallel}}{C}-\overset{\overset{\displaystyle |}{\underset{\underset{\displaystyle CH_3}{|}}{C}}}{CH}-CH_3$

    (c) $^-O\overset{\overset{\displaystyle O}{\parallel}}{C}-\overset{\overset{\displaystyle |}{\underset{\underset{\displaystyle CH_3}{|}}{C}}}{CH}-CH_3$

16. (a)

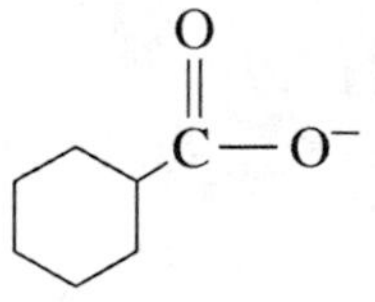

    (b) $^-O-\overset{\overset{\displaystyle O}{\parallel}}{C}-CH_2CH_2-\overset{\overset{\displaystyle O}{\parallel}}{C}-O^-$

    (c) $CH_3-CH_2-CH_2-\overset{\overset{\displaystyle O}{\parallel}}{C}-OH$

17. (a) propanal,

$$CH_3-CH_2-\overset{\overset{\displaystyle O}{\parallel}}{CH}$$

(b) butanone,

$$\text{CH}_3-\overset{\displaystyle O}{\overset{\|}{\text{C}}}-\text{CH}_2-\text{CH}_3$$

(c) butanal,

$$\text{CH}_3-\text{CH}_2-\text{CH}_2-\overset{\displaystyle O}{\overset{\|}{\text{CH}}}$$

18. (a) formaldehyde (methanal) $\overset{\displaystyle O}{\overset{\|}{\text{CH}_2}}$

(b) propanal, $\text{CH}_3-\text{CH}_2-\overset{\displaystyle O}{\overset{\|}{\text{CH}}}$

(c) propanal, $\text{CH}_3-\text{CH}_2-\overset{\displaystyle O}{\overset{\|}{\text{CH}}}$

19. (a) An aldehyde group, $-\overset{\displaystyle H}{\overset{|}{\text{C}}}=\text{O}$, must be present to give a positive Tollens test.

(b) The visible evidence for a positive Tollens test is the formation of a silver mirror on the inner walls of a test tube.

(c) $\text{CH}_3\overset{\displaystyle H}{\overset{|}{\text{C}}}=\text{O} + 2\,\text{Ag}^+ \xrightarrow[\text{H}_2\text{O}]{\text{NH}_3} \text{CH}_3\text{COO}^-\,\text{NH}_4^+ + 2\,\text{Ag}(s)$ (unbalanced equation)
  silver mirror

20. (a) An aldehyde group, $-\overset{\displaystyle H}{\overset{|}{\text{C}}}=\text{O}$, must be present to give a positive Fehling test.

(b) The visible evidence for a positive Fehling test is the formation of brick red Cu_2O, which precipitates during the reaction.

(c) $\text{CH}_3\overset{\displaystyle H}{\overset{|}{\text{C}}}=\text{O} + 2\,\text{Cu}^{2+} \xrightarrow[\text{H}_2\text{O}]{\text{NaOH}} \text{CH}_3\text{COO}^-\,\text{Na}^+\,\text{Cu}_2\text{O}(s)$ (unbalanced equation)
  brick red

21. (a) The aldehyde is oxidized in the Tollens test:

$\text{CH}_3\text{CH}_2\text{COO}^-\,\text{NH}_4^+$

(b) The ketone is reduced:

$$\text{CH}_3\overset{\displaystyle \text{OH}}{\overset{|}{\text{CH}}}\text{CH}_2\text{CH}_2\text{CH}_2\text{CH}_3$$

(c) The aldehyde is oxidized in the Benedict test:

$$CH_3-\underset{\underset{\displaystyle CH_3}{|}}{CH}-COO^-\ Na^+$$

(d) The aldehyde is reduced:

$$CH_3CH_2CH_2CH_2CH_2CH_2CH_2OH$$

22. (a) The aldehyde is reduced:

$$CH_3CH_2\underset{\underset{\displaystyle CH_3}{|}}{CH}CH_2OH$$

(b) The aldehyde is oxidized in the Benedict test:

$$CH_3\overset{\overset{\displaystyle O}{\|}}{C}O^-\ Na^+$$

(c) The ketone is reduced:

$$CH_3\underset{\underset{\displaystyle OH}{|}}{CH}CH_3$$

(d) The aldehyde is oxidized in the Tollens test:

$$CH_3CH_2\underset{\underset{\displaystyle CH_2CH_3}{|}}{CH}-\overset{\overset{\displaystyle O}{\|}}{C}O^-\ NH_4^+$$

23. HCN adds to a carbonyl functional group to form a cyanohydrin. In turn, the cyanohydrin can be reacted with water to make an α-hydroxy acid.

(a)
$$CH_3-\underset{\underset{\displaystyle CH_3}{|}}{\overset{\overset{\displaystyle OH}{|}}{C}}-COOH$$

(b)
$$CH_3CH_2CH_2-\underset{\underset{\displaystyle }{}}{\overset{\overset{\displaystyle OH}{|}}{C}H}-COOH$$

(c)
$$CH_3CH_2\underset{\underset{\displaystyle CH_3}{|}}{CH}-\underset{\underset{\displaystyle CH_3}{|}}{\overset{\overset{\displaystyle OH}{|}}{C}}-COOH$$

24. HCN adds to a carbonyl functional group to form a cyanohydrin. In turn, the cyanohydrin can be reacted with water to make an α-hydroxy acid.

(a) $HOCH_2COOH$

(b) 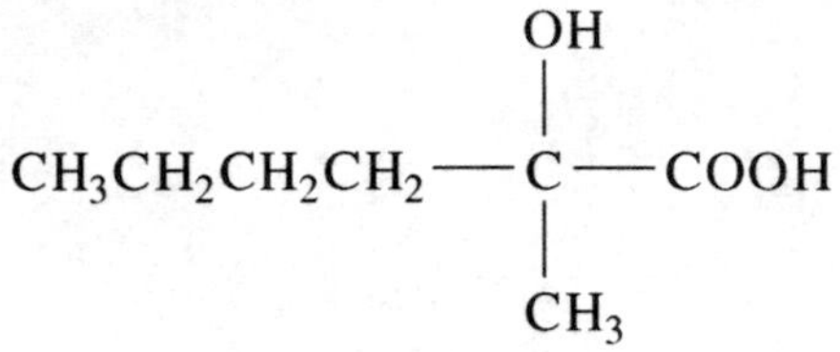
$$CH_3CH_2CH_2CH_2-\overset{\overset{\displaystyle OH}{|}}{\underset{\underset{\displaystyle CH_3}{|}}{C}}-COOH$$

(c)
$$CH_3-\overset{\overset{\displaystyle CH_3}{|}}{\underset{\underset{\displaystyle CH_3}{|}}{C}}-\overset{\overset{\displaystyle OH}{|}}{CH}-COOH$$

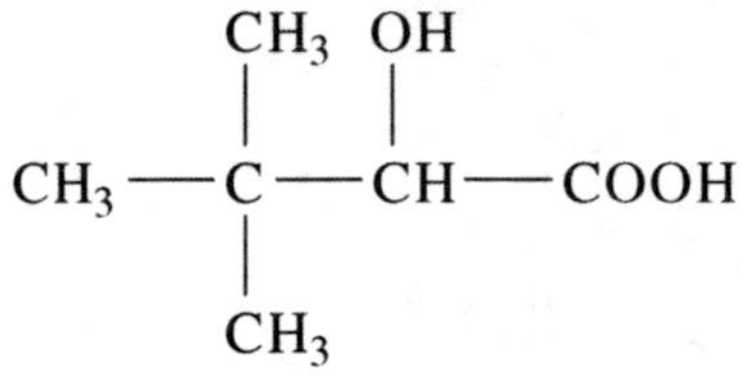

25. Aldol condensation

(a) butanal

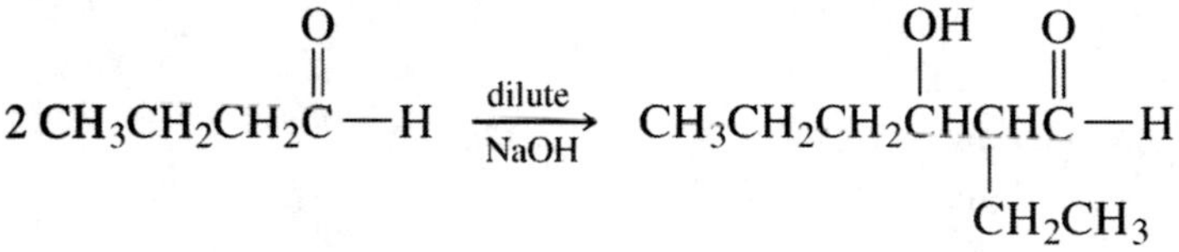

$$2\ CH_3CH_2CH_2\overset{\overset{\displaystyle O}{\|}}{C}-H \xrightarrow[NaOH]{dilute} CH_3CH_2CH_2\overset{\overset{\displaystyle OH}{|}}{CH}\underset{\underset{\displaystyle CH_2CH_3}{|}}{CH}\overset{\overset{\displaystyle O}{\|}}{C}-H$$

(b) 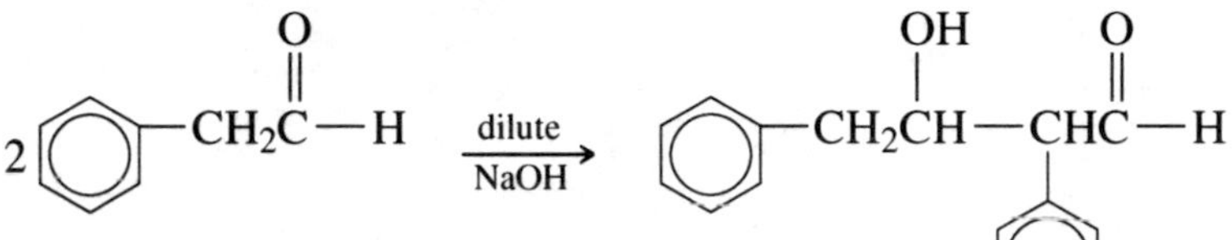 2 (phenyl)$-CH_2\overset{\overset{\displaystyle O}{\|}}{C}-H \xrightarrow[NaOH]{dilute}$ (phenyl)$-CH_2\overset{\overset{\displaystyle OH}{|}}{CH}-CH\overset{\overset{\displaystyle O}{\|}}{C}-H$ (phenyl)

phenylethanal

26. Aldol condensation

(a) 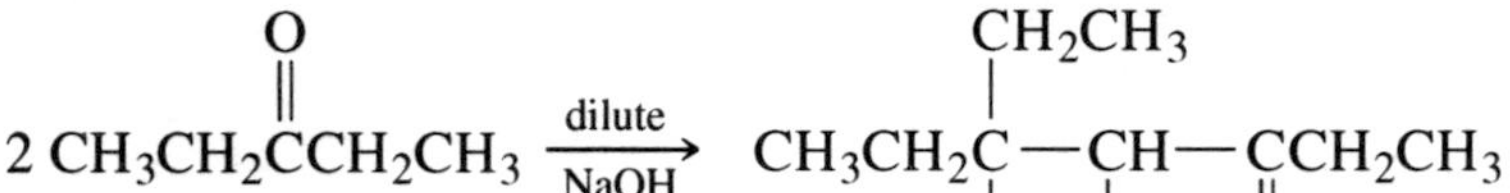 $2\ CH_3CH_2\overset{\overset{\displaystyle O}{\|}}{C}CH_2CH_3 \xrightarrow[NaOH]{dilute} CH_3CH_2\overset{\overset{\displaystyle CH_2CH_3}{|}}{C}-CH-\overset{\|}{C}CH_2CH_3$

3-pentanone
$$\underset{HO\quad CH_3\ \ O}{}$$

(b) 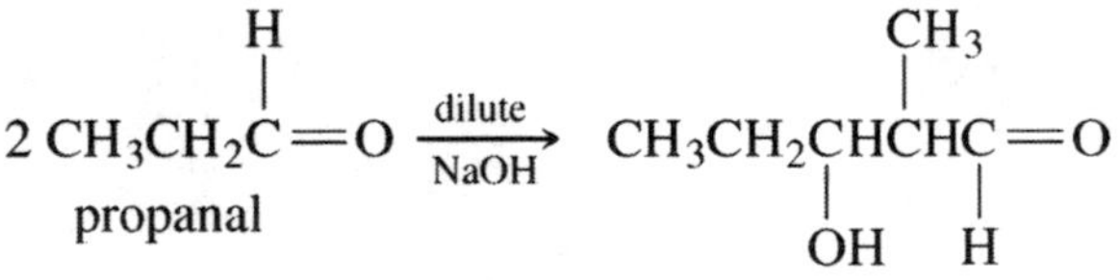 $2\ CH_3CH_2\overset{\overset{\displaystyle H}{|}}{C}=O \xrightarrow[NaOH]{dilute} CH_3CH_2CH\overset{\overset{\displaystyle CH_3}{|}}{CH}C=O$

propanal
$$\underset{OH\quad H}{}$$

27. The completed equations are:

(a) $CH_3\overset{O}{\overset{\|}{C}}CH_3$ + $\underset{OH\ OH}{CH_2CH_2}$ $\underset{}{\overset{dry\ HCl}{\rightleftharpoons}}$ $\underset{CH_3}{\overset{CH_3}{}}C\overset{OCH_2}{\underset{OCH_2}{}}$ + H_2O

(b) $CH_3CH_2\overset{H}{\underset{}{C}}{=}O$ + CH_3CH_2OH $\overset{H^+}{\rightleftharpoons}$ $CH_3CH_2\overset{OH}{\underset{H}{C}}-OCH_2CH_3$

(c) $\underset{CH_3}{CH_3CHCH_2CH(OCH_3)_2}$ $\overset{H_2O}{\underset{H^+}{\rightarrow}}$ $\underset{CH_3}{CH_3CHCH_2\overset{O}{\overset{\|}{C}}-H}$ + $2\ CH_3OH$

28. The completed equations are:

(a) $CH_3CH_2\overset{H}{\underset{}{C}}{=}O$ + $CH_3CH_2CH_2OH$ $\overset{dry\ HCl}{\rightleftharpoons}$ $CH_3CH_2\overset{OH}{\underset{H}{C}}-OCH_2CH_2CH_3$

(b) cyclohexanone + CH_3OH $\overset{H^+}{\rightleftharpoons}$ 1-methoxycyclohexanol (OH, OCH_3)

(c) $CH_3CH_2CH_2(OCH_3)_2$ $\overset{H_2O}{\underset{H^+}{\rightarrow}}$ $CH_3CH_2CH_2\overset{H}{\underset{}{C}}{=}O$ + $2\ CH_3OH$

29. Sequence of reactions:

(a) $CH_3\overset{}{\underset{O}{\overset{\|}{C}}}CH_3$ + HCN $\longrightarrow$ $\underset{OH}{\overset{CH_3}{CH_3C}-CN}$

(b) $\underset{OH}{\overset{CH_3}{CH_3C}-CN}$ + H_2O $\longrightarrow$ $\underset{OH}{\overset{CH_3}{CH_3C}-COOH}$

(c) $2\ \underset{OH}{\overset{CH_3}{CH_3C}-COOH}$ + $CH_3\overset{O}{\overset{\|}{C}}-H$ $\overset{dry\ HCl}{\longrightarrow}$ $CH_3CH\overset{O-\overset{CH_3}{\underset{CH_3}{C}}-COOH}{\underset{O-\overset{CH_3}{\underset{CH_3}{C}}-COOH}{}}$

30. Sequence of reactions:

(a) C₆H₅—CH(=O with H) + HCN $\xrightarrow{OH^-}$ C₆H₅—CH(OH)—CN

$$\text{(a)} \quad \text{C}_6\text{H}_5\overset{H}{\underset{}{\text{C}}}{=}\text{O} \;+\; \text{HCN} \xrightarrow{\text{OH}^-} \text{C}_6\text{H}_5\overset{\text{OH}}{\underset{}{\text{CH}}}{-}\text{CN}$$

$$\text{(b)} \quad \text{C}_6\text{H}_5\overset{\text{OH}}{\underset{}{\text{CH}}}{-}\text{CN} \;+\; \text{H}_2\text{O} \xrightarrow{\text{H}^+} \text{C}_6\text{H}_5\overset{\text{OH}}{\underset{}{\text{CHCOOH}}}$$

$$\text{(c)} \quad \text{C}_6\text{H}_5\overset{\text{OH}}{\underset{}{\text{CHCOOH}}} \xrightarrow[\text{H}_2\text{SO}_4]{\text{K}_2\text{Cr}_2\text{O}_7} \text{C}_6\text{H}_5\overset{\text{O}}{\underset{}{\text{C}}}{-}\text{COOH}$$

31. $\text{HOCH}_2\text{CH}_2\text{CH}_2\overset{\text{H}}{\underset{}{\text{C}}}{=}\text{O} \xrightarrow{\text{H}^+}$ (tetrahydrofuran ring with H and OH)

32.

$$\text{CH}_3\overset{\text{O}}{\underset{}{\text{C}}}{-}\text{H} \qquad \text{ethanal}$$

33. Four aldol condensation products from a mixture of ethanal (E) and propanal (P) are possible.

EE   PP   EP   PE

EE is $\text{CH}_3\underset{\text{OH}}{\text{CHCH}_2}\overset{\text{O}}{\text{C}}{-}\text{H}$

PP is $\text{CH}_3\text{CH}_2\underset{\text{OH}}{\text{CHCH}}\overset{\text{CH}_3}{\underset{\text{O}}{\text{C}}}{-}\text{H}$

EP is $\text{CH}_3\underset{\text{OH}}{\text{CHCHCH}}\overset{\text{CH}_3}{\underset{\text{O}}{}}$

PE is $\text{CH}_3\text{CH}_2\underset{\text{OH}}{\text{CHCH}_2}\overset{\text{O}}{\text{C}}{-}\text{H}$

34. A hemiacetal forms when an alcohol adds to a carbonyl group.

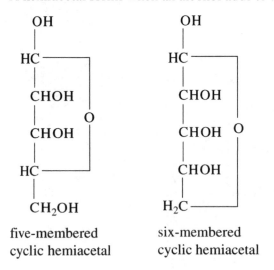

five-membered
cyclic hemiacetal

six-membered
cyclic hemiacetal

35. (a) $\overset{\overset{\displaystyle H}{|}}{CH_3CH_2C}{=}O$ and $\overset{\overset{\displaystyle O}{||}}{CH_3CCH_3}$.

The Tollens test (silver mirror) or Fehling test (red Cu_2O) will give positive results with propanal but not with acetone.

(b) $\overset{\overset{\displaystyle H}{|}}{CH_3CH_2C}{=}O$ and $CH_2{=}\overset{\overset{\displaystyle H}{|}}{CHC}{=}O$

Bromine will decolorize immediately with the second compound (propenal) but not with propanal.

(c) 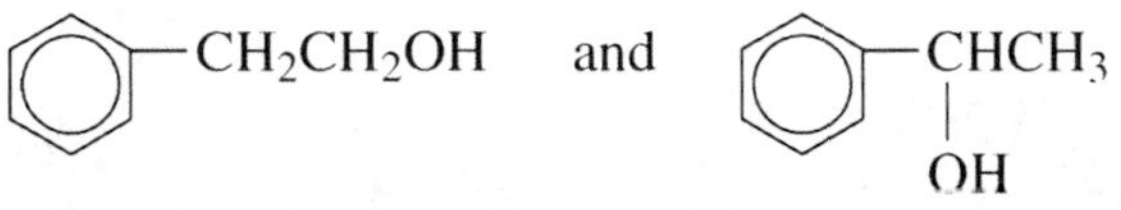 ⬡—CH_2CH_2OH and   ⬡—$\overset{\displaystyle CHCH_3}{\underset{\displaystyle OH}{|}}$

Oxidize both compounds: 2-phenylethanol will give 2-phenylethanal and 1-phenylethanol will give methyl phenyl ketone. 1-phenylethanal will give a positive Tollens or Benedict test and methyl phenyl ketone will not give a positive test.

36. $\overset{\overset{\displaystyle H}{|}}{CH_3C}{=}O + HCN \longrightarrow CH_3\overset{\overset{\displaystyle H}{|}}{\underset{\underset{\displaystyle OH}{|}}{C}}{-}CN \xrightarrow[H^+]{H_2O} CH_3\overset{\underset{\underset{\displaystyle OH}{|}}{}}{CH}{-}COOH$

lactic acid

37. 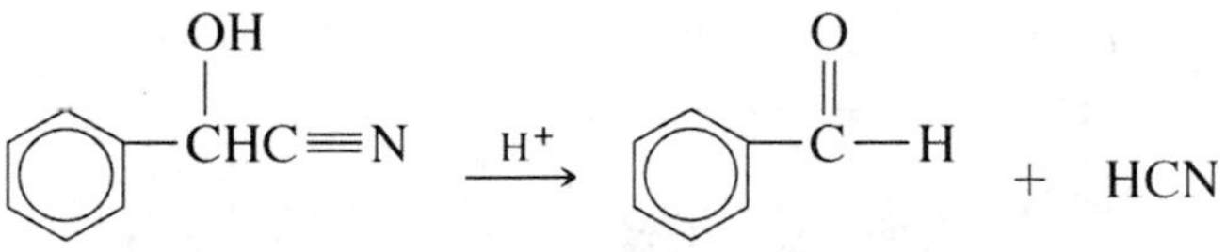 ⬡—$\overset{\overset{\displaystyle OH}{|}}{CHC}{\equiv}N \xrightarrow{H^+}$ ⬡—$\overset{\overset{\displaystyle O}{||}}{C}{-}H$ $+$ HCN

benzaldehyde

38. Pyruvic acid is changed to lactic acid by a reduction reaction.

39. The alcohols which should be oxidized to give these ketones.

(a) 3-pentanone: 3-pentanol $CH_3CH_2\overset{\underset{\underset{\displaystyle OH}{|}}{}}{CH}CH_2CH_3$

(b) methyl ethyl ketone: 2-butanol $CH_3\overset{\underset{\underset{\displaystyle OH}{|}}{}}{CH}CH_2CH_3$

(c) 4-phenyl-2-butanone: 4-phenyl-2-butanol   ⬡—$CH_2CH_2\overset{\underset{\underset{\displaystyle OH}{|}}{}}{CH}CH_3$

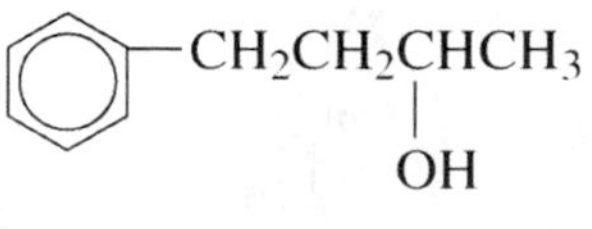

40. The following are alll the isomeric aldehydes and ketones with the formula, $C_5H_{10}O$:

$$CH_3CH_2CH_2CH_2\overset{\overset{\displaystyle O}{\|}}{C}H \qquad CH_3CH_2CH_2\overset{\overset{\displaystyle O}{\|}}{C}CH_3 \qquad CH_3CH_2\overset{\overset{\displaystyle O}{\|}}{C}CH_2CH_3$$

$$CH_3\overset{\overset{\displaystyle CH_3}{|}}{C}HCH_2\overset{\overset{\displaystyle O}{\|}}{C}H \qquad CH_3CH_2\overset{\displaystyle CH}{\underset{\overset{\displaystyle |}{CH_3}}{}}\overset{\overset{\displaystyle O}{\|}}{C}H \qquad CH_3\overset{\overset{\displaystyle O}{\|}}{C}\overset{\displaystyle CH}{\underset{\overset{\displaystyle |}{CH_3}}{}}CH_3$$

$$CH_3-\overset{\overset{\displaystyle CH_3}{|}}{\underset{\overset{\displaystyle |}{CH_3}}{C}}-\overset{\overset{\displaystyle O}{\|}}{C}H$$

41. Ketones will not react in the Fehling test. The following are all the ketones with the formula, $C_6H_{12}O$:

$$CH_3CH_2CH_2CH_2\overset{\overset{\displaystyle O}{\|}}{C}CH_3 \qquad CH_3CH_2CH_2\overset{\overset{\displaystyle O}{\|}}{C}CH_3CH_3 \qquad CH_3CH_2\overset{\overset{\displaystyle O}{\|}}{C}\overset{\displaystyle CH}{\underset{\overset{\displaystyle |}{CH_3}}{}}CH_3$$

$$CH_3\overset{\overset{\displaystyle O}{\|}}{C}CH_2\overset{\displaystyle CH}{\underset{\overset{\displaystyle |}{CH_3}}{}}CH_3 \qquad CH_3-\overset{\overset{\displaystyle O}{\|}}{C}-\overset{\overset{\displaystyle CH_3}{|}}{\underset{\overset{\displaystyle |}{CH_3}}{C}}-CH_3 \qquad CH_3\overset{\overset{\displaystyle O}{\|}}{C}\overset{\displaystyle CH}{\underset{\overset{\displaystyle |}{CH_3}}{}}CH_2CH_3$$

42. (a) 1. This step is an aldol condensation that occurs in dilute NaOH.
    2. This step oxidizes the aldehyde to a carboxylic acid. Common oxidizing conditions are set by the Tollens test (Ag^+, NH_3, H_2O) or the Fehling/Benedict test (Cu^{2+}, NaOH, H_2O).

    (b) 1. This step forms a cyanohydrin and the conditions require hydroxide (OH^-).
    2. This step reacts the cyanohydrin with water to form an α-hydroxy acid. Conditions require acid (H^+) as well as water.

    (c) 1. This step is an aldol condensation that occurs in dilute NaOH.
    2. This step reduces the aldehyde to form a primary alcohol. Common reducing conditions involve heat (Δ), hydrogen gas, and a nickel catalyst.

43. $$CH_3\overset{\overset{\displaystyle H}{|}}{C}{=}O \quad \left(CH_2\underset{\overset{\displaystyle |}{OH}}{CH}-CH_2\underset{\overset{\displaystyle |}{OH}}{CH}-CH_2\underset{\overset{\displaystyle |}{OH}}{CH}-CH_2\underset{\overset{\displaystyle |}{OH}}{CH}\right)_n$$

44. In phenol, three positions, ortho, ortho, and para to the OH group are used in the reaction to form a thermosetting polymer. However, in p-cresol the para position is occupied by a methyl group and cannot

react with formaldehyde. This leaves the *p*-cresol molecule as a bifunctional monomer, resulting in a linear, thermoplastic polymer.

45. (a)

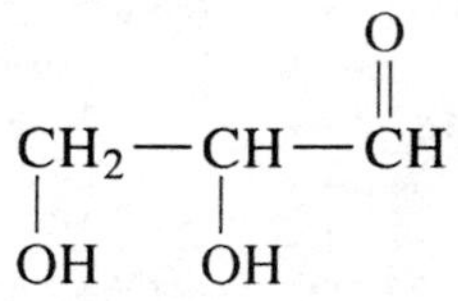

$$CH_2 - CH - \overset{\overset{\displaystyle O}{\|}}{C}H$$
$$\quad | \qquad |$$
$$\quad OH \quad OH$$

2,3-dihydroxypropanal

$$CH_2\overset{\overset{\displaystyle O}{\|}}{C} - CH_2$$
$$\quad | \qquad |$$
$$\quad OH \quad OH$$

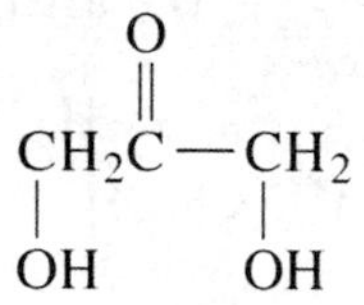

1,3-dihydroxypropanone

(b)

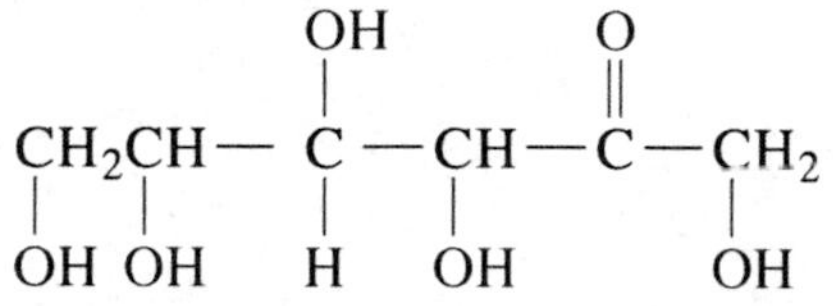

$$CH_2CH - \overset{\overset{\displaystyle OH}{|}}{C} - CH - \overset{\overset{\displaystyle O}{\|}}{C} - CH_2 \quad \text{and} \quad CH_2 \overset{\overset{\displaystyle OH}{|}}{C} - \overset{\overset{\displaystyle OH}{|}}{C} - \overset{\overset{\displaystyle O}{\|}}{C}H$$
$$\quad | \quad | \qquad | \qquad \qquad | \qquad \qquad\qquad | \qquad |$$
$$\quad OH \ OH \quad H \ \ OH \quad\quad OH \qquad\qquad HO - CH_2 \ CH_2OH$$

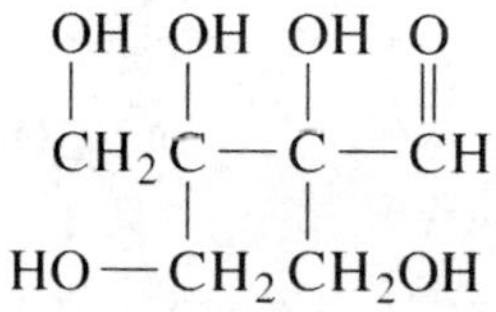

CARBOXYLIC ACIDS AND ESTERS

SOLUTIONS TO REVIEW QUESTIONS

1. (a) CH_3COOH

   (b) 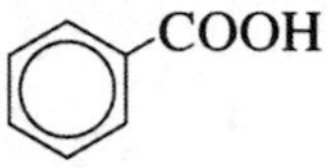 C_6H_5—COOH

   (c) $CH_3CHCOOH$ with OH

   (d) $CH_3CHCOOH$ with NH_2

   (e) CH_2CH_2COOH with Cl

   (f) $COOH$—$COOH$

   (g) $CH_2=CHCOOH$

   (h) CH_2C—OCH_3 with $\parallel O$

   (i) CH_3CN

   (j) CH_3COONa

   (k) CH_3C—Cl with $\parallel O$

   (l)
$$CH_2-OCC_{17}H_{35}$$
$$CH-OCC_{17}H_{35}$$
$$CH_2-OCC_{17}H_{35}$$

   (m) $CH_3(CH_2)_{16}COONa$

   (n) $HO-P(=O)(OH)-OH$

   (o) $HO-P(=O)(OH)-OCH_3$

2. The butyric acid solution would be expected to have the more objectionable odor because salts normally exhibit little or no odor. Salts are ionic and therefore have low volatility. For example, dilute solutions of acetic acid (vinegar) have considerable odor, but sodium acetate does not.

3. Which has the greater solubility in water?
   (a) Propanoic acid is more soluble than methyl propanoate.
   (b) Sodium palmitate is more soluble than palmitic acid.
   (c) Sodium stearate is more soluble than barium stearate.
   (d) Sodium phenoxide is more soluble than phenol.

4. Unlike alkanes, carboxylic acids hydrogen bond tightly to each other. Thus, they have much higher boiling points than alkanes.

5. (a) The major difference between fats and oils is that fats are solids at room temperature, oils are liquids. The fatty acids in the molecules are mostly saturated in fats; more unsaturated in oils.

(b) Soaps are the sodium salts of high molar mass fatty acids. Syndets are synthetic detergents and occur in several different forms. They have cleansing action similar to soaps, but have different structural and solubility characteristics, such as being soluble in hard water. Some syndets also contain long hydrocarbon chains.

(c) Hydrolysis is the breaking apart of an ester in the presence of water to form an alcohol and a carboxylic acid. Mineral acids or digestive enzymes are used to speed the hydrolysis process. Saponification breaks the ester apart using sodium hydroxide to form an alcohol and a salt of a carboxylic acid.

6. A sour taste is associated with acids. Citrus fruits and rhubarb are examples of foods that have a sour, acid taste.

7. Like a soap, a detergent contains a grease-soluble component and a water-soluble component. The grease-soluble component of many molecules dissolves in the grease film. The water-soluble components attract water, causing small droplets of grease-bearing dirt to break loose and float away.

8. The principal advantage of synthetic detergents over soap is that the syndets do not form insoluble precipitates with the ions in hard water (Ca^{2+}, Mg^{2+}, Fe^{3+}).

9. Aspirin acts in the body as an antipyretic, an analgesic, and as an antiinflammatory agent.

10. In a condensation reaction one of the products formed is a small molecule such as water. In polyester formation between an alcohol and a carboxylic acid, water is formed in addition to the ester (polyester).

11. Polyesters are polymers of molecules linked together by ester bonds. The ester bonds are susceptible to hydrolysis (being broken by reaction with water). When a polyester suture dissolves, the polymer's ester bonds are broken.

12. (a) Phosphoric acid anhydrides are used by living cells to store metabolic energy. The "energy currency" of the cell is adenosine triphosphate (ATP) that contains two phosphoric acid anhydrides.

(b) The phosphoric acid ester (or phosphate ester) has many important biochemical functions. Often a phosphate ester is used to "tag" a biochemical, labeling it for specific biological uses.

13. Partial hydrogenation of oils is no longer commonly used by the food industry because this process produces some *trans* fats. These fats are known to increase the risk of atherosclerosis.

14. MSG stands for monosodium glutamate. It is the sodium salt of the common amino acid, glutamic acid. The sodium salt is formed in an acid-base reaction between glutamate and a base such as sodium hydroxide.

15. Both compounds have a hydrophilic structural component. Capric acid contains a nine-carbon hydrocarbon chain which decreases its solubility in water. Formic acid does not have a hydrocarbon chain.

16. An ester is formed by bonding an hydroxyl group (alcohol) to an acid with the release of water. This compound meets the definition of ester because it combines phosphoric acid with ethanol.

$$
\underset{\substack{\text{from} \\ \text{phosphoric} \\ \text{acid}}}{\underbrace{HO-}}\overset{\displaystyle \overset{O}{\|}}{\underset{\displaystyle |}{P}}\underset{\substack{\text{from} \\ \text{ethanol}}}{\underbrace{-OCH_2CH_3}}
$$

17. Hard water contains much higher concentrations of Ca^{2+}, Mg^{2+}, and Fe^{3+} ions than soft water. Soaps precipitate (forming a soap scum) with these ions while synthetic detergents do not.

SOLUTIONS TO EXERCISES

1. (a) lactic acid
   (b) succinic acid
   (c) phthalic acid (*o*-phthalic acid)
   (d) propionic acid

2. (a) *m*-toluic acid
   (b) β-chloropropionic acid
   (c) butyric acid
   (d) glutaric acid

3. (a) 2,3-dichloropropanoic acid
   (b) hexanoic acid
   (c) 2-methylbutanoic acid
   (d) propanedioic acid

4. (a) propanoic acid
   (b) 2,4-dimethylpentanoic acid
   (c) butanedioic acid
   (d) 2-hydroxybutanoic acid

5. (a)
$$\overset{\displaystyle OH}{\underset{\displaystyle |}{CH_3CHCH_2COOH}}$$

   (b) $CH_3CH_2CH_2CH_2COOH$

   (c) $CH_3CH_2COO^-\ K^+$

   (d) 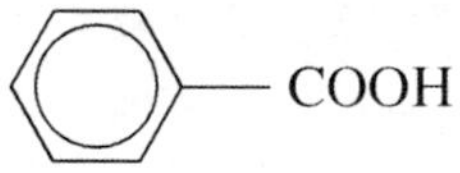 ⬡— COOH

   (e) $CH_3COO^-\ N\overset{+}{a}$

   (f)
$$\overset{\displaystyle CH_3}{\underset{\displaystyle |}{\underset{\displaystyle \underset{\displaystyle |}{CH_3}}{CH_3CH_2CH_2CCH_2COOH}}}$$

   (g)
   CH_3
   ⬡— COOH

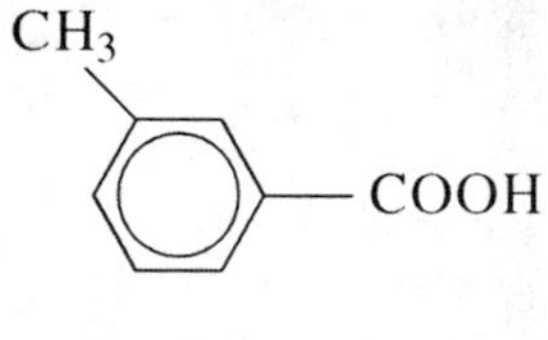

6. (a) $HOOCCH_2COOH$

   (b)
$$\overset{\displaystyle OH\quad\ CH_2CH_3}{\underset{\displaystyle |\qquad\ |}{CH_3CH_2CHCH_2CHCOOH}}$$

   (c)
$$\overset{\displaystyle Cl}{\underset{\displaystyle |}{CH_3CHCOOH}}$$

   (d) 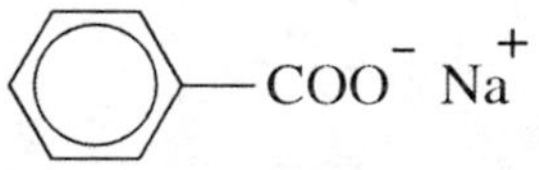 ⬡— $COO^-\ Na^+$

(e) HOOC — 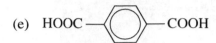 — COOH

(f) CH_3CH_2COOH

(g) $CH_3CH_2CH_2COO^-\overset{+}{Li}$

7. (a) Acetic acid is a common name. The IUPAC name is ethanoic acid.

   (b) 2-Bromo-1-hexanoic acid is incorrect because the carboxyl functional group should not be numbered. The correct IUPAC name is 2-bromohexanoic acid.

   (c) In α-chloropropanoic acid, the Greek letter is common nomenclature. The correct IUPAC name is 2-chloropropanoic acid.

8. (a) In β-bromobutanoic acid, the Greek letter is common nomenclature. The correct IUPAC name is 3-bromobutanoic acid.

   (b) Lactic acid is a common name. The IUPAC name is 2-hydroxypropanoic acid.

   (c) 3-Ethyl-5-pentanoic acid incorrectly includes a number for the carboxyl functional group. The correct IUPAC name is 3-ethylpentanoic acid.

9. Increasing pH means increasing basicity

$$HCl \; < \; CH_3COOH \; < \; \text{(phenol with OH)} \; < \; NaCl \; < \; NH_3 \; < \; NaOH$$

Increasing pH ⟶

10. Decreasing pH means increasing acidity

$$KOH \; > \; NH_3 \; > \; KBr \; > \; \text{(phenol with OH)} \; > \; HCOOH \; > \; HBr$$

Decreasing pH ⟶

11. (a) succinic acid   (c) acetic acid
    (b) butanoic acid   (d) butanoic acid

12. (a) ethanoic acid   (c) ethanoic acid
    (b) propanoic acid   (d) malonic acid

13. IUPAC and common names

    (a) methyl propenoate   methyl acrylate
    (b) ethyl butanoate   ethyl butyrate
    (c) phenyl-2-hydroxybenzoate   phenyl salicylate

14. IUPAC and common names

    (a) methyl methanoate   methyl formate
    (b) propyl benzoate   propyl benzoate
    (c) ethyl propanoate   ethyl propionate

15. Structural formulas

(a) $\overset{\overset{\displaystyle O}{\|}}{HC}-OCH_3$

(c) Ph$-\overset{\overset{\displaystyle O}{\|}}{C}-OCH_2CH_3$

(b) $CH_3CH_2CH_2\overset{\overset{\displaystyle O}{\|}}{C}-OCH_2CH_2CH_2CH_3$

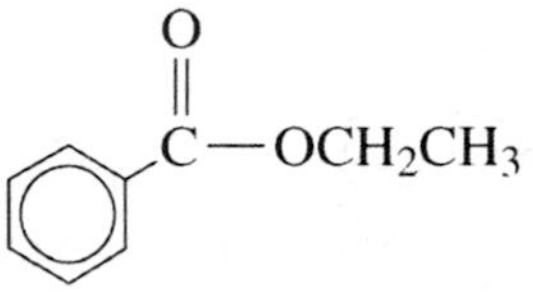

16. Structural formulas

(a) $CH_3\overset{\overset{\displaystyle O}{\|}}{C}-OCH_2CH_2CH_3$

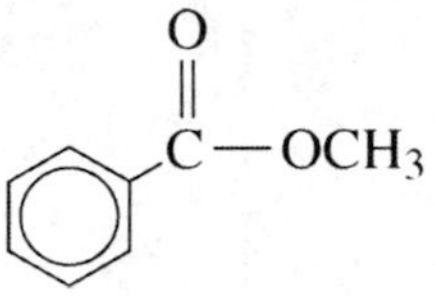

(c) $CH_3CH_2CH_2CH_2CH_2\overset{\overset{\displaystyle O}{\|}}{C}-OCH_2CH_3$

(b) Ph$-\overset{\overset{\displaystyle O}{\|}}{C}-OCH_3$

17. (a) Compounds are different: methyl propanoate and ethyl ethanoate
(b) Compounds are the same
(c) Compounds are different: methyl 2-methylpropanoate and isopropyl ethanoate

18. (a) Compounds are different: methyl benzoate and phenyl acetate
(b) Compounds are different: methyl acetate and methyl propanoate
(c) Compounds are the same

19. Structural formula for the ester that when hydrolyzed would yield:

(a) methanol and acetic acid $\qquad CH_3\overset{\overset{\displaystyle O}{\|}}{C}-OCH_3$

(b) ethanol and formic acid, $\qquad HC\overset{\overset{\displaystyle O}{\|}}{}-OCH_2CH_3$

(c) 2-propanol and benzoic acid $\qquad$ Ph$-\overset{\overset{\displaystyle O}{\|}}{C}-O-\underset{\underset{\displaystyle CH_3}{|}}{CH}-CH_3$

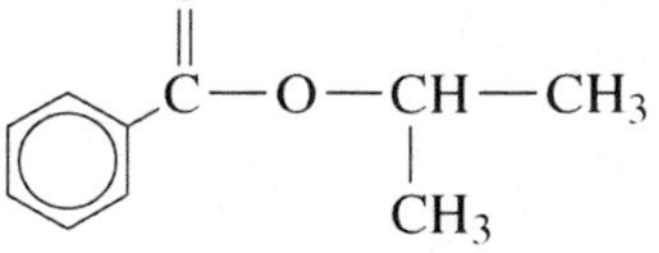

20. Structural formula for the ester that when hydrolyzed would yield:

(a) methanol and propanoic acid

$$CH_3CH_2\overset{\displaystyle O}{\overset{\|}{C}}\text{—}OCH_3$$

(b) 1-octanol and acetic acid

$$CH_3\overset{\displaystyle O}{\overset{\|}{C}}\text{—}OCH_2(CH_2)_6CH_3$$

(c) ethanol and butanoic acid

$$CH_3CH_2CH_2\overset{\displaystyle O}{\overset{\|}{C}}\text{—}OCH_2CH_3$$

21. Structural formulas for the reactants that will yield the following esters:

(a) methyl palmitate

$$CH_3OH + CH_3(CH_2)_{14}COOH$$

(b) phenyl propionate

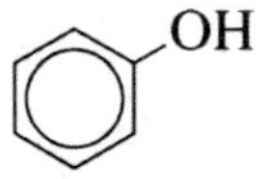

⬡$-OH$ $+ CH_3CH_2COOH$

(c) dimethyl succinate

$$CH_3OH + HOOCCH_2CH_2COOH$$

22. Structural formulas for the reactants that will yield the following esters:

(a) isopropyl formate

$$CH_3\underset{\underset{\displaystyle OH}{|}}{CH}CH_3 + HCOOH$$

(b) diethyl adipate

$$CH_3CH_2OH + HOOC(CH_2)_4COOH$$

(c) benzyl benzoate

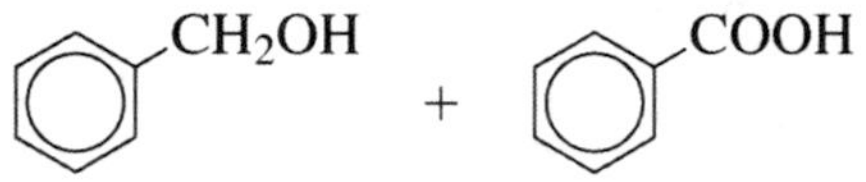

⬡$-CH_2OH$ $+$ ⬡$-COOH$

23. (a) These reagents ($Na_2Cr_2O_7$, H_2SO_4) oxidize the aldehyde to a carboxylic acid.

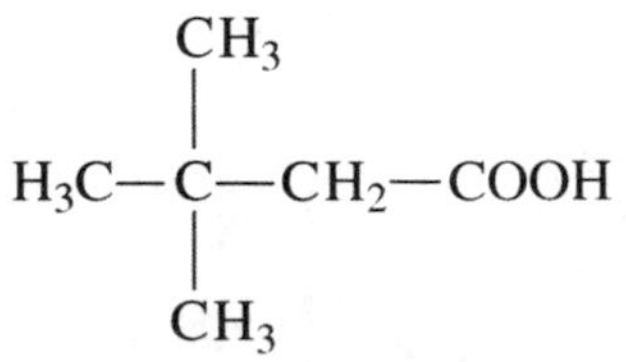

$$H_3C\text{—}\underset{\underset{\displaystyle CH_3}{|}}{\overset{\overset{\displaystyle CH_3}{|}}{C}}\text{—}CH_2\text{—}COOH$$

3,3-dimethylbutanoic acid

(b) These reagents (Na$_2$MnO$_4$, NaOH, Δ) oxidize the alkyl side chain to form the salt of benzoic acid.

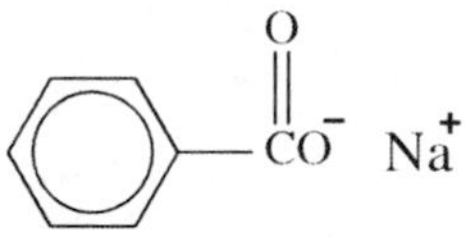

sodium benzoate

(c) This is an acid-base reaction that forms the salt of the carboxylic acid.

CH$_3$CH$_2$CH$_2$COO$^-$ K$^+$

potassium butanoate

24. (a) These reagents (Na$_2$MnO$_4$, NaOH, Δ) oxidize the alkyl side chain to form the salt of benzoic acid.

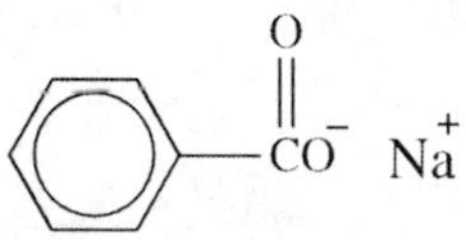

sodium benzoate

(b) These reagents (Na$_2$Cr$_2$O$_7$, H$_2$SO$_4$) oxidize the primary alcohol to a carboxylic acid.

CH$_3$CH$_2$CH$_2$COOH

butanoic acid

(c) This is an acid-base reaction that forms the salt of the carboxylic acid.

CH$_3$CH$_2$COO$^-$ Na$^+$

sodium propanoate

25. (a) A carboxylic acid reacts with a carboxylic acid chloride to yield a carboxylic acid anhydride.

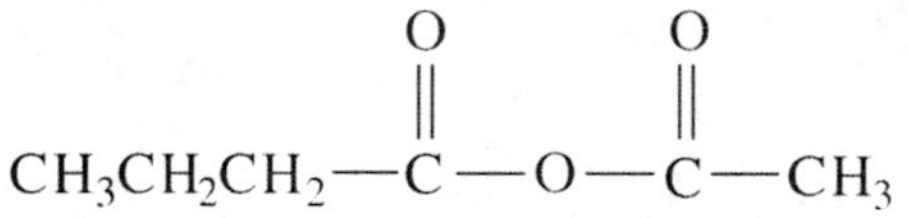

CH$_3$CH$_2$CH$_2$—C(O)—O—C(O)—CH$_3$

(b) A nitrile reacts with water to form a carboxylic acid.

CH$_3$COOH

(c) A carboxylic acid chloride reacts with an alcohol to form an ester.

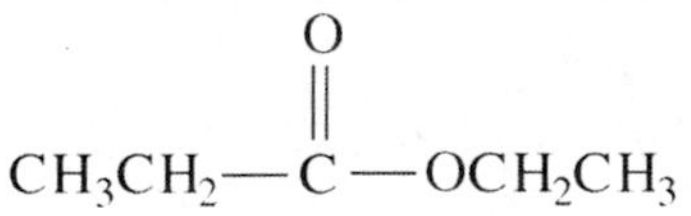

CH$_3$CH$_2$—C(O)—OCH$_2$CH$_3$

26. (a) A nitrile reacts with water to form a carboxylic acid.

CH$_3$CH$_2$COOH

(b) A carboxylic acid chloride reacts with an alcohol to form an ester.

$$CH_3-\overset{\overset{\displaystyle O}{\|}}{C}-OCH_2CH_2CH_3$$

(c) A carboxylic acid reacts with an alcohol to form an ester.

$$H-\overset{\overset{\displaystyle O}{\|}}{C}-OCH_2CH_3$$

27. Simple test to distinguish between:
   (a) Sodium benzoate and benzoic acid: Sodium benzoate is water soluble; benzoic acid is not.
   (b) Maleic acid and malonic acid. Maleic acid has a carbon-carbon double bond, it will readily add and decolorize bromine; malonic acid will not decolorize bromine.

28. Simple tests to distinguish between:
   (a) Benzoic acid and ethyl benzoate; benzoic acid is an odorless solid; ethyl benzoate is a fragrant liquid.
   (b) Succinic acid and fumaric acid: fumaric acid has a carbon-carbon double bond and will readily add and decolorize bromine; succinic acid will not decolorize bromine.

29. (a) $CH_3CH_2COOH + NaOH \longrightarrow CH_3CH_2COO^- Na^+ + H_2O$

   (b) $CH_3CH_2COOH + CH_3OH \longrightarrow CH_3CH_2\overset{\overset{\displaystyle O}{\|}}{C}OCH_3 + H_2O$

30. (a) $CH_3\overset{\overset{\displaystyle CH_3}{|}}{C}HCH_2COOH + SOCl_2 \longrightarrow CH_3\overset{\overset{\displaystyle CH_3}{|}}{C}HCH_2\overset{\overset{\displaystyle O}{\|}}{C}Cl + SO_2 + HCl$

   (b) $CH_3\overset{\overset{\displaystyle CH_3}{|}}{C}HCH_2COOH + CH_3CH_2CH_2OH \longrightarrow CH_3\overset{\overset{\displaystyle CH_3}{|}}{C}HCH_2\overset{\overset{\displaystyle O}{\|}}{C}OCH_2CH_2CH_3 + H_2O$

31. (a) These reagents (NaOH, H_2O, Δ) cause saponification of the ester.

   $CH_3OH + CH_3CH_2COO^- \overset{+}{Na}$

   (b) These reagents (H^+, H_2O) cause hydrolysis of the ester.

   $CH_3CH_2OH + CH_3COOH$

32. (a) These reagents (H^+, H_2O) cause hydrolysis of the ester.

   $CH_3\overset{\overset{\displaystyle OH}{|}}{C}HCH_3 + CH_3CH_2COOH$

(b) These reagents (NaOH, H_2O, Δ) cause saponification of the ester.

$$CH_3OH + CH_3CH_2CH_2CH_2CH_2COO^- \overset{+}{N}a$$

33. (a) $CH_3CH_2CH_2\overset{\displaystyle O}{\overset{\|}{C}}OCH_3 + H_2O \xrightarrow{\text{HCl}} CH_3CH_2CH_2\overset{\displaystyle O}{\overset{\|}{C}}OH + CH_3OH$

(b) $H\overset{\displaystyle O}{\overset{\|}{C}}OCH_2CH_3 + NaOH \xrightarrow[\Delta]{H_2O} H\overset{\displaystyle O}{\overset{\|}{C}}O^- Na^+ + CH_3CH_2OH$

34. (a) $CH_3CH_2\overset{\displaystyle O}{\overset{\|}{C}}OCH_2CH_3 + KOH \xrightarrow[\Delta]{H_2O} CH_3CH_2\overset{\displaystyle O}{\overset{\|}{C}}O^- K^+ + CH_3CH_2OH$

(b) $CH_3CH_2CH_2\overset{\displaystyle O}{\overset{\|}{C}}OCH_2CH_2CH_3 + H_2O \xrightarrow{\text{HCl}} CH_3CH_2CH_2\overset{\displaystyle O}{\overset{\|}{C}}OH + CH_3CH_2CH_2OH$

35. (a) These reagents (H_2O, H^+) cause hydrolysis of the diacylglycerol.

$$\begin{array}{l} CH_2OH \\ | \\ CHOH \quad + \ 2\,CH_3(CH_2)_{10}COOH \\ | \\ CH_2OH \end{array}$$

(b) These reagents (H_2, Ni) reduce the double bond.

$$CH_3(CH_2)_{14}COOH$$

(c) These reagents (H_2, copper chromite, Δ, pressure) cause hydrogenolysis of the triacylglycerol.

$$\begin{array}{l} CH_2OH \\ | \\ CHOH \quad + \ 3\,CH_3(CH_2)_{12}CH_2OH \\ | \\ CH_2OH \end{array}$$

36. (a) These reagents (H_2, Ni) reduce the double bond.

$$CH_3(CH_2)_{16}COOH$$

(b) These reagents (H_2, copper chromite, Δ, pressure) cause hydrogenolysis of the triacylglycerol.

$$\begin{array}{l} CH_2OH \\ | \\ CHOH \quad + \ 3CH_3(CH_2)_{10}CH_2OH \\ | \\ CH_2OH \end{array}$$

(c) These reagents (H_2O, H^+) cause hydrolysis of the triacylglycerol.

$$\begin{array}{l} CH_2OH \\ | \\ CHOH \quad + 3CH_3(CH_2)_{14}COOH \\ | \\ CH_2OH \end{array}$$

37.

$$\begin{array}{l} H_2C-O-\overset{\overset{\displaystyle O}{\|}}{C}(CH_2)_{12}CH_3 \\ | \\ HC-O-\overset{\overset{\displaystyle O}{\|}}{C}(CH_2)_{12}CH_3 \quad + 3\,H_2O \\ | \\ H_2C-O-\overset{\overset{\displaystyle O}{\|}}{C}(CH_2)_{12}CH_3 \end{array} \xrightarrow{H^+} \begin{array}{l} H_2C-OH \\ | \\ HC-OH \\ | \\ H_2C-OH \end{array} + 3\,CH_3(CH_2)_{12}\overset{\overset{\displaystyle O}{\|}}{C}OH$$

38.

$$\begin{array}{l} H_2C-O-\overset{\overset{\displaystyle O}{\|}}{C}(CH_2)_{14}CH_3 \\ | \\ HC-O-\overset{\overset{\displaystyle O}{\|}}{C}(CH_2)_{14}CH_3 \quad + 6\,H_2 \\ | \\ H_2C-O-\overset{\overset{\displaystyle O}{\|}}{C}(CH_2)_{14}CH_3 \end{array} \xrightarrow[\Delta,\ \text{pressure}]{\text{copper chromite}} \begin{array}{l} H_2C-OH \\ | \\ HC-OH \\ | \\ H_2C-OH \end{array} + 3\,CH_3(CH_2)_{14}CH_2OH$$

39. $CH_3(CH_2)_{12}COONa$ would be more useful than $CH_3(CH_2)_{12}COOH$ as a cleansing agent in soft water. Both have a long hydrocarbon chain which would dissolve in the fat, but the acid is not water soluble, whereas the sodium salt is soluble in water.

40. Sodium lauryl sulfate would be more effective as a detergent in hard water than sodium propyl sulfate because the hydrocarbon chain is only three carbons long in the latter, not long enough to dissolve grease well. Sodium lauryl sulfate is effective in both hard and soft water.

41. Only (a), hexadecyltrimethyl ammonium chloride would be a good detergent in water. It is cationic.

42. Only (c) $CH_3(CH_2)_{10}CH_2O(CH_2CH_2O)_7CH_2CH_2OH$, would be a good detergent in water. It is nonionic.

43. (a)

$$HO-\overset{\overset{\displaystyle O}{\|}}{\underset{\underset{\displaystyle OH}{|}}{P}}-OH + HOCH_2CH_3 \xrightarrow{H^+} HO-\overset{\overset{\displaystyle O}{\|}}{\underset{\underset{\displaystyle OH}{|}}{P}}-OCH_2CH_3 + H_2O$$

(b)

$$HO-\overset{\overset{\displaystyle O}{\|}}{\underset{\underset{\displaystyle OH}{|}}{P}}-OCH_2CH_3 + CH_3CH_2CH_2OH \xrightarrow{H^+} HO-\overset{\overset{\displaystyle O}{\|}}{\underset{\underset{\displaystyle OCH_2CH_2CH_3}{|}}{P}}-OCH_2CH_3 + H_2O$$

44. (a)

$$HO-\overset{\overset{\displaystyle O}{\|}}{\underset{\underset{\displaystyle OH}{|}}{P}}-OH + 3\ CH_3OH \xrightarrow{H^+} CH_3O-\overset{\overset{\displaystyle O}{\|}}{\underset{\underset{\displaystyle OCH_3}{|}}{P}}-OCH_3 + 3\ H_2O$$

(b)

$$HO-\overset{\overset{\displaystyle O}{\|}}{\underset{\underset{\displaystyle OH}{|}}{P}}-OH + CH_3OH \xrightarrow{H^+} CH_3O-\overset{\overset{\displaystyle O}{\|}}{\underset{\underset{\displaystyle OH}{|}}{P}}-OH + H_2O$$

$$CH_3O-\overset{\overset{\displaystyle O}{\|}}{\underset{\underset{\displaystyle OH}{|}}{P}}-OH + CH_3CH_2\overset{\underset{\displaystyle OH}{|}}{C}HCH_3 \xrightarrow{H^+} CH_3O-\overset{\overset{\displaystyle O}{\|}}{\underset{\underset{\displaystyle OH}{|}}{P}}-O\overset{\overset{\displaystyle CH_3}{|}}{C}HCH_2CH_3$$

45. Grams and moles of sodium benzoate ($NaC_7H_5O_2$):

$$0.001 \times 1\ \text{lb} \times \frac{454\ \text{g}}{1\ \text{lb}} = 0.5\ \text{g sodium benzoate}$$

$$0.5\ \text{g NaC}_7\text{H}_5\text{O}_2 \times \frac{1\ \text{mol}}{145.1\ \text{g}} = 3 \times 10^{-3}\ \text{mol NaC}_7\text{H}_5\text{O}_2$$

46.

salicylic acid   acetic anhydride   aspirin (acetylsalicylic acid)   acetic acid

47. The ester is propyl propanoate $CH_3CH_2\overset{\underset{\displaystyle O}{\|}}{C}OCH_2CH_2CH_3$

Compound A is an acid; B is an alcohol. If B is oxidized to an acid which is the same as A, then A and B both have the same carbon structure, three carbon atoms each. The acid A must be propanoic acid, CH_3CH_2COOH. The alcohol can be 1-propanol or 2-propanol. Only 1-propanol can be oxidized to propanoic acid which is the same as compound A.

48. Phosgene ($COCl_2$) contains two acid chloride functional groups. When phosgene (acid chlorides) reacts with bisphenol A (alcohols), esters are formed.

49. Polyesters are polymers of molecules linked together by ester bonds. The ester bonds are susceptible to hydrolysis (being broken by reaction with water). When a polyester suture dissolves, the polymer's ester bonds are broken.

50. $$CH_3COOH \rightleftharpoons CH_3COO^- + \overset{+}{H}$$

    98.7%   1.3%

    Of the total concentration of $0.1M$, $(0.013)(0.1M)$ is in the form of acetate. The acetate concentration is $0.0013M$.

51. (a) There are two phosphoric acid esters and no phosphoric acid anhydrides.

    (b) There is one phosphoric acid ester and one phosphoric acid anhydride.

52. Complete hydrogenation will convert the carbon-carbon double bonds to single bonds.

53.
$$CH_2-O-\overset{\overset{\displaystyle O}{\|}}{C}(CH_2)_{10}CH_3$$

$$CH-O-\overset{\overset{\displaystyle O}{\|}}{C}(CH_2)_{14}CH_3$$

$$CH_2-O-\overset{\overset{\displaystyle O}{\|}}{C}(CH_2)_7CH=CH(CH_2)_7CH_3$$

There are two other triacylglycerols containing all three of these acids. Each of the three acids can be attached to the middle carbon of the glycerol.

54. Names and formulas of products are

(a)
CH_2OH $\qquad$ $CH_3(CH_2)_{10}COOH$ $\qquad$ lauric acid

$CHOH$ $\quad$ and $\quad$ $CH_3(CH_2)_{14}COOH$ $\qquad$ palmitic acid

CH_2OH $\qquad$ $CH_3(CH_2)_7CH=CH(CH_2)_7COOH$ $\quad$ oleic acid
glycerol

(b)
CH_2OH $\qquad$ $CH_3(CH_2)_{10}CH_2OH$ $\qquad$ 1-dodecanol (lauryl alcohol)

$CHOH$ $\quad$ and $\quad$ $CH_3(CH_2)_{14}CH_2OH$ $\qquad$ 1-hexadecanol (cetyl alcohol)

CH_2OH $\qquad$ $CH_3(CH_2)_{16}CH_2OH$ $\qquad$ 1-octadecanol (stearyl alcohol)
glycerol

(c)
CH_2OH $\qquad$ $CH_3(CH_2)_{10}COOK$ $\qquad$ potassium laurate

$CHOH$ $\quad$ and $\quad$ $CH_3(CH_2)_{14}COOK$ $\qquad$ potassium palmitate

CH_2OH $\qquad$ $CH_3(CH_2)_7CH=CH(CH_2)_7COOK$ $\quad$ potassium oleate
glycerol

(d)
$$CH_2-O-\overset{\overset{\displaystyle O}{\|}}{C}(CH_2)_{10}CH_3$$

$$CH-O-\overset{\overset{\displaystyle O}{\|}}{C}(CH_2)_{14}CH_3$$

$$CH_2-O-\overset{\overset{\displaystyle O}{\|}}{C}(CH_2)_{16}CH_3$$

lauroylpalmitoylsteroylglycerol

55.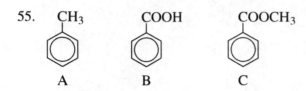

CH₃ ... A

COOH ... B

COOCH₃ ... C

56.

$$CH_3(CH_2)_{12}\overset{O}{\overset{\|}{C}}-OCHCH_3$$

(with CH₃ below the OCH) isopropyl myristate

57. $C_{10}H_{12}O_2 \longrightarrow C_7H_8O + C_3H_6O_2$

A (has a   B (alcohol)   C (acid)
benzene ring)

$$CH_3CH_2\overset{O}{\overset{\|}{C}}-OCH_2\!\!-\!\!\bigcirc$$
A

$\bigcirc\!-\!CH_2OH$
B

CH_3CH_2COOH
C

(1) C cannot be a benzene ring. If C is an acid, it must be propanoic acid. (2) If B is an alcohol and has a benzene ring, it must be benzyl alcohol. Putting B and C together gives the ester A.

58. Smell both samples. Butanoic acid has an unpleasant rancid odor; ethyl butanoate has a pleasant odor of pineapple.

59. Each molecule of triolein requires three molecules of H_2 (one for each double bond)

(a) $(1.00\,kg\ triolein)\left(\frac{10^3\,g}{kg}\right)\left(\frac{1\ mol\ triolein}{885.4\ g\ triolein}\right)\left(\frac{3\ mol\ H_2}{1\ mol\ triolein}\right)\left(\frac{22.4\ L}{1\ mol\ H_2}\right) = 75.9\ L\ H_2$

(b) $(1.00\,kg\ triolein)\left(\frac{891.5\ g\ tristearin}{885.4\ g\ triolein}\right) = 1.01\ kg\ tristearin$

60. The statement is false. When methyl propanoate is hydrolyzed, propanoic acid and methanol are formed.

61. Esters of formula $C_5H_{10}O_2$

$$\overset{O}{\overset{\|}{HC}}-OCH_2CH_2CH_2CH_3$$

$$\overset{O}{\overset{\|}{HC}}-OCH_2\overset{CH_3}{\overset{|}{CH}}CH_3$$

$$\overset{O}{\overset{\|}{HC}}-O\overset{CH_3}{\overset{|}{CH}}CH_2CH_3$$

$$\overset{O}{\overset{\|}{HC}}-O-\overset{CH_3}{\underset{CH_3}{\overset{|}{\underset{|}{C}}}}-CH_3$$

$$CH_3\overset{O}{\overset{\|}{C}}-OCH_2CH_2CH_3$$

$$CH_3\overset{O}{\overset{\|}{C}}-O\overset{CH_3}{\overset{|}{CH}}CH_3$$

$$CH_3CH_2\overset{O}{\overset{\|}{C}}-OCH_2CH_3$$

$$CH_3CH_2CH_2\overset{O}{\overset{\|}{C}}-OCH_3$$

$$CH_3\overset{|}{\underset{CH_3}{CH}}\overset{O}{\overset{\|}{C}}-OCH_3$$

62. (a) Mass percent of oxygen in dacron.

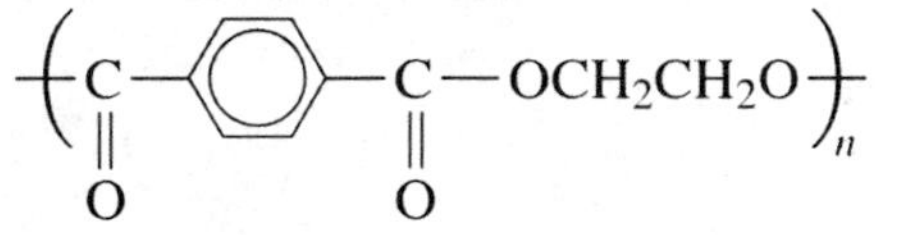

Mass of one unit of dacron = 192.2 g
Mass of oxygen in one unit of dacron = 64.00 g

$$\frac{64.00 \text{ g O}}{192.2 \text{ g}} \times 100 = 33.30\% \text{ O}$$

(b) Molar mass of 105 units

$$(105 \text{ units})\left(\frac{192.2 \text{ g}}{\text{unit}}\right) = 2.02 \times 10^4 \text{ g}$$

63. An alkyd polyester would most likely be thermosetting because glycerol is trifunctional and would thus allow cross linking between chains in forming the ester polymer.

64. Synthesis of:

(a) acetic acid

$$CH_3CH_2OH \xrightarrow[H_2SO_4]{K_2Cr_2O_7} CH_3COOH$$

(b) ethyl acetate

$$CH_3COOH + CH_3CH_2OH \underset{}{\overset{H+}{\rightleftharpoons}} CH_3\overset{O}{\overset{\|}{C}}OCH_2CH_3$$

(c) β-hydroxybutyric acid

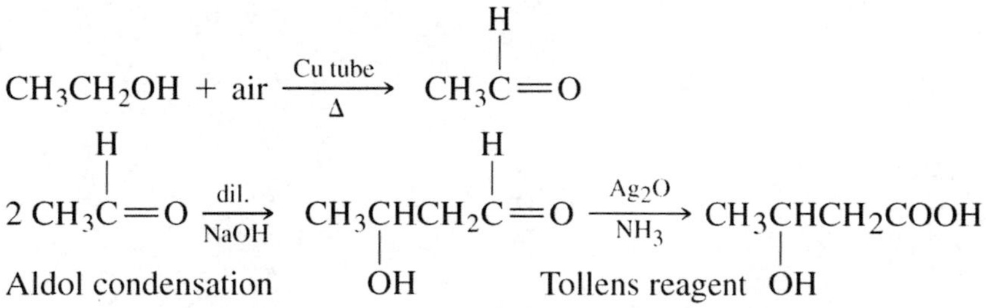

$$CH_3CH_2OH + air \xrightarrow[\Delta]{Cu\ tube} CH_3\overset{H}{\underset{}{C}}=O$$

$$2\ CH_3\overset{H}{C}=O \xrightarrow[NaOH]{dil.} CH_3\underset{OH}{CHCH_2}\overset{H}{C}=O \xrightarrow[NH_3]{Ag_2O} CH_3\underset{OH}{CHCH_2}COOH$$

Aldol condensation   Tollens reagent

65. (a) 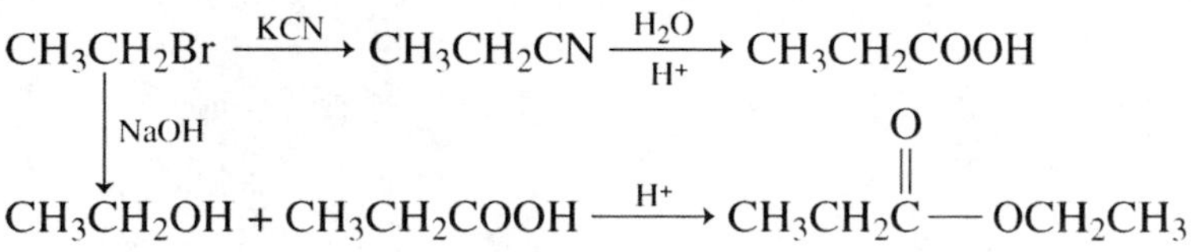 $CH_3CH_2Br \xrightarrow{KCN} CH_3CH_2CN \xrightarrow[H+]{H_2O} CH_3CH_2COOH$

$\downarrow NaOH$

$CH_3CH_2OH + CH_3CH_2COOH \xrightarrow{H+} CH_3CH_2\overset{O}{\overset{\|}{C}}OCH_2CH_3$

(b) $CH_3CH_2Br \xrightarrow{NaOH} CH_3CH_2OH \xrightarrow[H^+]{K_2Cr_2O_7} CH_3\overset{\overset{\displaystyle O}{\|}}{C}-H$

$CH_3\overset{\overset{\displaystyle O}{\|}}{C}-H + CH_3\overset{\overset{\displaystyle O}{\|}}{C}-H \xrightarrow[NaOH]{dilute} CH_3\underset{\underset{\displaystyle OH}{|}}{CH}CH_2\overset{\overset{\displaystyle O}{\|}}{C}-H$

Aldol condensation

$\xrightarrow{-H_2O}$

$CH_3CH_2CH_2CH_2OH \xleftarrow[Ni]{H_2} CH_3CH=CH\overset{\overset{\displaystyle O}{\|}}{C}-H$

66.

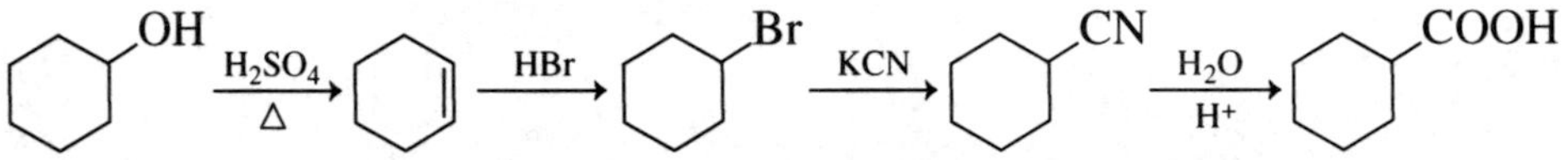

67. This polymer is formed as acids react with alcohols. Each reaction forms an ester splitting out a water molecule in the process. Thus, this is a condensation reaction.

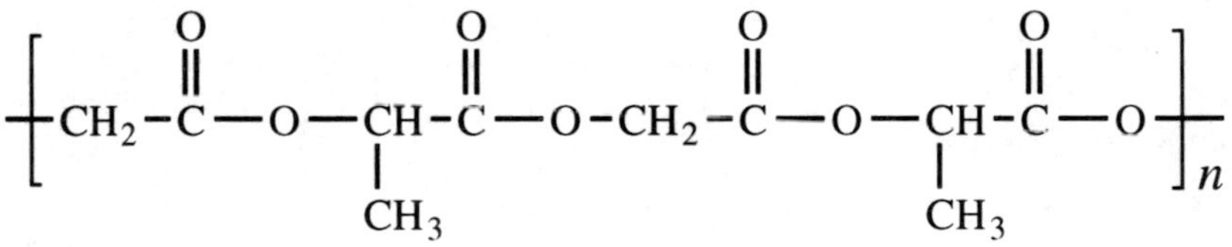

This polymer has been formed with alternating glycolic acid and lactic acid monomers. Other arrangements are possible.

CHAPTER 25

AMIDES AND AMINES: ORGANIC NITROGEN COMPOUNDS

SOLUTIONS TO REVIEW QUESTIONS

1. The lowest boiling point amide in the table is methanamide (formamide), b.p. = 210°C.

$$\overset{\overset{\displaystyle O}{\|}}{HC}-NH_2$$

The highest boiling point amide is *N*-phenylethanamide (acetanilide), b.p. = 304°C.

$$CH_3\overset{\overset{\displaystyle O}{\|}}{C}-\overset{\overset{\displaystyle H}{}}{N}-\bigcirc$$

The larger size of *N*-phenylethanamide contributes significantly to its higher boiling point.

2. Procaine contains an amine. When the amine reacts with HCl it forms a salt that is much more soluble in water.

3. (a) With reference to Section 25.6, acetylcholine contains a quaternary (4°) amine.
   (b) With reference to Section 25.8, cocaine contains a tertiary (3°) amine.
   (c) With reference to Section 25.7, serotonin contains a primary (1°) and a secondary (2°) amine.
   (d) With reference to Section 25.8, methamphetamine contains a secondary (2°) amine.

4. Amides: Unsubstituted amides (except formamide) are solids at room temperature. Many are odorless and colorless. Low molar-mass amides are water soluble. Solubility in water decreases as the molar mass increases. Amides are neutral compounds. The NH_2 group is capable of hydrogen bonding.

   Amines: Low molar-mass amines are flammable gases with an ammonia-like odor. Aliphatic amines up to six carbon atoms are water soluble. Many amines have a "fishy" odor and many have very foul odors. Aromatic amines occur as liquids and solids. Soluble aliphatic amines give basic solutions. Aromatic amines are less soluble in water and less basic than aliphatic amines. The NH_2 group is capable of hydrogen bonding.

5. Unlike esters, unsubstituted amides can hydrogen bond to each other. Thus, they have a higher melting point than esters of similar molar mass.

6. (a) Heterocyclic compounds are those in which all the atoms in the ring are not alike.
   (b) The number of heterocyclic rings in each of the compounds is:
      (i) purine, 2   (ii) histamine, 1   (iii) methadone, 0   (iv) nicotine, 2

7. Ammonia is a toxic, basic, water-soluble compound which can increase the pH of the blood and the urine and would be painful to pass through bodily tissues. However, ammonia is converted in the liver to the neutral diamide, urea, which is water soluble and is excreted in the urine.

8. A "condensation polymer" is made when monomers connect together with the loss of a small molecule, most often, water.

$$HOOC-(CH_2)_4-COOH + H_2N-(CH_2)_6-NH_2 \longrightarrow$$

$$\left[\begin{matrix} O \\ \parallel \\ C \end{matrix} -(CH_2)_4- \begin{matrix} O \\ \parallel \\ C \end{matrix} -NH-(CH_2)_6-NH \right]_n + H_2O$$

9. The nitrogen in a compound that has four groups bonded to it is positively charged and is called a quaternary ammonium nitrogen. The compound is called a quaternary ammonium salt.

10. An amine must have at least one hydrogen atom bonded to the nitrogen atom to react with an acid chloride. A tertiary amine does not meet this requirement.

11. The nitrogen in cocaine is an amine nitrogen. If it was an amide nitrogen it would be bonded to a carbonyl group.

12. The name of the compound is *N*-methyl-2-methylbutanamine. Methyl (a) is included in the name butamine; methyl (b) is part of 2-methyl; methyl (c) is included in *N*-methyl.

SOLUTIONS TO EXERCISES

1. (a)

$$\text{(benzene ring)}-\underset{\underset{O}{\|}}{C}-NH_2$$

(c) $CH_3CH_2-\underset{\underset{O}{\|}}{C}-NHCH_3$

(b) $CH_3CH_2\underset{\underset{CH_3}{|}}{\overset{\overset{CH_3}{|}}{C}}-\underset{\underset{O}{\|}}{C}-NH_2$

(d) $CH_3\underset{\overset{CH_3}{|}}{CH}-\underset{\underset{O}{\|}}{C}-NH_2$

2. (a) $CH_3CH_2CH_2CH_2-\underset{\underset{O}{\|}}{C}-\underset{\overset{CH_3}{|}}{N}-CH_3$

(b) $CH_3-\underset{\underset{O}{\|}}{C}-NH_2$

(c) $CH_3CH_2-\underset{\overset{CH_3}{|}}{CH}-\underset{\overset{CH_3}{|}}{CH}-\underset{\underset{O}{\|}}{C}-NH_2$

(d) $\text{(benzene ring)}-\underset{\underset{O}{\|}}{C}-NH-CH_2CH_3$

3. (a) common, formamide; IUPAC, methanamide
   (b) IUPAC, *N,N*-dimethyl-3,3-dimethylbutanamide
   (c) IUPAC, *N*-butylethanamide

4. (a) IUPAC, benzamide
   (b) IUPAC, *N,N*-dibutylethanamide
   (c) common, *N*-methylacetamide, IUPAC, *N*-methylethanamide

5. The larger amides are less water soluble. Thus, the largest amide, *N,N*-diethyl-3-methylhexanamide is the least water soluble. The intermediate sized amide, 3-methylhexanamide will be of intermediate solubility and the smallest amide, acetamide, will be the most water soluble.

6. The larger amides are less water soluble. Thus, the largest amide, benzamide, is the least water soluble. The intermediate sized amide, propanamide, will be of intermediate solubility and the smallest amide, formamide, will be the most water soluble.

7.

8.

9. Organic products

(a) $CH_3\overset{O}{\underset{||}{C}}-OH + CH_3NH_3^+$　　　　(b) $CH_3\overset{O}{\underset{||}{C}}-NHCH(CH_3)_2$

(c) $H_2NCH_2CH_2CH_2CH_2COONa$

10. Organic products

(a)
$$
\underset{CH_3C-N(CH_2CH_3)_2}{\overset{O}{\parallel}}
$$

(b)
$$
\text{C}_6\text{H}_5\text{CH}_2\text{COOH} \quad + \quad CH_3\overset{+}{N}H_2CH_3
$$

(c)
$$
\underset{\overset{|}{CH_3}}{CH_3CHCH_2}\overset{O}{\overset{\parallel}{C}}-NHCH_3
$$

11. (a)
$$
CH_3CH_2CH_2\overset{O}{\overset{\parallel}{C}}-OH \; + \; NH_2CH_3
$$

(b)
$$
CH_3CH_2\overset{O}{\overset{\parallel}{C}}-OH \; + \; NH_2CH_2CH_3
$$

(c)
$$
H\overset{O}{\overset{\parallel}{C}}-NH-C_6H_{11} \; + \; HCl \; + \; H_2O
$$

12. (a)
$$
CH_3\overset{O}{\overset{\parallel}{C}}-NHCH_2CH_3 \; + \; NaOH
$$

(b)
$$
CH_3CH_2\overset{O}{\overset{\parallel}{C}}-OH \; + \; CH_3NHCH_3
$$

(c)
$$
CH_3\overset{O}{\overset{\parallel}{C}}-NH_2 \; + \; HCl \; + \; H_2O
$$

13. Structures of amines with formula $C_4H_{11}N$.

$$
\underset{1^\circ}{CH_3CH_2CH_2CH_2NH_2} \qquad \underset{\underset{1^\circ}{\overset{|}{CH_3}}}{CH_3CHCH_2NH_2} \qquad \underset{\underset{1^\circ}{\overset{|}{NH_2}}}{CH_3CH_2CHCH_3}
$$

$$
\underset{\underset{1^\circ}{\overset{|}{CH_3}}}{\overset{\overset{CH_3}{|}}{CH_3CNH_2}} \qquad \underset{2^\circ}{CH_3CH_2CH_2NHCH_3} \qquad \underset{2^\circ}{(CH_3)_2CHNHCH_3}
$$

$$
\underset{2^\circ}{CH_3CH_2NHCH_2CH_3} \qquad \underset{\underset{3^\circ}{\overset{|}{CH_3}}}{\overset{\overset{CH_3}{|}}{CH_3CH_2NCH_3}}
$$

14. Structures of amines with formula C_3H_9N.

$$CH_3CH_2CH_2NH_2 \qquad CH_3\underset{\underset{\displaystyle NH_2}{|}}{C}HCH_3 \qquad CH_3CH_2NHCH_3 \qquad CH_3\underset{\underset{\displaystyle CH_3}{|}}{N}CH_3$$

$\qquad\quad 1° \qquad\qquad\qquad 1° \qquad\qquad\qquad\quad 2° \qquad\qquad\qquad\quad 3°$

15. Classification of amines

    (a)   primary   (b)   tertiary   (c)   primary   (d)   both primary

16. Classification of amines

    (a)   secondary   (b)   tertiary   (c)   secondary   (d)   tertiary

17. The triethylamine solution in 1.0 M NaOH would have the more objectionable odor because it would be in the form of the free amine, while in the acid solution the amine would form a salt that will have little or no odor.

18. The isopropylamine solution in 1.0 M KOH would have the more objectionable odor because it would be in the form of the free amine, while in the acid solution the amine would form a salt that will have little or no odor.

19. Names

    (a) CH_3NHCH_3 dimethylamine

    (b) 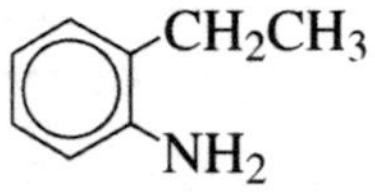   o-ethylaniline

    (c) 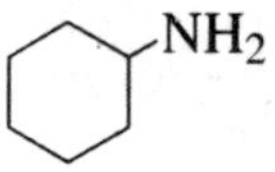   cyclohexanamine

    (d) $(C_2H_5)_4N^+ I^-$ tetraethylammonium iodide

    (e) 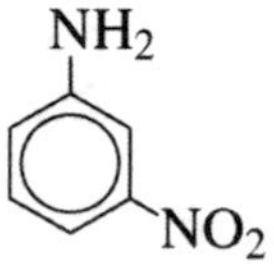   m-nitroaniline

    (f) $(CH_3CH_2)_2N$— 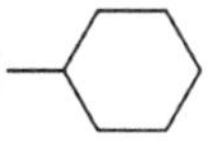   cyclohexyldiethylamine

20. Names

    (a) 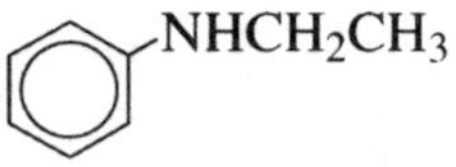   N-ethylaniline

(b) $CH_3CH_2CHCH_3$
$\quad\quad\quad\quad|$
$\quad\quad\quad\quad NH_2$

2-butanamine

(c)

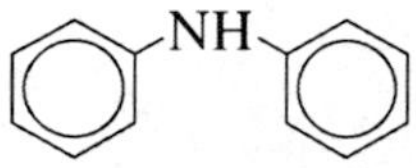

diphenylamine

(d) $CH_3CH_2NH_3^+Br^-$

ethylammonium bromide

(e)

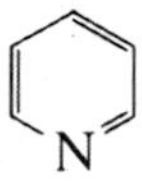

pyridine

(f) $CH_3\overset{\overset{\displaystyle O}{\|}}{C}-NHCH_2CH_3$

N-ethylacetamide

21. Structural formulas

(a) $CH_3CH_2NHCH_3$

(b)

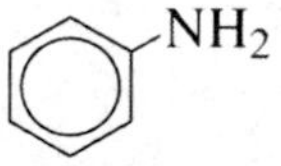

(c) $H_2NCH_2CH_2CH_2CH_2NH_2$

(d) $CH_3CH_2NCH(CH_3)_2$
$\quad\quad\quad\quad|$
$\quad\quad\quad\quad CH_3$

(e)

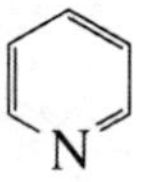

(f) $(CH_3CH_2)_3\overset{+}{N}HCl^-$

22. Structural formulas

(a) $CH_3CH_2CH_2CH_2NCH_2CH_2CH_2CH_3$
$\quad\quad\quad\quad\quad\quad\quad\quad|$
$\quad\quad\quad\quad\quad\quad\quad CH_2CH_2CH_2CH_3$

(b)

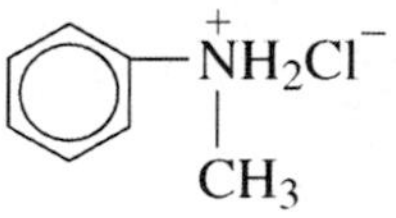

(c) $CH_3CH_2 \overset{+}{N}H_3Cl^-$

(d) 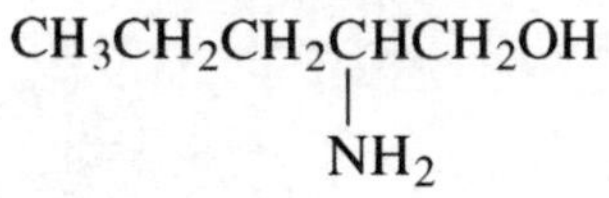 $CH_3CH_2CH_2CHCH_2OH$
$\qquad\qquad\quad |$
$\qquad\qquad\ NH_2$

(e) 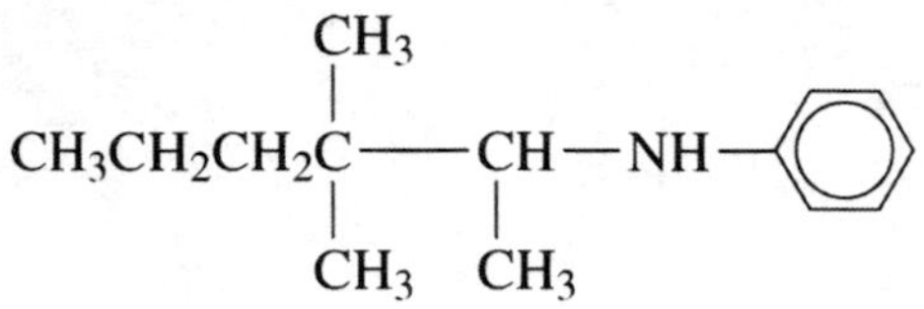 $CH_3CH_2CH_2\overset{\displaystyle CH_3}{\underset{\displaystyle CH_3}{\overset{|}{\underset{|}{C}}}}{-}\overset{}{\underset{\displaystyle CH_3}{\overset{|}{\underset{|}{CH}}}}{-}NH{-}\bigcirc$

(f) $H_2NCH_2CH_2NH_2$

23. (a) Butylamine has the following structure:
$CH_3CH_2CH_2CH_2NH_2$
Its IUPAC name is 1-butanamine.

(b) *tert*-butylamine has the following structure:

$CH_3-\overset{\displaystyle CH_3}{\underset{\displaystyle CH_3}{\overset{|}{\underset{|}{C}}}}-NH_2$

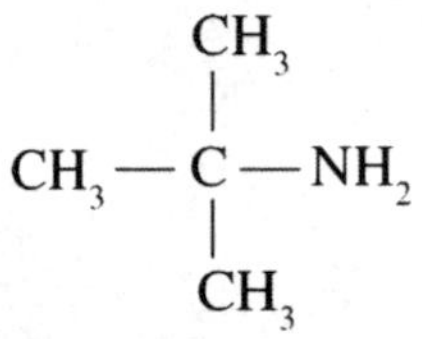

Its IUPAC name is 2-methyl-2-propanamine.

(c) Formamide has the following structure:

$\overset{\displaystyle O}{\overset{\|}{HC}}-NH_2$

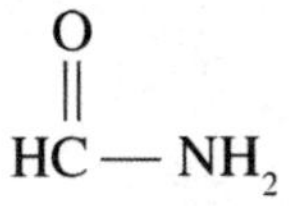

Its IUPAC name is methanamide.

24. (a) Isopropylamine has the following structure:

$\overset{\displaystyle CH_3}{\overset{|}{CH_3CHNH_2}}$

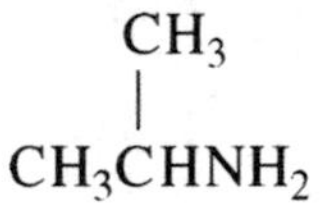

Its IUPAC name is 2-propanamine.

(b) Diethylamine has the following structure:

$CH_3CH_2NHCH_2CH_3$
Its IUPAC name is N-ethylethanamine.

(c) Acetamide has the following structure:

$\overset{\displaystyle O}{\overset{\|}{CH_3C}}-NH_2$

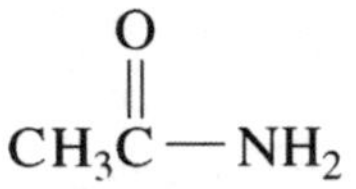

Its IUPAC name is ethanamide.

25. (a) This reaction adds methyl groups to a primary amine to form a tertiary amine.

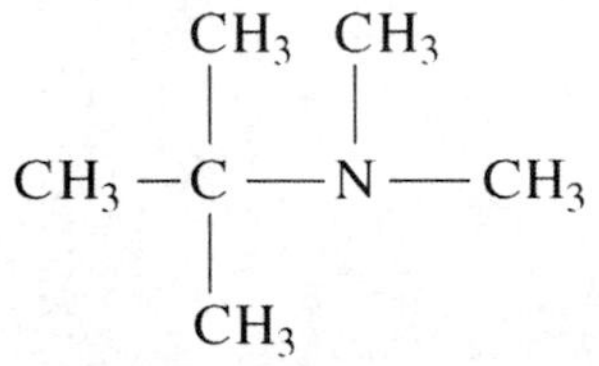

$$CH_3 - \underset{\underset{CH_3}{|}}{\overset{\overset{CH_3}{|}}{C}} - \underset{\overset{|}{CH_3}}{N} - CH_3$$

N,N-dimethyl-2-methyl-2-propanamine

(b) This reaction reduces the nitrile to form an amine.

CH_3NH_2

menthanamine (methylamine)

(c) This reaction reduces the amide to form an amine.

$CH_3CH_2NHCH_2CH_3$

N-ethylethanamine (diethylamine)

26. (a) This reaction reduces an amide to form an amine.

$(CH_3)_3N$

N,N-dimethylmethanamine (trimethylamine)

(b) This reaction reduces a nitro group to form a substituted aniline.

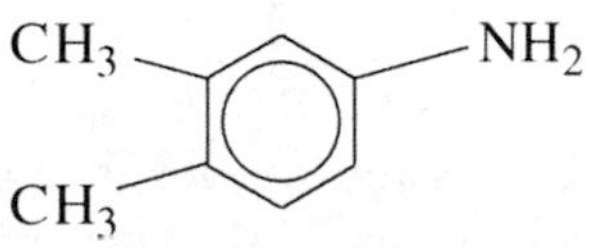

3,4-dimethylaniline

(c) This reaction adds a methyl group to a primary amine to form a secondary amine.

$CH_3CH_2NHCH_3$

N-methylethanamine (methylethylamine)

27. Organic molecule A

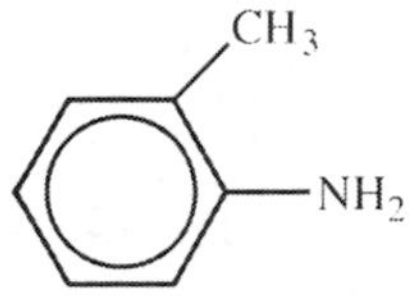

o-methylaniline

Organic molecule B

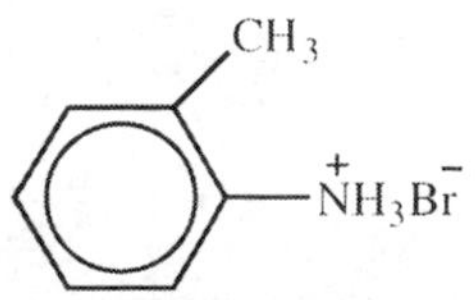

o-methylanilinium bromide

28. Organic molecule A

$$H_3C-\underset{\underset{CH_3}{|}}{CH}-CH_2-NH_2$$

2-methylpropanamine

Organic molecule B

$$H_3C-\underset{\underset{CH_3}{|}}{CH}-CH_2-\overset{+}{N}H_3\overset{-}{O}H$$

2-methylpropanammonium hydroxide

29. Amines are bases while carboxylic acids and phenols are acids. Alcohols and amides are neither acids nor bases.

(a) amine, base	(b) amide, neither	(c) amide, neither
(d) carboxylic acid, acid	(e) alcohol, neither	(f) alcohol, neither

30. Amines are bases while carboxylic acids and phenols are acids. Alcohols and amides are neither acids nor bases.

(a) alcohol, neither	(b) carboxylic acid, acid	(c) amide, neither
(d) amine, base	(e) amine, base	(f) phenol, acid

31. The main reason why trimethylamine has a lower boiling point than propylamine and ethylmethylamine is that trimethylamine cannot hydrogen bond because it has no hydrogen atoms bonded to the nitrogen atom. The other two amines do have hydrogen atom(s) bonded to the nitrogen atom and their molecules can hydrogen bond, which results in higher boiling points.

32. Biogenic amines are derived from amino acids and act as neurotransmitters and hormones in animals. In contrast, alkaloids are basic compounds derived from plants that show physiological activity.

33. This calculation requires Avogadro's number, 6.02×10^{23} dopamine molecules/mole.

$$(5 \times 10^{-18} \text{ moles})(6.02 \times 10^{23} \text{ dopamine molecules/mole}) = 3 \times 10^6 \text{ dopamine molecules}$$

About three million dopamine molecules are sent from one neuron to the next.

34. (a)

$$\underset{}{CH_3\overset{\overset{O}{\|}}{C}-Cl} + 2CH_3NH_2 \longrightarrow CH_3\overset{\overset{O}{\|}}{C}-NHCH_3 + CH_3\overset{+}{N}H_3\,\overset{-}{C}l$$

(b) $CH_3\overset{\overset{O}{\|}}{C}-NHCH_3 \xrightarrow{\text{LiAlH}_4} CH_3CH_2NHCH_3$

(c) $CH_3CH_2NHCH_3 + CH_3Br(1 \text{ mole}) \longrightarrow CH_3CH_2N(CH_3)_2$

35. There is one amine and two amides in ampicillin.

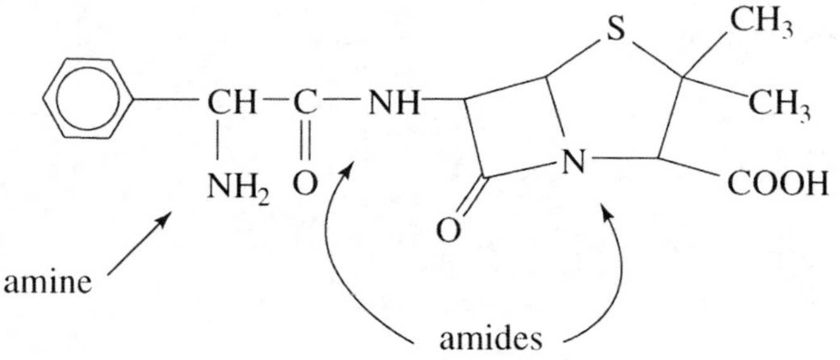

36. Lemon juice, being acidic, will react with the basic amines forming salts, which are soluble and can be washed away with water.

37. 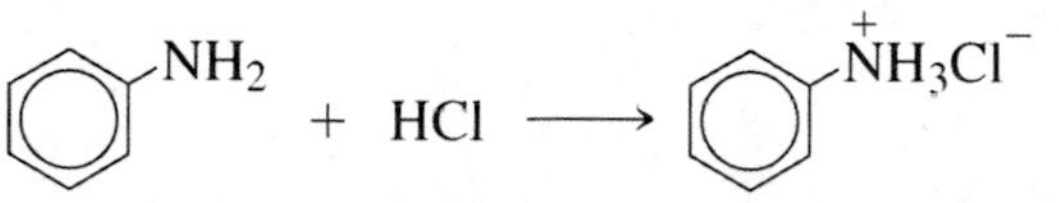

38. $H_2N—CH_2CH_2CH_2CH_2—NH_2$

Putrescine has two primary amines.

39. Drugs are given as ammonium salts because the salts are soluble in water.

40. 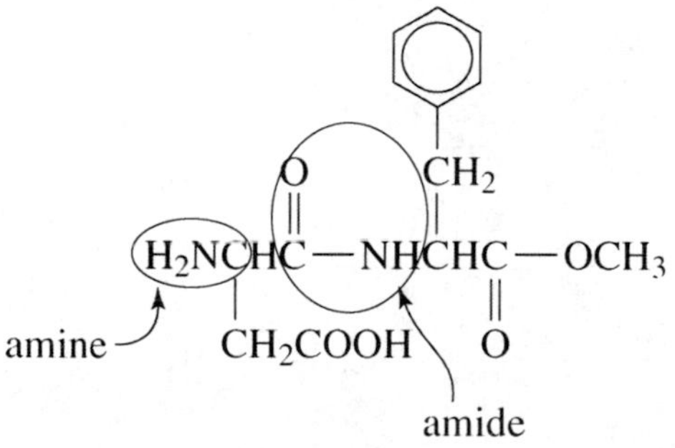

Aspartame

41. 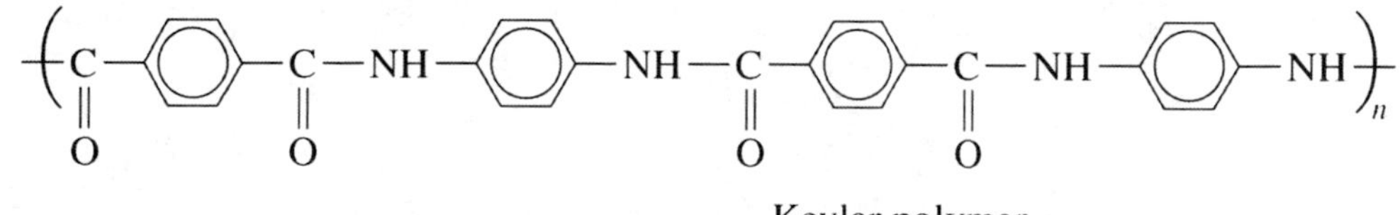

Kevlar polymer

42. (a) $HO—\overset{O}{\overset{\|}{C}}—(CH_2)_8\overset{O}{\overset{\|}{C}}—OH$ and $H_2N(CH_2)_6NH_2$

(b) 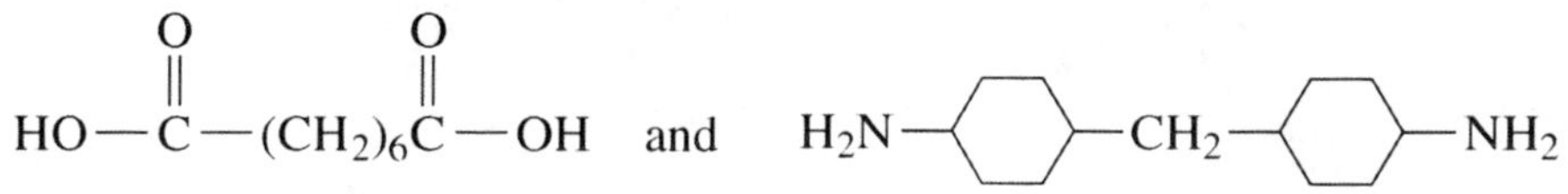

43. (a) $CH_3CH_2OH \xrightarrow[140°C]{H_2SO_4} CH_2=CH_2 \xrightarrow{Br_2} BrCH_2CH_2Br$

$\xrightarrow{KCN} NCCH_2CH_2CN \xrightarrow[Ni]{H_2} H_2NCH_2CH_2CH_2CH_2NH_2$

1, 4-butanediamine

(b) $CH_3\overset{O}{\underset{||}{C}}{-}Cl + NH_3 \longrightarrow CH_3\overset{O}{\underset{||}{C}}{-}NH_2 \xrightarrow{LiAlH_4} CH_3CH_2NH_2$

$CH_3\overset{O}{\underset{||}{C}}{-}Cl + CH_3NH_2 \longrightarrow CH_3\overset{O}{\underset{||}{C}}{-}NHCH_3 \xrightarrow{LiAlH_4} CH_3CH_2NHCH_3$

$CH_3\overset{O}{\underset{||}{C}}{-}Cl + (CH_3)_2NH \longrightarrow CH_3\overset{O}{\underset{||}{C}}{-}N(CH_3)_2 \xrightarrow{LiAlH_4} CH_3CH_2N(CH_3)_2$

CHAPTER 26

STEREOISOMERISM

SOLUTIONS TO REVIEW QUESTIONS

1. A chiral carbon atom is one to which four different atoms or groups are attached and is a center of asymmetry in a molecule. In the following three compounds, the chiral carbon atoms are marked with an asterisk. (These are merely three examples; there are an infinite number of compounds which contain one chiral carbon atom.)

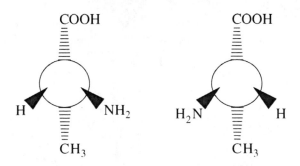

$$CH_3\overset{*}{C}H(OH)CH_2OH$$

2. When the axes of two pieces of polaroid film are parallel, you have maximum brightness of the light passing through both. When one piece has been rotated by 90° the polaroid appears black, indicating very little light passing through.

3. A necessary and sufficient condition for a compound to show enantiomerism is that the compound not be superimposable on its mirror image.

4. Enantiomers are nonsuperimposable mirror image isomers. Diastereomers are stereoisomers that are not enantiomers (not mirror image isomers).

5. Two enantiomers of the amino acid, alanine:

6. (±)-tartaric acid is an equal mixture of the two enantiomers, (+)-tartaric acid and (−)-tartaric acid and, so, has no net optical activity.

(−)-tartaric acid   (+)-tartaric acid

Meso-tartaric acid is a structure that is superimposable on its own mirror image and, thus, has no optical activity.

meso-tartaric acid

7. Each chiral carbon allows for two stereoisomers. The compound has three chiral carbons (marked with * as follows):

$$CH_3\overset{*}{C}HCl\overset{*}{C}HBrCH_2CCl_2\overset{*}{C}HClCH_3$$

The number of stereoisomers $= 2^n$, where n = number of chiral carbons.
For this compound, number of stereoisomers $= 2^3 = 8$.

8. Physical properties of a pair of enantiomers

	(+) 2-methyl-l-butanol	(−) 2-methyl-l-butanol
specific rotation	+5.76°	−5.76°
boiling point	129°C	129°C
density	0.819 g/mL	0.819 g/mL

9. Most chiral molecules are stereospecific in their biological activity. Therefore a racemic mixture of a drug provides only half the bioactive material prescribed. By using a single isomer of a compound, the dosage can be cut in half and possible side effects can be avoided from its enantiomer.

10. The specific rotation for the racemic mixture will be zero.

11. Run a sample of each compound in a polarimeter, which will show you which is (+) and which is (−) lactic acid.

12. If the following compound were meso it would not show optical rotation.

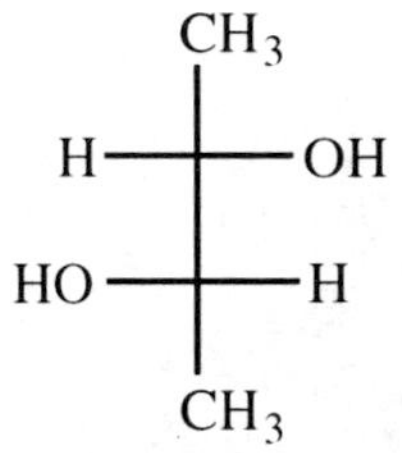

$$
\begin{array}{c}
CH_3 \\
H \!-\!\!\!-\!\!\!-\!\!\!- OH \\
HO \!-\!\!\!-\!\!\!-\!\!\!- H \\
CH_3
\end{array}
$$

13. Glucose has four chiral carbon atoms.

   $2^n = 2^4 = 16$ optical isomers of glucose

14. (a) and (d) are meso compounds. For (a), make two interchanges of the groups on carbons 2 or 3 and you have a structure which is meso. (d) is meso as written. There is a horizontal plane of symmetry through H and Br.

SOLUTIONS TO EXERCISES

1. specific rotation $[\alpha] = \dfrac{\text{observed rotation in degrees}}{\left(\begin{array}{c}\text{length of}\\\text{sample tube in decimeters}\end{array}\right)\left(\begin{array}{c}\text{sample concentration in}\\\text{grams per milliliter}\end{array}\right)}$

$$[\alpha] = \frac{+156°}{(20\,\text{dm})(0.55\,\text{g/mole})} = 14°$$

2. specific rotation $[\alpha] = \dfrac{\text{observed rotation in degrees}}{\left(\begin{array}{c}\text{length of}\\\text{sample tube in decimeters}\end{array}\right)\left(\begin{array}{c}\text{sample concentration in}\\\text{grams per milliliter}\end{array}\right)}$

$$[\alpha] = \frac{-216°}{(1\,\text{dm})(2.35\,\text{g/mole})} = -92°$$

3. Chiral objects can't be superimposed on their mirror image. The following are chiral: (a) the letter p; (c) a spiral staircase; (d) a left foot.

4. Chiral objects can't be superimposed on their mirror image. The following are chiral: (b) a human body; (c) a 1/4″ bolt; (d) the letter n.

5. Chiral carbons are bonded to four different groups. (a) one chiral carbon; (b) two chiral carbons; (c) two chiral carbons; (d) two chiral carbons.

6. Chiral carbons are bonded to four different groups. (a) no chiral carbons; (b) two chiral carbons; (c) one chiral carbon; (d) three chiral carbons.

7. (a) 2,4-Dibromohexane has two chiral carbons and is optically active.

$$\begin{array}{cc}\text{Br} & \text{Br}\\ | & |\\ \end{array}$$
$$\text{CH}_3\text{CHCH}_2\text{CHCH}_2\text{CH}_3$$

(b) 2,3-Dimethylhexane has one chiral carbon and is optically active.

$$\begin{array}{c}\text{CH}_3\\ |\\ \end{array}$$
$$\text{CH}_3\text{CHCHCH}_2\text{CH}_2\text{CH}_3$$
$$\begin{array}{c} |\\ \text{CH}_3\end{array}$$

(c) 3-chlorohexane has one chiral carbon and is optically active.

$$\begin{array}{c}\text{Cl}\\ |\\ \end{array}$$
$$\text{CH}_3\text{CH}_2\text{CHCH}_2\text{CH}_2\text{CH}_3$$

(d) 2,5-Dimethylhexane has no chiral carbons and is not optically active.

$$CH_3 \qquad CH_3$$
$$\quad | \qquad\qquad |$$
$$CH_3CHCH_2CH_2CHCH_3$$

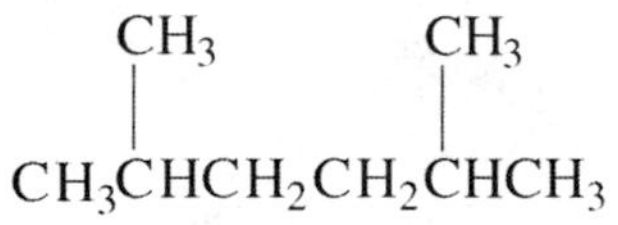

8. (a) 2,3,5-Trimethylhexane has one chiral carbon and is optically active.

$$CH_3 \qquad CH_3$$
$$\quad | \qquad\qquad |$$
$$CH_3CHCHCH_2CHCH_3$$
$$\qquad |$$
$$\qquad CH_3$$

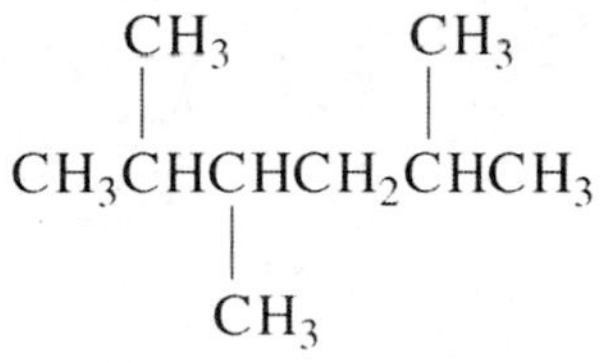

(b) 1-Iodohexane has no chiral carbons and is not optically active.

$$CH_2ICH_2CH_2CH_2CH_2CH_3$$

(c) 3-Ethylhexane has no chiral carbons and is not optically active.

$$CH_2CH_3$$
$$\quad |$$
$$CH_3CH_2CHCH_2CH_2CH_3$$

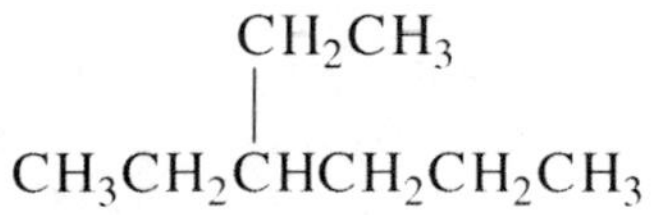

(d) 3-Chloro-2-iodohexane has two chiral carbons and is optically active.

$$CH_3CHICHClCH_2CH_2CH_3$$

9. (a) The sugar ribose has three chiral carbons (marked with *).

$$O$$
$$\|$$
$$CH$$
$$|$$
$$*CHOH$$
$$|$$
$$*CHOH$$
$$|$$
$$*CHOH$$
$$|$$
$$CH_2OH$$

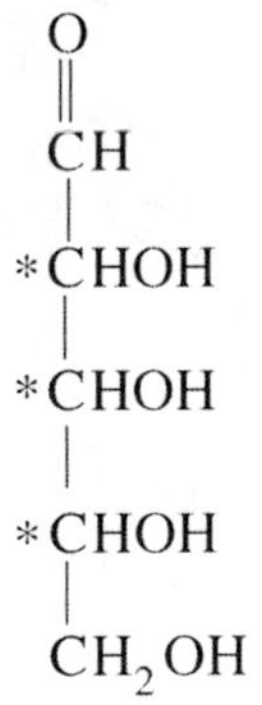

(b) The metabolite oxaloacetic acid has no chiral carbons.

(c) The amino acid threonine has two chiral carbons (marked with *).

$$CH_3$$
$$\quad |$$
$$*CHOH$$
$$\quad |$$
$$H_2N-CH-COOH$$
$$\qquad *$$

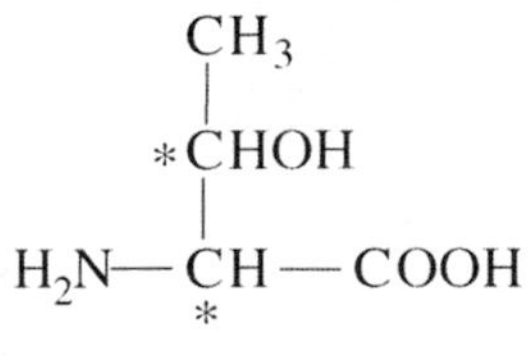

10. (a) The sugar 2-deoxyribose has two chiral carbons (marked with *).

```
       O
       ||
       CH
       |
       CH₂
       |
     *CHOH
       |
     *CHOH
       |
       CH₂OH
```

(b) The metabolite malic acid has one chiral carbon (marked with *).

```
       COOH
        |
      *CHOH
        |
       CH₂
        |
       COOH
```

(c) The amino acid isoleucine has two chiral carbons (marked with *).

```
        CH₃
         |
        CH₂
         |
       *CH—CH₃
         |
  H₂N——CH—COOH
         *
```

11. (a)
```
        CH₃
        *|
   H———————NH₂
         |
       CH₂CH₃
```
(c)
```
        COOH
         |
   CH₃———————NHCH₃
         |
        CH₃
```

(b)
```
        COOH
         |
   HO———————COOH
         |
         H
```
(d)
```
       CH₂CH₂CH₃
        *|
   H———————Cl
         |
        CH₃
```

12. (a)

CH₂NH₂
|
H—*|—CH₃
|
NHCH₃

(c)

COOH
|
CH₃—*|—OH
|
CH₂CH₃

(b)

CH₃
|
HO——CH₃
|
CH₃

(d)

CH₂CH₂CH₃
|
Cl—*|—CH₂CH₃
|
CH₃

13. The two projection formulas (A) and (B) are the same compound, for it takes two changes to make (B) identical to (A).

Cl
|
Br—C—H
|
F
(A)

Br
|
H—C—Cl
|
F
(B)

H
|
Br—C—Cl
|
F
1st change in (B)
(H and Br)

Cl
|
Br—C—H
|
F
2nd change in (B)
(H and Cl)

14. The two projection formulas (A) and (B) are the same compound, for it takes two changes to make (B) identical to (A).

H
|
Br—C—F
|
CH₃
(A)

CH₃
|
Br—C—H
|
F
(B)

H
|
Br—C—CH₃
|
F
1st change in (B)
(H and CH₃)

H
|
Br—C—F
|
CH₃
2nd change in (B)
(F and CH₃)

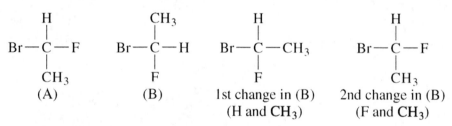

15. (−)-lactic acid is

COOH
|
H—C—OH
|
CH₃

(a), (e), and (f) are (−)-lactic acid

(+)-lactic acid is

COOH
|
HO—C—H
|
CH₃

(b), (c), and (d) are (+)-lactic acid

16. (+)-alanine is

COOH
|
H₂N————H
|
CH₃

(b), (c), and (d) are (+)-alanine

(−)-alanine is

COOH
|
H————NH₂
|
CH₃

(a), (e), and (f) are (−)-alanine

17.

```
           CH₂OH
           |
    HO —————— H
           |
    H —————— Br
           |
    HO —————— H
           |
           CH₃
```

18.

```
           CH₂OH
           |
    Cl —————— H
           |
    H —————— Br
           |
    H —————— Cl
           |
          COOH
```

19. Diastereomers are all stereoisomers that are not mirror images of

```
        COOH                          COOH            COOH
        |                             |               |
  H₂N — C — H     diastereomers =  H — C — NH₂   H₂N — C — H
        |                             |               |
    H — C — OH                    H — C — OH    HO — C — H
        |                             |               |
       COOH                          COOH            COOH
```

20. Diastereomers are all stereoisomers that are not mirror images of

```
        COOH                         COOH           COOH
        |                            |              |
   H — C — Cl     diastereomers =  H — C — Cl   Cl — C — H
        |                            |              |
   H — C — Cl                   Cl — C — H     H — C — Cl
        |                            |              |
       COOH                         COOH           COOH
```

21. Each chiral carbon allows for two stereoisomers. Galactose has four chiral carbons.
    number of stereoisomers = 2^n where n = number of chiral carbons
    number of stereoisomers for galactose = $2^4 = 16$ stereoisomers

22. Each chiral carbon allows for two stereoisomers. Ribose has three chiral carbons.
number of stereoisomers = 2^n where n = number of chiral carbons
number of stereoisomers for ribose = $2^3 = 8$ stereoisomers

23. All possible stereoisomers of the following compounds, with enantiomers and meso compounds labeled.

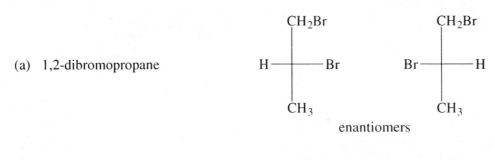

(a) 1,2-dibromopropane

enantiomers

(b) 2-butanol

enantiomers

(c) 3-chlorohexane

enantiomers

There are no meso compounds in 24(a), (b), or (c).

24. All possible stereoisomers of the following compounds with enantiomers and meso compounds labeled.

(a) 2,3-dichlorobutane

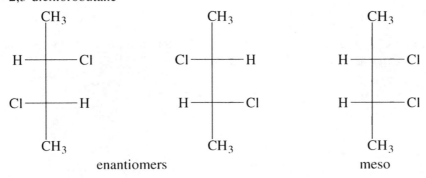

enantiomers   meso

(b) 2,4-dibromopentane

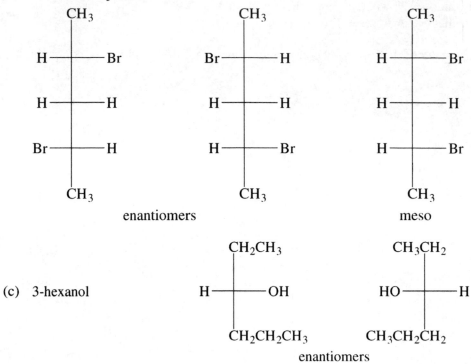

enantiomers   meso

(c) 3-hexanol

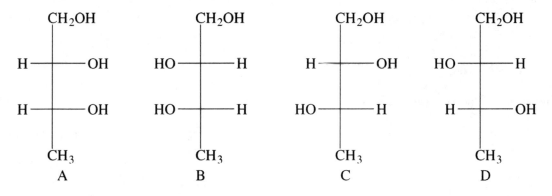

enantiomers

There are no meso compounds in 24 (c).

25. All the stereoisomers of 1,2,3-trihydroxybutane:

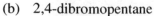

Compounds A and B, and C and D are pairs of enantiomers. There are no meso compounds. Pairs of diastereomers are A and C, A and D, B and C, and B and D.

26. All the stereoisomers of 3,4-dichloro-2-methylpentane:

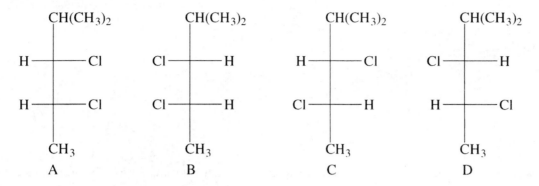

Compounds A and B, and C and D are pairs of enantiomers. There are no meso compounds. Pairs of diastereomers are A and C, A and D, B and C, and B and D.

27. The four stereoisomers of 2-hydroxy-3-pentene:

CH₃	CH₃	CH₃	CH₃
H—C—OH	HO—C—H	H—C—OH	HO—C—H
H—C	H—C	H—C	H—C
‖	‖	‖	‖
H—C	H—C	C—H	C—H
CH₃	CH₃	CH₃	CH₃
cis	*cis*	*trans*	*trans*

The two *cis* compounds are enantiomers and the two *trans* compounds are enantiomers.

28. The four stereoisomers of 2-chloro-3-hexene.

CH₃	CH₃	CH₃	CH₃
H—C—Cl	Cl—C—H	H—C—Cl	Cl—C—H
H—C	H—C	H—C	H—C
‖	‖	‖	‖
H—C	H—C	C—H	C—H
CH₂CH₃	CH₂CH₃	CH₂CH₃	CH₂CH₃
cis	*cis*	*trans*	*trans*

The two *cis* compounds are enantiomers and the two *trans* compounds are enantiomers.

29. Assume (+)-2-bromopentane is

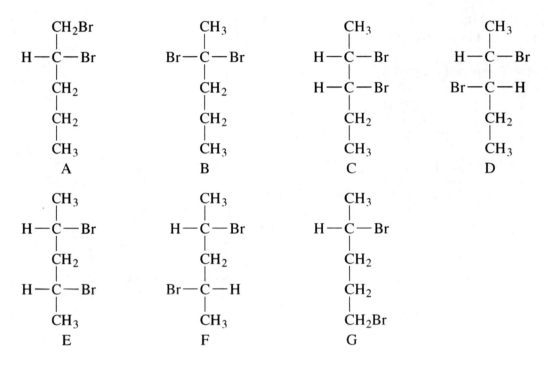

CH₃
|
H—C—Br
|
CH₂
|
CH₂
|
CH₃

All possible isomers formed when (+)-2-bromopentane is further brominated to dibromopentanes:

CH₂Br
|
H—C—Br
|
CH₂
|
CH₂
|
CH₃
A

CH₃
|
Br—C—Br
|
CH₂
|
CH₂
|
CH₃
B

CH₃
|
H—C—Br
|
H—C—Br
|
CH₂
|
CH₃
C

CH₃
|
H—C—Br
|
Br—C—H
|
CH₂
|
CH₃
D

CH₃
|
H—C—Br
|
CH₂
|
H—C—Br
|
CH₃
E

CH₃
|
H—C—Br
|
CH₂
|
Br—C—H
|
CH₃
F

CH₃
|
H—C—Br
|
CH₂
|
CH₂
|
CH₂Br
G

Compounds A, C, D, F, G are optically active; B has no chiral carbon atom; E is a meso compound.

30. Assume (+)-2-chlorobutane is

CH₃
|
CH₂
|
H—C—Cl
|
CH₃

All possible isomers formed when (+)-2-chlorobutane is further chlorinated to dichlorobutane:

<pre>
    CH₃   CH₂Cl   CH₃   CH₃   CH₃
    | | | | |
    CH₂   CH₂   CH₂   H—C—Cl   Cl—C—H
    | | | | |
H—C—Cl   H—C—Cl   Cl—C—Cl   H—C—Cl   H—C—Cl
    | | | | |
    CH₂Cl   CH₃   CH₃   CH₃   CH₃
     A   B   C   D   E
</pre>

Compounds A, B, and E are optically active; C does not have a chiral carbon atom; D is a meso compound.

31. Neither of the products, 1-chloropropane or 2-chloropropane, has a chiral carbon atom so neither product would rotate polarized light.

32. If 1-chlorobutane and 2-chlorobutane were obtained by chlorinating butane, and then distilled, they would be separated into the two fractions, because their boiling points are different. 1-chlorobutane has no chiral carbon, so would not be optically active. 2-chlorobutane would exist as a racemic mixture (equal quantities of enantiomers) because substitution of Cl for H on carbon-2 gives equal amounts of the two enantiomers. Distillation would not separate the enantiomers because their boiling points are identical. The optical rotation of the two enantiomers of the 2-chlorobutane fraction would exactly cancel, and thus would not show optical activity.

33. Compounds (a) and (d) are meso. Make two changes on C-3 in compound (a) to prove that it is meso.

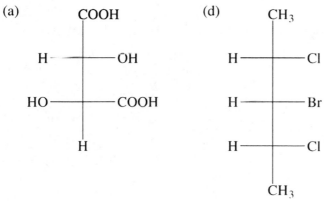

34. Compound (d) is meso.

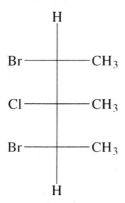

35. (a) enantiomers
    (b) nonisomers
    (c) diastereomers

36. (a) nonisomers
    (b) enantiomers
    (c) diastereomers

37. If four different groups were attached to a central carbon atom in a planar arrangement, it would not rotate polarized light because there would be a plane of symmetry in the molecule. No such plane of symmetry is possible when the four different groups are arranged in a tetrahedral structure.

38. (a) and (b)

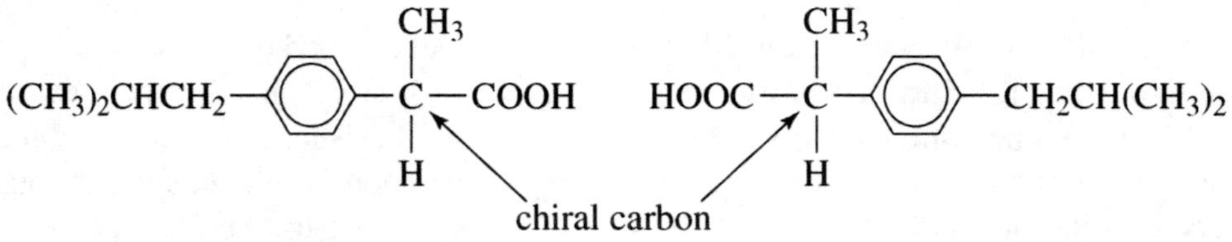

39. (a) A chiral primary alcohol of formula $C_5H_{12}O$.

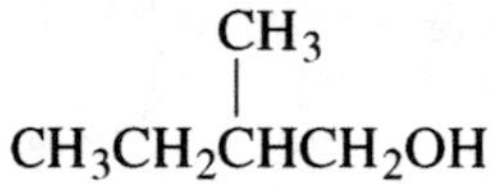

    (b) A compound with three primary alcohol groups is chiral, and has the formula $C_6H_{14}O_3$.

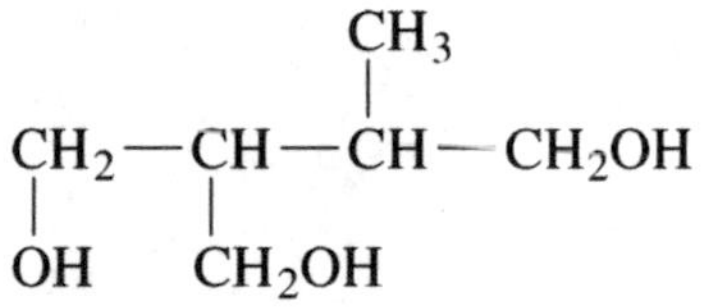

40. (a)

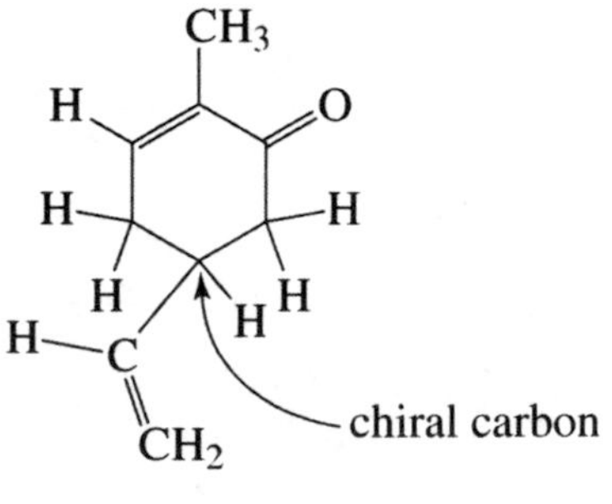

caraway

    (b) The spearmint molecule is the optical isomer of the caraway molecule and differs from it in structure at the chiral carbon atom.

41. Ephedrine has two chiral carbons and can have four stereoisomers. This number is calculated using $2^n = 2^2 = 4$.

42. A chiral carbon is bonded to four different groups. Tyrosine has one chiral carbon.

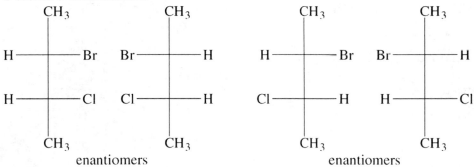

43. Based on its name, dextromethorphan, this drug is probably the dextrorotatory isomer. It rotates plane-polarized light in a clockwise direction. The symbol, (+), would be used to indicate this drug's optical activity.

44. Stereoisomer structures

    (a) 2-bromo-3-chlorobutane

              enantiomers   enantiomers

    There are no meso compounds.

(b) 2,3,4-trichloro-1-pentanol

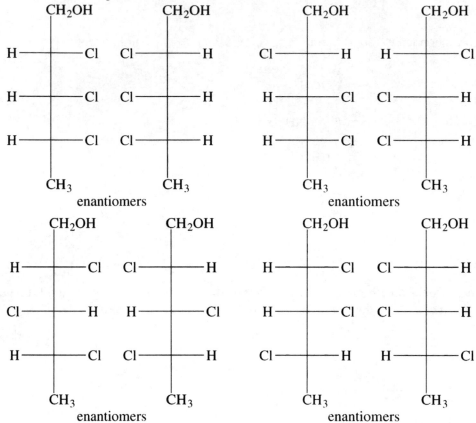

enantiomers   enantiomers

enantiomers   enantiomers

There are no meso compounds.

45. Meso structures for alkanes
(a) C$_8$H$_{18}$ (b) C$_9$H$_{20}$

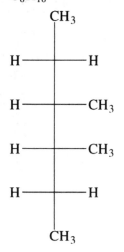

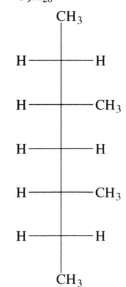

(c) $C_{10}H_{22}$

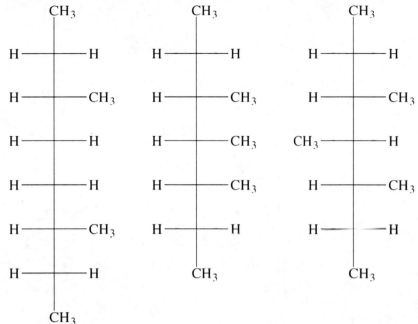

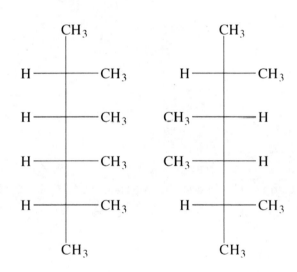

46. Optically active alcohols of $C_6H_{14}O$ (one enantiomer of each structure).

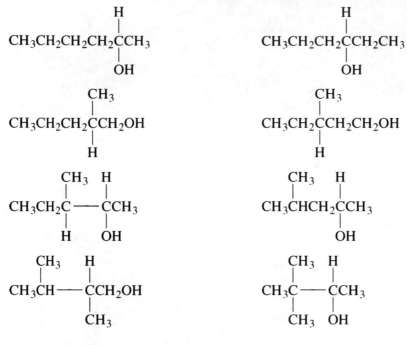

47.

$$\begin{array}{ccc}
\text{COOH} & \text{O} & \text{COOH} \\
 & \| & \\
\text{H}_2\text{N}\!-\!\!\bigoplus\!\!- & \text{C}\!-\!\text{NH}_2 & \text{H}_2\text{N}\!-\!\!\bigoplus\!\!- \\
 & -\!\!\bigoplus\!\!-\text{NH}_2 & \\
\text{CH}_3 & \text{COOH} & \text{CH}_3
\end{array}$$

The similarity is that the chiral carbon in each compound is bonded to an NH_2, a $COOH$, and an H group.

48. A compound of formula $C_3H_8O_2$:

(a) is chiral; contains two OH groups $CH_3\underset{\underset{\displaystyle OH}{|}}{CH}CH_2OH$

(b) is chiral; contains one OH group $CH_3-O-\underset{\underset{\displaystyle OH}{|}}{CH}CH_3$

(c) is achiral; contains two OH groups $\underset{\underset{\displaystyle OH}{|}}{CH_2}CH_2\underset{\underset{\displaystyle OH}{|}}{CH_2}$

CHAPTER 27

CARBOHYDRATES

SOLUTIONS TO REVIEW QUESTIONS

1. In general, the carbohydrate carbon oxidation state determines the carbon's metabolic energy content. The more oxidized a carbon is, the less energy it can provide in biological systems.

2. The notations D and L in the name of a carbohydrate specify the configuration on the last chiral carbon atom (from C-1) in the Fischer projection formula. If the —OH is written to the right of that carbon the compound is a D-carbohydrate. If the —OH is written to the left it is an L-carbohydrate. For example, D-glyceraldehyde and L-glyceraldehyde differ only at the chiral C—OH; D-glyceraldehyde has the —OH on the right while L-glyceraldehyde has the —OH on the left.

3. The notations (+) and (−) in the name of a carbohydrate specify whether the compound rotates the plane of polarized light to the right (+) or to the left (−).

4. Galactosemia is the inability of infants to metabolize galactose. The galactose concentration increases markedly in the blood and also appears in the urine. Galactosemia causes vomiting, diarrhea, enlargement of the liver, and often mental retardation. If not recognized a few days after birth it can lead to death.

5. There are four pairs of epimers among the D-aldohexoxes in Figure 27.1. They are: allose and altrose; glucose and mannose; gulose and idose; and galactose and talose. Glucose and mannose are epimers at carbon 2.

6. A carbohydrate forms a five-member or six-member heterocyclic ring (one oxygen atom, the rest carbon atoms). If it forms a five-member ring, it is termed a furanose, after the compound furan. If it forms a six-membered ring it is termed a pyranose, after the compound pyran.

Furan

C_4H_4O

Pyran

C_5H_6O

7. α-D-glucopyranose and β-D-glucopyranose differ in the configuration at the number 1 carbon in the cyclic structure. In the open-chain structure, carbon 1 is the aldehyde group, and is not chiral. In the cyclic structure that carbon contains a hemiacetal structure, which is chiral. When the ring forms, carbon 1 can have two configurations leading to the two structures called α and β.

In the Haworth structure for glucose the —OH on carbon 1 is written down for the α structure and up for the β structure.

8. The cyclic forms of monosaccharides are hemiacetals, because the number one carbon has an ether and an alcohol group; whereas a glycoside is an acetal which contains two ether linkages.

9. Mutarotation is the phenomenon by which the α or β form of a sugar, when in solution, will undergo change to reach an equilibrium mixture, not necessarily 50%-50%, of the two forms. To achieve this equilibrium, the chain must open up and then reclose. On closing, it has the possibility of closing in either the α or β form as the equilibrium mixture is achieved.

10. Major sources:
    (a) sucrose: sugar beets and sugar cane
    (b) lactose: milk
    (c) maltose: sprouting grain and partially hydrolyzed starch

11. The following parts are related to the eight D-aldohexoses shown in the text (Figure 27.1):
    (a) If each of the aldohexoses is oxidized by nitric acid to dicarboxylic acids, allose and galactose would become meso forms.

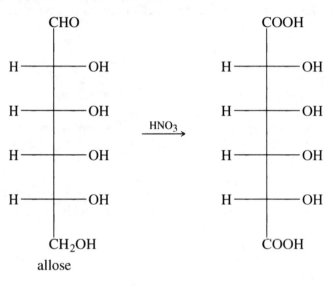

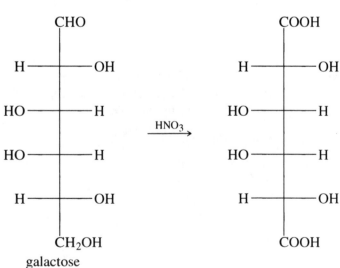

(b) Names and structures of enantiomers of D-altrose and D-idose:

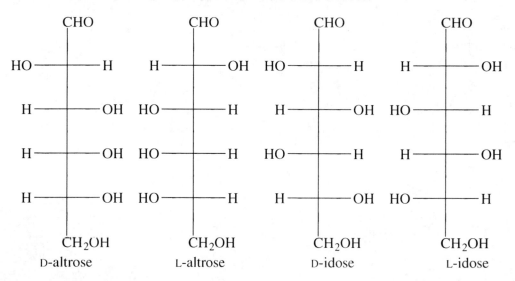

| CHO | CHO | CHO | CHO |
| D-altrose | L-altrose | D-idose | L-idose |

12. Invert sugar is sweeter than sucrose because it is a 50–50 mixture of fructose and glucose. Glucose is somewhat less sweet than sucrose, but fructose is much sweeter, so the mixture is sweeter.

13. Sucralose differs from sucrose in two important ways:
    (a) sucralose has three chlorines in place of three hydroxyl groups;
    (b) sucralose has a different configuration at one chiral carbon compared with sucrose. These two differences make sucralose a non-nutritive sweetener.

14. Sugar alcohols are poorly absorbed by the small intestine and, thus, can contribute only few dietary calories.

15. Based on Figure 27.3, the alpha glucose anomer gives a specific optical rotation of +112°.

16. Amylose is a linear polymer that forms a helix in solution. In contrast, amylopectin is a branching polymer that forms a tree-like shape in solution.

17. Glycoaminoglycans protect our joints from osteoarthritis by acting as a shock absorber and by keeping the bones from rubbing against each other.

18. Glucose provides metabolic energy via two important metabolic processes. They are glycolysis and the citric acid cycle.

19. Blood types A and B differ in one place. In the fourth pyranose ring at C-2, blood type B has an OH group, while blood type A has an amide group.

20. Branch chains of glucose in amylopectin are linked to the main chain by a single α-(1,6) linkage. All the other glucose units are linked by α-(1,4) linkages. Clearly, upon hydrolysis, many more maltose molecules than isomaltose molecules will be formed. (See Figure 27.7.)

21. Yes. L-glucose is a reducing sugar. The linear structure has an aldehyde group which is reducible by Benedict solution; or an hemiacetal structure which opens to form an aldehyde structure.

SOLUTIONS TO EXERCISES

1. There are no chiral carbons in dihydroxyacetone.

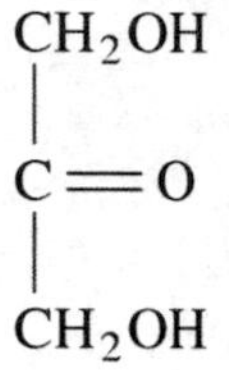

$$CH_2OH$$
$$|$$
$$C = O$$
$$|$$
$$CH_2OH$$

2. D-glyceraldehyde   L-glyceraldehyde

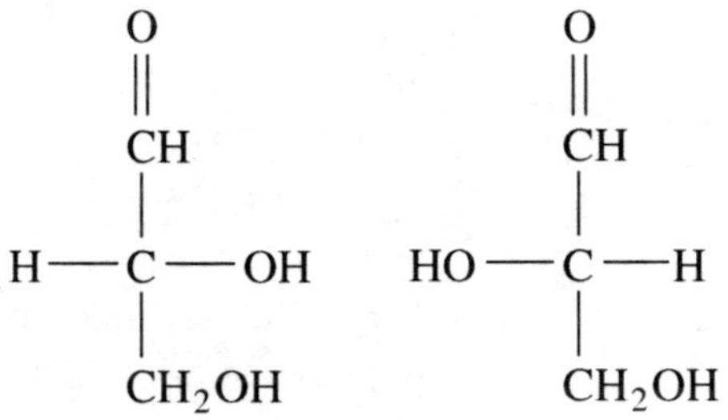

$$\begin{array}{cc}
O & O \\
\| & \| \\
CH & CH \\
| & | \\
H-C-OH & HO-C-H \\
| & | \\
CH_2OH & CH_2OH
\end{array}$$

3. D-aldopentose. The chiral carbon furthest from the aldehyde group has the —OH on the right in the Fischer projection formula.

4. L-aldopentose. The chiral carbon furthest from the aldehyde group has the —OH on the left in the Fischer projection formula.

5. A common carbohydrate with the formula, $C_4H_8O_4$, can be either

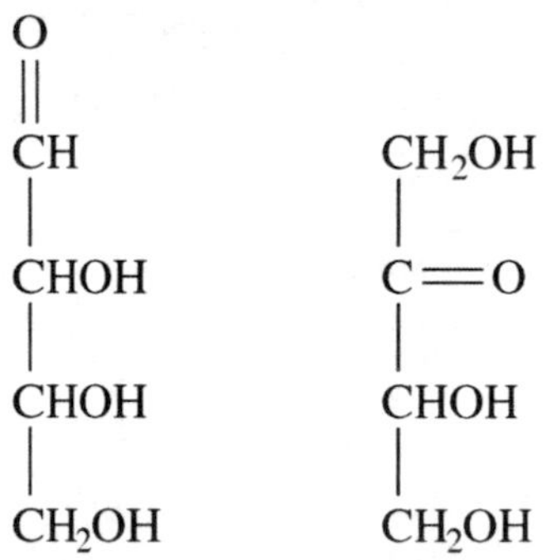

$$\begin{array}{cc}
O & \\
\| & CH_2OH \\
CH & | \\
| & C = O \\
CHOH & | \\
| & CHOH \\
CHOH & | \\
| & CH_2OH \\
CH_2OH & \\
\text{aldotetrose} & \text{ketotetrose}
\end{array}$$

but not a ketopentose. The correct answer is the aldotetrose since it has two chiral carbons.

6. A common carbohydrate with the formula, $C_6H_{12}O_6$, can be either

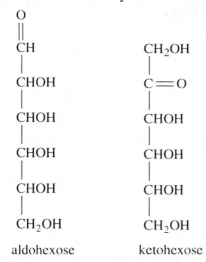

$$
\begin{array}{cc}
\overset{\displaystyle O}{\overset{\displaystyle \|}{CH}} & CH_2OH \\
| & | \\
CHOH & C{=}O \\
| & | \\
CHOH & CHOH \\
| & | \\
CHOH & CHOH \\
| & | \\
CHOH & CHOH \\
| & | \\
CH_2OH & CH_2OH \\
\text{aldohexose} & \text{ketohexose}
\end{array}
$$

but not an aldopentose. The correct answer is the ketohexose since it has 3 chiral carbons.

7. An enantiomer is a mirror-image stereoisomer. In the Fischer formula, the positions of the atoms around chiral carbons must be carefully shown. For achiral carbons, the positions of the atoms are not critical.

$$
\begin{array}{c}
\overset{\displaystyle O}{\overset{\displaystyle \|}{CH}} \\
| \\
HO{-}C{-}H \\
| \\
H{-}C{-}OH \\
| \\
H{-}C{-}OH \\
| \\
HO{-}C{-}H \\
| \\
CH_2OH
\end{array}
$$

8. An enantiomer is a mirror-image stereoisomer. In the Fischer formula, the positions of the atoms around chiral carbons must be carefully shown. For achiral carbons, the positions of the atoms are not critical.

$$
\begin{array}{c}
CH_2OH \\
| \\
C{=}O \\
| \\
H{-}C{-}OH \\
| \\
HO{-}C{-}H \\
| \\
H{-}C{-}OH \\
| \\
CH_2OH
\end{array}
$$

9. (a) An epimer is a stereoisomer that differs at one chiral carbon. For example, the following epimer differs from D-galactose at the carbon adjacent to the carbonyl carbon.

$$
\begin{array}{c}
\overset{\displaystyle O}{\underset{\displaystyle \parallel}{}} \\
CH \\
| \\
HO-C-H \\
| \\
HO-C-H \\
| \\
HO-C-H \\
| \\
H-C-OH \\
| \\
CH_2OH
\end{array}
$$

(b) An enantiomer is a mirror-image stereoisomer.

$$
\begin{array}{c}
\overset{\displaystyle O}{\underset{\displaystyle \parallel}{}} \\
CH \\
| \\
HO-C-H \\
| \\
H-C-OH \\
| \\
H-C-OH \\
| \\
HO-C-H \\
| \\
CH_2OH
\end{array}
$$

(c) A diastereomer is any stereoisomer that is not a mirror image. For example, the following sugar differs from D-galactose at all chiral carbons except the bottom-most <u>and</u> it is not a mirror image of D-galactose.

$$
\begin{array}{c}
\overset{\displaystyle O}{\underset{\displaystyle \parallel}{}} \\
CH \\
| \\
HO-C-H \\
| \\
H-C-OH \\
| \\
H-C-OH \\
| \\
H-C-OH \\
| \\
CH_2OH
\end{array}
$$

10. (a) An epimer is a stereoisomer that differs at one chiral carbon. For example, the following epimer differs from D-fructose at the carbon just below the carbonyl carbon.

```
        CH₂OH
         |
        C = O
         |
    H — C — OH
         |
    H — C — OH
         |
    H — C — OH
         |
        CH₂OH
```

(b) An enantiomer is a mirror-image stereoisomer.

```
        CH₂OH
         |
        C = O
         |
    H — C — OH
         |
   HO — C — H
         |
   HO — C — H
         |
        CH₂OH
```

(c) A diastereomer is any stereoisomer that is not a mirror image. For example, the following sugar differs from D-fructose at all chiral carbons except the bottom-most and it is not a mirror image of D-fructose.

```
        CH₂OH
         |
        C = O
         |
    H — C — OH
         |
   HO — C — H
         |
    H — C — OH
         |
        CH₂OH
```

11. Since D-glucose and D-mannose are epimers, they differ at only one chiral center. Their structures are almost identical.

12. Since D-galactose and D-mannose are diastereomers, they are non-mirror image stereoisomers. They may differ at one or more chiral carbons but they cannot be enantiomers.

13. (a)

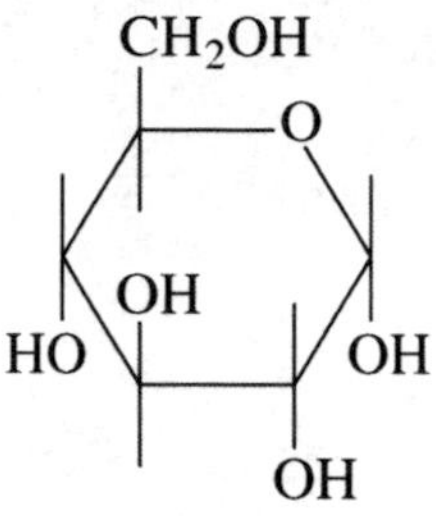

(c)

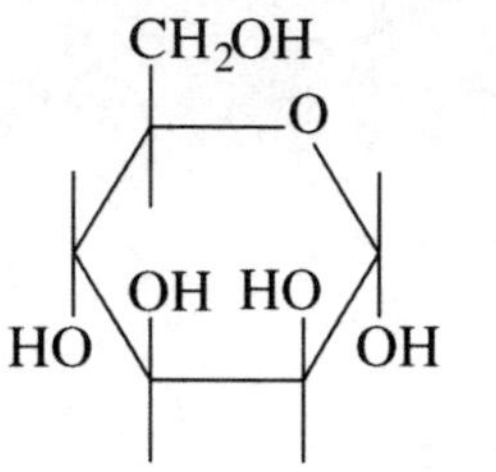

(b)

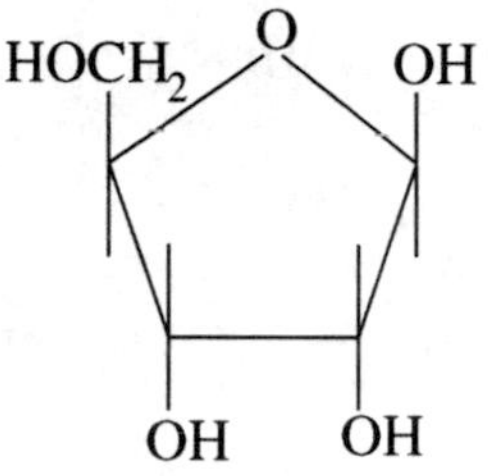

14. (a)

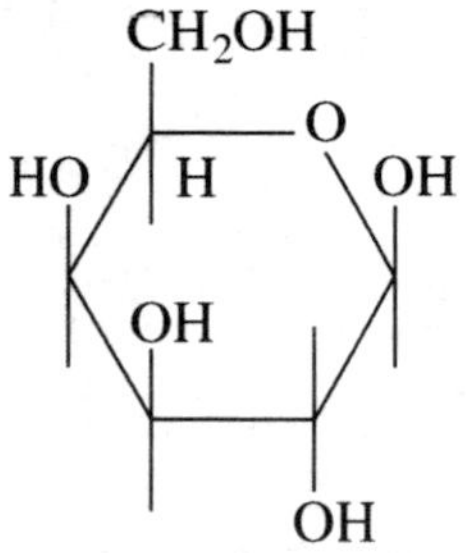

(b)

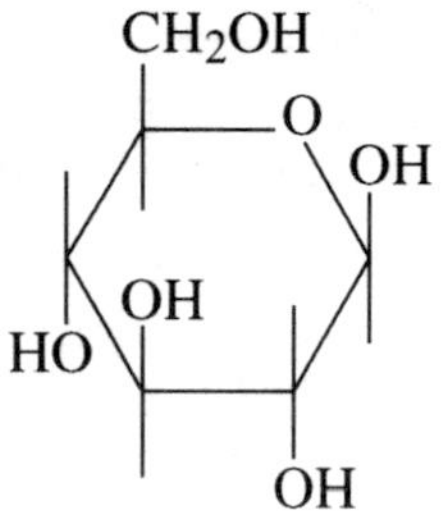

(c)

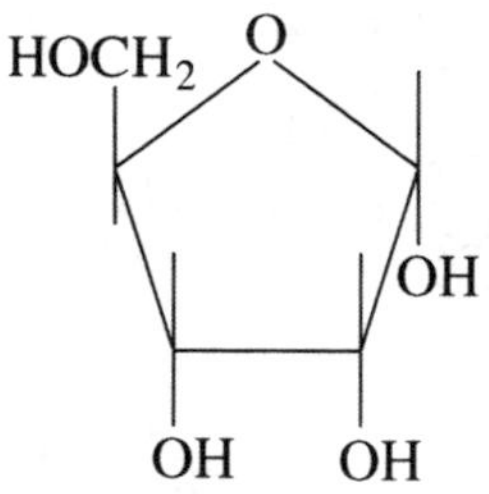

15. The hemiacetal carbon is the only carbon in this structure that is directly bonded to two oxygens.

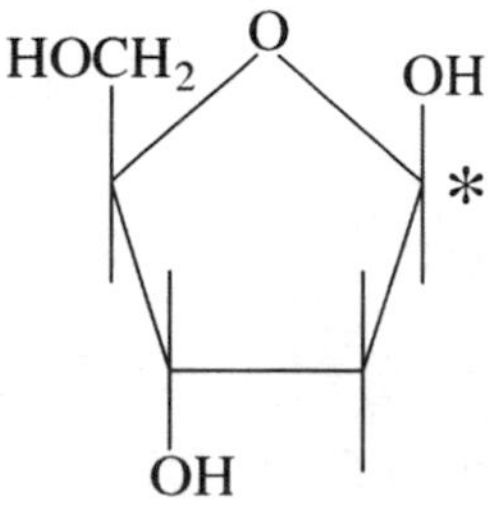

16. The hemiketal carbon is the only carbon in this structure that is directly bonded to two oxygens.

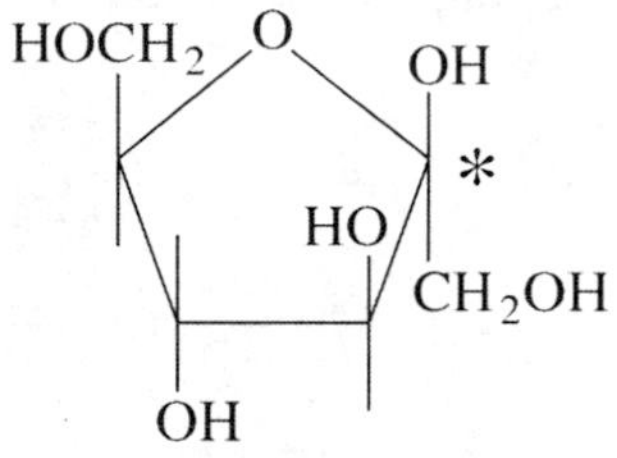

17. The monosaccharide composition of:
    (a) sucrose: one glucose and one fructose unit
    (b) glycogen: many glucose units
    (c) amylose: many glucose units
    (d) maltose: two glucose units

18. The monosaccharide composition of:
    (a) lactose: one glucose and one galactose unit
    (b) amylopectin: many glucose units
    (c) cellulose: many glucose units
    (d) sucrose: one glucose and one fructose unit

19. Both cellobiose and isomaltose are disaccharides composed of two glucose units. However, in cellobiose monosaccharides are linked by a β-1,4-acetal bond while for isomaltose the linkage is α-1,6.

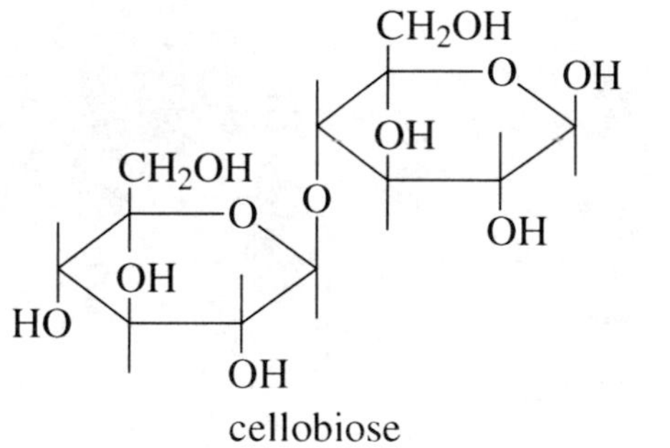

cellobiose

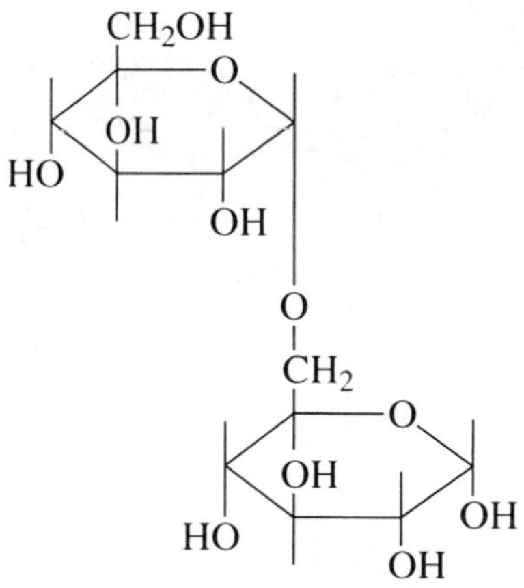

isomaltose

20. Both maltose and isomaltose are disaccharides composed of two glucose units. The glucose units in maltose are linked by an α-1,4-glycosidic bond while the glucose units in isomaltose are linked by an α-1,6-glycosidic bond.

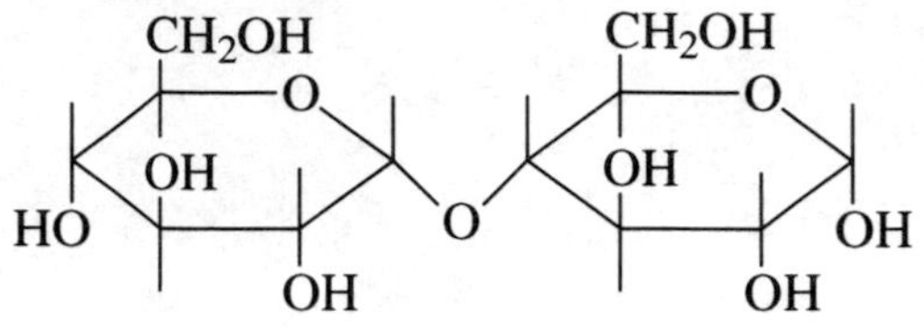

maltose

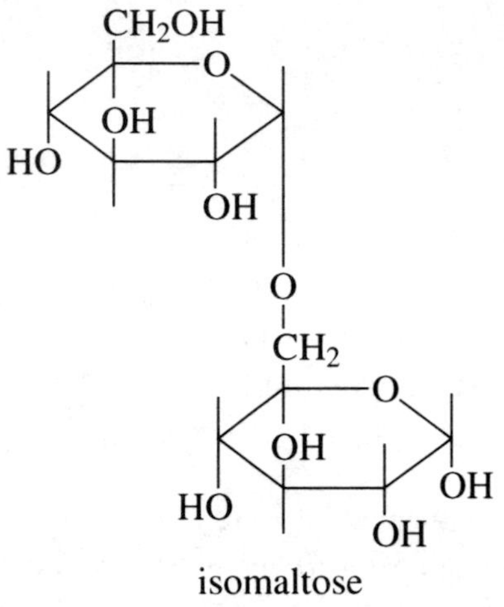

isomaltose

21. Lactose will show mutarotation; sucrose will not. The hemiacetal structure in lactose will open allowing mutarotation. Since sucrose has an acetal structure and no hemiacetal, it will not undergo mutarotation.

22. Both maltose and isomaltose will show mutarotation. Both disaccharides contain a hemiacetal structure which will open allowing mutarotation.

23.

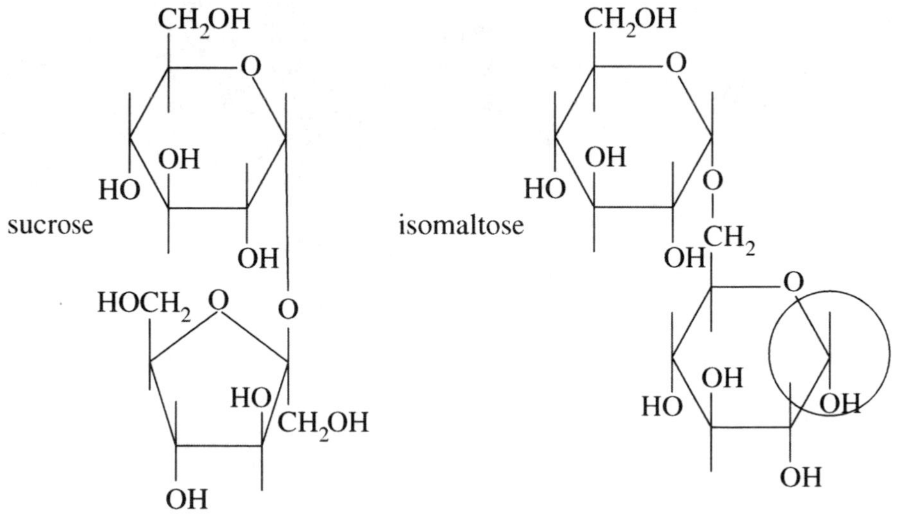

sucrose   isomaltose

24.

maltose

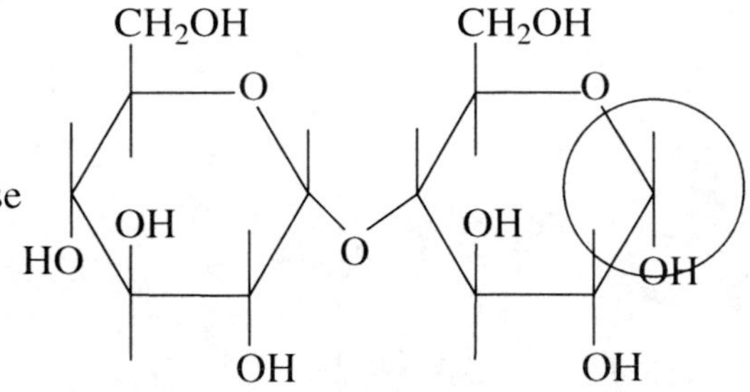

lactose

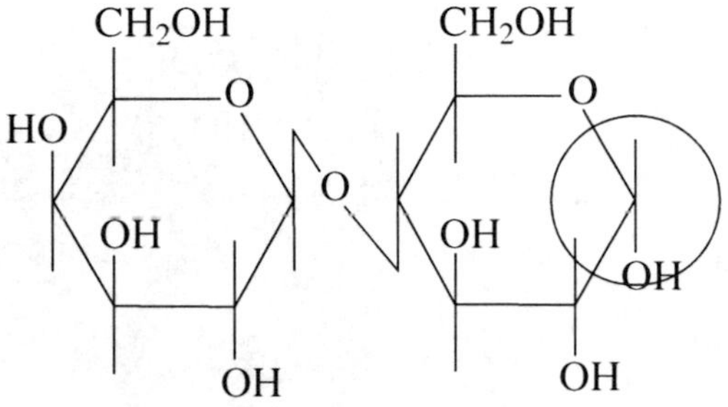

25. isomaltose, α-D-glucopyranosyl-(1,6)-α-D-glucopyranose; sucrose, α-D-glucopyranosyl-(1,2)-(β-D-fructofuranose.

26. maltose, α-D-glucopyranosyl-(1,4)-α-D-glucopyranose; lactose, β-D-galactopyranosyl-(1,4)-α-D-glucopyranose.

27. (a)

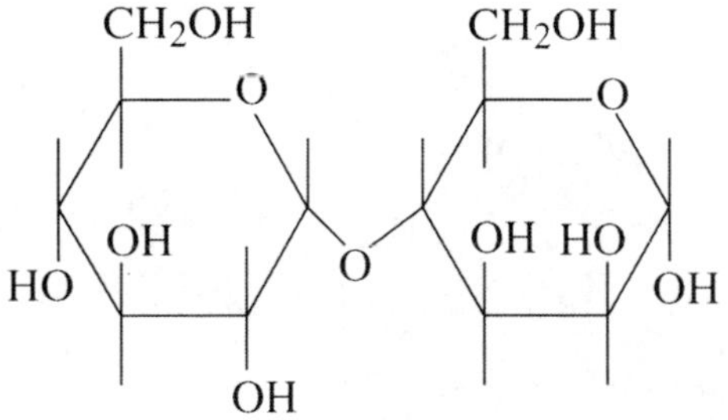

(b)

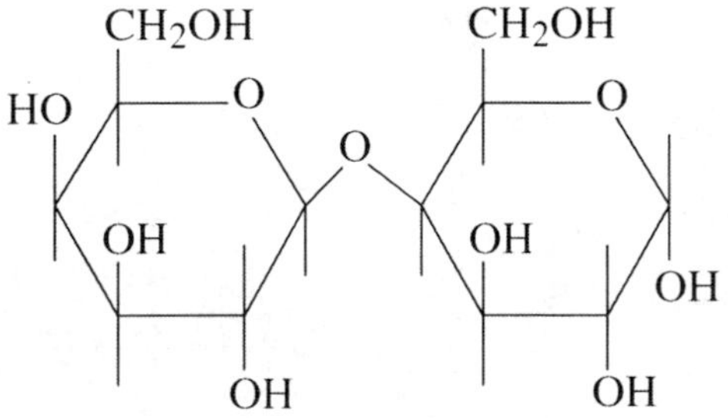

28. (a)

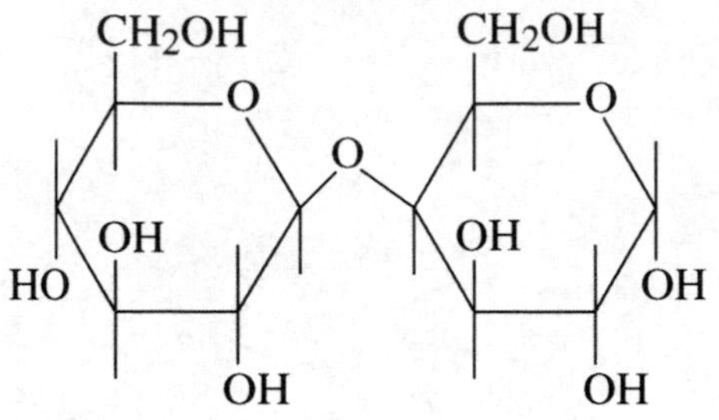

(b)

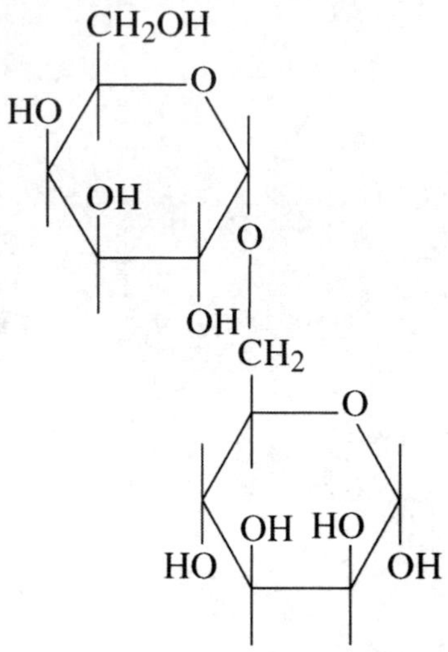

29.

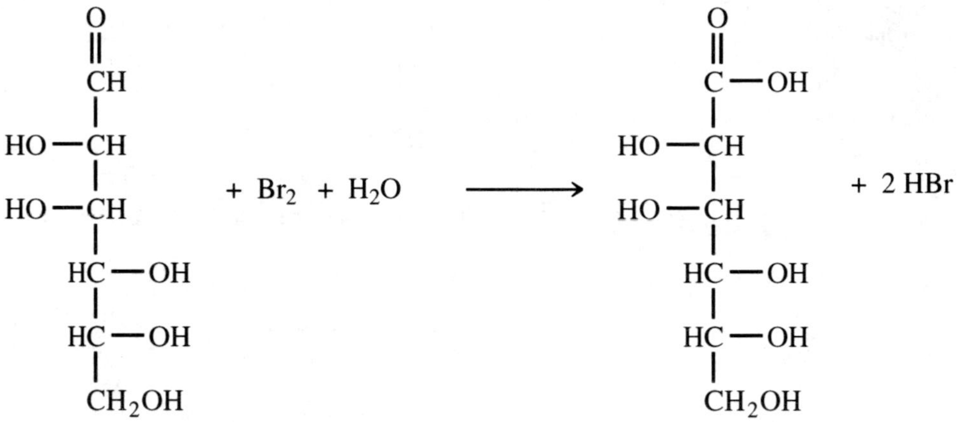

$+ Br_2 + H_2O \longrightarrow$

$+ 2 HBr$

30.

$+ Br_2 + H_2O \longrightarrow$

$+ 2 HBr$

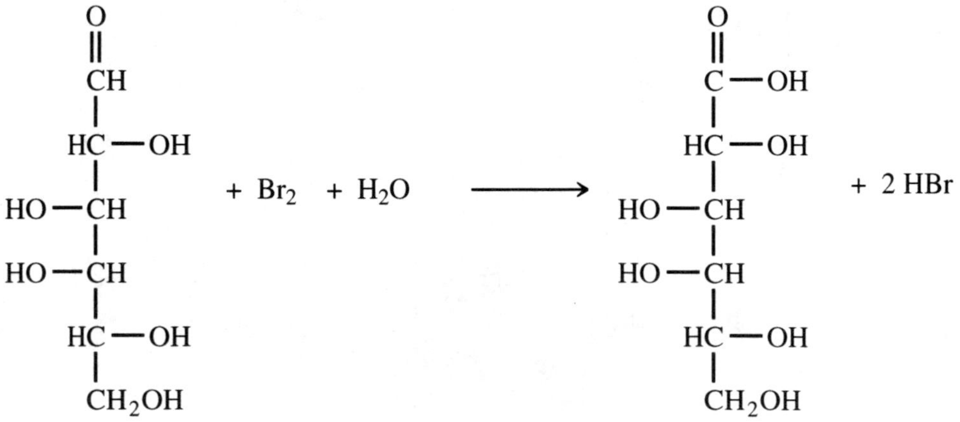

31. H_2/Pt reduces the aldehyde group on D-ribose to a primary alcohol.

$$
\begin{array}{c}
\boxed{CH_2OH} \\
| \\
H-C-OH \\
| \\
H-C-OH \\
| \\
H-C-OH \\
| \\
CH_2OH
\end{array}
$$

32. H_2/Pt reduces the aldehyde group on D-mannose to a primary alcohol.

$$
\begin{array}{c}
\boxed{CH_2OH} \\
| \\
HO-C-H \\
| \\
HO-C-H \\
| \\
H-C-OH \\
| \\
H-C-OH \\
| \\
CH_2OH
\end{array}
$$

33. Warm, dilute nitric acid oxidizes both primary alcohols and aldehyde groups to carboxylic acids.

$$
\begin{array}{c}
\boxed{COOH} \\
| \\
HO-C-H \\
| \\
HO-C-H \\
| \\
H-C-OH \\
| \\
H-C-OH \\
| \\
\boxed{COOH}
\end{array}
$$

34. Warm, dilute nitric acid oxidizes both primary alcohols and aldehyde groups to carboxylic acids.

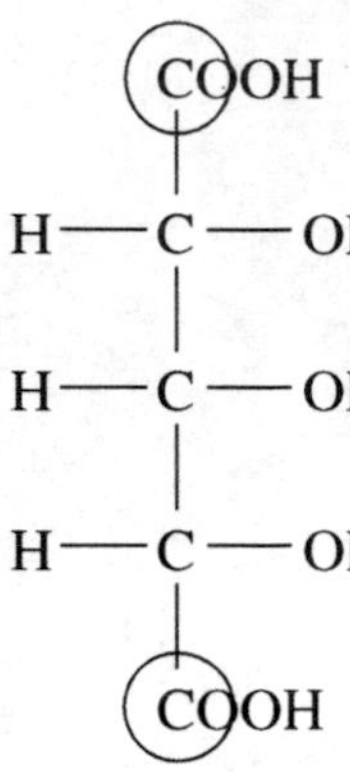

```
        (COOH)
          |
   H —— C —— OH
          |
   H —— C —— OH
          |
   H —— C —— OH
          |
        (COOH)
```

35. (a) Amylopectin and glycogen share the same structure except that glycogen has more 1,6-links and, thus, more branches.

   (b) D-glucose and D-galactose share the same structure except for the configuration at the C-4 chiral carbon.

   (c) Lactose and maltose are both disaccharides with a 1,4 linkage. Each has one α-D-glucopyranose unit. Lactose has a β-D-galactopyranose for the other unit while maltose has an α-D-glucopyranose.

36. (a) D-ribose and D-ribulose share the same structure except that D-ribose is an aldose while D-ribulose is a ketose (and, thus, has one less chiral carbon).

   (b) Maltose and isomaltose share the same structure except that maltose contains a (1,4) acetal bond while isomaltose contains a (1,6) acetal bond.

   (c) Cellulose and amylose share the same structure except that cellulose has β-anomer glucose units while amylose has α-anomer glucose units.

37.

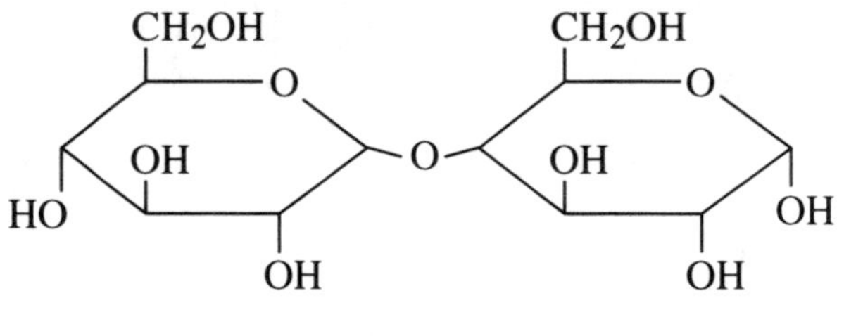

maltose

38.

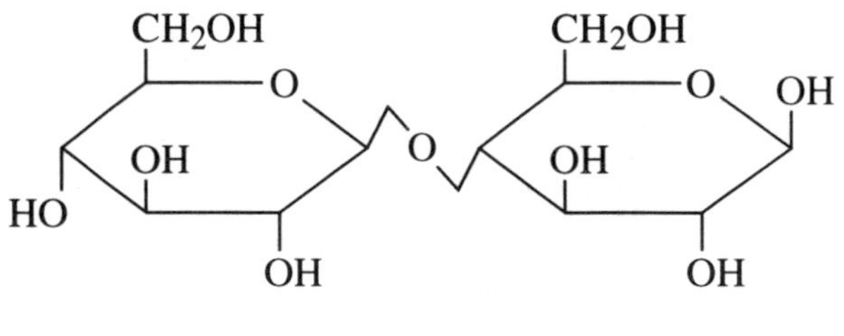

cellobiose

39.

40.

41. Glucose is called blood sugar because it is the most abundant carbohydrate in the blood and is carried by the bloodstream to all parts of the body.

42. Two advantages of aspartame over sucrose: (a) aspartame sweetens without added calories; (b) aspartame does not cause dental caries. Two advantages of neotame over aspartame: (a) neotame does not release phenylalanine, an amino acid that is dangerous to people with phenylketonuria; (b) neotame is intensely sweet and can be used in very small amounts.

43. Glucose molecules change from one anomer to the other in solution. This process is called mutarotation. A solution with a specific optical rotation of $+18.7°$ contains primarily the β-anomer of glucose and, as the optical rotation becomes more positive, more of the α-anomer is forming.

44. When D-galactose reacts to form a cyanohydrin, the carbonyl group reacts with HCN. This reaction is called addition because cyanide adds to the carbonyl carbon while H adds to the carbonyl oxygen. The double bond is lost in this process. Two isomers form in this reaction.

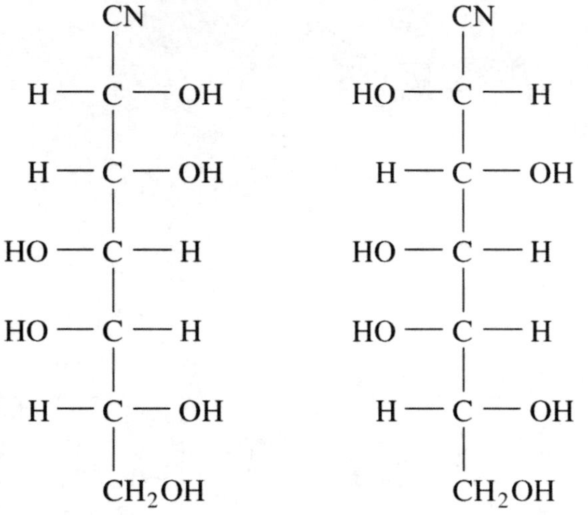

45. (a) D-2-deoxyribose has two chiral carbons (marked with asterisks in the following structure).

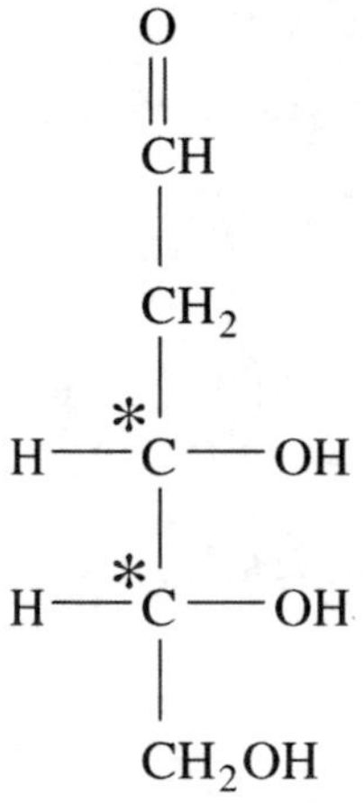

(b) When D-2-deoxyribose forms the ringed structure, α-D-2-deoxyribofuranose, a new chiral carbon forms at the hemiacetal position. α-D-2-deoxyribofuranose has three chiral carbons as marked by asterisks in the following structure:

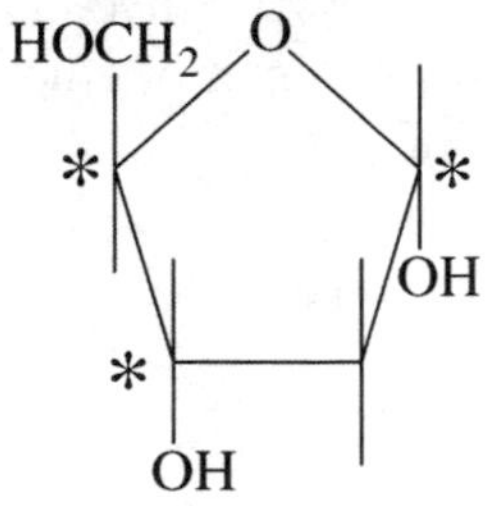

46. (a) The sugar acid could be most easily derived from β-D-mannopyranose.

(b)

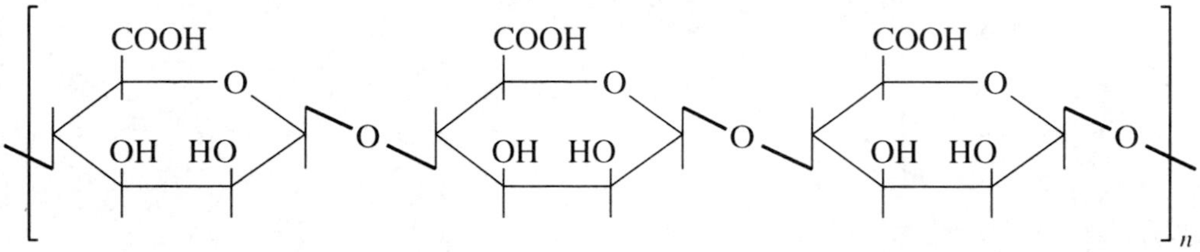

47. Compound A must be a reducing disaccharide because it produces a reddish color in the Benedict test. Sucrose is nonreducing with the Benedict test so compound A must be maltose.

48. A nonreducing disaccharide composed of two molecules of α-D-galactopyranose can have no hemiacetal structures. Thus, the hemiacetal structure of one α-D-galactopyranose must be used to form the glycosidic link to the hemiacetal structure of the other α-D-galactopyranose unit forming an acetal structure.

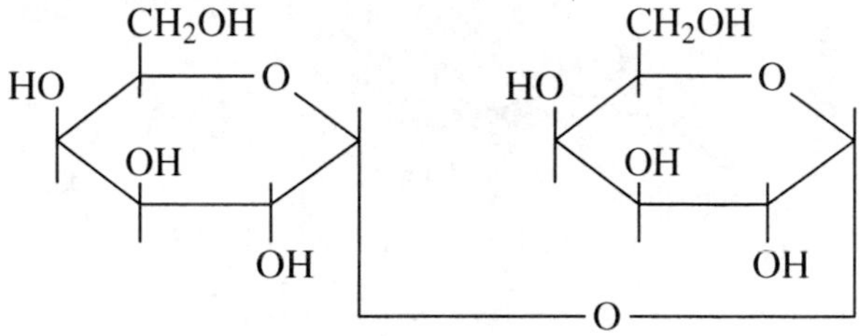

49. Cellulose, amylose, and amylopectin are polymers of glucose. Cellulose exists in the form of fibers, is not digestible by humans, and therefore remains in the digestive tract as fibers. Amylose and amylopectin are digested to glucose which is dissolved into the bloodstream.

50. (a) D-galactose and D-glucose differ only at carbon 4. Thus, D-galactose must be changed at carbon 4 to be converted to D-glucose.

(b) D-galactose is an *epimer* of D-glucose.

51. No, the classmate should not be believed. Although D-glucose and D-mannose are related as epimers, it is pairs of *enantiomers* which have equal and opposite optical rotation.

52. β-D-glucopyranose differs at two chiral carbons from the structure shown (α-D-galactopyranose). To convert α-D-galactopyranose to β-D-glucopyranose, the α-anomer must be changed to the β-anomer and the hydroxyl at C-4 must be changed from up to down. The structure for β-D-glucopyranose follows:

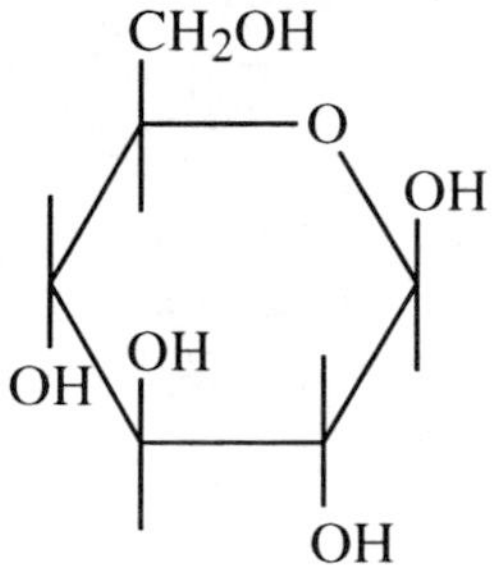

53. (a)

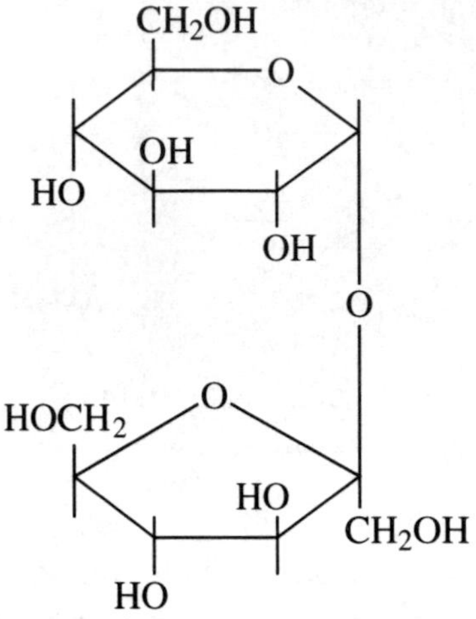

(b)

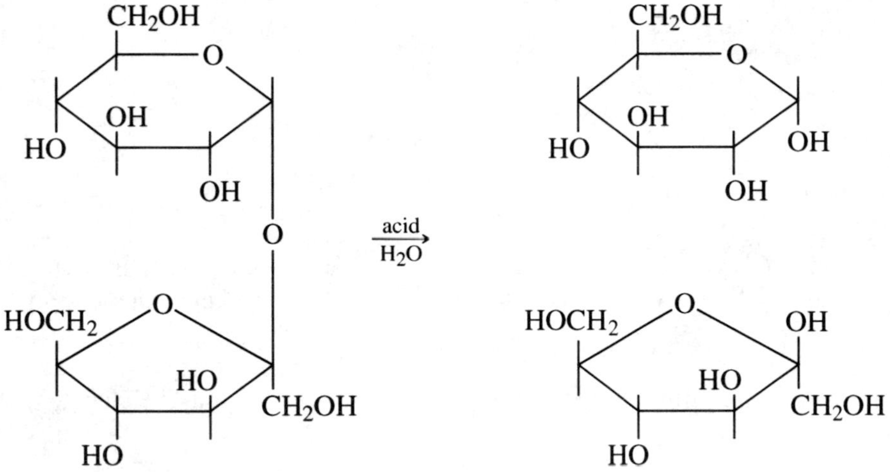

acid / H_2O

(c) Fructose is sweeter than sucrose. When sucrose is broken down to its two component monosaccharides, glucose and fructose, the candy becomes sweeter.

CHAPTER 28

LIPIDS

SOLUTIONS TO REVIEW QUESTIONS

1. The lipids, which are dissimilar substances, are arbitrarily classified as a group on the basis of their solubility in fat solvents and their insolubility in water.

2. Butyric acid is infinitely soluble in water while palmitic acid is water insoluble. Each molecule has a carboxyl functional group. The molecules differ in size, butyric acid is four carbons long while palmitic acid has sixteen carbons. Palmitic acid's greater size causes it to be much more hydrophobic and insoluble in water.

3. Eicosanoids are called "local hormones" because (a) they take effect close to where they are synthesized and (b) they are destroyed before they can move far away from the point of synthesis.

4. The omega numbering system for fatty acids starts with the carbon furthest from the carboxyl group (commonly a CH_3 group). The first double bond from this end is at carbon 3 in an omega-3 fatty acid and at carbon 6 in an omega-6 fatty acid.

5. Most metabolic energy is derived from foodstuffs by carbon oxidation. The average carbon in a fatty acid is less oxidized than the average carbon in a carbohydrate. Thus, fatty acids can be oxidized more extensively and will provide more energy per carbon. Furthermore, fatty acids have a much higher percentage carbon by mass (75%) than carbohydrates (40%). Fatty acids yield about twice the energy per gram as compared with carbohydrates.

6. NSAID stands for non-steroidal anti-inflammatory drug. This name partly describes the drug's structure (non-steroidal) and partly describes the drug's action (anti-inflammatory). These drugs inhibit cyclooxygenase and block the formation of prostaglandins, prostacyclins, and thromboxanes. By blocking synthesis of these "local hormones," NSAIDs effectively control pain, fever, and inflammation.

7. A low daily dose of aspirin reduces the risk of stroke by inhibiting synthesis of the eicosanoids.

8. A membrane lipid must be (a) partially hydrophobic to act as a barrier to water and (b) partially hydrophilic, so that the membrane can interact with water along its surface.

9. Phospholipids are mainly produced in the liver.

10. The four classes of eicosanoids are prostaglandins, prostacyclins, thromboxanes, and leukotrienes.

11. The omega-3 fatty acids can substitute for arachidonic acid and block its action as a trigger for heart attack and stroke.

12. In general, a membrane lipid will have both hydrophilic and hydrophobic parts. Sphingomyelin can serve as a membrane lipid because its phosphate and choline groups are hydrophilic and its two long carbon chains are hydrophobic.

$$
\begin{array}{c}
\text{OH} \\
|\\
\text{CHCH}{=}\text{CH(CH}_2)_{12}\text{CH}_3 \quad \text{hydrophobic} \\
|\\
\text{hydrophobic} \quad\; \text{RC}{-}\text{NH}{-}\text{CH} \qquad\quad \text{CH}_3\\
\end{array}
$$

hydrophobic   RC—NH—CH   O   CH₃

CH₂—O—P—O—CH₂CH₂—N⁺—CH₃   hydrophilic

O⁻   CH₃

sphingomyelin

13. Atherosclerosis is the deposition of cholesterol and other lipids on the inner walls of the large arteries. These deposits, called plaque, accumulate, making the arterial passages narrower and narrower. Blood pressure increases as the heart works to pump sufficient blood through the restricted passages. This may lead to a heart attack, or the rough surface can lead to coronary thrombosis.

14. Dietary cholesterol is transported first to the liver where it is bound to other lipids and proteins to form the very low density lipoprotein (VLDL). This aggregate moves through the blood stream delivering lipids to various tissues. As lipids are removed, the VLDL is converted to a low-density lipoprotein (LDL). Cells needing cholesterol can absorb LDL. Cholesterol is transported back to the liver by the high-density lipoprotein (HDL).

15. Dietary fish oils provide fatty acids which inhibit formation of thromboxanes, compounds which participate in blood clotting.

16. HDL is a cholesterol scavenger, picking up this steroid in the serum and returning it to the liver.

17. Because the interior of a lipid bilayer is very hydrophobic, molecules with hydrophilic character only cross the lipid bilayer with difficulty. A lipid bilayer acts as a barrier to water-soluble compounds.

18. Liposomes can package drugs so that only the target organ/tissue is exposed to the drug's effects.

19. A micelle forms when lipids such as fatty acids are mixed with water. The micelle is a sphere whose surface is composed of the hydrophilic ends of the lipids while the hydrophobic portions form the micelle's interior. In contrast, a liposome or vesicle forms when lipids like phospholipids or sphingolipids are mixed with water. The vesicle is a hollow sphere whose surface is formed by a lipid bilayer. A water solution is contained within.

20. Palmitoleic acid (m.p. = 0.5°C) has a lower melting point than palmitic acid (m.p. = 63°C). Palmitoleic acid is an unsaturated fatty acid while palmitic acid is saturated. Palmitoleic acid's *cis* double bond makes it difficult for the molecules to stack together and means this fatty acid will melt at a lower temperature than palmitic acid.

21. Up to 80% of the total body cholesterol is metabolically produced. The statin drugs block metabolic production of cholesterol.

22. They are all derived from arachidonic acid and they all have 20 carbon atoms.

23. Three ester groups.

$$CH_3CH_2O - \overset{\overset{\displaystyle O}{\|}}{\underset{\underset{\displaystyle OCH_2CH_3}{|}}{P}} - OCH_2CH_3$$

triethyl phosphate

SOLUTIONS TO EXERCISES

1.

$$CH_2O—\overset{\overset{\displaystyle O}{\|}}{C}—(CH_2)_7\overset{CH=CH}{\underset{CH_2}{}}\overset{CH=CH}{\underset{(CH_2)_4CH_3}{}}$$
$$|$$
$$CHOH$$
$$|$$
$$CH_2OH$$

(Linoleic acid can be located at any one of the three hydroxyls of glycerol.)

$$CH_2O—\overset{\overset{\displaystyle O}{\|}}{C}—(CH_2)_7\overset{CH=CH}{\underset{CH_2}{}}\overset{CH=CH}{\underset{(CH_2)_4CH_3}{}}$$
$$|$$
$$CHOH \quad O$$
$$| \qquad \|$$
$$CH_2O—C—(CH_2)_7\overset{CH=CH}{\underset{CH_2}{}}\overset{CH=CH}{\underset{(CH_2)_4CH_3}{}}$$

(Linoleic acid can be located at any two of the three hydroxyls of glycerol.)

2.

$$CH_2O—\overset{\overset{\displaystyle O}{\|}}{C}—(CH_2)_7\overset{CH=CH}{\underset{CH_2}{}}\overset{CH=CH}{\underset{CH_2}{}}\overset{CH=CH}{\underset{CH_2CH_3}{}}$$
$$|$$
$$CHOH$$
$$|$$
$$CH_2OH$$

(Linolenic acid can be located at any one of the three hydroxyls of glycerol.)

$$CH_2O—\overset{\overset{\displaystyle O}{\|}}{C}—(CH_2)_7\overset{CH=CH}{\underset{CH_2}{}}\overset{CH=CH}{\underset{CH_2}{}}\overset{CH=CH}{\underset{CH_2CH_3}{}}$$
$$|$$
$$CHOH \quad O$$
$$| \qquad \|$$
$$CH_2O—C—(CH_2)_7\overset{CH=CH}{\underset{CH_2}{}}\overset{CH=CH}{\underset{CH_2}{}}\overset{CH=CH}{\underset{CH_2CH_3}{}}$$

(Linolenic acid can be located at any two of the three hydroxyls of glycerol.)

3. A triacylglycerol containing one unit each of palmitic, stearic, and oleic acids:

$$CH_2O—\overset{\overset{\displaystyle O}{\|}}{C}(CH_2)_{14}CH_3 \qquad \text{palmitic acid}$$
$$|$$
$$CHO—\overset{\overset{\displaystyle O}{\|}}{C}(CH_2)_{16}CH_3 \qquad \text{stearic acid}$$
$$|$$
$$CH_2O—\overset{\overset{\displaystyle O}{\|}}{C}(CH_2)_7CH=CH(CH_2)_7CH_3 \qquad \text{oleic acid}$$

There would be two other triacylglycerols possible from these same components. Since the top and bottom attachments are equivalent, it only matters which acid is attached to the middle carbon of glycerol.

4. A triacylgylcerol containing two units of palmitic acid and one unit of oleic acid:

$$CH_2O-\overset{\overset{\displaystyle O}{\|}}{C}(CH_2)_{14}CH_3 \qquad \text{palmitic acid}$$

$$CHO-\overset{\overset{\displaystyle O}{\|}}{C}(CH_2)_{14}CH_3 \qquad \text{palmitic acid}$$

$$CH_2O-\overset{\overset{\displaystyle O}{\|}}{C}(CH_2)_7CH-CH(CH_2)_7CH_3 \qquad \text{oleic acid}$$

There is one other possible triacylglycerol with the same components; this triacylglycerol would have palmitic acid units at both ends with the oleic acid unit bound to the middle carbon of glycerol.

5. Yes. Hydrophobic *("water fearing")* molecules are relatively large and non-polar. A triacylglycerol is glycerol esterified with three fatty acids while a monoacylglycerol only contains one fatty acid. The triacylglycerol's extra size makes it much more hydrophobic.

6. No. Hydrophobic *("water fearing")* molecules are relatively large and non-polar. A triacylglycerol is glycerol esterified to three fatty acids while a phospholipid only contains two fatty acids. Additionally, phospholipids contain functional groups that attract water (e.g., phosphate). A phospholipid is smaller and more polar than a triacylglycerol, and, thus, is not more hydrophobic.

7. When palmitic acid reacts with glycerol, an ester forms linking the two molecules together.

$$CH_3(CH_2)_{14}COOH \;+\; \begin{matrix} CH_2OH \\ | \\ CHOH \\ | \\ CH_2OH \end{matrix} \longrightarrow \begin{matrix} CH_2O-\overset{\overset{\displaystyle O}{\|}}{C}(CH_2)_{14}CH_3 \\ | \\ CHOH \\ | \\ CH_2OH \end{matrix} \;+\; H_2O$$

the palmitic acid may be connected at any one of the three glycerol hydroxyl groups

8. When two stearic acids react with glycerol, two esters form linking the two stearic acid molecules to the glycerol.

$$2\ CH_3(CH_2)_{16}COOH + \begin{array}{c} CH_2OH \\ | \\ CHOH \\ | \\ CH_2OH \end{array} \longrightarrow \begin{array}{c} CH_2O-\overset{\overset{\displaystyle O}{\|}}{C}(CH_2)_{16}CH_3 \\ | \\ CHOH \\ | \\ CH_2O-\overset{\overset{\displaystyle O}{\|}}{C}(CH_2)_{16}CH_3 \end{array} +\ 2\ H_2O$$

the stearic acids may be connected at any two of the three glycerol hydroxyl groups

9. $CH_3(CH_2)_{16}\overset{\overset{\displaystyle O}{\|}}{C}-O-CH_2(CH_2)_{24}CH_3$

10. $CH_3(CH_2)_{14}\overset{\overset{\displaystyle O}{\|}}{C}-O-CH_2(CH_2)_{28}CH_3$

11. $\begin{array}{c} CH_2O-\overset{\overset{\displaystyle O}{\|}}{C}-(CH_2)_{18}CH_3 \\ | \\ CHOH \\ | \\ CH_2OH \end{array}$

(The arachidic acid may be located at any one of the three hydroxyls of glycerol.)

12.

$$\begin{array}{c} CH_2O-\overset{\overset{\displaystyle O}{\|}}{C}-(CH_2)_7\overset{\displaystyle CH=CH}{\diagup \qquad \diagdown}(CH_2)_7CH_3 \\ | \\ CHOH \\ | \\ CH_2OH \end{array}$$

(The oleic acid may be at any of the three hydroxyls of glycerol.)

13. The phospholipid structure is:

$$
\begin{array}{l}
\quad\quad\quad\quad O \\
\quad\quad\quad\quad \| \\
CH_2O-C(CH_2)_{14}CH_3 \\
| \\
\quad\quad\quad\quad O \\
\quad\quad\quad\quad \| \\
CHO-C(CH_2)_{14}CH_3 \\
| \\
\quad\quad\quad\quad O \\
\quad\quad\quad\quad \| \\
CH_2O-P-OCH_2CH_2NH_3^{+} \\
\quad\quad\quad\quad | \\
\quad\quad\quad\quad O^{-}
\end{array}
$$

The phosphoric acid and ethanolamine must be linked to the bottom glycerol carbon. Since a typical phospholipid contains two fatty acid units, a palmitic acid unit must be linked to the top glycerol carbon and another to the middle glycerol carbon.

14. The phospholipid structure is:

$$
\begin{array}{l}
\quad\quad\quad\quad O \\
\quad\quad\quad\quad \| \\
CH_2O-C(CH_2)_{16}CH_3 \\
| \\
\quad\quad\quad\quad O \\
\quad\quad\quad\quad \| \\
CHO-C(CH_2)_{16}CH_3 \\
| \\
\quad\quad\quad\quad O \quad\quad\quad\quad CH_3 \\
\quad\quad\quad\quad \| \quad\quad\quad\quad | \\
CH_2O-P-OCH_2CH_2N^{+}-CH_3 \\
\quad\quad\quad\quad | \quad\quad\quad\quad | \\
\quad\quad\quad\quad O^{-} \quad\quad\quad\quad CH_3
\end{array}
$$

The phosphoric acid and choline units must be linked to the bottom glycerol carbon. Since a typical phospholipid contains two fatty acid units, one stearic acid unit must be linked to the top glycerol carbon and a second stearic acid unit must be linked to the middle glycerol carbon.

15. The sphingolipid structure is:

$$
\begin{array}{l}
OH \\
| \\
CHCH=CH(CH_2)_{12}CH_3 \\
| \\
\quad\quad\quad\quad O \\
\quad\quad\quad\quad \| \\
CHNH-C(CH_2)_7CH=CH(CH_2)_7CH_3 \\
| \\
\quad\quad\quad\quad O \quad\quad\quad\quad CH_3 \\
\quad\quad\quad\quad \| \quad\quad\quad\quad | \\
CH_2O-P-OCH_2CH_2N^{+}-CH_3 \\
\quad\quad\quad\quad | \quad\quad\quad\quad | \\
\quad\quad\quad\quad O^{-} \quad\quad\quad\quad CH_3
\end{array}
$$

The phosphate and choline units are linked to the bottom glycerol carbon of sphingosine (when the molecule is written as above). The oleic acid unit is then linked to the nitrogen to form the sphingolipid.

16. The sphingolipid structure is:

$$
\begin{array}{l}
\overset{\displaystyle OH}{\underset{\displaystyle |}{}} \\
CHCH{=}CH(CH_2)_{12}CH_3 \\
| \qquad\qquad\quad O \\
| \qquad\qquad\quad \| \\
CHNH{-}C(CH_2)_{16}CH_3 \\
| \qquad\quad O \\
| \qquad\quad \| \\
CH_2O{-}P{-}OCH_2CH_2\overset{+}{N}H_3 \\
\qquad\quad | \\
\qquad\quad O^-
\end{array}
$$

The phosphate and ethanolamine units are linked to the bottom sphingosine carbon (when the molecule is written as above). The stearic acid unit is then linked to the nitrogen to form the sphingolipid.

17. Sphingomyelin is composed of sphingosine, one fatty acid, a phosphate, and choline.

$$
\begin{array}{l}
\overset{\displaystyle OH}{\underset{\displaystyle |}{}} \\
HC{-}CH{=}CH(CH_2)_{12}CH_3 \\
| \qquad\quad H \quad O \\
| \qquad\quad | \quad\; \| \\
HC{-}N{-}C(CH_2)_{14}CH_3 \\
| \qquad\quad O \\
| \qquad\quad \| \\
CH_2O{-}P{-}OCH_2CH_2\overset{+}{N}(CH_3)_3 \\
\qquad\quad | \\
\qquad\quad O^-
\end{array}
$$

18. A triacylglycerol (triglyceride) consists of three fatty acids esterified to glycerol.

$$
\begin{array}{l}
\qquad\qquad O \\
\qquad\qquad \| \\
CH_2O{-}C(CH_2)_{14}CH_3 \\
| \qquad\quad O \\
| \qquad\quad \| \\
CHO{-}C(CH_2)_{14}CH_3 \\
| \qquad\quad O \\
| \qquad\quad \| \\
CH_2O{-}C(CH_2)_{14}CH_3
\end{array}
$$

19. A glycolipid consists of sphingosine, one fatty acid, and carbohydrate.

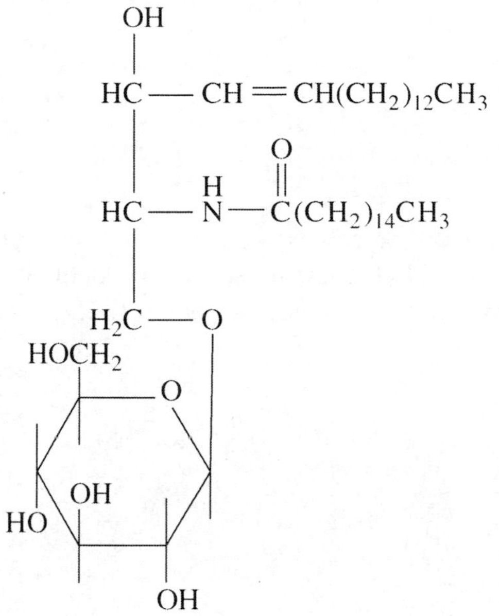

$$
\begin{array}{l}
\underset{|}{\overset{OH}{|}} \\
HC-CH=CH(CH_2)_{12}CH_3 \\
\end{array}
$$

OH
|
HC — CH = CH(CH₂)₁₂CH₃

        O
    H   ‖
HC — N — C(CH₂)₁₄CH₃
|
H₂C — O
HOCH₂
         O

    OH
HO

    OH

20. Phosphatidylethanolamine contains glycerol, two fatty acids, a phosphate, and ethanolamine.

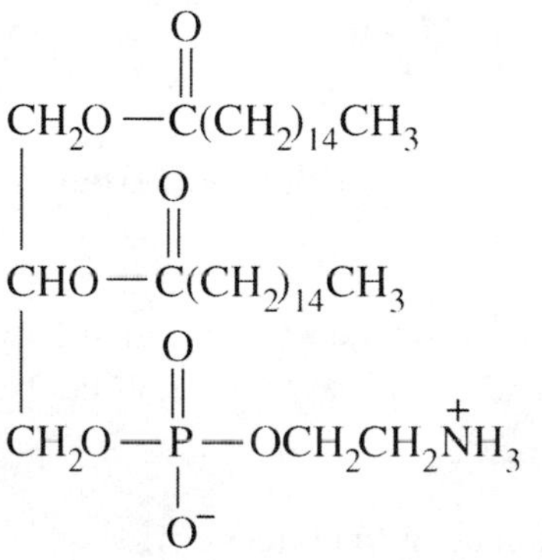

$$
\begin{array}{l}
\overset{O}{\overset{\|}{}} \\
CH_2O-C(CH_2)_{14}CH_3 \\
|\quad\;\; O \\
|\quad\;\; \| \\
CHO-C(CH_2)_{14}CH_3 \\
|\quad\;\; O \\
|\quad\;\; \| \\
CH_2O-P-OCH_2CH_2\overset{+}{N}H_3 \\
\quad\;\;\; | \\
\quad\;\; O^-
\end{array}
$$

21. LDL (low-density lipoprotein) differs from VLDL (very low density lipoprotein) in that
    (a) VLDL has a lower density than LDL.
    (b) VLDL is formed in the liver while LDL is formed from VLDL as the lipoproteins circulate in the blood.
    (c) VLDL is larger than LDL.

22. HDL (high-density lipoprotein) differs from LDL (low-density lipoprotein) in that
    (a) LDL has a lower density than HDL.
    (b) LDL delivers cholesterol to peripheral tissues while HDL scavenges cholesterol and returns it to the liver.
    (c) LDL is often described as "bad" cholesterol because it causes atherosclerosis while HDL is described as "good" cholesterol because it aids in removing cholesterol from the body.

23.

$$\underset{\substack{|\\ \text{CH}-\text{NH}_2\\ |\\ \text{CH}_2\text{OH}\\ \text{sphingosine}}}{\overset{\substack{\text{OH}\\ |\\ \text{CHCH}=\text{CH(CH}_2)_{12}\text{CH}_3}}{}}$$

$$\underset{\substack{|\\ \text{CH}-\text{OH}\\ |\\ \text{CH}_2\text{OH}\\ \text{monoacylglycerol}}}{\overset{\substack{\text{O}\\ ||\\ \text{CH}_2-\text{O}-\text{C}-\text{R}}}{}}$$

OH

$\text{CH}-\text{NH}_2$

CH_2OH

sphingosine

$\text{CHCH}=\text{CH(CH}_2)_{12}\text{CH}_3$

O
||
$\text{CH}_2-\text{O}-\text{C}-\text{R}$

$\text{CH}-\text{OH}$

CH_2OH

monoacylglycerol

Sphingosine is similar to the monoacylglycerol in that (a) both molecules contain a long hydrophobic chain, (b) the sphingosine amino group reacts with a fatty acid as does the secondary alcohol of the monoacylglycerol and, (c) for both compounds, the primary alcohol can react further with either acids (to form esters) or sugars (to form acetals).

24.

OH

$\text{CHCH}=\text{CH(CH}_2)_{12}\text{CH}_3$

O
||
$\text{CHNH}-\text{C}-\text{R}$

CH_2OH

sphingosine and fatty acid unit

O
||
$\text{CH}_2\text{O}-\text{C}-\text{R}$

O
||
$\text{CHO}-\text{C}-\text{R}$

CH_2OH

diacylglycerol

(a)   Both compounds have two long hydrophobic chains.

(b)   For both compounds, the primary alcohol can react further with either acids (to form esters) or sugars (to form acetals).

25.   Sodium ions will move from a region of high concentration to a region of low concentration as they move from a $0.1\,M$ solution across a membrane to a $0.001\,M$ solution. This process does not require energy and can be accomplished by facilitated diffusion.

26.   Phosphate ion will move from a region of low concentration to a region of high concentration as it moves from a $0.1\,M$ solution across a membrane to $0.5\,M$ solution. This process requires energy and must be accomplished by active transport.

27.   Lipoxygenase is an enzyme that starts the process to make leukotrienes.

28.   Cyclooxygenase is an enzyme that starts the process to make prostaglandins, thromboxanes, or prostacyclins.

29.   (a)   phosphatidylethanolamine; stearic acid and oleic acid
       (b)   sphingomyelin; palmitic acid

30.   (a)   triacylglycerol (triglyceride); palmitic acid, oleic acid, linoleic acid
       (b)   glycolipid; stearic acid

31.   Glycolipid. A glycolipid contains sphingosine that bonds to a fatty acid via an amide link and a carbohydrate (e.g., D-glucose).

32. Diacylglycerol (diglyceride). A diacylglycerol contains only glycerol and two fatty acids (linked to the glycerol by ester bonds).

33. Ibuprofen (an NSAID) inhibits the cyclooxygenase enzyme. This NSAID blocks the oxidation of arachidonic acid to form prostaglandins, thus, blocking inflammation.

34. Molecule A should be more hydrophobic than molecule B. Molecule A is larger and less polar, both features that contribute to hydrophobicity. Also, molecule A is in the hydrocarbon category—a class of molecules that are hydrophobic.

35. Omega numbering counts from the end furthest from the carboxyl group and indicates the position of the first double bond.
    (a) From Table 28.1, a common 18-carbon, omega-6 fatty acid is linoleic acid:

$$\overset{1}{CH_3CH_2CH_2CH_2CH_2} \diagup \overset{6}{CH}=CH \diagdown \diagup CH=CH \diagdown$$
$$CH_2 \qquad CH_2CH_2CH_2CH_2CH_2CH_2CH_2COOH$$

an omega-6 fatty acid

From Table 28.1, a common 20-carbon, omega-3 fatty acid is eicosapentaenoic acid:

$$\overset{1}{CH_3CH_2} \diagup \overset{3}{CH}=CH \diagdown CH_2 \diagup CH=CH \diagdown CH_2 \diagup CH=CH \diagdown CH_2 \diagup CH=CH \diagdown CH_2 \diagup CH=CH \diagdown CH_2CH_2CH_2COOH$$

an omega-3 fatty acid

    (b) Eicosanoids are formed from omega-3 and omega-6 fatty acids. These local hormones coordinate various cellular responses including blood pressure changes, blood clotting, and immune responses.

36. This meal is changed by increasing the fat content from 10 grams to 15 grams, an increase of 5 grams. Since each gram of fat yields an average of 9.5 Cal, an increase of 5 grams equates to an increase of 47.5 Cal (5 grams $\times$ 9.5 Cal/g). Thus, this meal will now contain 234 Cal + 47.5 Cal = 282 Cal.

37. Formula for beeswax

$$CH_3(CH_2)_{14}\overset{\overset{\textstyle O}{\|}}{C}-O-(CH_2)_{29}CH_3$$

38. $CH_3(CH_2)_7CH=CH(CH_2)_7COOH$
    oleic acid or 9-octadecenoic acid

39. original triacylglycerol

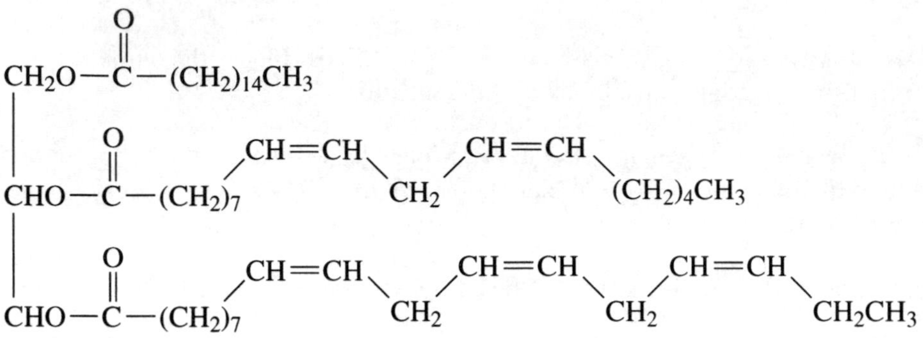

$$CH_2O-\overset{\overset{\displaystyle O}{\|}}{C}-(CH_2)_{14}CH_3$$

$$CHO-\overset{\overset{\displaystyle O}{\|}}{C}-(CH_2)_7\overset{CH=CH}{\diagdown}\underset{CH_2}{}\overset{CH=CH}{\diagup\diagdown}(CH_2)_4CH_3$$

$$CHO-\overset{\overset{\displaystyle O}{\|}}{C}-(CH_2)_7\overset{CH=CH}{\diagup\diagdown}\underset{CH_2}{}\overset{CH=CH}{\diagup\diagdown}\underset{CH_2}{}\overset{CH=CH}{\diagup\diagdown}CH_2CH_3$$

(The order in which the fatty acids are connected to glycerol may vary.)

final product

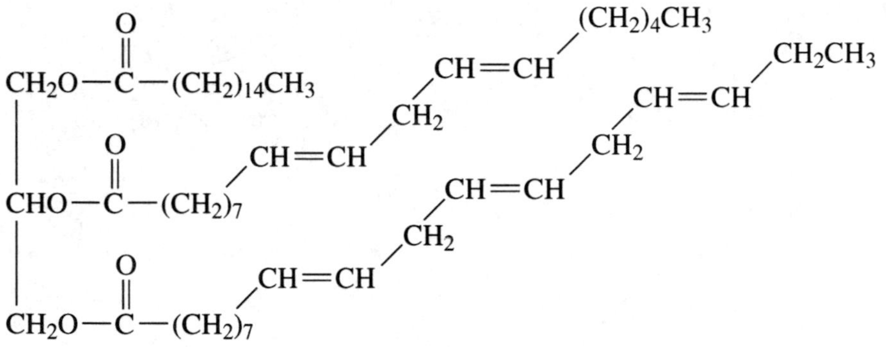

$$CH_2O-\overset{\overset{\displaystyle O}{\|}}{C}-(CH_2)_{14}CH_3$$

$$CHO-\overset{\overset{\displaystyle O}{\|}}{C}-(CH_2)_7\;CH=CH\;\underset{CH_2}{}\;CH=CH\;(CH_2)_4CH_3\;\;CH=CH\;\underset{CH_2}{}\;CH=CH\;CH_2CH_3$$

$$CH_2O-\overset{\overset{\displaystyle O}{\|}}{C}-(CH_2)_7\;CH=CH\;\underset{CH_2}{}\;CH=CH$$

(The order in which the fatty acids are connected to glycerol may vary.)

CHAPTER 29

AMINO ACIDS, POLYPEPTIDES, AND PROTEINS

SOLUTIONS TO REVIEW QUESTIONS

1. The designation α (alpha) means that the amine group in common amino acids is connected to the carbon immediately adjacent to the carboxylic acid. The designation, L, means that the common amino acids all have a specific configuration around the α-carbon. The amine group is on the left when the amino acids are written in a standard Fischer projection formula.

2. CH_3CH_2—CH—CH—COOH $\qquad$ H_2N-$(CH_2)_4$-CH—COOH

    $\qquad\qquad$ | $\quad$ | $\qquad\qquad\qquad\qquad\qquad$ |

    $\qquad\qquad$ CH_3 $\quad$ NH_2 $\qquad\qquad\qquad\qquad\qquad$ NH_2

    $\qquad$ I, isoleucine $\qquad\qquad\qquad\qquad\qquad$ K, lysine

3. The amino acid, lysine (K), has a pH of 9.7 at its isoelectric point. This pH fits into the range of 7.8 to 10.8 found for basic amino acids.

4. The zwitterion form of isoleucine (I) follows:

    CH_3CH_2—CH—CH—$\overline{COO}$

    $\qquad\qquad$ | $\quad$ |$_+$

    $\qquad\qquad$ CH_3 $\quad$ NH_3

    This is a zwitterion because the compound is ionized but the charges cancel so that the overall charge is zero at its isoelectric point.

5. Amino acids are amphoteric because the carboxyl group can react with a base to form a salt, or the amine group can react with an acid to form a salt. They are optically active because the alpha carbon is chiral, except for glycine. They commonly have the L configuration at carbon two, as in L-serine.

6. At its isoelectric point, a protein molecule must have an equal number of positive and negative charges.

7. (a) Primary structure. The number, kind, and sequence of amino acid units comprising the polypeptide chain making up a molecule.
    (b) Secondary structure. Regular three-dimensional structure held together by the hydrogen bonding between the oxygen of C=O groups and the hydrogen of the N—H groups in the polypeptide chains.
    (c) Tertiary structure. The distinctive and characteristic three-dimensional conformation or shape of a protein molecule.
    (d) Quaternary structure. The three-dimensional shape formed by an aggregate of protein subunits found in some complex proteins.

8. The sulfur-containing amino acid, cysteine, has the special role in protein structure of creating disulfide bonding between polypeptide chains which helps control the shape of the molecule.

9. Collagen is a good structural protein because its three-dimensional structure is held together strongly. Three protein strands are coiled in left-handed helices and then wrapped together in a right-handed helix. This cable construction resists stretching.

10. Ferritin is a good iron storage protein because its three-dimensional structure provides a sack for holding iron atoms. This protein is made up of many subunits that together form a hollow sphere (a quaternary structure) within which the iron is stored.

11. Hydrolysis breaks the peptide bonds, thus disrupting the primary structure of the protein. Denaturation involves alteration or disruption of the secondary, tertiary, or quaternary but not of the primary structure of proteins.

12. Amino acids containing a benzene ring give a positive xanthoproteic test (formation of yellow-colored reaction products). Among the common amino acids, these would include phenylalanine, tryptophan, and tyrosine.

13. The visible evidence observed in the:
    (a) Xanthoproteic test gives a yellow-colored reaction product when a protein containing a benzene ring is reacted with concentrated nitric acid.
    (b) Biuret test gives a violet color when dilute $CuSO_4$ is added to an alkaline solution of a peptide or a protein.
    (c) Ninhydrin test gives a blue solution with all amino acids except proline and hydroxyproline, both of which produce a yellow solution when ninhydrin is added to an amino acid.
    (d) In the Lowry Assay test a dark violet-blue color is produced when a protein contains tyrosine and tryptophan amino acids.
    (e) In the Bradford Assay test a deep blue color develops when a protein binds to the dye Coomassie Brilliant Blue.

14. Protein column chromatography uses a column packed with polymer beads (solid phase) through which a protein solution (liquid phase) is passed. Proteins separate based on differences in how they react with the solid phase. The proteins move through the column at different rates and can be collected separately.

15. (a) Thin layer chromatography is a way of separating substances based on a differential distribution between two phases, the liquid phase and the solid phase.
    (b) A strip (or sheet) is prepared with a thin coating (layer) of dried alumina or other adsorbent. A tiny spot of solution containing a mixture of amino acids is placed near the bottom of the strip. After the spot dries, the bottom edge of the strip is placed in a suitable solvent. The solvent ascends in the strip, carrying the different amino acids upwards at different rates. When the solvent front nears the top, the strip is removed from the solvent and dried.
    (c) Ninhydrin is the reagent used to locate the different amino acids on the strip.

16. In ordinary electrophoresis the rate of movement of a protein depends on its charge and size. In SDS electrophoresis a detergent, sodium dodecyl sulfate, is added to the protein solution, which masks the differences in protein charges, leaving the separation primarily due to the size of the various proteins.

17. Simple proteins contain only amino acids in their structure. Conjugated proteins contain another substance in their structure.

18. All amino acids (except one) have a chiral carbon atom. Since all proteins contain amino acids, all proteins will rotate plane polarized light.

19. Another name for a peptide-linkage is an amide bond.

20. The primary structure of a protein is not changed during denaturation.

21. The ribbon structure represents the primary structure without the side chains. The spacefilling structure represents the entire protein structure.

SOLUTIONS TO EXERCISES

1.

$$\begin{array}{c} \text{COOH} \\ | \\ \text{H}_2\text{N} - \text{C} - \text{H} \\ | \\ \text{CH}_2\text{OH} \end{array} \qquad \begin{array}{c} \text{COOH} \\ | \\ \text{H} - \text{C} - \text{NH}_2 \\ | \\ \text{CH}_2\text{OH} \end{array}$$

L-serine   D-serine

The primary alcohol group causes serine to be hydrophilic and, thus, this amino acid prefers to be on the surface of proteins where it can interact with water.

2.

$$\begin{array}{c} \text{COOH} \\ | \\ \text{NH}_2 - \text{C} - \text{H} \\ | \\ \text{CH}_2 \end{array} \qquad \begin{array}{c} \text{COOH} \\ | \\ \text{H} - \text{C} - \text{NH}_2 \\ | \\ \text{CH}_2 \end{array}$$

L-phenylalanine   D-phenylalanine

The benzene ring causes phenylalanine to be hydrophobic and, thus, this amino acid prefers to be inside proteins and away from water.

3. Basic. The amine functional group allows the amino acid side chain to accept a proton under physiological conditions (pH of about 7). Thus, this amino acid would be classed as basic and also as "positively-charged."

4. Polar, uncharged. The amide functional group causes the amino acid side chain to be polar but uncharged.

5. Pro = proline

$$\overset{+}{\text{NH}_2} \qquad \text{COO}^-$$

6. Gln = glutamine.

$$\begin{array}{c} \text{O} \\ \| \\ \text{NH}_2\text{CCH}_2\text{CH}_2\text{CHCOO}^- \\ | \\ \overset{+}{\text{NH}_3} \end{array}$$

7. For phenylalanine:
   (a) zwitterion formula   (b) formula in 0.1 M H$_2$SO$_4$ (c) formula in 0.1 M NaOH

   CH$_2$CHCOO$^-$ CH$_2$CHCOOH   CH$_2$CHCOO$^-$
   NH$_3^+$ NH$_3^+$ NH$_2$

8. For tryptophan:
   (a) zwitterion formula   (b) formula in 0.1 M H$_2$SO$_4$ (c) formula in 0.1 M NaOH

   CH$_2$CHCOO$^-$ CH$_2$CHCOOH   CH$_2$CHCOO$^-$
   NH$_3^+$ NH$_3^+$ NH$_2$

9. H$_2$N—C—NHCH$_2$CH$_2$CH$_2$CHCOOH + CH$_3$—CHCOOH ⟶
   ‖ | |
   NH   NH$_2$ NH$_2$

   O   CH$_3$
   ‖ |
   H$_2$N—C—NHCH$_2$CH$_2$CH$_2$CHCNHCHCOOH
   ‖ |
   NH   NH$_2$

   CH$_3$
   |
   OH   O   CHOH
   | ‖ |
10. CH$_2$COOH + CH$_3$CHCHCOOH ⟶ CH$_2$CNHCHCOOH + H$_2$O
    | | |
    NH$_2$ NH$_2$ NH$_2$

11. At a very acidic pH, the dipeptide will carry two positive charges. The amine end of the dipeptide and the side chain of the arginine will be protonated.

    O   CH$_3$
    ‖ |
    H$_2$N—C—NHCH$_2$CH$_2$CH$_2$CHCNHCHCOOH
    ‖$_+$ |$_+$
    NH$_2$ NH$_3$

12. At a very basic pH, the carboxyl end of the dipeptide will lose a proton and will carry a negative charge.

$$\begin{array}{c} CH_3 \\ | \\ O \quad CHOH \\ \| \quad | \\ CH_2CNHCHCOO^- \\ | \\ NH_2 \end{array}$$

13. The two dipeptides containing serine and alanine:

$$\begin{array}{cc} CH_2OH \quad CH_3 \\ | \quad\quad | \\ NH_2CHC\!-\!NHCHCOOH \\ \| \\ O \end{array}$$
peptide bond

Ser-Ala

$$\begin{array}{cc} CH_3 \quad CH_2OH \\ | \quad\quad | \\ NH_2CHC\!-\!NHCHCOOH \\ \| \\ O \end{array}$$
peptide bond

Ala-Ser

14. The two dipeptides containing glycine and threonine:

$$\begin{array}{c} CH_3 \\ | \\ CHOH \\ | \\ NH_2CH_2C\!-\!NHCHCOOH \\ \| \\ O \end{array}$$
peptide bond

Gly-Thr

$$\begin{array}{c} CH_3 \\ | \\ CHOH \\ | \\ NH_2CHC\!-\!NHCH_2COOH \\ \| \\ O \end{array}$$
peptide bond

Thr-Gly

15. All the possible tripeptides containing one unit each of glycine, phenylalanine, and leucine:

GFL   GLF   FGL
FLG   LGF   LFG

16. All the possible tripeptides containing one unit each of tyrosine, aspartic acid, and alanine:

YDA   YAD   DYA
DAY   AYD   ADY

17. Hydrolysis breaks the peptide bonds. One water molecule will react with each peptide bond, a hydrogen atom attaches to the nitrogen to complete the amino group and an -OH group attaches to the carboxyl carbon. The tripeptide, Ala-Phe-Asp, will hydrolyze to yield the following:

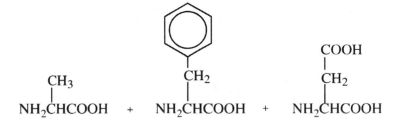

18. Hydrolysis breaks the peptide bonds. One water molecule will react with each peptide bond, a hydrogen atom attaches to the nitrogen to complete the amino group and an -OH group attaches to the carboxyl carbon. The tripeptide, Ala-Glu-Tyr, will hydrolyze to yield the following:

$$
\underset{\substack{| \\ NH_2CHCOOH}}{CH_3} \quad + \quad \underset{\substack{| \\ CH_2 \\ | \\ CH_2 \\ | \\ NH_2CHCOOH}}{COOH} \quad + \quad \underset{\substack{| \\ CH_2 \\ | \\ NH_2CHCOOH}}{OH-C_6H_4}
$$

19.
$$
-\underset{\substack{| \\ H}}{\overset{\substack{H \\ |}}{C}}-\underset{\substack{\| \\ O}}{C}-\underset{\substack{| \\ H}}{N}-\underset{\substack{| \\ H}}{\overset{\substack{CH_3 \\ |}}{C}}-
$$

20.
$$
-\underset{\substack{| \\ H}}{\overset{\substack{H \\ |}}{C}}-\underset{\substack{\| \\ O}}{C}-\underset{\substack{| \\ H}}{N}-\underset{\substack{| \\ H}}{\overset{\substack{CH_3 \\ |}}{C}}-
$$

21. Tertiary protein structure is usually held together by bonds between amino acid side chains. Serine side chains will hydrogen bond to each other:

$$
-CH_2O-H \cdots \overset{\substack{H \\ |}}{O}CH_2-
$$
hydrogen bond

22. Tertiary protein structure is usually held together by bonds between amino acid side chains. At pH = 7, the lysine side chain will contain a positive charge, $\overset{+}{N}H_3CH_2CH_2CH_2CH_2-$, and the aspartic acid side chain will contain a negative charge, $^-OOCCH_2-$. These two side chains will be held together by an ionic bond:

$$
-CH_2CO\overset{-}{O} ----- \overset{+}{N}H_3CH_2CH_2CH_2CH_2-
$$
ionic bond

23. Elastin serves a structural function; it is an integral part of artery vessel walls. The elasticity of this protein allows arteries to flex with changes in blood pressure. Elastin is also a fibrous protein which means that it has an especially strong structure, a characteristic important to its function.

24. Epidermal growth factor (EGF) has a regulatory function. Cells change their metabolism in response to EGF. This protein signals cells to grow and differentiate.

25. (a) There are three alpha helices in trypsin.
    (b) The space-filling model shows trypsin to be a compact, roughly spherical molecule. Trypsin is a globular protein.

26. (a) The ribbon structure of the epidermal growth factor shows four strands of beta-pleated sheet.
    (b) The space-filling model shows the epidermal growth factor (EGF) to be a compact protein. EGF is a globular protein.

27. The tripeptide, Gly-Ala-Thr, will
    (a) react with $CuSO_4$ to give a violet color. The tripeptide has the required two peptide bonds.
    (b) not react to give a positive xanthoproteic test because there are no benzene ring compounds in this tripeptide.
    (c) react with ninhydrin to give a blue solution. (Contains the required amino acids for reaction.)

28. The tripeptide, Gly-Ser-Asp, will
    (a) react with $CuSO_4$ to give a violet color. The tripeptide has the required two peptide bonds.
    (b) not react to give a positive xanthoproteic test because there are no benzene ring amino acids in this tripeptide.
    (c) react with ninhydrin to give a blue solution. (Contains the required number of amino acids for reaction.)

29. $$\left(\frac{1 \text{ mol Fe}}{1 \text{ mol cytochrome c}}\right)\left(\frac{55.85 \text{ g Fe}}{\text{mol Fe}}\right)\left(\frac{100. \text{ g cytochrome c}}{0.43 \text{ g Fe}}\right) = 1.3 \times 10^4 \frac{g}{mol}$$

The molar mass of cytochrome c is 1.3×10^4 g/mol

30. $$\left(\frac{4 \text{ mol Fe}}{1 \text{ mol hemoglobin}}\right)\left(\frac{55.85 \text{ g Fe}}{\text{mol Fe}}\right)\left(\frac{100. \text{ g hemoglobin}}{0.33 \text{ g Fe}}\right) = 6.8 \times 10^4 \frac{g}{mol}$$

The molar mass of hemoglobin is 6.8×10^4 g/mol

31. The amino acid sequence of the heptapeptide is:

```
                Gly - Phe - Leu
    Phe - Ala - Gly       Leu - Ala - Tyr
    Phe - Ala - Gly - Phe - Leu - Ala - Tyr
```

32. The amino acid sequence of the heptapeptide is:

```
                      Phe - Gly - Tyr
              Ala - Leu - Phe
    Phe - Ala - Ala
    Phe - Ala - Ala - Leu - Phe - Gly - Tyr
```

33. This newly discovered protein is probably a structural-support protein. The high percentage of beta-pleated sheet means that there are many hydrogen bonds holding the protein together in a very stable structure.

34. This newly discovered protein is probably not a structural-support protein because it is globular in shape and has secondary structure (beta pleated sheet) at its core. Thus, it is more likely to be a binding protein.

35. A domain is a compact piece of the overall protein structure that is relatively small (about the size of myoglobin, for example). A protein with a molar mass of about 452,000 g/mole is likely to have many domains.

36. A domain is a compact piece of protein structure of about 20,000 g/mole. The newly discovered protein with two domains is more likely to have a molar mass between 40,000 and 60,000 g/mole.

37. Alpha keratins have a high percentage of the alpha helix secondary structure. The alpha helix is like a spring in that it is stretchable, so hair is stretchable.

38. The silk protein, fibroin, has a high percentage of the secondary structure, beta-pleated sheet. The secondary structure is like a sheet of paper in that it is flexible but not stretchable. Thus, fibroin is not easily stretched.

39. The amino acid sequence of the nonapeptide is:

```
Arg - Pro
      Pro - Pro
            Pro - Gly - Phe
                        Phe - Ser
                              Ser - Pro - Phe
                                          Phe - Arg
_____
Arg - Pro - Pro - Gly - Phe - Ser - Pro - Phe - Arg
```

40. A binding site has attractive forces and a specific shape that selectively binds a particular biochemical.

41. Most fibrous proteins provide structural support. For example, connective tissue (e.g., tendons), skin, and blood vessel walls all are strengthened by fibrous proteins.

42. Transport proteins carry chemicals from one place to another. In general, motion proteins move cells or organisms from one place to another. Hemoglobin is an example of a transport protein that moves oxygen from the lungs to the tissues. Myosin is an example of a motion protein that allows muscles to contract.

43. The steroisomers of threonine:

44. The immunoglobulin hypervariable regions allow the body to produce millions of different immunoglobulins. Each immunoglobulin has a unique antigen binding site and two hypervariable regions with distinct amino acid sequences.

45. (a) The structure of alanine at pH = 9.0 would be $\overset{\displaystyle CH_3}{\underset{|}{}}$ NH_2CHCOO^-

(b) The structure of lysine at pH = 9.0 would be $\overset{+}{N}H_3CH_2CH_2CH_2CH_2CHCOO^-$
$$\underset{NH_3^+}{|}$$

(c) The net charge on lysine at pH = 9.0 would be positive [see the structure of lysine in part (b)].

46. Nineteen dipeptides can be written with glycine on the N-terminal side. Another nineteen are possible with glycine on the C-terminal end. Finally, one dipeptide can be written with two glycines giving a total of thirty-nine dipeptides.

47. Vasopressin will have a higher isoelectric point than oxytocin. Vasopressin has two different amino acids as compared with oxytocin, a phenylalanine instead of an isoleucine and an arginine instead of a leucine. Thus, vasopressin has one additional basic amino acid (arginine) which will cause the vasopressin isoelectric point to be higher than the oxytocin isoelectric point.

48. Leucine   Alanine   Glutamic Acid

$(CH_3)_2CHCH_2CHCOOH$ $CH_3CHCOOH$ $HOOCCH_2CH_2CHCOOH$
$\qquad\quad\underset{NH_2}{|}$ $\quad\underset{NH_2}{|}$ $\qquad\qquad\quad\underset{NH_2}{|}$

Glutamic acid is the only one of these three amino acids with two polar bonds in its side chain. Thus, glutamic acid will be the most polar.

49. (a)

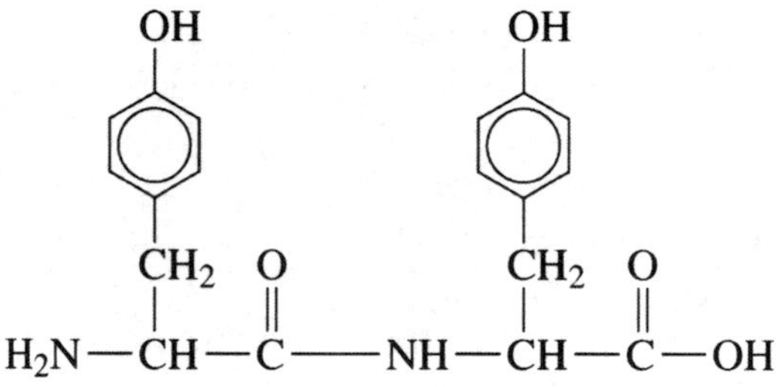

(b)

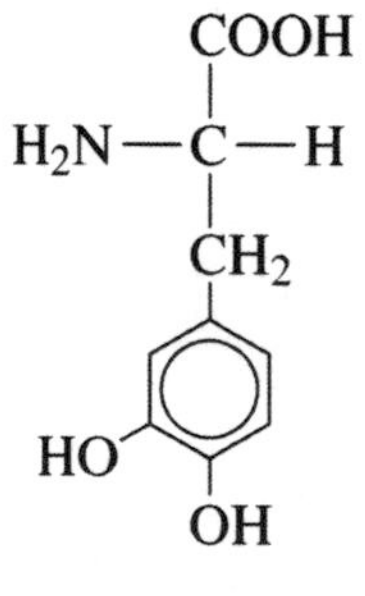

L-dopa 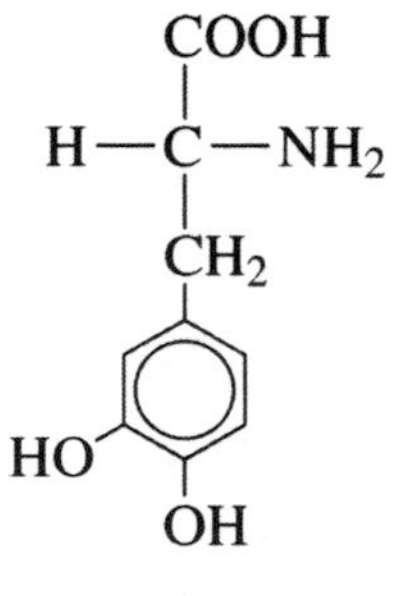   D-dopa

(c) Dopamine does not have a chiral carbon and thus, does not exist as a pair of stereoisomers.
(d) Norepinephrine contains a primary amine while epinephrine contains a secondary amine.

CHAPTER 30

ENZYMES

SOLUTIONS TO REVIEW QUESTIONS

1. Activation energy is the energy barrier to chemical reaction and is measured as the difference between the reactant(s) energy level and the transition state energy level. Enzymes lower the activation energy barrier and, thus, increase biochemical reaction rates.

2. Enzymes are proteins. As noted in the answer to question 1, enzymes serve as catalysts for chemical reactions in the body.

3. A coenzyme is the nonprotein part of a conjugated enzyme. An apoenzyme is the protein part of a conjugated enzyme.

4. Lactase catalyzes the reaction of lactose with water that splits the disaccharide into its component monosaccharides, glucose and galactose. This enzyme is in the IUB (International Union of Biochemistry) class of hydrolases because it hydrolyzes a carbohydrate.

5. An enzyme that converts D-mannose to D-glucose is causing an isomerization. This enzyme is in the IUB (International Union of Biochemistry) class of isomerases because it catalyzes an interconversion of stereoisomers.

6. By definition, a catalyst decreases the activation energy barrier for a reaction. This will decrease the energy difference between reactants and the transition state. It will not change the energy difference between reactants and products.

7. The induced-fit model proposes that the enzyme active site changes its shape due to substrate binding. If the glucose isomerase active site is exactly complementary to the shape of the reactant, glucose, then there is no need for the active site shape to change. This enzyme would probably not follow the induced-fit model.

8. When substrates bind to an active site, they may be converted more easily to products because (1) the substrates are close together (proximity catalysis); (2) the substrates are oriented to best react (productive binding hypothesis); (3) when the substrates bind to the active site, they change shape to be more like products (strain hypothesis).

9. Both an enzyme substrate and an enzyme inhibitor may bind to the active site. However, a substrate will react to form product while an inhibitor will only block the active site from further reaction.

10. Pectinase is used to remove the peeling from oranges and grapefruit. The pectinase penetrates the peeling and dissolves the white stringy material that attaches the peeling to the fruit. The peeling can then be easily removed from the fruit.

11. Proteases digest proteins. The proteases in detergents digest protein stains allowing the stains to be washed away.

12. Specific proteases are injected into stroke victims to degrade plasminogen and, thereby, dissolve blood clots.

13. An enzyme activator causes the enzyme to be a better catalyst. An enzyme substrate is the molecule upon which the enzyme works—the reactant.

14. The statin drugs inhibit a key enzyme in cholesterol biosynthesis and, thus, lower blood cholesterol levels.

15. An enzyme has specificity to react with that particular substrate.

SOLUTIONS TO EXERCISES

1. reaction rate $= \left(\dfrac{0.005\,M}{3.5\,\text{min}}\right)\left(\dfrac{1\,\text{min}}{60\,\text{s}}\right) = 2 \times 10^{-5}\,M/s$

2. reaction rate $= \left(\dfrac{0.02\,M}{8\,\text{min}}\right)\left(\dfrac{1\,\text{min}}{60\,\text{s}}\right) = 4 \times 10^{-5}\,M/s$

3. (a) $\dfrac{0.03\,M}{5\,\text{min}} = 0.006\,M/\text{min}$

   (b) $\left(\dfrac{0.006\,M}{\text{min}}\right)\left(\dfrac{1000\,mM}{M}\right)\left(\dfrac{1\,\text{min}}{60\,\text{s}}\right) = 0.1\,mM/s$

4. (a) $\dfrac{0.15\,M}{37\,\text{min}} = 0.0041\,M/\text{min}$

   (b) $\left(\dfrac{0.0041\,M}{\text{min}}\right)\left(\dfrac{10^6\,\mu M}{M}\right) = 4100\,\mu M/\text{min}$

5. $\dfrac{0.028\,M}{2\,\text{min}} = \dfrac{0.014\,M}{\text{min}}$

6. $\left(\dfrac{0.75\,M}{4.0\,\text{hr}}\right)\left(\dfrac{1\,\text{hr}}{60\,\text{min}}\right) = 3.1 \times 10^{-3}\,M/\text{min}$

7. $\left(\dfrac{0.014\,M}{\text{min}}\right)\left(\dfrac{1\,\text{min}}{60\,\text{s}}\right) = 0.00023\,M/s$ or $2.3 \times 10^{-4}\,M/s$

8. $\left(\dfrac{0.0031\,M}{\text{min}}\right)\left(\dfrac{1000\,mM}{M}\right) = 3.1\,mM/\text{min}$

9. $\left(\dfrac{2470\,\text{molecules}}{360\,\text{s}}\right) = 6.9\,\text{molecules}/s$

10. $\dfrac{5.2 \times 10^4\,\text{molecules}}{500\,\text{s}} = 104\,\text{molecules}/s$

11. The turnover number for lysozyme shows that this enzyme converts 0.5 reactant to products every second. Thus, in 1 min,

    (0.5 reactant converted to products/s)$\left(\dfrac{60\,\text{s}}{\text{min}}\right)$(1 min) = 30 reactants converted to products

12. The turnover number for pepsin shows that this enzyme converts 1.2 reactants to products every second. Thus, in 5 min,

    (1.2 reactants converted to products/s) $\left(\dfrac{60\,\text{s}}{\text{min}}\right)$(5 min)(3 pepsin molecules) $= 1 \times 10^3$ reactants converted to products by three pepsin molecules

13. A protease catalyzes the breaking of a peptide bond by bring together the two reactants, water and a protein. The reaction is made quicker because the enzyme attracts both reactants to the active site—proximity catalysis.

14. The hexokinase active site attracts both glucose and ATP. By bringing the two reactants together, the reaction is made quicker—proximity catalysis.

15. For proximity catalysis, active site attraction brings the reactants together; for the productive binding hypothesis, active site attraction orients the reactants for optimum reaction.

16. For proximity catalysis, active site attraction brings the reactants together; for the strain hypothesis, active site attraction forces the reactant shape to change toward the product shape.

17. All proteins, including enzymes, lose their three-dimensional shape at high temperatures. This denaturation causes enzymes to lose their catalytic ability and results in a slowing of the enzyme-catalyzed reaction.

18. Very low pH destroys the natural 3-D enzyme shape (denaturation) and the enzyme becomes inactive.

19. Enzyme A: the reaction rate goes up much more gradually for Enzyme B because of the lower attraction of this enzyme for substrate.

20. Enzyme A: the reaction rate for Enzyme A goes to a much higher value because this enzyme has a higher V_{max}.

21. Salivary amylase probably has a pH optimum of about 6 because it is found in the mouth (pH = 6). The activity of salivary amylase will therefore decrease at pH values higher and lower than 6. In the stomach (pH = 2), the salivary amylase activity will be very low.

22. Since pepsin is produced in the stomach, this enzyme probably has a pH optimum around 2. When pepsin moves into the small intestine (pH = 7), the enzyme's activity decreases.

23. Since pepsin is produced in the stomach, this enzyme probably has a pH optimum around 2. The pepsin activity should decrease as the antacid neutralizes stomach acid and increases the pH.

24. Trypsin probably has a pH optimum around 7. This will allow the enzyme to be optimally active in its natural environment, the small intestine.

25. Condition A has the same reactant concentration and activation energy as Condition B. However, Condition A has a higher enzyme concentration. Since there is more catalyst in Condition A, the reaction will go faster.

26. Condition B is the same as Condition A except that the activation energy is lower (15 kcal/mole for Condition B versus 20 kcal/mole for Condition A). Condition B's lower activation energy will allow for a faster reaction.

27. This must be *proximity catalysis* because the enzyme attracts and brings the reactants together.

28. Although ribose has four hydroxyls that could react, only the hydroxyl at C-2 does react with ATP. The enzyme must orient ribose so that only the C-2 hydroxyl reacts. This is an example of the *productive binding hypothesis*.

29. Glutamine is the product of this liver metabolic pathway. Thus, glutamine control of this pathway must be "feedback". Since an increase in glutamine concentration causes the metabolic pathway to slow down, this control must be feedback inhibition. Feedback inhibition means that when a large amount of product has been formed, the beginning of a process will slow down.

30. Glutamine is a starting material (reactant) for the liver metabolic pathway which produces urea. Thus, glutamine control of this pathway must be "feedforward." Since an increase in glutamine concentration causes the metabolic pathway to speed up, this control must be feedforward activation.

31. Cellulases are used to digest cellulose containing materials. Paper producers use cellulases to complete the breakdown of wood chips to wood pulp, which is used to manufacture paper products.

32. Blood clots in the brain are the major cause of strokes. Specific proteases are used to dissolve these blood clots to alleviate the cause of a stroke.

33. A lipase is present in laundry detergents to digest fatty residues.

34. Amylase is present in laundry detergents to digest starchy residues.

35. Lactase splits lactose into glucose and galactose, a reaction that no longer occurs in the digestive tract of lactose-intolerant individuals. Once glucose and galactose are formed, digestion is easy.

36. Sucrase converts sucrose to a mixture of glucose and fructose. Since fructose is sweeter than sucrose, a sweeter solution is created.

37. The turnover number measures the number of substrate molecules converted to product by one enzyme molecule under optimum conditions. Chymotrypsin can convert one glycine-containing substrate to product every twenty seconds while the enzyme can convert two hundred L-tyrosine-containing substrates to products each second. Chymotrypsin is a much more efficient catalyst for the L-tyrosine-containing substrate.

38. Feedback inhibition means that when a large amount of product has been formed, the beginning of a process will slow down. Thus, feedback inhibition protects against overproduction. "Feedback activation" means that when a large amount of product is formed, the beginning of the process will accelerate. A state of overproduction will disrupt the normal activity of the cell due to overproduction of products and might lead to cell death.

39.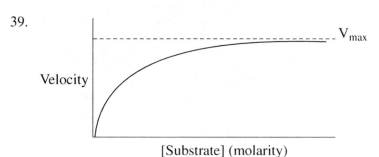

When the V_{max} is reached, an increase in substrate concentration will not increase the velocity.

40. The enzyme lactase catalyzes hydrolysis of the disaccharide lactose to produce two monosaccharides, D-glucose and D-galactose.

41. The induced-fit model requires a flexible active site because the active site must conform to the reactant shape as it binds. In contrast, the strain hypothesis requires a rigid active site. According to this hypothesis, the reactant is forced to change shape as it is pulled against a differently-shaped and firm active site.

CHAPTER 31

NUCLEIC ACIDS AND HEREDITY

SOLUTIONS TO REVIEW QUESTIONS

1. The five nitrogen bases found in nucleotides:

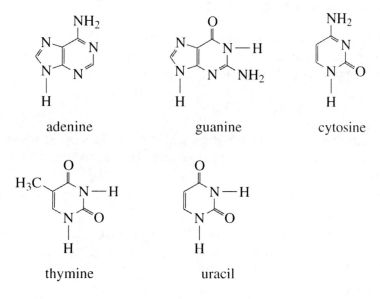

adenine   guanine   cytosine

thymine   uracil

2. A nucleoside is a purine or pyrimidine base linked to a sugar molecule, usually ribose or deoxyribose. A nucleotide is a purine or pyrimidine base linked to a ribose or deoxyribose sugar, which in turn is linked to a phosphate group.

3. There are three structural differences between DNA and RNA.
   (a) In RNA the sugar molecule is always ribose. In DNA, the sugar molecule is always deoxyribose, which has H instead of OH at carbon number two.
   (b) Both molecules use a mixture of four nitrogen bases. Both use cytosine, adenine, and guanine. In DNA, the fourth base is thymine. In RNA, the fourth base is uracil.
   (c) DNA exists as a double helix whereas RNA is a single strand of nucleotides.

4. The major function of ATP in the body is to store chemical energy, and to release it when called upon to carry out many of the complex reactions that are essential to most of our life processes. An equilibrium exists with ADP.

$$\text{ATP} + H_2O \underset{\text{energy storage}}{\overset{\text{energy utilization}}{\rightleftharpoons}} \text{ADP} + P_i + 35 \text{ kJ}$$

5. The structure of DNA as proposed by Watson and Crick is a double-stranded helix. Each strand has a backbone of alternating phosphate and deoxyribose units. Each deoxyribose unit has one of the four nitrogen bases attached, but coming off the backbone, not part of the backbone. These nitrogen bases thus link to their complementary nitrogen base by hydrogen bonds on the other strand of the double helix.

6. The ratio of adenine to thymine in the human liver is 1.00. This ratio would be expected because adenine and thymine are a complementary base pair in double-stranded DNA. That is, for every adenine on one DNA strand, there will be a corresponding thymine on the other strand.

7. Genes must code for 20 different amino acids. If a codon consisted of only one nucleotide, then only four different codons would be possible. A codon of two nucleotides allows for 16 different codons (4×4), still not enough codons for 20 different amino acids. In nature, a codon consists of three nucleotides because this allows 64 different codons ($4 \times 4 \times 4$)—more than enough for 20 different amino acids.

8. DNA replication starts with one double stranded DNA (the template strands) and produces two double-stranded DNA products. Each product contains one template strand and one newly-synthesized strand. Since each product has conserved one template strand out of two, replication is said to be "semiconservative."

9. A brief outline of the biosynthesis of proteins:
   (a) A DNA strand produces a complementary mRNA strand which leaves the nucleus and travels to the cytoplasm where it becomes associated with a cluster of ribosomes, binding to five or more ribosomes.
   (b) With the aid of an enzyme, the proper amino acid attaches to a tRNA molecule by an ester linkage.
   (c) The amino acids are brought to the protein synthesis site by tRNA.
   (d) The initiation of a polypeptide chain always uses the mRNA codon AUG or GUG, which ties to the tRNA anticodon UAC. This code brings N-formylmethionine for procaryotic cells. The amino group is blocked by the formyl group, leaving the carboxyl group available to react with the amino group of the next amino acid.
   (e) The next tRNA, bringing an amino acid, comes in to the mRNA and links up, anticodon to codon. The peptide linkage is then made between amino acids.
   (f) The first tRNA is ejected, and a third enters the ribosome.
   (g) The polypeptide chain terminates at a nonsense or termination-codon. The protein molecule breaks free.

10. Initiation of protein synthesis requires a special codon and starts with a single amino acid. In contrast, elongation requires an unfinished protein chain and, depending on the mRNA, may use any of the standard codons.

11. Termination of protein synthesis occurs when a "nonsense" or termination codon appears. Since no tRNA's recognize these codons, protein synthesis stops. In contrast, elongation requires an unfinished protein chain and, depending on the mRNA, may use any of the standard codons.

12. A codon is a triplet of three nucleotides, and each codon specifies one amino acid. The cloverleaf model of transfer RNA has an anticodon loop consisting of seven unpaired nucleotides. Three of these make up the anticodon, which is complementary to, and hydrogen-bonds to the codon on mRNA.

13. The role of N-formylmethionine in procaryotic protein synthesis is to start the polypeptide chain so it goes in the right direction. It can only build from the carboxyl end. After the synthesis, the N-formylmethionine breaks loose from the protein.

14. From time to time a new trait appears in an individual that is not present in either parents or ancestors. These traits which are generally the result of genetic or chromosomal changes are called mutations.

15. A "DNA fingerprint" is a pattern of tagged DNA fragments on an electrophoretic gel which can be used to identify possible suspects.

16. In genetic engineering, scientists insert specific genes into the genome of a host cell and program it to produce new and different proteins. Protein engineering uses the techniques of genetic engineering in microorganisms to make proteins with new primary structures and, thus, new functions. These new proteins often have an industrial application.

17. siRNA stands for small interfering ribonucleic acid. This RNA is "interfering" because it hydrogen bonds to, and triggers destruction of, messenger RNA.

18. Apoptosis is cellular self-destruction. If a normal cell is not functioning correctly, it can start a coordinated sequence of steps that lead to cell death. Apoptosis maintains an organism's health at the expense of individual cells. Often cancer cells grow because apoptosis has been inactivated. If this programmed cell death could be restarted, the cancer cells would be eliminated.

19. A mutation often causes a defective gene. Gene therapy seeks to replace the defective gene with a normal version, eliminating the mutation.

20. (a) ADP has two phosphate groups; ATP has three phosphate groups.
    (b) ATP has more potential energy.

21.

1,2-diazine   1,4-diazine

SOLUTIONS TO EXERCISES

1.

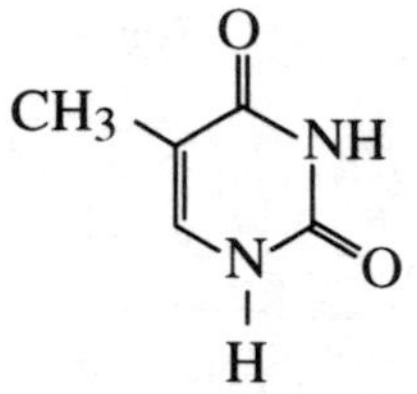

thymine

2.

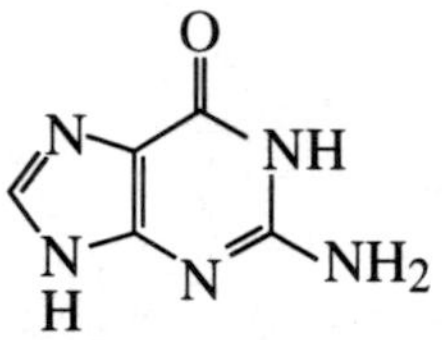

guanine

3. The letters are associated with compound names as follows:
   (a) A, adenosine
   (b) AMP, adenosine-5′-monophosphate
   (c) dADP, deoxyadenosine-5′-diphosphate
   (d) UTP, uridine-5′-triphosphate

4. The letters are associated with compound names as follows:
   (a) G, guanosine
   (b) GMP, guanosine-5′-monophosphate
   (c) dGDP, deoxyguanosine-5′-diphosphate
   (d) CTP, cytidine-5′-triphosphate

5. Structural formulas

   (a)   A   (b)   AMP

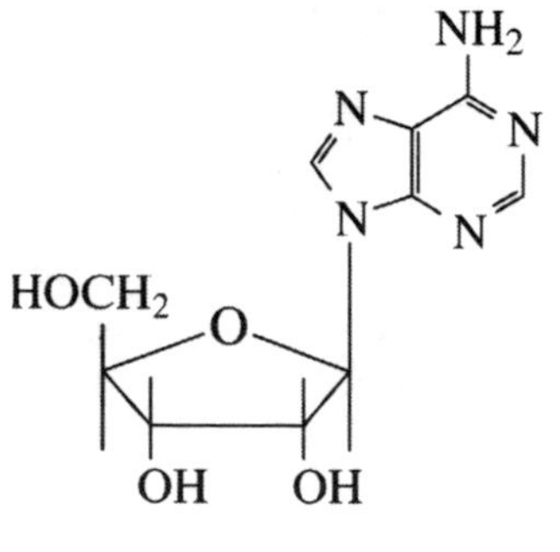

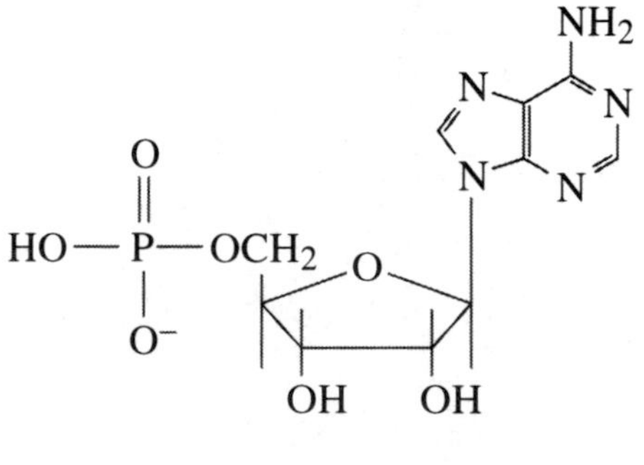

(c) CDP   (d) dGMP

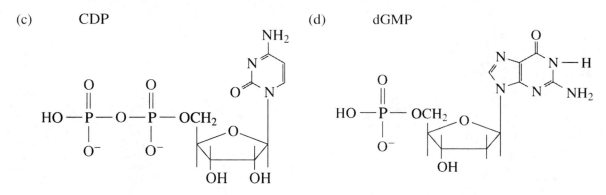

6. Structural formulas

(a) U   (b) UMP

(c) CTP   (d) dTMP

7. (a) This molecule is a nucleotide and is named cytidine-5′-monophosphate.

(b) This molecule is a nucleoside and is named deoxyadenosine.

8. (a) This molecule is a nucleoside and is named deoxyguanosine.

(b) This molecule is a nucleotide and is named thymidine-2′-monophosphate.

9.

10.

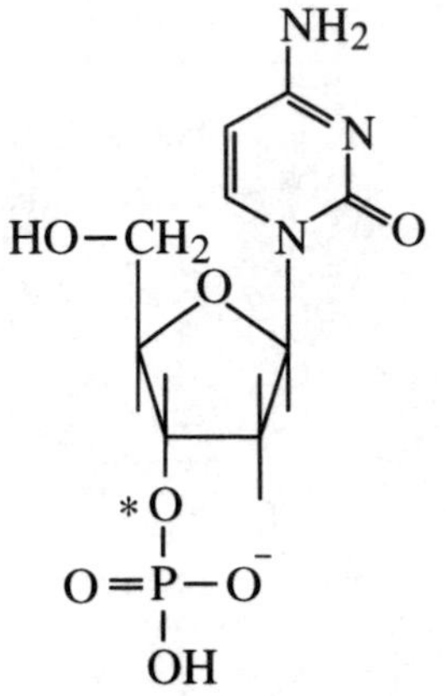

11. Two different small nucleic acids can be made from dCMP and dGMP:

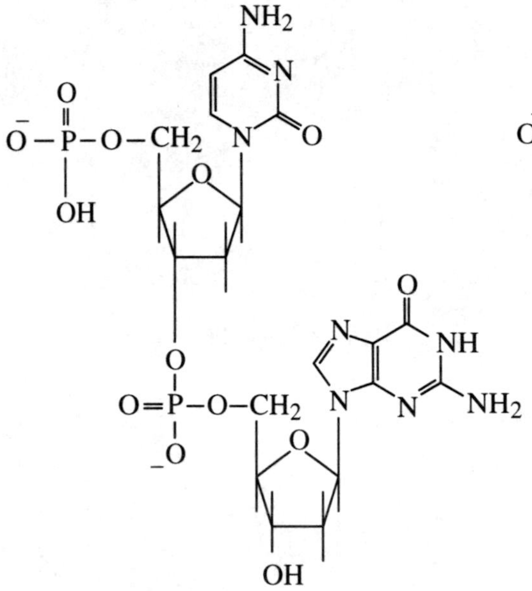

or

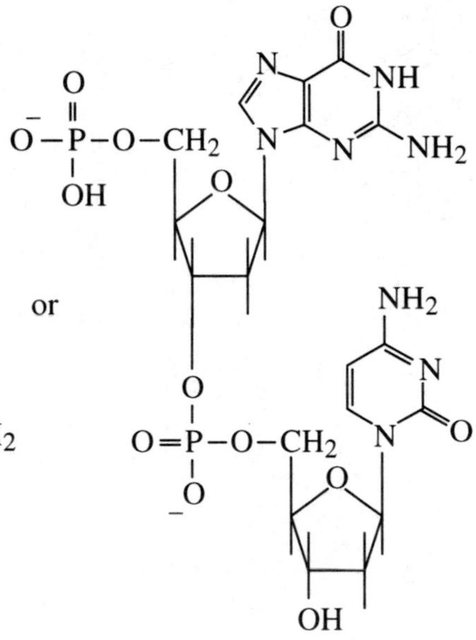

12. Two different small nucleic acids can be made from AMP and UMP:

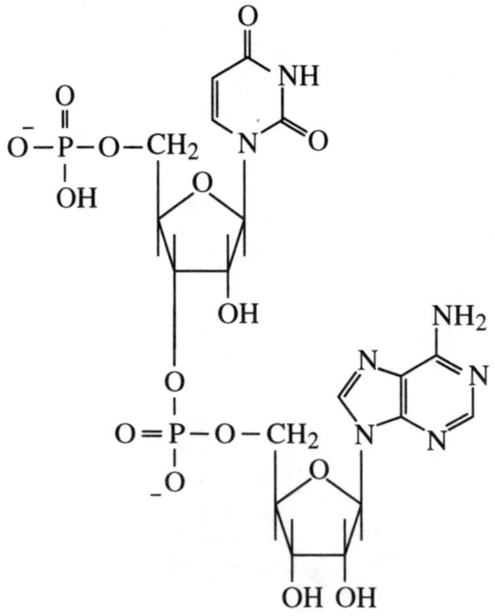

or

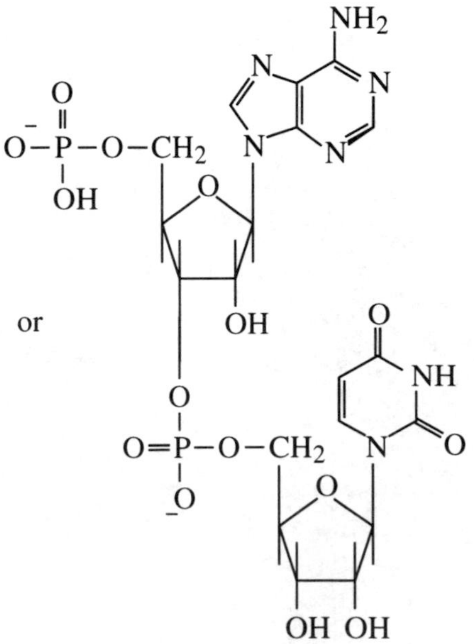

13.

adenine   uracil

14.

guanine   cytosine

15. tRNA acts as an "adapter" molecule. It (1) carries a specific amino acid to the ribosome and (2) matches its anticodon to a specific codon on the mRNA. In this way specific amino acids are added to a growing polypeptide chain following the message encoded in the RNA.

16. mRNA provides the information or "message" that dictates the protein amino acid sequence (the 1° structure).

17. Replication is the biological process of making DNA using a DNA template while the transcription process refers to the making of RNA using a DNA template.

18. Transcription is the process of making RNA using a DNA template while translation is a process for making protein using a mRNA template.

19. tRNA binds specific amino acids and brings them to the ribosome for protein synthesis. mRNA carries genetic information from DNA to the ribosome and serves as a template for protein synthesis.

20. Both mRNA and rRNA can be found in the ribosomes. rRNA serves as part of the ribosome structure while mRNA serves as a template for protein synthesis.

21. For a DNA sequence, TCAATACCCGCG,
    (a) the complementary mRNA will be: AGUUAUGGGCGC.
    (b) the sequence of amino acids coded by the DNA will be: Ser-Tyr-Gly-Arg

22. A segment of DNA strand consists of GCTTAGACCTGA.
    (a) The order in the complementary mRNA will be: CGAAUCUGGACU.
    (b) The sequence of amino acids coded by the DNA is Arg-Ile-Trp-Thr.

23. Transcription makes a polymer of nucleotides by forming phosphate ester bonds to connect the nucleotides to each other. The phosphate ester combines a phosphoric acid with an alcohol.

24. Translation makes a polymer of amino acids by forming amide bonds to connect the amino acids to each other. The amide bond combines an amine with a carboxylic acid.

25. Translation termination occurs when the ribosome reaches a "nonsense" or termination codon along the mRNA. No tRNA (in normal cells) has the anticodon to match the termination codon and so no more amino acids are added to the newly synthesized protein chain. The peptidyl-tRNA connection is broken and the protein chain is released from the ribosome.

26. Translation initiation occurs when the ribosome reaches a special AUG or GUG codon along the mRNA. Since there is commonly more than one AUG or GUG codon, the ribosome must use other information to choose the special AUG or GUG. This codon is the starting point for protein synthesis and is bound by either a special tRNA carrying N-formylmethionine (in procaryotes) or a tRNA carrying methionine (in eucaryotes).

27. mRNA codons and corresponding tRNA anticodons follow:

    (a) GUC: anticodon is CAG   (c) UUU: anticodon is AAA
    (b) AGG: anticodon is UCC   (d) CCA: anticodon is GGU

28. mRNA codons and corresponding tRNA anticodons follow:

    (a) CGC: anticodon is GCG   (c) GAU: anticodon is CUA
    (b) ACA: anticodon is UGU   (d) UUC: anticodon is AAG

29. siRNA (small interfering ribonucleic acid) and miRNA (microribonucleic acid) are found in the cytoplasm. Both commonly block translation (protein synthesis). miRNA stops translation by blocking correct functioning of the ribosome. In contrast, siRNA stops translation by causing destruction of the mRNA.

30. Both snRNA and snoRNA are small RNAs that aid posttranscriptional modification of RNA in the nucleus. snoRNA specifically helps modify rRNA in the nucleolus while snRNA is found in other parts of the nucleus and modifies other RNA.

31. This procedure would best be described as genetic engineering (c). The scientists are seeking to add new genetic material to allow more production of the nerve growth factor protein.

32. (b) protein engineering. This science works to modify proteins by changing individual amino acids in the protein's primary structure. These specific changes are caused by selective changes in the genetic code.

33. The original mRNA sequence codes for amino acids, . . . Leu-Pro-Thr . . . The mutation changes the first amino acid to Phe, so the sequence becomes . . . Phe-Pro-Thr . . . This mutation can change the function of the protein.

34. The original mRNA sequence codes for the amino acids . . . Val-Gln-Lys . . . The mutation changes the second codon to a termination or nonsense codon UAA. The protein will end at this point which may cause a major disruption of the protein function.

35. "Recombinant" means the DNA has genes that have been rearranged to contain new or different hereditary information.

36. "Nonsense" means the codon does not code for an amino acid. Instead, this codon causes translation to stop. (It is also known as a termination codon.)

37. Termination of protein synthesis means ending the polymerization of amino acids. Polymerization requires a continuous supply of tRNAs carrying specific amino acids. Each tRNA matches its anticodon to a specific codon on the mRNA. During termination a codon occurs in the mRNA that matches no tRNA. Without a tRNA the polymerization process cannot continue and protein synthesis ends.

38. The genome is the sum of all hereditary material contained in a cell. The genome contains many genes. Each gene is a segment of DNA that controls the formation of a molecule of RNA.

39. This thymine isomer now has the appropriate H-bonding groups to connect with guanine instead of adenine.

40. A substitution mutation will change one codon. An insertion mutation will shift the sequence position of all bases following the insertion by one. Each base will take the position of its neighbor. For example, if base $\boxed{C}$ is inserted into the sequence . . . UUC ACG GCC . . . , every codon will change, . . . U $\boxed{C}$ U CAC GGC C . . . In general, every codon following an insertion is changed.

41. For the mRNA segment, UUUCAUAAG,
    (a) the coded amino acids are: Phe-His-Lys.
    (b) the sequence of DNA for this mRNA is: AAAGTATTC.

42. In DNA fingerprinting, DNA polymerase is used to copy very small samples of DNA so that enough will be available to separate and visualize on gel electrophoresis.

43. No, RNA would not be predicted to have a 1:1 ratio of guanine to cytosine. In DNA, these bases have a 1:1; ratio because they are complementary and, thus, are found in equal amounts in the double-stranded nucleic acid. Since RNA is most often found as a single-stranded molecule, there is no requirement that the ratio of guanine to cytosine be set.

CHAPTER 32

NUTRITION

SOLUTIONS TO REVIEW QUESTIONS

1. Nutrients are components of food that provide for body growth, maintenance, and repair. Food is the material we eat.

2. The "energy allowance" represents the recommended dietary energy supply needed to maintain good health.

3. Marasmus is a chronic calorie deficiency while kwashiorkor represents a protein deficiency condition.

4. Adequate Intakes (AIs) are listed as the DRIs for very young infants (0.0–0.5 yrs). The AI is an experimentally determined average requirement and is used when a Recommended Dietary Allowance (RDA) has not been determined.

5. The DRI for calcium peaks at an age range of 9–18 for both males and females. This is an age range where the human body is growing rapidly and adding much new bone mass.

6. Essential amino acids cannot be synthesized by the human body. *Essential* means these amino acids must be supplied in the diet.

7. The essential fatty acids have 18 or more carbon atoms per molecule and are polyunsaturated.

8. Animal proteins generally are a source of all 20 common amino acids while vegetable proteins may be deficient in one or several of the amino acids.

9. The water-soluble vitamins are C. The B vitamins are biotin and folic acid. The fat-soluble vitamins include A, D, E, and K.

10. Vitamin D functions as a regulator of calcium metabolism; Vitamin K enables blood clotting to occur normally; Vitamin A functions to furnish the pigment that makes vision possible. Many vitamins act as coenzymes.

11. Major elements are needed in relatively large quantities by the body while only small amounts of the trace elements are required.

12. Calcium is a major constituent of bones and teeth and is also important in nerve transmission.

13. Iron is essential for oxygen and electron transport as well as hemoglobin replenishment. Anemia is the disease associated with an iron deficiency.

14. (a) Beriberi results from a vitamin B_1 (thiamine) deficiency.
    (b) Pellagra is the result of a deficiency of niacin (nicotinic acid and nicotinamide).
    (c) Rickets results from a vitamin D (ergocalciferol, D_2; cholecalciferol, D_3) deficiency.

15. Both vitamins and trace elements are *essential* nutrients. That is, they must be in the diet for good health. They are also similar in that they are only required in small amounts.

16. The U.S. Food and Drug Administration is responsible for regulating the use of food additives.

17. Enzymes present in the digestive juices are: **saliva:** salivary amylase; **gastric juice:** pepsin, and gastric lipase; **pancreatic juice,** trypsinogen, chymotrypsinogen, procarboxypeptidase, amylopsin, steapsin; **bile:** no enzymes; **intestinal juice:** sucrase, maltase, lactase, aminopeptidase, dipeptidase, nucleases, phosphatase, and intestinal lipase.

18. Chyme is the material of liquid consistency found in the stomach consisting of food particles reduced to small size resulting from gastric digestion of food.

19. Intestinal mucosal cells produce many enzymes needed to complete digestion, e.g., disaccharidases, aminopeptidases, and dipeptidases.

20. GRAS means "generally recognized as safe."

21. Sodium benzoate is termed an antimicrobial and is used to prevent food spoilage. Calcium silicate is described as an anticaking agent and is used to keep powders and salt free flowing.

22. (a) RDA stands for Recommended Dietary Allowance. It is the daily dietary uptake that is sufficient to meet the requirements of nearly all individuals in a specific age and gender group.
    (b) EER stands for Estimated Energy Requirement. It is the average dietary energy intake that is predicted to maintain an energy balance for good health.
    (c) AI stands for Adequate Intake. It is an experimentally determined average dietary requirement (used when an RDA is not available).

23. The four food groups that should provide the major portion of the diet are grains, vegetables, fruits, and milk products.

24. According to the food pyramid, grains should be the food group in largest proportion in the diet and oils should be the food group in smallest proportion.

25. All three kinds of foods are digested in the small intestines: fats, proteins, and carbohydrates.

26. The enzymes functioning in the small intestines are optimally active in an alkaline environment.

27. The duodenum is located in the first section of the small intestines after the stomach.

28. Excess glucose is converted to glycogen and is stored in the liver.

SOLUTIONS TO EXERCISES

1. (a) (23 g protein) + (0 g carbohydrates) + (19 g fat) = 42 g

   (b) (52 mg sodium) + (297 mg potassium) + (23 mg magnesium) + (2.5 mg iron)
       +(4.7 mg zinc) + (9.0 mg calcium) = 388 mg

   (c) (0.015 mg Vitamin A) + (0.092 mg thiamin) + (0.218 mg riboflavin)
       +(3.2 mg niacin) + (0.330 mg Vitamin B_6) + (0.007 mg folic acid)
       +(0.002 mg Vitamin B_{12}) = 3.9 mg

2. (a) (70 g protein) + (25 g carbohydrate) + (37 g fat) = 132 g

   (b) (770 mg sodium) + (564 mg potassium) + (68 mg magnesium) + (3.5 mg iron)
       +(2.7 mg zinc) + (56 mg calcium) = 1464 mg

   (c) (0.0565 mg Vitamin A) + (0.322 mg thiamin) + (0.408 mg riboflavin)
       +(29.5 mg niacin) + (1.2 mg Vitamin B_6) + (0.016 mg folic acid)
       +(0.00082 mg Vitamin B_{12}) = 31.5 mg

3. (a) (9 kcal/g)(19 g) = 171 kcal from fat in the sirloin steak

   (b) (171 kcal from fat)(100)/(2600 kcal/day) = 6.6% of the EER

4. (a) (9 kcal/g)(37 g) = 333 kcal from fat in breaded chicken breast

       (4 kcal/g)(25 g) = 100 kcal from carbohydrates in the breaded chicken breast.

   (b) (333 kcal + 100 kcal)(100)/(2200 kcal/day) = 20% of the teacher's EER.

5. The zinc DRI for males over 70 years old is 11 mg/day. The steak provides 4.7 mg.

$$(4.7 \text{ mg})(100)/(11 \text{ mg/day}) = 43\% \text{ of the zinc DRI}$$

6. The iron DRI for females from 19 to 30 years old is 18 mg/day. The breaded chicken breast provides 3.5 mg.

$$(3.5 \text{ mg})(100)/(18 \text{ mg/day}) = 19\% \text{ of the iron DRI.}$$

7. The thiamine DRI for females from 31 to 50 years old is 1.1 mg/day. The steak provides 92 μg of thiamine.

$$(92 \text{ μg})(1 \text{ mg}/1000 \text{ μg}) = 0.092 \text{ mg}$$
$$(0.092 \text{ mg thiamine})(100)/(1.1 \text{ mg/day}) = 8.4\% \text{ of the thiamine DRI}$$

8. The riboflavin DRI for males from 14 to 18 years old is 1.3 mg/day. The breaded chicken breast provides 408 μg riboflavin, 31% of the riboflavin DRI.

$$(408 \text{ μg})(1 \text{ mg}/1000 \text{ μg}) = 0.408 \text{ mg}$$
$$(0.408 \text{ mg riboflavin})(100)/(1.3 \text{ mg/day}) = 31\% \text{ of the riboflavin DRI}$$

9. The macronutrient DRIs for a male 31–50 years old are as follows: protein, 56 g/day; carbohydrate, 130 g/day; fat, ND. The steak provides 23 g protein and 0 g carbohydrate.

$$(23 \text{ g protein})(100)/(56 \text{ g/day}) = 41\% \text{ of protein DRI}$$
$$(0 \text{ g carbohydrate})(100)/(130 \text{ g/day}) = 0\% \text{ of carbohydrate DRI}$$

10. The macronutrient DRIs for a female 51–70 years old are as follows: protein, 46 g/day; carbohydrate, 130 g/day; fat, ND. The breaded chicken breast provides 70 g protein and 25 g carbohydrate.

$$(70 \text{ g protein})(100)/(46 \text{ g/day}) = 150\% \text{ of protein DRI}$$
$$(25 \text{ g carbohydrate})(100)/(130 \text{ g/day}) = 19\% \text{ of carbohydrate}$$

11. The trace elements in the steak are iron and zinc. The total mass of trace elements is 7.2 mg.

12. Major elements: Na, K, Mg, Ca
    Total major element mass: 1458 mg

13. Three reasons to chew food well:
    (a) To break the food down into small pieces for easy digestion in the stomach.
    (b) To begin the hydrolysis of starch by the salivary amylase in saliva.
    (c) To mix the food with the lubricant, mucin, so it can easily pass through the esophagus into the stomach.

14. When food enters the stomach, it triggers the release of gastric juice, which contains, among other things, hydrochloric acid and the digestive enzymes pepsin and lipase. The main digestive function of the stomach is the partial digestion of proteins aided by the enzyme pepsin, and also aided by the acid which helps denature proteins.

15. The three classes of nutrients which do not commonly supply energy for the cells are (a) vitamins, (b) minerals, and (c) water.

16. The three classes of nutrients which commonly supply energy for the cells are proteins, carbohydrates and lipids.

17. Starch, as a complex carbohydrate, provides dietary carbohydrate in a form which is slowly digested, enabling the body to control distribution of this energy nutrient.

18. Cellulose is important as dietary fiber. Although it is not digested, cellulose absorbs water and provides dietary bulk which helps maintain a healthy digestive tract.

19. Saliva is slightly acidic with a pH below 7.

20. Pancreatic juice is slightly basic with a pH above 7.

21. Butylated hydroxytoluene (BHT) is an antioxidant food additive that prevents changes in color and flavor of the potato chips.

22. Lecithin acts as an emulsifying agent.

23. Pancreatic amylase digests carbohydrates.

24. Pepsin digests proteins.

25. Bread, cereal, rice, and pasta (complex carbohydrates) should be the most abundant foods in our diet.

26. Vegetables should be the second most abundant food group in our diet.

27. Vitamin C (ascorbic acid) is the vitamin that is missing in the disease, scurvy.

28. Vitamin B_1 (thiamin) is the vitamin that is lost in beri-beri.

29. One serving of tomato soup provides 25% of the % Daily Value of Vitamin C. Four servings of soup are required for 100% Daily Value of Vitamin C.

    (100% Daily Value)/(25% Daily Value/serving) = 4 servings

30. The tomato soup serving size is reported as 1 cup (240 g). One serving (1 cup) provides 2 g of fiber. Thus, two servings (2 cup) are needed for 4 g of fiber.

    (4 g fiber)/(2 g fiber/cup) = two cups

31. Twenty-seven grams of total carbohydrate are supplied by one serving of tomato soup. Fiber accounts for 7% of the total carbohydrate.

    (2 g fiber/serving)/(27 g total carbohydrate/serving)(100%) = 7%

32. There is 0.5g of fat in one serving of this tomato soup. Saturated fat represents 0% of the total fat.

    (0 g saturated fat/serving)/(0.5 g total fat/serving)(100%) = 0%

33. Milk contains (a) lipids, (b) minerals (calcium), (c) water, and (d) carbohydrates.

34. 9 kg of body fat represents about 80,000 kcal of energy. Complete starvation for about 30 days would cause a net loss of about 9 kg of fat. Thus, this new diet is unreasonable.

35. A complete protein contains all essential amino acids. There are ten essential amino acids that cannot be synthesized by human cells and that must be supplied in the diet.

36. Fat (9 kcal/g) contains more than twice as much energy as a carbohydrate (4 kcal/g). Thus, a tablespoon of butter will greatly increase the calorie content of a medium sized baked potato.

37. Galactose is a component of the carbohydrate lactose which is found in milk. Thus, milk must commonly be changed to eliminate lactose from the diet of a baby who has galactosemia.

38. Nutritionists recommend less than 300 mg of dietary cholesterol per day. One large egg (270 mg cholesterol) represents almost the total daily recommended amount (about 90% of the RDA).

39. (a) $C_{18}H_{30}O_2$
    (b) $CH_3CH_2CH{=}CHCH_2CH{=}CHCH_2CH{=}CH(CH_2)_7COOH$
    (c) 9,12,15-octadecatrienoic acid

40. (a) $\left[\dfrac{5\,\text{fat Cal per serving}}{130\,\text{total Cal per serving}}\right](100\%) = 4\%$

(b) $(3\,\text{g protein per serving})(2\,\text{servings}) = 6\text{g protein}$

(c) $\left[\dfrac{(1\,\text{serving per day})(130\,\text{total Cal per serving})}{1000\,\text{Cal per day}}\right](100\%) = 13\%$

(d) $\left[\dfrac{(27\,\text{g total carbohydrates per serving})(1\,\text{serving per day})}{375\,\text{g total carbohydrates per 2500 Cal daily diet}}\right](100\%) = 7\%$

BIOENERGETICS

SOLUTIONS TO REVIEW QUESTIONS

1. The IUPAC name is 2,3,4,5-tetrahydroxypentanal.

2.

The class of linkage is an ester.

3. Organic nutrients contain potential metabolic energy in reduced carbon atoms. The more reduced the carbon, the more potential energy the atom holds. Also, the more oxidized the carbon, the less potential energy is present. Oxalic acid has two carbons that each have three bonds to oxygen. (Each carbon has an oxidation number of +3.) These carbons are already very oxidized and, therefore, contain little potential metabolic energy. Your friend is wrong.

4. Reactive oxygen species are harmful to living organisms. The most common reactive oxygen species include hydrogen peroxide (H_2O_2), superoxide (O_2^-), and the hydroxyl radical (OH).

5. Oxidative stress occurs when levels of reactive oxygen species rise to harmful levels inside a cell.

6. ADP has one phosphate anhydride bond. It is high in energy partly because of the repulsion between adjacent negative charges.

7. ATP is known as the "common energy currency" of the cell because energy from many different catabolic processes is stored in this molecule. In turn, most anabolic pathways draw energy from ATP.

8. An oxidation-reduction coenzyme (redox coenzyme) is a reusable organic compound which helps an enzyme carry out an oxidation-reduction reaction.

9. The nicotinamide ring of NAD^+ becomes reduced.

10. Oxidative phosphorylation uses the mitochondrial electron transport system to directly produce ATP from oxidation-reduction reactions. Substrate-level phosphorylation does not use the electron transport system, but, instead involves transfer of a phosphate group from a substrate to ADP to form ATP.

11. Oxidative phosphorylation takes place in the mitochondria.

12. Both chloroplasts and mitochondria are organelles which are bounded by two membranes. In both cases they contain much folded internal membrane (which facilitates oxidation-reduction reactions).

13. Chloroplast pigments trap light to provide energy for photosynthesis.

14. The overall photosynthetic reaction in higher plants is as follows:

$$6\,CO_2 + 6\,H_2O + 2820\,kJ \longrightarrow C_6H_{12}O_6 + 6\,O_2$$

15. Myosin protein is part of the muscle fiber and lies alongside actin protein. The myosin structure is partly bent like a hinge. The ATP reaction causes this "hinge" to open and close which moves the myosin along the actin. As these two proteins slide along each other the muscle fiber contracts.

16. The FAD formula is in the oxidized state.

SOLUTIONS TO EXERCISES

1. The oxidation state of carbon in CH_3OH is -2. The oxidation state of carbon in CO_2 is $+4$. The carbon in CH_3OH is in a lower oxidation state than in CO_2, and, therefore, will deliver more energy than CO_2 in redox reactions.

2. The average oxidation state of carbon in ethanal is -1. The average oxidation state of carbon in acetic acid is 0. The carbons in ethanal, being in a lower oxidation state, will deliver more energy in biological redox reactions.

3. The more oxidized the carbon, the less potential metabolic energy the atom holds. The carbon dioxide carbon has four bonds to oxygen (an oxidation number of $+4$) and is maximally oxidized. This carbon carries essentially no potential metabolic energy.

4. The carbons in a fat are more reduced than those in a carbohydrate. These more reduced carbons can be oxidized more extensively and, thus, can provide more metabolic energy.

5. Dioxygen gains four electrons when it is reduced to two molecules of water. If dioxygen gains only two electrons, the dangerous reactive oxygen species (ROS), hydrogen peroxide (H_2O_2), is formed.

6. Dioxygen gains four electrons when it is reduced to two molecules of water. If dioxygen gains only one electron, the dangerous reactive oxygen species (ROS), superoxide (O_2^-), is formed.

7. Oxidative phosphorylation forms ATP for the cell. *Oxidative* means that redox reactions supply the energy to form the ATP.

8. "Substrate" refers to an compound that is an intermediate in a metabolic pathway. During substrate-level phosphorylation, substrates contribute phosphates to ADP, making ATP.

9. The equation is unbalanced. An additional proton is always used when NADH is converted to NAD^+.

10. Both oxidized forms of the coenzymes (NAD^+ and FAD) are on the same side of the equation. For a reaction to occur one reduced coenzyme (NADH or $FADH_2$) must be paired with one oxidized coenzyme.

11. The conversion of glucose to carbon dioxide is catabolic. A larger molecule (glucose) is converted to a smaller molecule (carbon dioxide) and the glucose carbons are progressively oxidized as they are converted to carbon dioxide.

12. The conversion of acetate to long-chain fatty acids is anabolic. A smaller molecule (acetate) is converted to a larger molecule (long-chain fatty acid) and the acetate carbons are progressively reduced as they are converted to long-chain fatty acid.

13. Photosynthesis is an anabolic process, not a catabolic one.

14. Oxidative phosphorylation creates ATP and is a catabolic process not an anabolic one.

15. Procaryotic cells contain no chloroplasts.

16. Chloroplasts carry out photosynthesis, not oxidative phosphorylation.

17. The chemical changes in the mitochondria are catabolic. The mitochondria convert many different larger molecules to smaller carbon dioxide, the carbons become progressively more oxidized, and, as a result, ATP is produced.

18. The chemical changes in the chloroplasts are anabolic. Smaller carbon dioxide molecules are converted to larger glucose molecules, the carbons become progressively more reduced, and this process requires an input of energy (from sunlight).

19. Three important characteristics of an anabolic process are: (1) simpler substances are built up into complex substances; (2) often carbons are reduced; (3) often cellular energy is consumed.

20. Three important characteristics of a catabolic process are: (1) complex substances are broken down into simpler substances; (2) often carbons are oxidized; (3) often cellular energy is produced.

21. The nicotinamide ring is the reactive center of NAD^+.

22. The flavin ring is the reactive center of FAD.

23. Since the conversion of NAD^+ to NADH is reduction, some pathway metabolites are being oxidized. A pathway that oxidizes carbon compounds is catabolic.

24. Since the conversion of $FADH_2$ to FAD is oxidation, some pathway metabolites are being reduced. A pathway that reduces carbon compounds is anabolic.

25. $2\,FADH_2 + O_2 \longrightarrow 2\,FAD + 2\,H_2O$

$$(2.38 \text{ mol } FADH_2)\left(\frac{1 \text{ mol } O_2}{2 \text{ mol } FADH_2}\right) = 1.19 \text{ mol } O_2$$

26. $2\,NADH + 2\,H^+ + O_2 \longrightarrow 2\,NAD^+ + 2\,H_2O$

$$(0.67 \text{ mol } NADH)\left(\frac{1 \text{ mol } O_2}{2 \text{ mol } NADH}\right) = 0.34 \text{ mol } O_2$$

27. $NAD^+ + H^+ + 2\,e^- \longrightarrow NADH$

$$(11.75 \text{ mol } NAD^+)\left(\frac{2 \text{ mol } e^-}{1 \text{ mol } NAD^+}\right) = 23.50 \text{ mol } e^-$$

28. $FAD + 2\,H^+ + 2\,e^- \longrightarrow FADH_2$

$$(0.092 \text{ mol } FAD)\left(\frac{2 \text{ mol } e^-}{1 \text{ mol } FAD^+}\right) = 0.18 \text{ mol } e^-$$

29. There is one high-energy phosphate anhydride bond in this compound. The anhydride is formed directly between two phosphate groups and is high in energy partly because of the repulsion between adjacent negative charges.

30. There are no high-energy phosphate anhydride bonds in this compound. The phosphates are connected to carbons by lower energy ester bonds.

31. NADH allows more ATP to be produced than $FADH_2$ does and, thus, NADH stores more energy. Three ATP are made when NADH is oxidized versus only two ATP from $FADH_2$.

32. Energy is required. Figure 33.10 shows that the electrons must increase in energy before converting $NADP^+$ to NADPH. The energy comes from light.

33. Since the phosphorylated substrate phosphoenolpyruvate contains only one high-energy phosphate bond, substrate-level phosphorylation can produce only one ATP.

34. Since the phosphorylated substrate 1,3-diphosphoglycerate contains only one high-energy phosphate bond, substrate-level phosphorylation can produce only one ATP.

35. Mitochondrial electron transport and oxidative phosphorylation produce 3 ATP per NADH and 2 ATP per $FADH_2$. Thus,

$$(4 \text{ NADH})\left(\frac{3 \text{ ATP}}{\text{NADH}}\right) + (2 \text{ FADH}_2)\left(\frac{2 \text{ ATP}}{\text{FADH}_2}\right) = 16 \text{ ATP}$$

36. Mitochondrial electron transport and oxidative phosphorylation produce 3 ATP per NADH and 2 ATP per $FADH_2$. Thus,

$$(2 \text{ NADH})\left(\frac{3 \text{ ATP}}{\text{NADH}}\right) + (3 \text{ FADH}_2)\left(\frac{2 \text{ ATP}}{\text{FADH}_2}\right) = 12 \text{ ATP}$$

37. According to the sliding filament model, the myosin protein acts as a "hinge" to cause muscle contraction. As part of this model (see Figure 33.8, i. Step 1), ATP binding opens the hinge and allows muscle relaxation. Without ATP, the muscle is locked in the contracted state—rigor mortis.

38. According to the sliding filament model, calcium initiates muscle contraction. By blocking calcium release, botulism toxin stops the initial step in muscle contraction. Myosin does not react with ATP and the muscles remain relaxed.

39. NAD^+ has two phosphate esters. Each ester bond connects a phosphate to a carbon.

40. FAD has one phosphate anhydride where the two phosphates are connected to each other.

41. Myosin moves into its extended structure in Figure 33.8a, part i. Myosin changes its shape due to the binding of the "large" ATP molecule.

42. Myosin moves into its "kinked" structure in Figure 33.8a, part iii. Myosin changes its shape when ATP is reacted with water and released.

43. No, photosynthesis is an anabolic process by which carbon atoms are reduced. (Carbon dioxide is converted to glucose.)

44. The myosin protein acts as a "hinge," opening and closing to cause the muscle to contract. The "hinge" opens when ATP binds. Then, a high-energy phosphate anhydride in ATP is broken. The products (ADP and phosphate) are easily released from myosin and the "hinge" springs shut causing muscle contraction.

45. No. Higher plant cells need mitochondria as well as chloroplasts in order to oxidize energy storage molecules to provide for the cellular energy needs.

46. Both ATP and NAD^+ contain a ribose which (1) is linked to an adenine at carbon one and (2) is linked to a phosphate at carbon five.

47. The photosynthetic dark reactions produce glucose and do not directly require light energy. In contrast, the photosynthetic light reactions directly use light energy to make ATP and reduced coenzymes.

48. The oxidation number sum for glyceraldehyde is more negative than the oxidation number sum for pyruvic acid. Glyceraldehyde is more reduced than pyruvic acid.
    (a) NADH is a reducing agent and will be a reactant not a product.
    (b) Since the pathway metabolite is being reduced, I would expect this pathway to be anabolic. Thus, ATP is probably a reactant, not a product.

CHAPTER 34

CARBOHYDRATE METABOLISM

SOLUTIONS TO REVIEW QUESTIONS

1. Carbohydrates are considered energy storage molecules because they contain biologically usable, reduced carbon atoms.

2. Glucose catabolism breaks down glucose to the smaller carbon dioxide molecule in an oxidative process. This produces energy for the cell and is catabolic. Photosynthesis produces glucose from carbon dioxide via a reductive path. Light energy is required, and this is an anabolic process.

3. Each mole of glucose will yield 2820 kJ of energy when oxidized. Thus, 3 moles of glucose will provide 3 moles × 2820 kJ/mole or 8460 kJ.

4. Enzymes are needed so that biochemical reactions will proceed fast enough to keep the cell alive. Metabolism is often controlled by controlling enzyme activity.

5. Initial muscle contraction requires a readily and quickly available form of energy, ATP. Once ATP stores have been depleted, muscle cells turn to muscle glycogen as a source of chemical energy.

6. The lungs support muscle contraction in two important ways. First, they supply oxygen that is required by the muscles for long-term energy production. Second, the lungs exhale carbon dioxide that is the end product of oxidative muscle metabolism.

7. The preparatory phase of the Embden-Meyerhof pathway reacts carbohydrates with ATP to form low-energy phosphate ester bonds. These will be "energized" later in the pathway to yield a net increase in ATP.

8. The liver is responsible for maintaining a constant blood-glucose level. Von Gierke disease blocks glycogenolysis. The liver cannot release much glucose to the blood and patients experience hypoglycemia. This low blood sugar means the patients fail to thrive.

9. The citric acid cycle oxidizes carbon compounds with the ultimate goal of producing ATP. However, ATP can only be produced in conjunction with two processes, electron transport and oxidative phosphorylation, that require molecular oxygen. The citric acid cycle is described as aerobic because it can only accomplish its ultimate goal (producing ATP) when molecular oxygen is available.

10. Hormones are chemical substances that act as control or regulatory agents in the body. Hormones are secreted by the endocrine, or ductless, glands directly into the bloodstream and are transported to various parts of the body to exert specific control functions.

11. (a) The glucose concentration in blood under normal fasting conditions is about 70 to 100 mg/100 mL of blood.

    (b) The blood glucose concentration considered as (1) hyperglycemic is 100 to 140 mg/100 mL of blood; and (2) hypoglycemic is 50 to 70 mg/100 mL of blood.

12. The renal threshold is the concentration of a substance in the blood above which the kidneys begin to excrete that substance.

13. In gluconeogenesis glucose is synthesized from non-carbohydrate sources: lactate, amino acids, and glycerol. These three substances are converted to Embden-Meyerhof pathway intermediates, which are then converted to glucose. Since the Embden-Meyerhof pathway is anaerobic, gluconeogenesis is considered an anaerobic process.

14. Disagree. Acetyl-CoA, with reduced carbons, enters the citric acid cycle (Fig. 34.4). During the cycle, chemical reactions occur in which oxidized carbons in the form of CO_2 are expelled.

15. An alternate name for the citric acid cycle is the Krebs cycle. The Krebs cycle is named for Hans A. Krebs, who discovered this metabolic path.

16. The addition of H_2O to the double bond in fumerate is a typical alkene reaction.

17. The loss of two hydrogen atoms from succinate to form fumerate means succinate has been oxidized.

SOLUTIONS TO EXERCISES

1. The student's brain uses 22% of 2400 kcal in one day.

   $(22\%)/100 = 0.22$

   $(0.22)(2400\,\text{kcal/day}) = 530\,\text{kcal/day}$ used by the brain

2. The older woman's heart uses 9% of 1900 kcal in one day.

   $(9\%)/100 = 0.09$

   $(0.09)(1900\,\text{kcal/day}) = 170\,\text{kcal/day}$ used by the heart

3. In Exercise 1, the brain is calculated to use 530 kcal/day. Glucose provides 3.7 kcal/g.

   $(530\,\text{kcal/day})/(3.7\,\text{kcal/g}) = 140\,\text{g/day}$ of glucose

4. In Exercise 2, the heart is calculated to use 170 kcal/day. Lactate provides 3.6 kcal/g.

   $(170\,\text{kcal/day})/(3.6\,\text{kcal/g}) = 47\,\text{g/day}$ of glucose

5. Emphysema will have two important impacts on skeletal muscle work. First, the lungs will provide only small amounts of molecular oxygen and skeletal muscle aerobic metabolism will be limited. Second, the lungs will exhale only small amounts of carbon dioxide. The blood will become more acidic which will limit muscle work.

6. The liver will lose its ability to maintain the blood glucose levels needed for skeletal muscle work. Additionally, the liver will not be able to remove the blood lactate derived from skeletal muscle work.

7. When the muscle takes glucose and makes lactate, this is catabolic metabolism. First, a larger molecule, glucose, is converted to a smaller molecule, lactate. Second, this metabolism is used to provide ATP for muscle work.

8. When the liver takes lactate and converts it to glucose this is anabolic metabolism. First, a smaller molecule, lactate, is converted to a larger molecule, glucose. Second, this process requires an input of ATP.

9. Glycolysis and the citric acid cycle must work together to convert glucose to carbon dioxide. This is a catabolic process because (a) glucose is broken down to smaller carbon dioxide, (b) the glucose carbons are oxidized completely to carbon dioxide, and (c) ATP is produced.

10. Gluconeogenesis and glycogenesis must work together to convert lactate to glycogen. This is an anabolic process as the smaller lactate is reacted to form the much larger glycogen. Also, ATP is required for this process.

11. ATP production in the citric acid cycle is substrate-level phosphorylation. A phosphate is transferred directly from a pathway substrate to ADP, forming ATP.

12. The Embden-Meyerhof pathway uses substrate-level phosphorylation to produce ATP. That is, following carbon oxidation a phosphate group is transferred from a substrate to ADP, forming ATP.

13. Although glucose is oxidized, molecular oxygen (O_2) is not used in the Embden-Meyerhof pathway. Thus, this pathway is anaerobic.

14. Although no molecular oxygen (O_2) is used in the citric acid cycle, reduced coenzymes (NADH and $FADH_2$) are produced. These reduced coenzymes must be oxidized by electron transport so that the citric acid cycle can continue. And, it is electron transport which uses molecular oxygen. Thus, the citric acid cycle depends on the presence of molecular oxygen and is considered to be aerobic.

15. Glycolysis produces 2 moles of lactate per mole of glucose. Therefore, 2.8 moles of glucose will yield 5.6 moles of lactate.

16. Glycolysis yields 2 moles of ATP per mole of glucose. Therefore, 0.85 moles of glucose will yield 1.7 moles of ATP.

17. The end products of the anaerobic catabolism of glucose in yeast cells are ethanol and carbon dioxide.

18. The end product of the anaerobic catabolism of glucose in muscle tissue is lactate.

19. The Embden-Meyerhof pathway is defined as catabolic because (a) glucose is broken down to smaller compounds, (b) the carbons of glucose are oxidized and, (c) this pathway causes production of ATP.

20. The gluconeogenesis pathway is considered to be anabolic because (a) smaller, noncarbohydrate precursors such as lactate are converted to the larger product, glucose, and (b) as this conversion takes place, the carbons become progressively more reduced.

21. Electron transport and oxidative phosphorylation are needed to convert the reduced coenzyme products of the citric acid cycle into oxidized coenzymes with ATP formation.

22. The glycolysis pathway uses substrate-level phosphorylation to produce ATP and, thus, does not need electron transport or oxidative phosphorylation.

23. One pass through the citric acid cycle uses one acetyl-CoA and yields 3 NADH, 1 $FADH_2$ and 1 ATP. Each NADH yields a maximum of three ATP in electron transport and oxidative phosphorylation. Each $FADH_2$ yields a maximum of two ATP in electron transport and oxidative phosphorylation. Thus, for each acetyl-CoA, the cell produces the following ATP:

> 1 ATP directly from the citric acid cycle
> 9 ATP from the three NADH produced in the citric acid cycle
> <u>2 ATP</u> from the one $FADH_2$ produced in the citric acid cycle
> 12 ATP/acetyl-CoA

For 0.6 moles of acetyl-CoA, the following moles of ATP will be made:

$$(0.6 \text{ moles acetyl-CoA})(12 \text{ moles ATP/mole acetyl-CoA}) = 7.2 \text{ moles ATP}$$

24. One pass through the citric acid cycle uses one acetyl-CoA and yields 3 NADH, 1 $FADH_2$, and 1 ATP. Each NADH yields a maximum of three ATP in electron transport and oxidative phosphorylation. Each $FADH_2$ yields a maximum of two ATP in electron transport and oxidative phosphorylation. Thus, for each acetyl-CoA, the cell produces the following ATP:

    1 ATP directly from the citric acid cycle
    9 ATP from the three NADH produced in the citric acid cycle
    2 ATP from the one $FADH_2$ produced in the citric acid cycle
    12 ATP/acetyl-CoA

    For 0.38 moles of acetyl-CoA, the following moles of ATP will be made:

    $$(0.38 \text{ moles acetyl-CoA})(12 \text{ moles ATP/mole acetyl-CoA}) = 4.6 \text{ moles ATP}$$

25. The glycolysis pathway causes the anaerobic breakdown of glucose to form lactate. In contrast, the glycogenolysis pathway breaks down glycogen to liberate glucose.

26. Gluconeogenesis synthesizes glucose from non-carbohydrate precursors while glycogenesis is the process that makes glycogen from glucose.

27. The one reaction in the Embden-Meyerhof pathway that is classed as oxidation-reduction converts glyceraldehyde-3-phosphate to 1,3-bisphosphoglycerate. The top carbon of glyceraldehyde-3-phosphate is oxidized from an aldehyde to a carboxyl group. Simultaneously, NAD^+ is reduced to NADH. Both oxidation and reduction occur in this reaction.

28. When pyruvate is converted to lactate, a carbonyl group is changed to an hydroxyl group. The carbon in these functional groups loses a bond to oxygen and gains a bond to hydrogen. So, the carbon is reduced. Simultaneously, NADH is converted to NAD^+, losing electrons and becoming oxidized.

29. Both epinephrine and glucagon cause an increase in blood glucose levels. Glucagon is made in the pancreas while epinephrine comes from the adrenal glands.

30. While insulin decreases blood glucose levels and stimulates glycogen synthesis, glucagon increases blood glucose levels and stimulates glycogen breakdown.

31. The steroid is progesterone.

32. The polypeptide hormone is vasopressin.

33. Solving this problem requires reference to the conversion factor, $4.184 \text{ J} = 1 \text{ cal}$.

    $(5.2 \text{ MJ/day})(10^6 \text{ J/MJ}) = 5.2 \times 10^6 \text{ J/day}$

    $(5.2 \times 10^6 \text{ J/day})(1 \text{ cal}/4.184 \text{ J}) = 1.2 \times 10^6 \text{ cal/day}$

    $(1.2 \times 10^6 \text{ cal/day})(1 \text{ kcal}/1000 \text{ cal}) = 1200 \text{ kcal/day}$

34. If glucose was oxidized to pyruvate in one step a large amount of energy would be released, more than the cell can handle at one time. Also, the ten steps of the Embden-Meyerhof pathway provide intermediate compounds that can be used for other metabolic purposes within the cell.

35. Insulin is not effective when given orally because it is a protein, and would be hydrolyzed to amino acids in the gastrointestinal tract.

36. If a large overdose of insulin was taken by accident, the blood glucose concentration would drop to a low level, probably in the hypoglycemic range. This could result in fainting, convulsions, and unconsciousness.

37. Epinephrine (adrenaline) is sometimes called the emergency or crisis hormone because it stimulates glycogenolysis, which raises the concentration of glucose in the blood, giving the body additional energy for an emergency or crisis situation.

38. $$CH_3\underset{\underset{\displaystyle OH}{|}}{C}HCOO^-$$

    lactate

    All three carbons of lactate are more reduced than the carbon in CO_2. Thus, lactate can provide more cellular energy upon further oxidation of the carbons.

39. After two hours in a glucose-tolerance test, a blood glucose level of 210 mg/100 mL blood is too high. Under normal circumstances, insulin would have returned the blood glucose to about 100 mg/100 mL blood. Thus, insulin control is not functioning correctly for this patient.

40. Yeast first converts pyruvate to acetaldehyde with the release of carbon dioxide. This is useful to bakers as it causes dough to rise. Second, yeast converts acetaldehyde to ethanol. This reaction allows brewers to produce alcoholic beverages.

41. All these compounds contain an adenosine diphosphate group in their structure.

42. Oxidation. The average oxidation number of carbon in glucose is 0. The oxidation number of carbon in carbon dioxide is $+4$. Thus, the conversion of glucose to carbon dioxide is oxidation.

43. (a) Glycolysis: The anaerobic catabolic pathway for conversion of glucose to lactate.
    (b) Gluconeogenesis: The metabolic pathway for the synthesis of glucose from non-carbohydrate sources.
    (c) Glycogenesis: The synthesis of glycogen from glucose.
    (d) Glycogenolysis: The hydrolysis, or breakdown of glycogen to glucose.

44. $$C_6H_{12}O_6 + 2\,NAD^+ + 2\,ADP + 2\,P_i \longrightarrow 2\,CH_3\overset{\displaystyle O}{\overset{\|}{C}}COO^- + 2\,NADH + 2\,ATP$$

    glucose$\qquad\qquad\qquad\qquad\qquad\qquad\qquad\qquad$pyruvate

45. (a)
$$CH_3CHCOO^-$$
$$|$$
$$OH$$
lactate

(b)
$$O$$
$$\parallel$$
$$CH_3CCOO^-$$
pyruvate

(c)
$$CH_2OPO_3^-$$

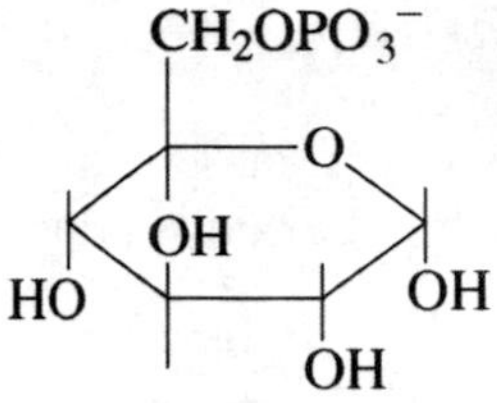

glucose-6-phosphate

(d)
$$CH_2OPO_3^-$$

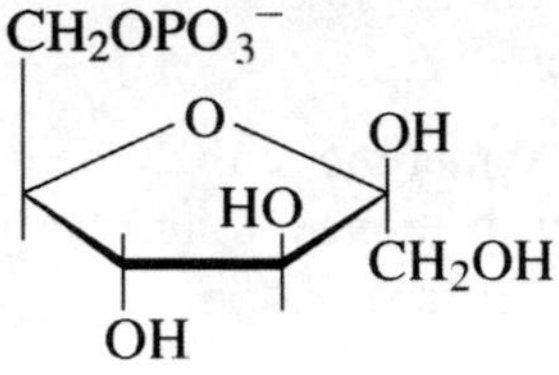

fructose-6-phosphate

(e)
$$CH_2OPO_3^-$$

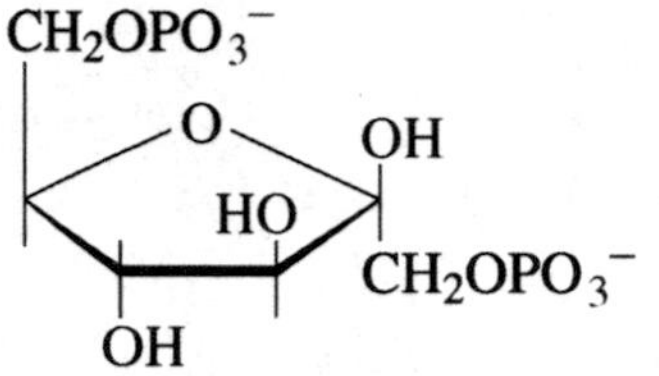

fructose-1,6-bisphosphate

(f)

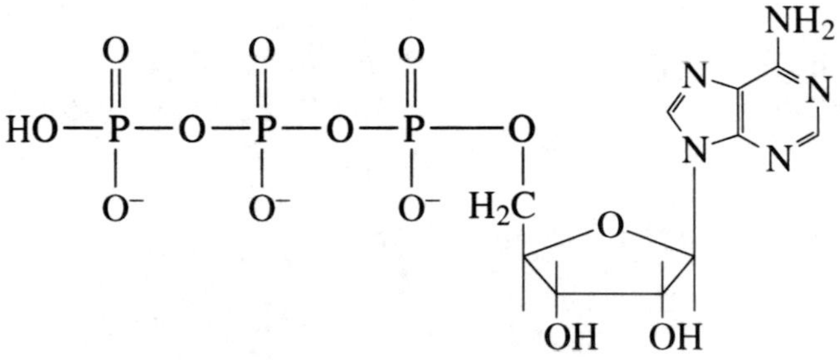

adenosine triphosphate
(ATP)

46. (a) (i)

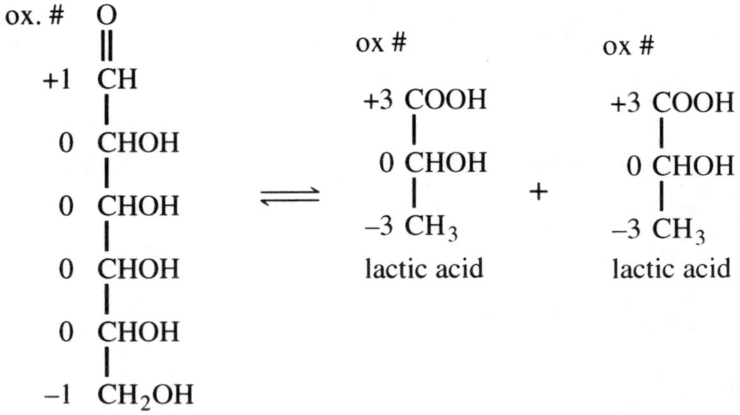

ox. #				ox #		ox #	
+1	CH=O			+3	COOH	+3	COOH
0	CHOH			0	CHOH	0	CHOH
0	CHOH	$\rightleftharpoons$		-3	CH$_3$	+	-3 CH$_3$
0	CHOH				lactic acid		lactic acid
0	CHOH						
-1	CH$_2$OH						

glucose

Average oxidation number for glucose carbons = oxidation number sum/6 carbons
$$= (+1+0+0+0+0-1)/6 \text{ carbons} = 0$$

Average oxidation number for lactic acid carbons = oxidation number sum/6 carbons
$$= 2(+3+0-3)/6 \text{ carbons} = 0$$

There is no net change in oxidation number.

(ii) Two ATP molecules are produced for each glucose under anaerobic conditions. Since glucose is composed of six carbon atoms,

$$2 \text{ ATP}/6 \text{ carbons} = 0.33 \text{ ATP/carbon}$$

(b) (i) All six carbons of glucose are converted to carbon dioxide. The oxidiation number of the carbon in carbon dioxide is +4. So, the average oxidation number changes from 0 to +4.

(ii) Thirty-two ATP are produced under aerobic conditions as glucose is converted to carbon dioxide. (See Section 34.3.) Since glucose is composed of six carbon atoms,

$$32 \text{ ATP}/6 \text{ carbons} = 5.33 \text{ ATP/carbon}$$

This comparison illustrates the general principle: more carbon oxidation allows formation of more ATP.

CHAPTER 35

METABOLISM OF LIPIDS AND PROTEINS

SOLUTIONS TO REVIEW QUESTIONS

1. For the typical adult, adipose (fat) tissue stores the most metabolic energy. Energy is stored in the fatty acids. These molecules (1) contain relatively reduced carbons and (2) have a high percentage of carbon. On a mass basis, fatty acids contain more than twice the energy of carbohydrates.

2. The liver is responsible for the urea cycle. First, this organ removes nitrogens from amino acids and makes urea, a non-toxic, nitrogen excretion product. Second, the liver uses the carbon skeletons from amino acids to make other molecules such as glucose and acetyl-CoA.

3. During starvation, muscle protein is degraded to its component amino acids. The nitrogen from these amino acids is excreted as urea and the carbon skeletons are used for energy.

4. During fasting when carbohydrates become scarce, the liver changes its metabolism to start making ketone bodies. These molecules are water-soluble like glucose but are metabolized like fat. Ketone bodies are the only other molecules besides glucose that can be used by the brain for energy.

5. In the oxidation of fatty acids, the carbon atom beta to the carboxyl group is oxidized forming a β-keto acid which is then cleaved between the α and β carbon atoms leaving a new fatty acid that is two carbons shorter than the original fatty acid.

$$R-\underset{\substack{\uparrow \\ \text{oxidation} \\ \text{of this} \\ \text{carbon atom}}}{\overset{\beta}{CH_2}}-\underset{\substack{\nwarrow \\ \text{cleavage here}}}{\overset{\alpha}{CH_2}}-\overset{\overset{\textstyle O}{\|}}{C}-OH$$

6. Ketone bodies are relatively soluble in the blood stream and can be transported to energy-deficient cells in time of need. Ketone bodies supply energy via β-oxidation and the citric acid cycle.

7. Ketosis is the accelerated production of ketone bodies leading to high blood levels of these compounds with a corresponding increase in blood acidity.

8. Fatty acid synthesis is not the reverse of β-oxidation. β-oxidation occurs in the mitochondria while fatty acids are synthesized in the cytoplasm. In fatty acid synthesis, the growing molecule is linked to an acyl carrier protein while in β-oxidation the fatty acid being broken down is linked to a coenzyme A. Finally, fatty acid synthesis uses some different reactions and a different coenzyme ($NADP^+$) which are not found in β-oxidation.

9. In the enzymatic oxidation of fatty acids from fats, the fatty acids are oxidized, ultimately forming acetyl-CoA. Acetyl-CoA enters the citric acid cycle to produce energy.

10. Soybeans enrich the soil because their roots serve as host to the symbiotic, nitrogen-fixing bacteria.

11. Transamination is the transfer of nitrogen to and from amino acids. Transamination allows transformation of one amino acid to another, supplying needed, specific amino acids for protein synthesis. Also, the amino acid carbon skeletons may be used for other metabolic purposes.

12. A glucogenic amino acid can be used to produce glucose. Most amino acids are glucogenic. A ketogenic amino acid can be used to make acetyl CoA. Only a few amino acids fall into this category.

13. Nitrogen is excreted as (a) ammonia by fish and (b) uric acid by birds.

14. The structure of urea is   and the structure of uric acid is

15. Three ATP molecules are used for every urea molecule produced. The production of urea is important because this molecule provides a nontoxic means for excreting nitrogen.

16. Acetyl-CoA is considered to be an important central intermediate in metabolism for the following reasons:
    (a) Complete catabolism of most energy, containing nutrients is achieved by conversion to acetyl-CoA and oxidation via the citric acid cycle.
    (b) Fats and amino acids can be synthesized from acetyl-CoA.
    (c) Carbohydrates, fats, and amino acids can be converted to acetyl-CoA.
    (d) Acetyl-CoA is the central intermediate in converting (1) carbohydrates to fats or amino acids, (2) fats to amino acids, or (3) amino acids to fats.

17. A negative nitrogen balance means the body is excreting more nitrogen than is being consumed in the diet. An adult who is starving consumes very little. Meanwhile, this person's muscle protein breaks down to amino acids. The nitrogen from these amino acids is excreted as urea and the carbon skeletons are used for energy. There is much more nitrogen that is excreted than is consumed.

18. Nitrogen is fixed by three general processes, (1) bacterial action, (2) high temperature, and (3) chemical fixation. The high temperatures of a lightning flash convert dinitrogen to nitrogen oxide. The most important chemical fixation is the Haber process for making ammonia from nitrogen and hydrogen. Neither of these processes is biochemical.

19. Ketone bodies are derived from fatty acids. Both ketone bodies and fatty acids are broken down in the beta-oxidation pathway.

SOLUTIONS TO EXERCISES

1. The two carbons that will be found in acetyl-CoA after one pass through the beta oxidation pathway by butyric acid are circled.

   $CH_3CH_2\overparen{CH_2COOH}$

2. The two carbons that will remain after two passes through the beta oxidation pathway by caproic acid are circled.

   $\overparen{CH_3CH_2}CH_2CH_2CH_2COOH$

   β-carbon
   ↓
3. $CH_3(CH_2)_{12}CH_2CH_2COOH$ palmitic acid

   β-carbon
   ↓
4. $CH_3(CH_2)_{10}CH_2CH_2COOH$ myristic acid

5. Lauric acid ($CH_3(CH_2)_{10}COOH$) has 12 carbons and will yield six molecules of acetyl-CoA.

6. Palmitic acid ($CH_3(CH_2)_{14}COOH$) has 16 carbons and will yield eight molecules of acetyl-CoA.

7. The obese man carries 40 kg of fat as an energy store. Every gram is worth 9 kcal of energy.

   $(40\,kg)(1000\,g/kg)(9\,kcal/g) = 360{,}000\,kcal$ in energy reserves.

   This obese man uses 2000 kcal/day.

   $(360{,}000\,kcal)/(2000\,kcal/day) = 180$ days or about 6 months on a fast to lose this fat.

8. The woman will use 1800 kcal/day over 3 days.
   $(1800\,kcal/day)(3\,days) = 5400\,kcal$ will be used over the three-day fast.

   Each kilogram of fat holds about 9000 kcal of metabolic energy.

   $(5400\,kcal)/(9000\,kcal/kg) = 0.60\,kg$ will be lost.

9. A high-protein diet requires the body to use protein for energy. This protein will be broken down to amino acids. The nitrogen will be removed from the amino acids and the carbon skeletons used for energy. Increased nitrogen removal means an increase in urea excretion.

10. The liver will increase urea production. During a fast, muscle protein is broken down to amino acids. The carbon skeletons of these amino acids are used for energy while the nitrogen is excreted as part of urea.

11. The acyl carrier protein (ACP) is similar to coenzyme A (CoA) in that
    (a) both molecules have —SH groups which form thioesters with carboxylic acids;
    (b) both molecules act as carriers for fatty acids during metabolism: ACP during anabolism and CoA during catabolism.

12. The acyl carrier protein (ACP) differs from coenzyme A (CoA) in that
    (a) ACP is a protein while CoA is a coenzyme;
    (b) ACP is used in fatty acid synthesis while CoA is used in the beta oxidation pathway.

13. Yes, the student's liver will be able to synthesize ketone bodies. Although the diet contains no ketogenic amino acids, ketone bodies can be formed from acetyl-CoA. Carbohydrates and fats, as well as proteins, can be used to make acetyl-CoA. So, the low-carbohydrate, high-fat, high-protein diet provides the nutrients needed to make ketone bodies.

14. A diet of protein and carbohydrate (but no fat) will not make the office worker "fat deficient." The human liver can convert both proteins and carbohydrates into fat.

15. During a wasting disease, muscle protein is broken down to amino acids. As the amino acids are metabolized, nitrogen is removed and excreted as urea. Because there is a net decrease in muscle protein, these patients will have a negative nitrogen balance.

16. After surgery, part of the healing process involves protein synthesis. More nitrogen will be consumed than excreted and a positive nitrogen balance will occur.

17. During transamination,

    (a) L-alanine yields
    $$\underset{\text{}}{CH_3\overset{\displaystyle O}{\overset{\|}{C}}COOH}$$

    (b) L-serine yields
    $$HOCH_2\overset{\displaystyle O}{\overset{\|}{C}}COOH$$

18. During transamination,

    (a) L-aspartic acid yields
    $$HOOCCH_2\overset{\displaystyle O}{\overset{\|}{C}}COOH$$

    (b) L-phenylalanine yields
    $$\text{⬡}-CH_2\overset{\displaystyle O}{\overset{\|}{C}}COOH$$

19. The portion of carbamoyl phosphate which is incorporated into urea is circled.

$$H_2N-\overset{\displaystyle O}{\overset{\|}{C}}-O-\overset{\displaystyle O}{\underset{\underset{\displaystyle O^-}{|}}{\overset{\|}{P}}}-O^-$$

20. The portion of L-aspartic acid which is incorporated into urea is circled.

$$COOH$$
$$|$$
$$CH_2$$
$$|$$
$$H_2N-CH-COOH$$

21. One mole of urea is produced for every mole of L-glutamic acid used by the urea cycle. Thus, 1.6 moles of urea will be made from 1.6 moles of L-glutamic acid. Three moles of ATP are used for every L-glutamic acid mole that goes into the urea cycle. Thus, 4.8 moles of ATP will be used to process 1.6 moles of L-glutamic acid.

22. One mole of urea is produced for every mole of L-aspartic acid used. 0.84 moles of aspartic acid will yield 0.84 moles of urea. One mole of L-glutamic acid is used for every mole of L-aspartic acid. 0.84 moles of aspartic acid uses 0.84 moles of glutamic acid.

23. No. The transamination process is not nitrogen fixation. Nitrogen fixation must start with elemental nitrogen while transamination transfers an amine group to an alpha-keto acid from a different amino acid.

24. Although this is part of the nitrogen cycle, it is not nitrogen fixation. The bacteria are starting with ammonia rather than elemental nitrogen as is required in nitrogen fixation.

25. As shown in Figure 35.6, proteins and fats can be converted to acetyl CoA. However, there can be no net synthesis of glucose from acetyl CoA (the conversion of glucose to acetyl CoA is a one-way process).

26. During starvation, most of the amino acids coming from protein breakdown are glucogenic and are used to make glucose. Late in starvation the body has shifted to using more ketone bodies and less glucose. Thus, muscle protein breakdown slows.

27. In step 3, the carbonyl group is reacted to form an hydroxyl group. The reactant's carbonyl carbon has two bonds to oxygen while the product's hydroxyl carbon has only one bond to oxygen (and one bond to hydrogen). Thus, the carbon is reduced.

28. β-hydroxybutyric acid is a ketone body which does not contain a ketone group.

$$CH_3CHCH_2COOH$$
$$|$$
$$OH$$

29. Beta-oxidation is a pathway with relatively few reactions, requiring FAD, NAD^+, ATP, and coenzyme A. This process forms no ATP directly but depends upon mitochondrial electron transport and oxidative phosphorylation to form ATP.

30. Malonyl-CoA is used in fatty acid synthesis (anabolism) but not in beta oxidation (catabolism). Thus, the lecture would probably focus on fatty acid anabolism.

31. Dehydrogenation refers to the removal of hydrogen. In step 2 of beta oxidation,

$$\underset{\displaystyle RCH_2CH_2CH_2\overset{\textstyle O}{\overset{\|}{C}}-SCoA}{} + FAD \longrightarrow \underset{\displaystyle RCH_2CH=CH\overset{\textstyle O}{\overset{\|}{C}}-SCoA}{} + FADH_2$$

The fatty acid carbons at the alpha and beta positions change their oxidation numbers from -2 to -1; they are oxidized.

32. $(1.0 \text{ mol palmitic acid}) \left(\dfrac{256.4 \text{ g}}{\text{mol palmitic acid}} \right) = 2.6 \times 10^2 \text{ g}$

   $\left(\dfrac{9 \text{ kcal}}{\text{g}} \right) (2.6 \times 10^2 \text{ g}) = 2.3 \times 10^3 \text{ kcal}$

33. Disagree. Ketosis (the increased production of ketone bodies) occurs when carbohydrates are in short supply. Thus, as long as carbohydrates remain the single most abundant dietary component, ketosis should not develop.

34. In metabolism, the majority of ATP is formed when acetyl-CoA is broken down in the citric acid cycle, electron transport, and oxidative phosphorylation.

   (1) One six-carbon hexose (glucose) will be converted to two molecules of pyruvate by the Embden-Meyerhof pathway which, in turn, are converted to two acetyl-CoA molecules.

   (2) A six-carbon fatty acid will yield three molecules of acetyl-CoA via the beta oxidation pathway.

   The six-carbon fatty acid yields more ATP than the six carbon hexose during catabolism primarily because more acetyl-CoA molecules are formed and can feed into the citric acid cycle, electron transport, and oxidative phosphorylation.

CPSIA information can be obtained at www.ICGtesting.com
Printed in the USA
BVOW00n1628211213

339395BV00030B/44/P